Plants: Evolution and Diversity

Plants are so much part of our environment that we often take them for granted, yet beautiful, fascinating and useful plants are everywhere, from isolated moss colonies on stone walls to vast complex communities within tropical rainforests. How did this array of form and habitat come about, and how do we humans interact with the plant kingdom? This book provides a refreshing and stimulating consideration of these questions and throws light in a new way on the complexity, ecology, evolution and development of plants and our relationship with them. Illustrated throughout with numerous line diagrams and beautiful colour photographs, the book provides a unique source of information about the fascinating lives that plants lead and the way in which our lives are inextricably linked to theirs. It will be particularly useful to those seeking a more ecological and process-oriented approach than is available in other textbooks.

MARTIN INGROUILLE is a senior lecturer at Birkbeck College, University of London, where he teaches botany, genetics, ecology, biogeography and evolution and researches the evolutionary ecology and systematics of plants. He has travelled widely during the course of his studies, including Europe, North and Central America, East Africa, South East Asia, Australia and New Zealand. He is author of the widely adopted textbook *Diversity and Evolution of Land Plants* (1992) from which this present book has been developed.

BILL EDDIE is a tutor at the University of Edinburgh where he teaches botany, ornithology, evolution and geopoetics as part of the University's Open Studies programme. His research focuses on the evolutionary ecology and systematics of plants and birds, and he has a particular interest in the history of botany, and in birds and wildlife in art. He has a long-term interest in the Campanulaceae, and did molecular research on this family at the University of Texas, Austin. He has travelled widely during the course of his career, including Europe, North and Central America, Central and Southern Africa, South East Asia, and Australia, and has held a four-year teaching post at the University of Papua New Guinea.

Plants: Evolution and Diversity

Martin Ingrouille
School of Biological and Chemical Sciences, Birkbeck College, University of London
Bill Eddie
University of Edinburgh

CAMBRIDGE
UNIVERSITY PRESS

CAMBRIDGE UNIVERSITY PRESS
Cambridge, New York, Melbourne, Madrid, Cape Town, Singapore, São Paulo

Cambridge University Press
The Edinburgh Building, Cambridge CB2 2RU, UK

Published in the United States of America by Cambridge University Press, New York

www.cambridge.org
Information on this title: www.cambridge.org/9780521790970

First published 2006

Printed in the United Kingdom at the University Press, Cambridge

A catalogue record for this book is available from the British Library

ISBN-13 978-0-521-79097-0 hardback
ISBN-10 0-521-79097-2 hardback

ISBN-13 978-0-521-79433-6 paperback
ISBN-10 0-521-79433-1 paperback

Contents

Preface

Dancing is surely the most basic and relevant of all forms of expression. Nothing else can so effectively give outward form to an inner experience. Poetry and music exist in time. Painting and architecture are a part of space. But only the dance lives at once in both space and time. In it the creator and the thing created, the artist and the expression, are one. Each participant is completely in the other. There could be no better metaphor for an understanding of the . . . cosmos.

Lyall Watson (*Gifts of Unknown Things*)

The metaphor of dance is a very apt way to portray the unfolding and increasing complexity of plant-life on Earth. The dance of plants is the dance of plant form in space and time. From a reductionist point of view, the conversion of solar energy is what plants are really all about, either at the level of the individual, or the community, or even in the characteristics of the plant-life of a given region. Form, is the physical expression of the energy captured and transformed by plants, and it provides the basis for all ecological relationships. It is not surprising then that, broadly speaking, the plants of tropical regions that have access to the greatest input of radiant energy also have the greatest exuberance, while those of energy- and nutrient-limited environments, such as alpine moorlands and bogs, have a more restricted range of body plans.

In the continuum of time the dance of plants is both developmental and evolutionary. From this perspective the unity of all life can be seen in its infinite diversity. No longer can organisms be viewed in isolation but must be seen in the context of environment – they **are** environment. The dancers are the plants and the music is their physical and biotic relations with their environment. They are simultaneously the creators and the created for they themselves contribute to the music.

As the orchestra of life tuned up, the first steps of sub-cellular and cellular structure and physiology were rehearsed. Initially it was a slow dance and the first notes of the evolution of life were the solar and thermal energy driving the chemistry of simple living organisms. The overture only hinted at what was to come and, for a long time, there was a simple melody where the principal players were not heard and the dancers were few, but even at an early stage the dance was one of innovation and improvisation. It was a dance of increasing sophistication accompanied by harmonies in a major key as plants arose. They were the first truly terrestrial organisms and they transformed the landscape making it habitable for other organisms.

The dance of plants is complex beyond our wildest dreams. Plants perform epic dances of cooperation and competition. They dance with their environment, adapting in step with it and modifying it, by cooling the air, changing atmospheric carbon dioxide concentration, providing oxygen, making soil and by altering the relative abundance of the biotic components. They dance with each other in complex communities, exploiting water, mineral nutrients and sunlight, each finding a place to grow. They dance with other organisms, avoiding or repelling herbivores, attracting and feeding pollinators and dispersers of seeds and fruits, and cooperating with fungi to exploit the soil's nutrients. There is an endless variation in the music and the dance, and the degree of complexity of their interrelationship.

The growth of plants from seed is the source of some powerful metaphors for human life but mostly plants do not have immediate impact on us in terms of their adaptive evolution and developmental processes. We appreciate them more for their beauty of form and colour, and grow them in our gardens and homes to lend harmony to our lives and as a reminder of wild nature. There may be more to 'phyto-psychology' than we realise. Humans have highly developed senses of colour and spatial order and there may be a connection here with our love of highly symmetrical plants such as cacti and succulents, or rosette plants such as African violets and primulas. Many bird-pollinated species such as fuchsias and columneas with their bright scarlet flowers, or herbs of the rainforest floor such as marantas with their strange metallic pigments, are perennial favourites in our homes.

Plants lack the spontaneity of animals, whose movements, grace, complex behaviour, and often intricate and bizarre colours and patterns attract us in profound yet familiar ways. Animals arouse our curiosity. They are like us in so many ways, yet are different, and this novelty requires investigation. Plants live in a different time dimension and television documentaries often resort to the use of time-lapse photography in order to 'animate' plants. This is perhaps unfortunate because it fails to convey the true nature of the relationship between the spatial and temporal organisation of the plant world.

While plants could also be said to lack the 'aloofness' that is so tantalising about wild animals, we can easily touch plants and we can imagine that they pose for our photographs, but they still remain somewhat alien. Their texture is not that of the animal, although we can be intrigued when some leaf textures seem fur-like. Plants appear to lack movement or, if they do move, we are bemused. We know they are formed by the conversion of radiant energy, but the nature of their nutrition remains mysterious, and when they occasionally devour insects we are amazed. They are living organisms but we cannot quite comprehend the nature of their experiences of the world, what it means to actually be a plant. Perhaps it is no great surprise that some of the earliest space invaders of science fiction were plant-like creatures, the triffids.

Ironically, plants are so much part of our environment that we also tend to take them for granted, and that is part of the problem for conservation. How do we become aware, how do we redirect our attention? To comprehend the grandeur of these organisms, a visit to the silent groves of coastal redwoods of California, the towering dipterocarps of a Bornean rainforest, or the remnant primeval kauri forests of New Zealand may be necessary. For others it requires the crazy kaleidoscopic colours of an alpine meadow, or a desert after rain, to take the breath away. But plants also impress us on a tiny scale. Some of the loveliest flowering plants are tiny ephemeral beauties that can be found only on the highest mountains. But, at this scale there is still much to be seen in our immediate, even urban, environment, especially the enchanting, if largely unsung, world of bryophytes. On an even smaller scale is the world of plants through the microscope. We can remember the first time we viewed the jewel-like appearance of moss leaf cells through a microscope, a truly wondrous sight.

Aesthetic appeal will probably have a more profound influence on the conservation of plants than economic arguments, and to encourage the conservation of the world's flora is one of the main aims of writing this book. For more than 30 years we have studied plants in laboratory and field and they have led us to some very exciting places as well as the more mundane. Even industrial slag-heaps have provided raw data for theories of plant adaptation. Beautiful and fascinating plants are everywhere, from the bryophyte communities of old walls, to the scattered plants holding a tenacious grip on the scree slopes of glaciated mountains, or the weedy fringe at the high-tide marks of sandy seashores. Even old derelict buildings can be a source of pleasure. When travelling in the middle of a city such as London one can see buddleias, growing in such incongruous sites. The wonder of being a botanist is that literally almost anywhere you can find something beautiful and fascinating. In the words of Alan Paton from his moving novel, *Cry, the Beloved Country*:

> *. . . the train passes through a world of fancy, and you can look through the misty panes at the green shadowy banks of grass and bracken. Here in their season grow the blue agapanthus, the wild watsonia, and the red-hot poker, and now and then it happens that one may glimpse an arum in a dell.*

The writing of this book has taken much longer than we intended, and many of our ideas have evolved in keeping with the progress of the book. Inevitably this meant more changes. Originally, our plan was to write a celebration of plant diversity as a successor to *Diversity and Evolution of Land Plants* (Ingrouille, 1992). However, it soon became obvious that there was a definite need for a new kind of approach, one that would go beyond the bounds of conventional textbooks, of which there are several excellent examples already available for students. The research for such a book meant that the material we acquired would fill several volumes, so painful decisions were made to cut the ever-expanding prodigy down to an acceptable size. Meanwhile, other events, including a lengthy research post overseas, intervened to delay publication even further.

Some of our personal views of plant-life and the ideas expounded in this book might be considered unorthodox by the standards of mainstream science. For example, we have aimed to bring into the foreground the work of botanists whose work no longer fits current orthodoxy, but whose views we believe still have value today. There are past masters, such as Goethe, Hoffmeister, Church, Arber, and Corner, and undoubtedly many others, to whom we are happy to pay our dues, as well as those whose works we have consulted for this book. In the words of John Bartlett,

> *I have gathered a posie of other men's flowers, and nothing but the thread that binds them is mine own.*

Generally speaking, we believe that science and art are but two ways of comprehending the world, two forms of creativity, and that the scientific method, particularly in the realm of botany, could be applied in a more phenomenological way, and even augmented by intuitive approaches. Like art, science provides a way of knowing, of making sense of the world, but the best scientists must go beyond the scientific method. Current scientific procedures and methodologies are inadequate to explain much of the complexities of plantlife, which often require subtle, broader-based holistic approaches. For example, we have always been struck by the similarity of forms throughout many unrelated plant families, be it at the level of gross morphology or confined to the flower. Such phenomena are usually explained away as instances of parallelism or convergence (or homoplasy, to use a currently popular term), and the explanation is always framed in Darwinian terms of adaptation and natural selection. However, we feel that there is a deeper, underlying law of form or morphogenesis that constrains expression of form to within certain boundaries, and which cannot be understood simply in terms of linear cause and effect. From a holistic perspective, the genome may also be portrayed as a self-organising network capable of producing new forms of order. In addition, the aesthetic dimension has undoubtedly great potential in promoting empathy for plants at the personal level as well as a more widespread conservation ethic.

It is unfortunate that, in this age of instant information, general botany and its long history are no longer taught, at least to the extent that we would prefer. How we react to plants and how we ultimately treat them is intimately bound up with our ways of regarding them. Western science, at least since the time of Descartes and Bacon, has promoted the idea that plants and other living organisms are objects (*res extensa*) existing in isolation from the subject observer (*res cogitans*). The disinterested objective method became the scientific method and a cornerstone of the philosophy of science. We strongly believe that this philosophy is flawed and has contributed to many of the difficulties facing science today.

Thus, initially it may be difficult for some students to get situated in this book, to see it in its entirety, for, at first sight, the combination of different approaches is apparent. We make no apologies for this because we do not reject the advances made in botany over the last four hundred years. There is no doubt that reductionist science has been singularly successful in elucidating much of our current knowledge of plantlife, particularly relating to anatomy and physiology, and in the fields of genetics, development and systematics. However, in the age of the expert, plant science courses in universities are often so narrowly specialised that we are in danger of losing sight of the plants altogether, and therefore we feel that certain new approaches or new perspectives are needed.

The traditional role of the amateur is the foundation upon which botany was built. Without disparaging the importance of modern computerised methods, and molecular and theoretical developments, we encourage a return to a broad approach to botany that would reinstate the importance of the amateur. Botany is an immense and deeply satisfying subject and one that we can attest to providing a lifetime of riches and rewards. It is therefore difficult for an undergraduate to get the flavour of botany in three or four short years, especially to develop a feeling for plants, and to understand the role of plants in diverse ecosystems.

Where possible, we have tried to keep abreast of the multifarious changes that have revolutionised so much of current biology in recent years. Chapter 1 has been strongly influenced by developments in complexity theory, including phenomena such as hypercycles and autopoiesis (see Kauffman, 1993). It was felt necessary to touch on such topics in order to give as complete a picture of the events leading to the early evolution of plant life, and for this reason we have also included many aspects of the evolution and diversity of the algae, although technically we would normally exclude them from the category 'plant'. There are several excellent and complementary texts on the biology, evolution and diversity of the algae that we recommend.

In Chapter 2, although we have basically adopted a conventional reductionist approach, we have tried to integrate this with some of the most recent ideas in plant morphology and developmental genetics, including the 'theory of morphospace'. Much of this chapter was influenced by the 'process morphology' of Rolf Sattler and his colleagues, although the philosophy behind this approach goes back to A. N. Whitehead (see Whitehead, 1929), in addition to more recent theories on developing and transforming dynamic systems and biological form (Webster and Goodwin, 1996). There is no doubt that morphology and developmental genetics has benefited from this trend away from a static typology to a more dynamic process-orientated approach but it has to be admitted that, by including the dimension of time, practical difficulties in the analysis, interpretation, and description of form are also introduced. This is particularly the case with respect to descriptive morphology and the use of homology in classification.

There have been changes in the world of evolutionary botany over the past 20 years. The familiar Neo-Darwinian paradigm is being augmented by views that see evolutionary change as a result of life's inherent tendency to create novelty, and which may or may not be accompanied by adaptations to changing environmental conditions. Some believe that we are in the process of a paradigm shift (in the sense of Kuhn) while others believe that the Neo-Darwinian paradigm is sufficient to explain evolution, or that it only needs some amendments. There is no doubt, however, that, in biology, we are witnessing a general move from a mechanistic world view to a systems view of life involving the triple helix of phenotype, genotype and environment.

In Chapters 3 and 4, we have tried to explore the processes of evolution and plant reproduction within an evolving Darwinian framework that gives more weight to phenotypic plasticity and the ability of plants to harmonise their form and life cycles with changing physical parameters, rather than to simply view plants in more orthodox terms of mutation and selection within populations. We have also tried to emphasise the recognition of both constraint and relaxation in form-making and the resultant phenomena of convergence and novelty, respectively. In addition, we have highlighted processes that might be pertinent to the evolution of plants, especially the founder effect on island populations, and those that may result in major genomic and morphological reorganisation. The reciprocal relation of space and time with form is central to Leon Croizat's panbiogeography and this approach to plant distribution has much to commend it rather than the viewpoint whereby organisms are treated a priori within the framework of a simple dispersalist model.

In Chapter 5, we have used the arrangements of plant families that have resulted from the most recent findings of molecular systematics. Of course, this may be a highly controversial and somewhat contradictory stance, especially in view of what we say about methodologies. However, we believe that this provides the student with the best means of gaining access to, and evaluating, current developments in plant systematics. Within the realm of plant systematics we take the view that cladistics and molecular methods are only several ways of handling data, and that a pluralistic approach involving time-honoured methods (e.g. morphology and biogeography) is essential.

In Chapter 6, which is an overview of the world's flora, we have deliberately taken an adaptationist approach knowing full-well the pitfalls of 'the adaptationist programme', which were so elegantly exposed by Gould and Lewontin in their seminal paper 'The Spandrels of San Marco and the Panglossian Paradigm: A Critique of the Adaptationist Programme' (1978). A naïve interpretation of functional morphology is certainly to be avoided but we feel that there is an overwhelming heuristic value in the adaptationist approach and, if soberly used, it can be an invaluable teaching aid and inspiration for students. Story-telling is fundamental to humans and can be the most effective way of inspiring an empathetic relationship with the plant world.

The earliest botanists were herbalists and plants were studied mainly for their culinary, curative and magical properties. In Chapter 7 we have emphasised some of the most important uses of plants by humans, in addition to some of the more worrying aspects of globalised food production and distribution. For example, in many western countries, the larger supermarkets now stock a diversity of fruits and vegetables from around the world that would rival some of the traditional fruit markets in places such as Malaysia and Thailand. One wonders what the effect of such large-scale imports will have on local economies and traditional crops. Today, plants sustain a multi-billion dollar global pharmacy industry, and a growing research and development programme for genetically modified crops, but there has been a backlash to all these so-called technological improvements to our food supplies. There has been a tremendous resurgence of interest in recent years in herbal medicines, vegetarianism and organic farming.

All this has been happening at a time when we are witnessing widespread disaffection with modernity. We are now more acutely aware of the impacts of technological/industrial activities on the climate, and on plant and animal life of the planet, as well as the gross inequalities in human societies, owing to an unrestrained desire for material wealth and consumer goods. We suggest that political answers to these problems are, in reality, only short-term solutions, and that we will only realise a paradigm shift to a more eco-centric way of living in harmony with the Earth when, as individuals, we adopt a transpersonal way of relating to other living organisms. This is the essence of the movement known as Deep Ecology that was first formulated by the Norwegian philosopher Arne Naess. A *poesis* of life or 'living poetically' is what we try to live up to in our relationship with living organisms and the environment.

The evolutionary dance of plants has taken place during the past 400 million years. In the past 10 000 years a different tune in a minor key has been heard as plants have begun a new dance with humans. They have been manipulated and transformed by us for food and materials, and have enabled human civilisations to evolve. Simultaneously we have also damaged and destroyed much of the plant life on Earth and rendered numerous species extinct or nearly so. The book is about the relationships between plants and humans, how we perceive them, form concepts of them, study and analyse them, and enjoy them, although it does not provide clear answers as to why we do this. In the third millenium we need to adopt a new philosophy for the planet we inhabit and all its unique life-forms if we are to survive. We have tried to steer clear of metaphysics, but maybe we also need to retain a sense of the mysteries of life, especially if we are to develop a sane and non-exploitational relationship with the Earth.

Evolution is the polestar of the biological sciences, and this book says a lot about the evolution of plants, but it goes beyond scientific concerns to embrace our intuitive processes, our aesthetic senses and the human ability to wonder and to imagine. According to Wordsworth, imagination is 'reason in its most exalted mood'. Therefore, we have given much emphasis to the visual aspect of plants, their form and colour, and have promoted a return to a more 'in-depth seeing' as exemplified by phenomenology. The phenomenological method tries to take into account the subjective feelings of the observer within a more dynamic framework of observation and concept-formation ('reciprocal illumination'). To do this effectively we also have to have some grounding in epistemology, and therefore we have provided an outline in Chapter 8 of the philosophical traditions that impinge on botany, as well as the major developments in its history. Phenomenology was essentially the way of Goethe, and consequently his much maligned and overlooked contributions to botany are given due consideration.

We hope that this book will provide a much-needed stimulus to the student of botany with an inquiring mind, particularly advanced undergraduates, but it is not designed solely as a university textbook. It is also aimed at all who enjoy plants for their form and beauty but want to delve deeper into their complexity, their ecology, evolution and development, and who, hopefully, will find inspiration and seek out other sources of knowledge. We have tried to bear in mind Corner's words about botany texts.

> *. . . the books that deal with general botany have grown so tediously compendious, so canalised in circuitous fertility, so thoroughly dull and dully thorough*

The interaction with plants can invoke feelings of empathy, but the sheer pleasure of discovery, of finding things out, can invoke feelings of revelation. We have tried to present the material in a way that will stimulate the reader to find pleasure and wonder in the world of plants, much of which is unknown, and probably will remain unknowable. At the end of each chapter we have listed only a fraction of our sources but, hopefully, these works should provide a gateway to the larger literature.

....

for the question is always

how

out of all the chances and changes

to select

the features of real signficance

so as to make

of the welter

a world that will last

and how to order

the signs and symbols

so they will continue

to form new patterns

developing into

new harmonic wholes

so to keep alive

in complexity

and complicity

with all of being -

there is only poetry.

(Kenneth White, 'Walking the Coast')

Chapter 1

Process, form and pattern

. . . an autopoietic system is a homeostat . . . a device for holding a critical systemic variable within physiological limits . . .: in the case of autopoietic homeostasis, the critical variable is the system's own organization. It does not matter, it seems, whether every measurable property of that organizational structure changes utterly in the system's process of continuing adaptation. It survives.

S. Beer, 1980

1.1 Living at the edge of chaos

This chapter provides a short history of the pre-biotic Earth and of organisms in the early stages of the evolution of life. It covers the origins of photosynthetic organisms, the setting of the stage for the evolution of plants and terrestrial ecosystems, and for the subsequent diversification of plants from the Silurian Period onwards. Key early events are the evolution of metabolism, including photosynthesis, of mechanisms of heredity and of cells. Later symbiotic associations between cells provide a much broader canvas for life-forms to diverge. Other important stages in the evolution of plants were the origin of multicellularity and subsequently the functional specialisation of cell types in the multicellular organism.

Process, form and pattern are three primary features of living systems. In this section we focus individually on each of these primary criteria of life. Process first, concentrating on the origin of the processes fundamental to life, and particularly to plants – photosynthesis. Then we focus on form, by describing some key aspects of the evolution of complex cells. Finally we look at pattern – cells together in multicellular organisms.

Using musical metaphors we trace in this section the origins of life from the white noise of chaos to the full symphony of life. The first notes of life are the complex molecules and beating out with the drum of metabolism. At first the noise is cacophonous as if the orchestra is tuning up, but with the origin of cellular life, coordinated metabolism arises, like snatches of melodies. Gradually at first, but then more and more speedily, as the rhythms of cellular life assert themselves, the first snatches of melody grow louder against the cacophonous background. Simple melodies are taken up and repeated in counterpoint as the seas and lakes become populated with living organisms, some complex and multicellular. Later symbiotic associations between cells, like the origin of musical harmony, provide a much broader potential for new life forms to diverge. The origin of multicellularity and subsequently the functional specialisation of cell types in the multicellular organism enrich the sound. At the margin of land and water some of these themes were to be taken up and elaborated by the first plants.

1.1.1 The pre-biotic Earth

The probability of life evolving is so small that is seems impossible, yet in the aeons that passed from the formation of the Earth the almost impossible became the probable. The key to understanding this distant past is in the present. All life is built on what has gone before and in order to understand how life evolved we must study the common metabolic processes that connect all living organisms, but we have to seek life's origins in processes of chemical evolution that occurred on the pre-biotic Earth.

Geological eras	Dates started (millions years ago)
Cenozoic	65
Mesozoic	250
Palaeozoic	570
Sinian	800
Riphean	1650
Animikean	2200
Huronian	2450
Randian	2800
Swazian	3500
Isuan	3800
Hadean	4650

The Earth is at least 5 billion years old and has been changing all the time. About 4.6 billion years ago, and for about 1 billion years thereafter, our planet was cooling and an atmosphere consisting of hydrogen and helium, and continental crust was forming. Then about 3.5 billion years ago the stage was set for the grandest chemical experiment, that was to create life.

At this stage the world was a huge laboratory test-tube and was constantly subjected to intense electrical storms, meteoric impacts and volcanic eruptions, and, because the Earth was not shielded by the oxygen-rich atmosphere that we have now, it was bombarded by ultra-violet (UV) and gamma radiation. There was a steady input of molecules from the out-gassing of volcanoes. There was also the input of complex molecules based on carbon (organic molecules) from meteorites. The steady intense energy of radiation and the cataclysms of storms and volcanic eruptions forced chemical elements to combine or compounds to break apart, setting off a myriad tiny fireworks, and sparked life into being. These chemical reactions were orderly, determined by the atomic structure of the elements and they happened again and again so that the products of particular reactions became more and more abundant.

It was hot because of high levels in the atmosphere of carbon dioxide (CO_2) and methane (CH_4) produced by volcanic activity. Hydrogen, hydrogen sulphide, hydrogen cyanide and formaldehyde were also present. These conditions have been replicated in the laboratory in

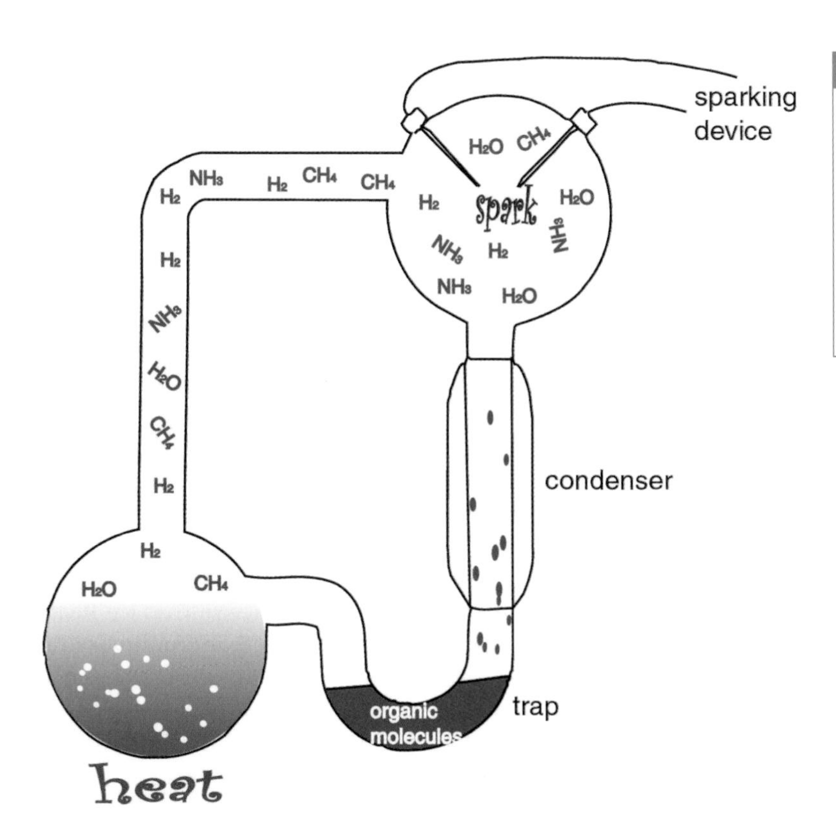

Figure 1.1. The Miller/Urey experiment. A continuous electric current was passed through an 'atmosphere' of methane (CH_4), ammonia (NH_3), hydrogen (H_2), and water (H_2O) to simulate lightning storms. After a week 10%–15% of the carbon was now in the form of organic compounds including 2% in amino acids.

the classic Miller/Urey experiment (Figure 1.1). Gradually more stable and more complex compounds were produced and accumulated but this was not yet life. For that a level of complexity had to be achieved that was self-sustaining and growing.

A vital component of the living mixture was the most important compound to accumulate at this early stage, water. It was almost the most simple molecule, made from a single oxygen atom and two hydrogen atoms. Together with other gases such as ammonia and methane, water formed in the atmosphere, and began to fill the pre-biotic ocean basins. The oceans were very warm, slightly acidic and rich in dissolved ferrous ions (Fe^{2+}), carbon dioxide (CO_2) and bicarbonate ions (HCO^-). A continuous process of chemical evolution led to a great diversity of molecular species that formed compounds possessing emergent properties not possessed by their constituent elements. For example, water has the properties of a liquid not possessed by either of the gases oxygen or hydrogen. Indeed water is a pretty unique liquid and life without water is only conceivable in science fiction.

Water has remarkable properties because although it is a very small molecule it has a very strong polarity from an uneven distribution of positive and negative charge, giving it a kind of stickiness. Consequently water molecules tend to join loosely together and stick to other charged atoms or molecules. Since the hydrogen atoms in the water molecule are involved this is called hydrogen bonding. Strong

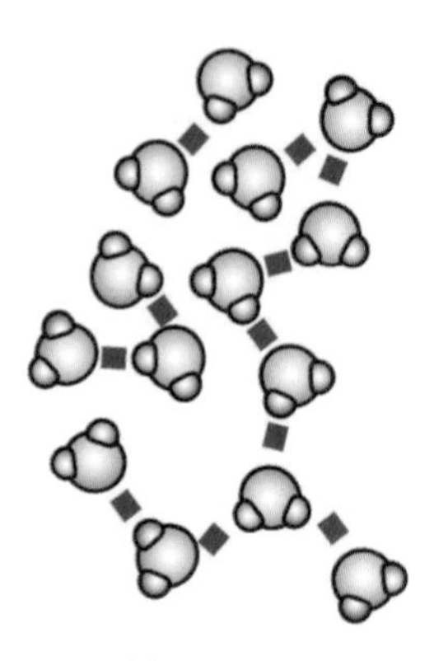

Figure 1.2. The asymmetric arrangement of hydrogen atoms leads to an unequal distribution of charge across the water molecule and attraction between the hydrogen atom of one molecule and the oxygen of another.

hydrogen bonding makes water an excellent solvent. In aqueous solution ionic compounds break down into their constituent ions each surrounded by a halo of water molecules.

Other polar molecules also dissolve readily in water. Water also takes part in many chemical reactions. By condensation large organic molecules, made up of a skeleton of carbon and hydrogen, are built up through the formation of a covalent bond and the elimination of water. Large organic molecules can also be broken down by the addition of water as covalent bonds are split by hydrolysis. As more complex compounds accumulated and became more concentrated, their formation and destruction established the first elements of living metabolism, the constant cycle of building and breaking, anabolism and catabolism, the work of life.

The stickiness of water also gives it remarkable physical properties. It has a high heat capacity so that it buffers aqueous systems from large temperature changes. In addition, as liquid water evaporates it cools the remaining liquid; and when it freezes the water molecules form an ice lattice taking up more space so that ice floats providing an insulating blanket. Water had a profound influence on the origin of life not only at the smallest scale, that of metabolism, by influencing chemical interactions between atoms and molecules, but also at the largest scale, that of the whole Earth, by buffering it from temperature extremes.

1.1.2 Complex molecules and self-organisation

The conditions on Earth before life began favoured the progressive evolution of complex molecules that had the ability to self-organise and replicate. These precursors of living chemical systems must have been stable, with the ability to correct replication errors. They must also have been capable of inheriting favourable replication errors. The ability to change over time became established, and, in this respect, these molecules are quite unlike non-living matter. Self-replication is a catalysed reaction, and catalytic cycles play an essential role in the metabolism of living organisms. In its simplest form, a living system may be modelled as an autocatalytic chemical cycle, but these self-organising molecules can hardly be called *living* because they are limited by factors that are independent of the catalytic process.

Living systems can maintain their existence in an energetic state that is relatively stable and far from thermodynamic equilibrium. They have been called dissipative structures by Ilya Prigogine. In contrast, thermodynamic equilibrium exists when all metabolic processes cease. These hypothesised dissipative systems must have possessed multiple feedback loops in the manner of catalytic cycles, what have been termed 'hypercycles' by Manfred Eigen. Hypercycles are those loops where each link is itself a catalytic cycle. Almost every pathway is linked to every other pathway in some way. As chemical instabilities originate the system is pushed farther and farther away from equilibrium until it reaches a threshold of stability. This hypothetical point is called the bifurcation point and it is at this stage that increased complexity and higher levels of organisation may emerge spontaneously.

If we apply the above ideas to living systems we can also say that living systems exist in a poised state far from equilibrium in that boundary region near 'the edge of chaos'. Evolution may favour living systems at the edge of chaos because these may be best able to coordinate complex interactions with the environment and evolve. In such 'poised' systems most perturbations have small consequences because of the system's homoeostatic nature but occasionally some cause larger cascades of change.

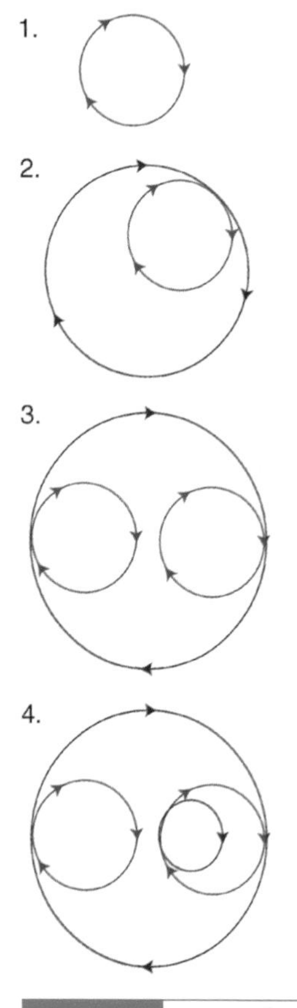

Figure 1.3. Four stages in the evolution of a hypothetical hypercycle: each loop represents a catalytic cycle like the citric acid cycle, or the production of a series of autocatalytic enzymes.

Living systems can be conceptualised as maintaining such hypercycles, thus allowing for evolutionary change without loss of the cyclic processes themselves. Living organisation is manifested therefore, not in the properties of its components, but in processes and relations between processes, as realised through its components, and in the context of the environment. Matter and energy continually flow through it but it maintains a stable form through self-organisation. This self-making characteristic of living systems has been termed 'autopoietic' by Humberto Maturana and Francisco Varela. Paraphrasing the cyberneticist Stafford Beer quoted at the beginning of the chapter, every measurable property of the system may change while it maintains itself. It is its continuation that is 'it'. Autopoiesis is a network of production processes in which the function of each component is to participate in the production or transformation of other components in the network. In this way the entire network continually 'makes itself'; the product of the operation is its own organisation. It becomes distinct from its environment through its own dynamics. It is in this context that we can recognise the three criteria of life: pattern, form and process.

Autopoiesis = the process by which an organisation produces itself.

One of the best examples of an autopoietic system is the complete set of genes in an organism, the genome, which forms a vast interconnected network, rich in feedback loops, where genes directly and indirectly regulate each other's activities. At its simplest in transcription and translation the DNA sequence of genes provides the template for an RNA sequence (transcription) that codes for a polypeptide (translation) that may be required for either the processes of transcription or translation, or even DNA replication. But it is much more complex than that. The genes are only a part of a highly interwoven network of multiple relationships mediated through repressors, depressors, exons, introns, jumping genes, enzymes and structural proteins, constantly changing, evolving.

The autopoietic gene system does not exist in isolation but as part of the autopoietic living cell. The bacterial cell is the simplest autopoietic system found in nature, though it is hugely complex. Simpler autopoietic structures with semi-permeable membranes (but lacking a protein component) may have been the first autopoietic systems before the evolution of the cell. The evolution of autopoiesis was undoubtedly a landmark in the history of the Solar System, but almost 1 billion years were to elapse before the evolution of the first cells and the beginning of life at about 2.5 billion years ago.

1.1.3 The RNA world

A protobiological system (called a 'chemoton' by Tibor Ganti) should consist of a minimum of three sub-systems: a membrane, a metabolic cycle, and some genetic material. In the development of primordial living systems some sort of compartmentalisation such as a vesicle was necessary.

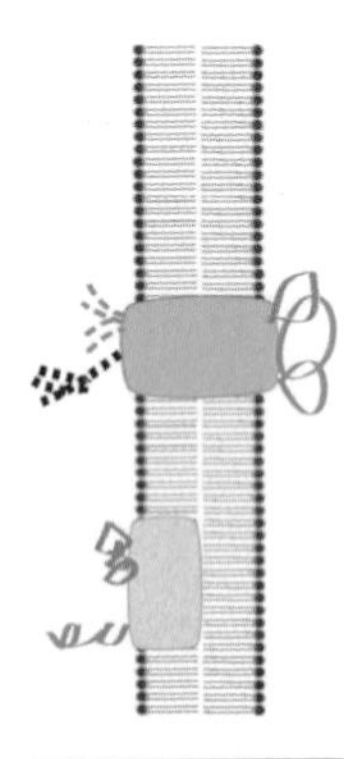

Figure 1.4. A bi-lipid membrane showing the hydrophilic heads situated on the surface of the membrane and the hydrophobic tails in the middle of the membrane. Various proteins float in or on the membrane.

Lipids and nucleic acids are complex organic molecules in which carbon-based chains form the main structural components. Carbon atoms have an outstanding capacity to combine with each other and with other kinds of atoms to produce an unlimited morphological diversity of molecules. A key feature must have been vesicles formed from fatty acids. Fatty acids are organic molecules with a long water-repellant (hydrophobic) hydrocarbon tail and a hydrophilic polar head. They orientate with their tails together and the heads towards water, and consequently form globules or two layered sheets called membranes. Membranes provide the outer layer of vesicles. At the earliest stages of life membrane-bound vesicles probably formed in shallow tidal pools as a consequence of repeated cycles of desiccation and rehydration. Only certain molecular species possessed the necessary characteristics for living systems; of forming membranes sufficiently stable and plastic to be effective barriers and to have changing properties for the diffusion of ions and molecules. Such membranes were necessary for the formation of organic molecules such as nucleotides that had the potential to act as catalysts and to replicate.

Because of some extra properties of the membrane, imparted by other molecular components floating in it, vesicles can contain a solution with a different chemical constitution to the surrounding aqueous solution. They are semi-permeable, completely permeable to water and some other small molecules, but less permeable to other molecules, so that they can encapsulate and keep large molecules concentrated.

Reactive molecules are called radicals. The appearance of autocatalytic networks of carbon-based radicals, containing one carbon atom (plus hydrogen, oxygen and nitrogen) and organic compounds such as sugars and acids could lead to the evolution of simple enzyme-free metabolic pathways. However, the synthesis of more complex potentially replicating chemical compounds is problematical. It is now thought that the early evolution of life was dominated by the nucleic acid RNA, and that the original genetic material was an RNA analogue. Like DNA, RNA is a series of four different nucleotide bases strung together; differences in the sequence of bases, the four-letter alphabet, gives limitless variation in the molecule, providing a language. RNA also has catalytic properties. For the evolution of RNA to occur, some sort of intermediary mechanism must have occurred within the vesicle, for example, a polynucleotide analogue of RNA could have been replicating within chemoton-like systems.

One key feature of the nucleic acids like RNA and DNA is their ability to splice together; parts of the molecule can be looped out or into the sequence of bases. The parts of the sequence excised are called introns and those spliced together exons. Thus, in the evolution of life before the emergence of bacteria, we envisage an 'RNA world' where some molecules are active enzymes, others contain introns and exons and convert themselves, either to RNA by self-splicing, or recombine to yield novel combinations by trans-splicing. Subsequently DNA took the replication and information-storing role, and proteins the catalytic role, and RNA was left as an intermediary. In our 'DNA world' proteins have taken over almost every catalytic activity.

In a chemical system change is likely to extinguish a chemical reaction, but a living system has the potential to change without destroying the circular processes that makes its components. There is change because self-replication is not perfect and slightly different but stable daughter molecules are sometimes produced, but the living system continues to replicate instead of spluttering to a halt. The system could evolve because some of these altered daughter molecules had an improved ability for autocatalysis as if they 'remembered' the changes that brought them about. This was the birth of inheritance. With the combination of self-regulating hypercycles and inheritance, the brake was taken off chemical evolution and new kinds of metabolism evolved.

Creativity, the generation of novelty, is a key property of all living systems. A special form of creativity is the generation of diversity through reproduction, from simple cell division to the highly complex dance of sexual reproduction. Driven by the creativity inherent in all living systems the life of the planet diversified in forms of ever-increasing complexity.

> Life emerged, I suggest, not simple, but complex and whole, and has remained complex and whole ever since – not because of a mysterious élan vital, but thanks to the simple profound transformation of dead molecules into an organization by which each molecule's formation is catalyzed by some other molecule in the organization. The secret of life, the wellspring of reproduction, is not to be found in the beauty of Watson–Crick pairing, but in the achievement of collective catalytic closure. So, in another sense, life – complex, whole, emergent – is simple after all, a natural outgrowth of the world in which we live.
>
> Stuart Kauffman, *At Home in the Universe*, Oxford University Press 1995 pp. 47–48.

1.1.4 How to recognise a living system

The age of the microcosm lasted (from about 3.5 billion years ago) for about 2 billion years, during which time many of the metabolic processes essential to life evolved. These processes include fermentation, nitrogen fixation, and oxygenic photosynthesis, the most important single metabolic innovation in the history of life on the planet. About 1.5 billion years ago self-regulation of the biosphere and an oxidising atmosphere were established, setting the stage for the evolution of macrocosmic life.

It is the traces of patterned cellular structure in rocks (and chemical processes in sediments and atmosphere) that provide the first hard evidence for the presence of life. The earliest traces date back to the early Archaean age 3500 million years ago from several parts of the world. The fossils are recognisable because they are composed of alternating dark and light layers of sediment. The fossil structures can be understood by reference to living stromatolites, 'living' rocks found in shallow water that grow in layers consisting of alternating

Figure 1.5. Stromatolites in Laguna, NE Mexico. Changes in lake level here exposed them above the surface of the water.

mats of photosynthetic microbes, cyanobacteria, and precipitated calcium carbonate. The cyanobacterial mats trap sediment and the photosynthetic activity of these microbes precipitates a layer of calcium carbonate on top. Eventually the microbes establish a new living layer on top of the calcium carbonate layer. The alternating light and dark bands of fossil stromatolites are the earliest evidence of a living process. The living examples, discovered only in the last century in Shark Bay in Australia, are often mentioned, but stromatolites are found in a few other places in the world such as Laguna in NE Mexico (Figure 1.5). The earliest kinds are cone-shaped fossil stromatolites (*Conophyton*) similar to living stromatolites from the hot springs of Yellowstone National Park in the USA.

It has been suggested that some fossil stromatolites may have a purely physical origin, but nevertheless microbial filaments of presumed cyanobacterial origin, from the Apex Basalt of Western Australia about 2700 million years old, have also been described. The presence of characteristic hydrocarbons such as 2 alpha-methylhopanes indicates the presence of cyanobacteria long before the atmosphere became oxidising. It is probably not coincidental that the first evidence for oxygen production is found around 2.8 billion years ago at about the time cyanobacteria were colonising shallower waters.

The evolution of oxygen producing photosynthesis was a pivotal event in the history of life on Earth because it permitted dramatically increased rates of carbon production, and a much wider range of metabolism associated with novel ecosystems. By changing the atmosphere to one that was rich in oxygen it set the stage for the evolution of aerobic organisms. However, it is likely that other organisms pre-date the cyanobacteria. Numerous bacterial species capable of metabolising sulphur are found near the root of the 'Tree of Life'. Many are active at very high temperatures and are commonly found in modern sulphide-rich hydrothermal systems, such as geysers and fuming deep-ocean vents. Here they utilise chemical energy trapped in the rocks from the time of the formation of the Earth. It is in these organisms that we must look for evidence about the first stages in the evolution of metabolism including photosynthesis, because they also include species that carry out photosynthesis but do not produce oxygen.

1.2 Process: the evolution of photosynthesis

Chemical energy trapped in the rocks is a kind of leftover from the very origins of the Earth. This energy is still utilised by some microorganisms, but life would have been very limited if it had been restricted to geysers or hydrothermal vents and sediments. Photosynthesis, by harnessing an inexhaustible supply of energy, vastly expanded the possibilities of life. Today plants and some kinds of plankton are the major photosynthetic organisms but the origins of photosynthesis must be sought in bacteria.

The fundamental chemical equation of plant photosynthesis is

$$6\ CO_2 + 12\ H_2O + \text{energy from sunlight} \rightarrow C_6H_{12}O_6 + 6O_2 + 6H_2O.$$

This kind of photosynthesis is oxygenic (releases oxygen). Carbon dioxide and water are combined in the presence of energy to make energy-storing sugars. Oxygen is released as a by-product. In fact photosynthesis occurs in two main stages. In the first light-dependent stage, light energy is used to form the energy-containing compound, ATP, and to produce chemical power, mainly in the form of a compound called NADPH. Fundamentally it does this by providing electrons to compounds thereby making them chemically reactive.

There are a number of distinct events in the first stage. Light is caught by an array of pigments, acting as an antenna, and the energy of the light photons raises electrons in the pigments to an excited state. The energy of excitation is transferred via intermediates to the reaction centre (RC). At the reaction centre energy is transduced into chemical energy by the donation of an electron to an electron acceptor, which is thereby chemically 'reduced'. Then, by a series of reactions associated with electron transport, molecules storing energy (ATP) and reducing power (NADPH) are formed. In the second stage of photosynthesis, the light-independent stage, ATP and NADPH are used to chemically link carbon dioxide covalently to an organic molecule, thereby creating a sugar. Sugars are suitable molecules for the transport and storage of energy and can be broken down later in respiration to release that energy.

Any hypothesis about the evolution of photosynthesis must explain how such a complex series of events might have arisen step by step. One possible starting point is in the origin of pigments that protected the earliest living organisms from the damaging effects of ultra-violet (UV) light.

1.2.1 Pigments

The portion of a pigment molecule that absorbs light and hence imparts colour is called a chromophore. At the earliest stages it is likely that pigments evolved in a purely protective role, providing protection from UV. The amount of UV radiation was considerably higher then because of the lack of UV-absorbing oxygen in the atmosphere. The radiation reaching the surface of the Earth included the potentially highly damaging short wavelengths (UV-C, wavelength 190–280 nm) that are now completely shielded out, as well as slightly less-damaging longer wavelengths (UV-B, 280–320 nm). Even today cyanobacteria produce a pigment in their sheath called scytonemin, which strongly absorbs UV-C radiation. The presence of this pigment may explain their ability to have colonised shallow marine environments prior to 2.5 billion years ago.

Absorption of a photon of light energy in a chromophore elevates electrons to an excited state. The energy must then be dissipated in a way that does not produce toxic photoproducts. It can occur in one of four different ways:

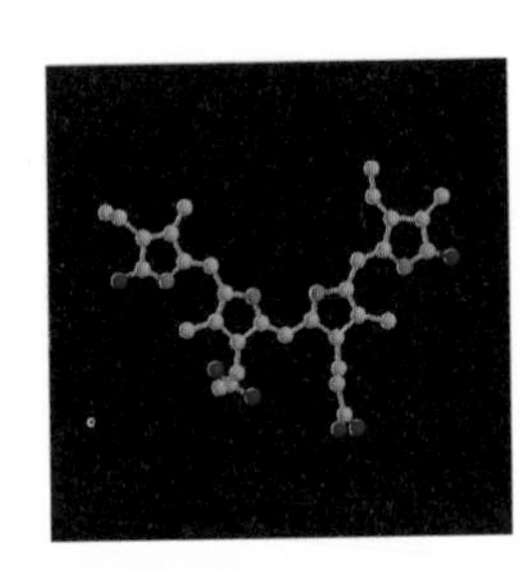

Figure 1.6. The pigment phycocyanobilin (ball and stick model: grey represents the hydrocarbon backbone, blue – nitrogen, red – oxygen).

- by emission of infra-red radiation, i.e. heat;
- by fluorescence;
- by transferring the excited electron state to a neighbouring molecule;
- by the receptor molecule becoming an electron donor.

For example the phycobilin pigments found in cyanobacteria and red algae (Rhodophyta) absorb strong light at different wavelengths and release it by fluorescing at a very narrow range of wavelengths.

Phycobiliproteins (= phycobilins) have a tetrapyrrole-based structure like haemoglobin. One kind is the bluish pigment phycocyanin that gives the cyanobacteria or blue-green algae their name. Another phycobilin called phycoerythrin makes the red algae, Rhodophyta, red. The absorbance spectra of phycocyanin and phycoerythrin pigments are shown in Figure 1.8.

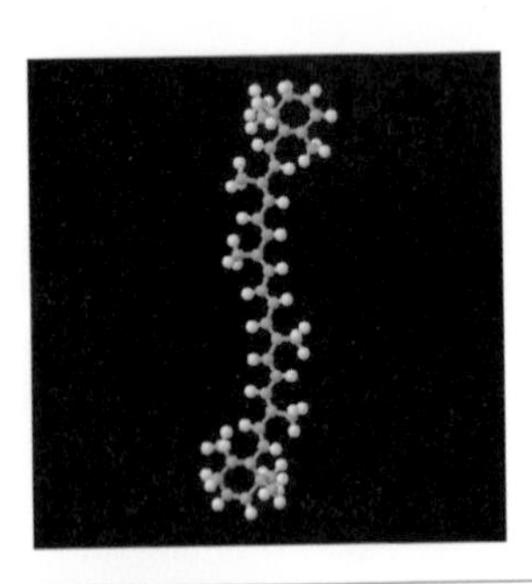

Figure 1.7. The pigment β-carotene.

Another class of pigments is the carotenoids of which β-carotene, the carrot pigment, is one. It absorbs blue light strongly and so looks orange. Others are red. Different carotenoid pigments absorb wavelengths between 400 and 550 nm. The carotenoids also have a protective role in plants though not only by shielding the cell. They seem to have gained another way of protecting the cell from damage because they scavenge toxic products such as superoxide (O_2^-) and singlet oxygen ($^1O_2^*$) that are created by absorbing light. Like many pigments, carotenoids have a ring-based structure but here with two six-carbon rings attached to either end of a long carbon chain. The carotene found in some green photosynthetic bacteria has a carbon ring at only one end. Carotenoids are soluble in lipids and are normally attached to the cell membrane or found in specialised vesicles (plastids) called chromoplasts.

Another interesting class of compounds that absorb light are the phytochromes. They are used by green plants as photoreceptors, signal-receiving molecules, directing their development depending on the quality of light. Phytochrome-like proteins may have

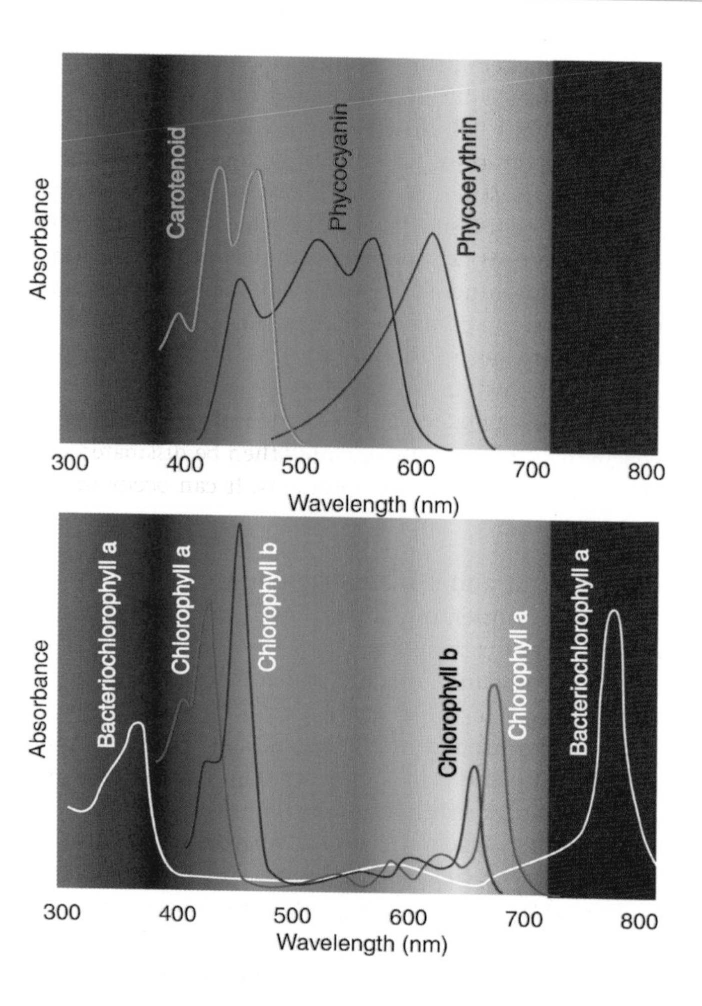

Figure 1.8. Absorption spectra of pigments involved in photosynthesis in various organisms, and the level of excitation achieved.

an ancient history pre-dating the origin of plants. For example they have been detected in non-photosynthetic bacteria, such as *Deinococcus radiodurans*, where they protect the bacterium from visible light. *Deinococcus* has a close evolutionary relationship with the cyanobacteria.

The most important photosynthetic pigments are chlorophylls but carotenoids and other pigments are also usually present and act to extend the light harvesting capabilities of the organism. They garner these different wavelengths and pass on the trapped energy to chlorophyll. Several types of chlorophyll have been identified and they all have a complex multiple ring structure, a porphyrin, like a tetrapyrrole but with magnesium at its centre. What makes chlorophylls such powerful photosynthetic pigments is the stable ring structure, around which electrons can move freely and be lost and gained easily. Different chlorophylls differ either in the form of one of the rings, as in bacteriochlorophyll compared to chlorophyll, or in the side chains, as in the different forms of chlorophyll called a, b, c, cs d, e and g.

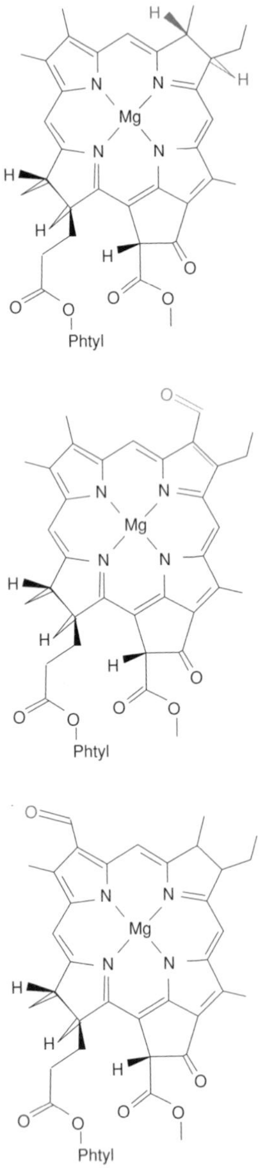

Figure 1.9. Chlorophyll pigments. Molecular structure showing how the pigments differ in the presence and position of oxygen, resulting in subtle changes in the absorption spectra of the molecules: chlorophyll a, chlorophyll b and bacteriochlorophyll.

These differences in chemical structure have the effect of modifying the wavelength at which different pigments, including the chlorophylls, absorb light (Figure 1.8) and the level of excitation achieved. This is particularly important in water or in shade because different wavelengths penetrate to different degrees. Water normally absorbs longer wavelength red light faster than the shorter blue wavelengths. The deepest living seaweeds are species of coralline red algae. Their ability to live and photosynthesise in only 0.05%–0.1% of surface irradiance is attributable to the pigment phycoerythrin, which is able to absorb in the middle ranges of the visible spectrum and then pass on the energy to chlorophyll. In shallower coastal water organic compounds from decomposing materials or released by vegetation absorb the blue wavelengths preferentially and therefore a different range of pigments are required.

1.2.2 Harvesting light and transferring energy

The first steps in the evolution of photosynthesis may have occurred by the photoreduction of carbon dioxide by iron rich clays to form the simple organic compounds, oxalate and formate. Iron remains an important component of the electron transport processes of living cells as part of cytochromes, which contain iron atoms held in place by a haem group; the iron atoms alternate between an oxidised ferric state Fe^{3+} and a reduced ferrous state Fe^{2+} as they lose or gain electrons. An earlier stage of the evolution of electron transport systems is indicated by the continued presence of non-haem bound iron–sulphur proteins. Ferredoxin, a small water-soluble iron–sulphur protein, passes reducing power from another iron–sulphur protein, the Rieske protein, to NADH, and is also an important elsewhere in electron transport. Pheophytin is another molecule involved in electron transport. It is a form of chlorophyll a in which magnesium is replaced by two hydrogen atoms.

Sulphur-containing (thio-) compounds were also important precursors in synthesis. For example acetyl thioesters polymerise to form the important electron acceptor molecule, quinone. Pheophytin passes electrons on to a quinone. Quinone is a molecule with a six-carbon ring. It is reduced to hydroxyquinone, but oxidised back to quinone when it passes these electrons on to the next part of the electron transport system. Molecular data indicate that the mechanism of photosynthesis in purple sulphur bacteria is the earliest evolved surviving type of photosynthesis. Light capture evolved from photoreduction in iron-rich clays through the use of phycobilins and carotenoids to chlorophyll pigments. Photosynthesis began in the UV and evolved through the absorption of blue, yellow, orange and red light as a consequence of bacteria colonising more productive upper layers of microbial mats where the sunlight intensity was greater. When pigments acting as sunblock did not just dissipate the energy they absorbed from sunlight, but utilised it, photosynthesis had originated. By this hypothesis, photosynthesis is one of the primary metabolic processes in the evolution of life.

In photosynthesis efficiency is gained by pigments being arranged in an antenna-like complex that funnels captured light energy to a reaction centre. In the first stages of transfer some energy is lost as heat. Different organisms have different antennae. Most cyanobacteria and Rhodophyta have phycobilins (phycobiliproteins) feeding electrons to chlorophyll a. Phycobilins are found aggregated together in a particular arrangement; one, called allophycocyanin, is attached to the photosynthetic membrane and surrounded by phycocyanin and phycoerythrin molecules. Plant chloroplasts have a photosynthetic antenna system with carotenoids instead of phycobilins feeding electrons to chlorophyll b, then to chlorophyll a, and then finally to another chlorophyll a molecule in the reaction centre. Some derived members of the cyanobacteria, called the Prochlorophytes, possess a plant-like pattern, including the possession of both chlorophyll a and b, and are without phycobilins. There is a greater diversity of chlorophyll(ide) pigments among groups of small planktonic algae than large sedentary algae. This diversity may be related to the lower degree of self-shading in the free-floating smaller organisms. They can use a wider range of pigments to exploit a spectrally more-diverse environment.

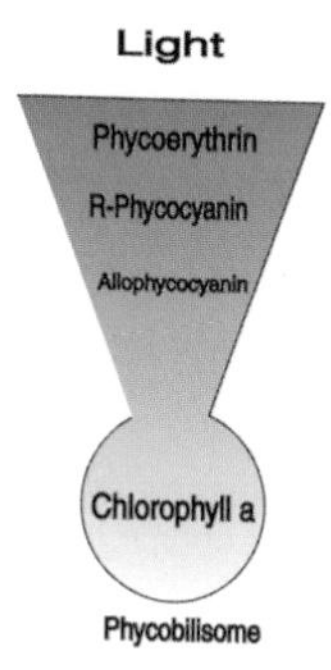

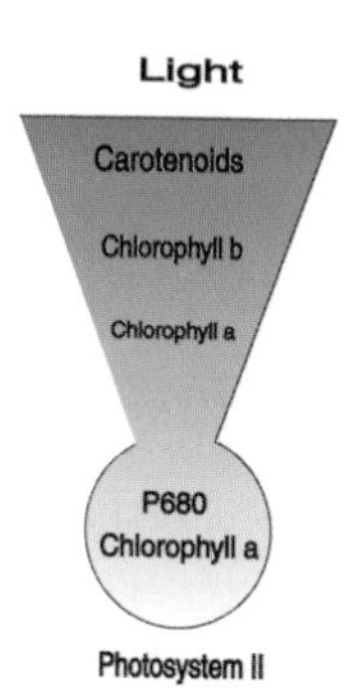

Figure 1.10. Absorption of light energy in a phycobilisome and Photosystem II of a chloroplast, showing how light is first absorbed by subsidiary pigments and the excitation passed on to chlorophyll.

In photosynthesis electrons boosted to an excited state by absorbing light are transferred from the chromophore to neighbouring molecules and with their transport down a chain of electron acceptors produce power, stabilised in forms utilisable by the cell. Photosynthesis is a process that can drive other chemical reactions. The transfer of energy by the transport of electrons permits, for example, the fixation of carbon dioxide into energy storing sugars, or the production of the energy storing compound ATP. An important electron carrier, the target molecule of the light reactions of photosynthesis, is the molecule nicotinamide-adenine dinucleotide phosphate ($NADP^+$). It is freely diffusible and when reduced by the gain of an electron to make NADPH it carries reduction potential to where it can be utilised.

There are two distinct kinds of reaction centres that differ in the form of their electron transport. They are so distinct that they may have evolved separately, although cyanobacteria and plants have both kinds. In one kind pheophytins and quinones act as intermediates and terminal electron acceptors, whereas the other kind uses iron–sulphur centres as terminal acceptors. The first kind is present in Photosystem II of plants, algae and cyanobacteria, and is the only one found in purple bacteria and green non-sulphur bacteria. The other kind is present in Photosystem I of plants, algae and cyanobacteria, and is also present in green sulphur bacteria and heliobacteria. Halophilic (salt loving) archaebacteria in anaerobic conditions carry out a different, and relatively inefficient, kind of photosynthesis

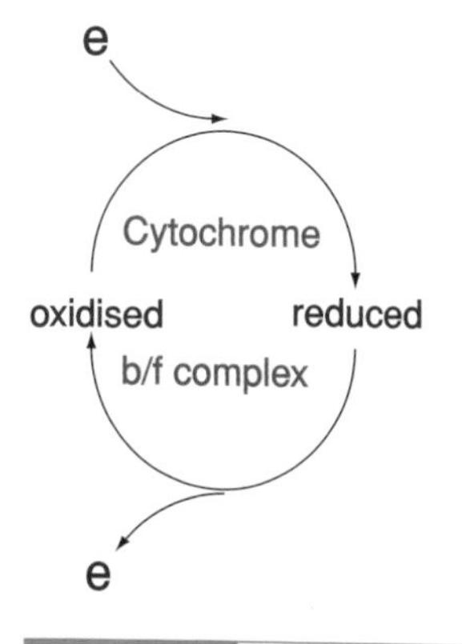

Figure 1.11. Electron transport by cytochrome.

utilising a purple pigment called bacteriorhodopsin. It may have evolved separately from other kinds of photosynthesis.

1.2.3 Anoxygenic photosynthesis

Bacteria, including photosynthetic bacteria, can be divided into two main kinds depending on their staining reaction to a procedure called Gram staining. Differences in staining are a measure of a fundamental difference in their cell walls. Heliobacteriaceae are the only Gram-positive photosynthetic bacteria. They are one of several kinds of photosynthetic bacteria that are anoxygenic, that is, they do not split water to provide electrons for photosynthesis but use other sources of electrons. The study of the Heliobacteriaceae and filamentous photosynthetic green bacteria is particularly useful for understanding the earliest stages in the evolution of photosynthesis. The Heliobacteriaceae include the genus *Heliobacter*. They have bacteriochlorophyll g, which closely resembles chlorophyll a, but absorbs wavelengths of light that can penetrate deep water. A photosynthetic reaction centre (RC-1) is embedded in the cytoplasmic membrane and contains only a core FeS (iron–sulphur cluster) and lacks an extensive peripheral antenna system. Heliobacteriaceae are strict anaerobes and reside in places like stagnant rice paddy fields and alkaline soils. Their cells are red-brown owing to the presence of a carotenoid pigment neurosporene. Although they are photosynthetic, gaining energy from light, they are heterotrophic because they cannot fix carbon dioxide but must utilise simple carbon compounds such as pyruvate, acetate and lactate as a carbon source. These simple carbon compounds are also their primary source of electrons. In the dark they can live by fermentation of pyruvate.

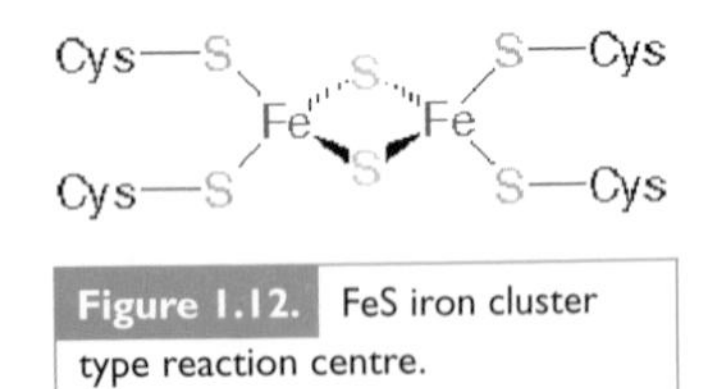

Figure 1.12. FeS iron cluster type reaction centre.

A similar kind of reaction centre (RC-1) is found in the green sulphur bacteria such as *Chlorobium*, though here the primary source of electrons is hydrogen sulphide. By oxidising hydrogen sulphide they produce sulphur and release electrons and hydrogen ions.

Green sulphur bacteria probably evolved in deep water where light levels are low and filtered by the organisms above, and where reduced sulphur compounds are also available. They have exceptionally large antenna arrays of 1000–1500 bacteriochlorophyll c molecules to each bacteriochlorophyll a molecule at the reaction centre. The pigments are packed into vesicles called chlorosomes attached to the cytoplasmic membrane. The green sulphur bacteria can make organic compounds through the fixation of CO_2 by a reductive tricarboxylic acid cycle.

Another type of reaction centre, called RC-2, which probably evolved from RC-1, is present in green filamentous bacteria such as *Chloroflexus*. It has pheophytin and a pair of quinones as early electron acceptors. *Chloroflexus* forms thick microbial mats in neutral or alkaline hot springs. It has bacteriochlorophyll a located in the chlorosomes. The early oceans were rich in sulphide and the use of hydrogen gas or hydrogen sulphide (H_2S) as the initial electron donor in *Chloroflexus* may date from that time. *Chloroflexus* is also sometimes called a green non-sulphur bacterium to contrast it with the green sulphur bacteria like *Chlorobium* from which it differs in many ways, not least in that it has a unique chemical pathway, the hydroxypropionate pathway, for carbon dioxide fixation.

Photosynthetic reaction centre RC-2 is also found in the two groups of purple bacteria: the purple sulphur bacteria (including the genus *Chromatium*) and the purple non-sulphur bacteria (the genus *Rhodospirillum*). The former utilise sulphide as a primary electron source and the latter utilise hydrogen. Two novel features are of particular interest in these organisms. Firstly carbon dioxide is fixed by the Calvin cycle, as it is in plants. Secondly these organisms are versatile. They can grow autotrophically by photosynthesis, and will do so in the light in anaerobic conditions, but they can also grow heterotrophically in aerobic conditions. In fact many of the components of the energy metabolism, the electron transport system, are the same or are very similar for both these activities. Paradoxically in the context of the evolution of plants, it is not the photosynthesis of purple bacteria, which is most interesting, but their aerobic metabolism. Purple non-sulphur bacteria are the probable ancestors of the mitochondria of eukaryotic organisms including plants.

The activities of these primitive anoxygenic photosynthetic bacteria are recorded in rocks of a great age because, by their activity, soluble ferrous iron was oxidised to the insoluble ferric state. The brown precipitate was preserved in rocks as 'banded iron formations' (BIF) that formed extensively in ocean sediments in the Archaean eon (Precambrian pre-2500 million years ago) and early Proterozoic eon (Precambrian 2500–590 million years ago). The banded iron formations are composed of silica-rich layers of fine grained quartz or chert interspersed by Fe_3O_4 (ferrous oxide) and Fe_2O_3 (ferric oxide) with about 30% iron content. Later the production of ferric precipitates was enhanced because of the greater concentration of oxygen in the atmosphere from the evolution of oxygenic photosynthesis along with the burying of organic carbon in sediments 'freeing' existing oxygen from CO_2.

1.2.4 Oxygenic photosynthesis

The cyanobacteria are the most important oxygenic photosynthetic bacteria. They have two photosystems: Photosystem I is related to the RC-1 containing photosynthesis most primitively seen in the heliobacteria, and Photosystem II is related to the RC-2 containing photosynthesis seen most primitively in the green filamentous bacteria. Probably this conjunction of photosystems occurred by gene transfer between distinct *Heliobacter* and *Chloroflexus* type organisms. It was a coupling that was to prove enormously successful, transforming the world because it permitted oxygenic photosynthesis.

It worked because the two photosystems acting in concert provide a double hit, boosting the energy level of electrons and thereby providing sufficient oxidising power to split the inexhaustible supply of water to provide a reductant without sacrificing the ability to use

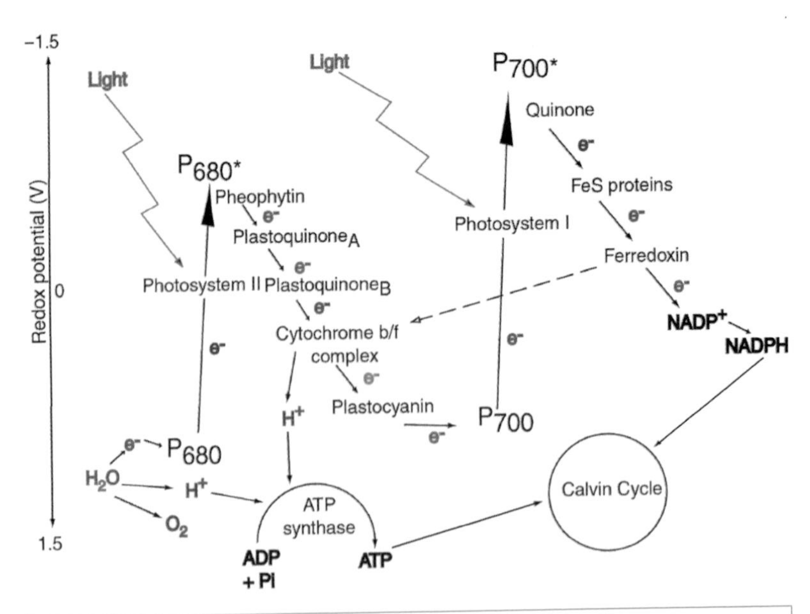

Figure 1.13. Oxygenic photosynthesis. Simplified diagram showing the light reactions of photosynthesis starting with the oxidation of water. Through the absorption of light in Photosystem II the electrons gained from water are elevated to an excited state. Subsequently they are transferred to Photosystem I. Here they are excited again by absorption of light and transferred ultimately to NADPH. The action of Photosystems II and I together is called non-cyclic photophosphorylation. The ATP and NADPH produced as a result are input into the Calvin cycle. The dashed line shows cyclic phosphorylation involving Photosystem I only and without the production of NADPH. The diagram does not attempt to be chemically balanced. Not all the details are shown. For example the cytochrome complex contains two cytochrome b and one cytochrome f molecules as well as a Rieske iron–sulphur protein.

photons in the red region of the spectrum. Oxgen is produced as a side-product. The first hit is from Photosystem II and results in non-cyclic photophosphorylation, the flow of electrons to Photosystem I with the production of ATP, and the splitting (oxidisation) of water (Figure 1.13). The second hit is from Photosystem I where electrons are excited again, and now they are transported to ultimately produce reducing power in the shape of NADPH. However, when sufficient reducing power is already present Photosystem I can carry out cyclic photophosphorylation to produce ATP.

The structure of fossil microbes from the Warrwoona Group in Western Australia from about 3500 million years ago is comparable to living cyanobacteria and is taken as evidence that oxygen producing photosynthesis had evolved by then. However, the close relationship between oxygenic and anoxygenic photosynthesis is emphasised by the activity of some cyanobacteria. Although they possess both photosystems, they are able to carry out anoxygenic photosynthesis by utilising Photosystem I alone to carry out cyclic photophosphorylation. In this case they oxidise H_2S to gain electrons

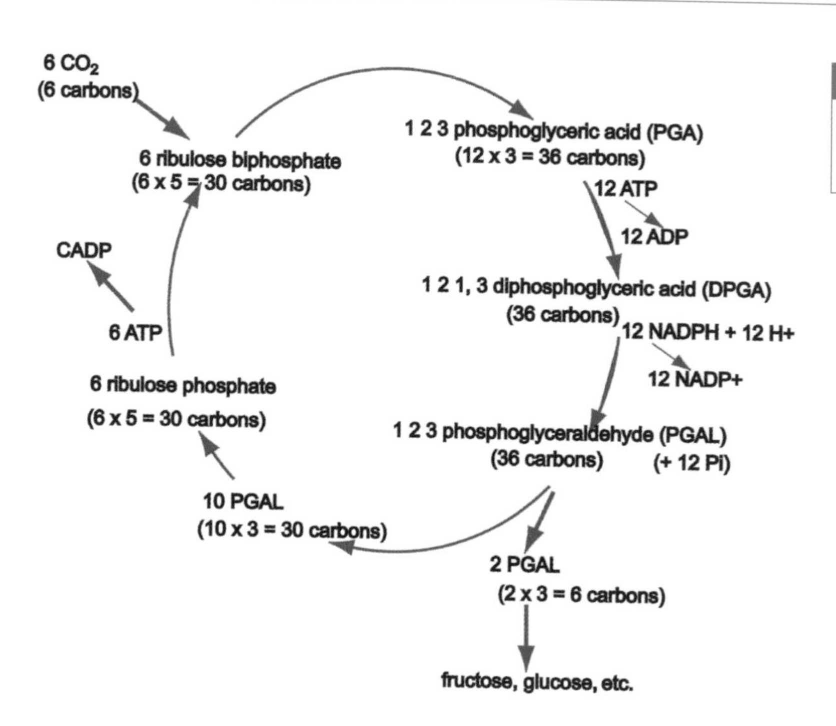

Figure 1.14. The Calvin cycle. The reducing power of NADPH and the energy from ATP is used to build sugars.

and produce sulphur in the same way as the sulphur bacteria. For example the cyanobacterium *Oscillatoria limnetica* lives in sulphide-rich saline ponds along with sulphur bacteria. Rather than giving off oxygen, globules of sulphur accumulate on the outside of its filaments.

1.2.5 Carbon fixation

Carbon dioxide is found in the atmosphere and dissolved in water. There are a number of different chemical pathways by which it is utilised or fixed to make organic compounds. The most important is a cyclical process called the Calvin cycle or Calvin–Benson cycle after the workers who discovered it (Figure 1.14). The first step is the covalent linking of the carbon in the carbon dioxide to a five-carbon compound, ribulose-1,5-bisphosphate (RuBP). This process is catalysed by the enzyme rubisco (D-ribulose-1,5-bisphosphate carboxylase/oxygenase). Rubisco makes up more than 15% of the protein in chloroplasts and may be the most abundant protein on Earth. It is also found in purple bacteria, cyanobacteria, chemolithotrophic bacteria and even some archaebacteria as well as plants. In many algae and in hornworts it is particularly associated with the pyrenoid, a region inside the chloroplast that forms part of a CO_2-concentrating mechanism.

The widespread occurrence of rubisco hints at an alternative function of rubisco at early stages of life on Earth when it may have acted as an oxygen detoxifier; in low CO_2 concentrations it catalyses a reaction in which oxygen is taken up, causing what is called

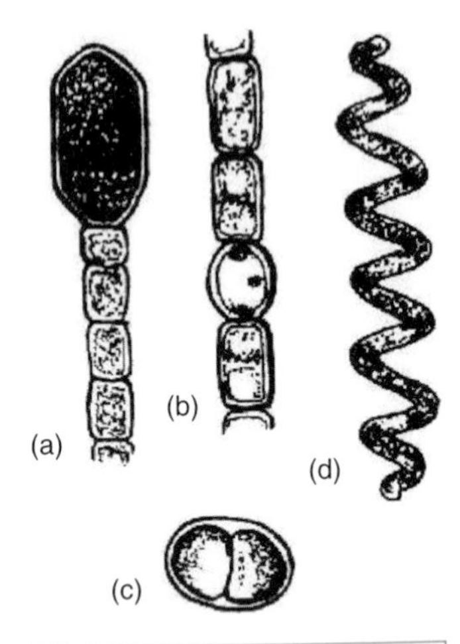

Figure 1.15. The diversity of Cyanobacteria. (a,b) *Anabaena*: the enlarged cell in (a) is an akinete (a cell that forms a resting stage); (b) a heterocyst (a cell associated with nitrogen fixation). (c) *Chlorococcus*. (d) *Spirulina* (drawn from http://vis-pc.plantbio.ohiou.edu/algaeindex.htm).

photorespiration. This can be very wasteful in plants, because in normal atmospheric conditions up to 50% of carbon fixed in photosynthesis may be reoxidised to CO_2, but it is an important capacity in anaerobic organisms.

1.2.6 The cyanobacteria and Prochlorophytes

The cyanobacteria are photosynthetic Gram-negative eubacteria that traditionally have been referred to as 'blue-green algae'. They are quite diverse, especially morphologically, which is unusual for Eubacteria. Over 150 genera and 1000 species have been described. The cyanobacteria have chlorophyll a and phycoerythrin (a phycobiliprotein) as primary pigments. Like chloroplasts, which are derived from them, they have a complex system of thylakoid membranes with associated spherical phycobilisomes to which the photosynthetic pigments are attached.

Cyanobacteria occupy a diverse range of extreme environments. Some species photosynthesise and grow in the high temperatures of hot springs and hyper-saline pools. They can also be found in the polar regions and at high altitude, surviving in snow and ice or in cracks in transparent rocks like quartz. They survive desiccation in deserts. Cyanobacteria of the order Chamaesiphonales (*Chamaesiphon*) occur in terrestrial and fresh-water habitats and are also epiphytic on mosses. In marine and fresh-water habitats they are important components of the plankton (e.g. *Trichodesmium* and *Microcystis*) and are often responsible for algal blooms. The Nostocales, exemplified by the genera *Nostoc* and *Scytonema* are found in soils, rocks and on tree trunks. As well as their importance as photosynthetic organisms, cyanobacteria are important ecologically because many can fix atmospheric nitrogen, and they are often symbiotically associated with plants.

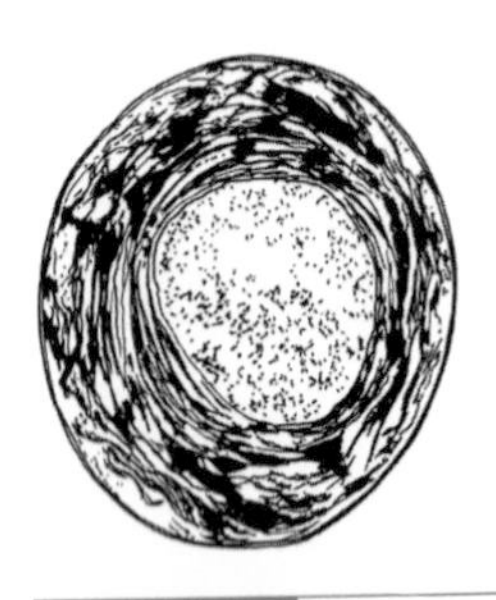

Figure 1.16. *Prochloron*, a representative of the small number of genera in the 'Prochlorophytes', Cyanobacteria that, like green algae and plants, have both chlorophyll a and b (from http://www-cyanosite.bio.purdue.edu/images/images.html).

Some workers recognise a group of planktonic photosynthetic bacteria called the 'Prochlorophytes', but DNA sequence data indicate that the three Prochlorophyte genera, *Prochloron*, *Prochlorococcus* and *Prochlorthrix*, have evolved separately and are not a single group distinguishable from other cyanobacteria. However, they are interesting because they are similar to plants as they have divinyl-chlorophylls a and b, which are very similar to plant chlorophyll a and b, and they lack phycoerythrin. However, although they have plant-like stacked thylakoid membranes, their own light-harvesting complex probably evolved as a response to the permanent iron-depleted conditions found in inter-tropical oceanic waters. 'Prochlorophytes' are very widespread in oceans and constitute up to 40% of the chlorophyll present in some regions.

Oxygen producing photosynthesis by the cyanobacteria and Prochlorophytes gradually enriched the atmosphere with oxygen. As a shielding ozone layer formed in the upper atmosphere, UV exposure declined. Paradoxically, although the damaging effects of UV light were reduced, the presence of highly reactive oxygen provided a different kind of challenge to living organisms. Some organisms were poisoned by it and survived only in the remaining anaerobic areas of deep stagnant water and waterlogged soil. Meanwhile many opportunities were created for aerobic organisms that had the mechanisms to mitigate the toxic effect of oxygen. An ecological transition was established in microbial mats between aerobic organisms growing on the surface of the soil, to more and more strictly anaerobic organisms, growing in deeper and deeper layers. It was the evolution of oxygen liberating photosynthesis by cyanobacteria and Prochlorophytes that provided an environment for dramatically increased rates of organic molecule production.

1.3 Form: the origin of complex cells

The evolution of cells permitted the localisation and isolation of potentially competing metabolic processes and a much more energy efficient metabolism. Increasingly complex metabolism evolved with the development of distinct membrane systems and intra-cellular compartmentalisation. For example membranes could become energy transducing by the location of electron transporters in separate places in them. Photosynthesis is only one activity that drives electrons across a membrane to establish an electrochemical potential. This potential is then used as the motive power for other activities.

It is especially in the boundary of the cell, in the cell membrane and any cell wall exterior to it, that a profound difference between three kinds or domains of living organism can be recognised: the Archaebacteria, the Eubacteria and the Eukarya.

1.3.1 Cell membranes and cell walls

The cell membrane is normally called the cytoplasmic membrane because it separates the living cytoplasm of the cell from the exterior environment. It is a phospholipid bilayer. The interior is hydrophobic and composed of long-chain fatty acids, and linked to it by an ester link is the outer part that has relatively hydrophilic glycerol and phosphate components. Embedded in the membrane, and sometimes passing right through it, are proteins that carry out many of the activities of the cell.

The cytoplasmic membrane does not just surround the cell: it is often highly folded inwardly, providing a greatly increased area for the localisation of other components of the cell. The chlorosomes of green sulphur Eubacteria are a good example of this but intra-cellular membrane systems (endomembrane systems), including intra-cellular membrane bound organelles are a particular feature of the eukaryote grade of living organism (Table 1.1).

Vesicles bud off the cytoplasmic membrane by endocytosis, capturing materials from outside the cell, or are part of an excretory system by carrying out exocytosis. The Golgi apparatus and the rough endoplasmic reticulum (rough ER) are important examples of an endomembrane system in eukaryotes. Another example

Table 1.1 Fundamental grades of organisation

Prokaryotes (Monera) (includes both Archaebacteria and Eubacteria)	Eukaryotes (Eukarya) (includes the four kingdoms: protists, plants, fungi and animals)
• No membrane-bound nucleus	• Membrane-bound nucleus
• DNA in circular chromosomes and without histones	• DNA complexed with histones in chromosomes
• Cell fission	• Mitosis and cytokinesis
• No cytoplasmic membrane-bound organelles (but mesosomes and membrane systems may be present)	• Cytoplasmic membrane-bound organelles (mitochondria, chloroplasts, Golgi apparatus, endoplasmic reticulum)

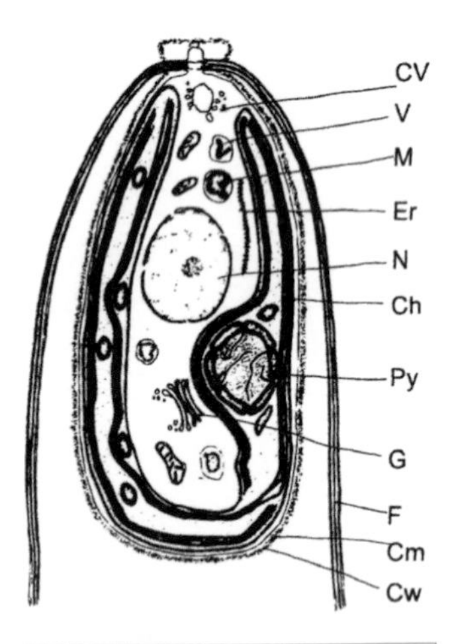

Figure 1.17. Drawing of a transmission electron micrograph of the cell of the green alga *Chlamydomonas* showing organelles and membranes systems: (Ch) chloroplast, (CV) contractile vacuole, (Er) endoplasmic reticulum, (G) Golgi body, (F) flagellum, (M) mitochondrion, (V) vesicle, (N) nucleus, (Py) pyrenoid, (Cm) cell membrane, (Cw) cell wall (from Lee, 1999).

is the extensive thylakoid membrane system of cyanobacteria and chloroplasts: thylakoids are stacks of flat membranes in which the photosynthetic pigments are located. In addition, most plant cells have one or more vacuoles; liquid-filled sacs surrounded by a membrane called the tonoplast. An example of the complex internal structure of a eukaryote is shown in Figure 1.17.

The cytoplasmic membrane is quite fluid but is stabilised in the Eukarya and methanotrophic Eubacteria by the presence of rigid flat sterol molecules that are absent from most prokaryotes, although some of these have similar molecules called hopanoids. The Archaebacteria have slightly different membranes from other organisms, perhaps because they have a tendency to occupy high-temperature environments that would disrupt a fluid cytoplasmic membrane: the interior fatty acids are linked to the glycerol part of the membrane by ether links and, in addition, some Archaebacteria have a membrane in which the interior hydrophobic part is stabilised as a monolayer.

The cytoplasmic membrane is also stabilised, in many organisms by the presence of a cell wall exterior to the membrane. The simplest cell walls are found in Gram-positive Eubacteria and Archaebacteria. They are called Gram-positive because during a particular staining regime, devised by the microbiologist Christian Gram, they retain a stain called crystal violet even when washed with ethanol. Their cell wall is thick and composed of 90% peptidoglycan. Gram-negative prokaryotes have a more complex multi-layered cell wall in which peptidoglycan makes up only 10%. Different kinds of cell walls are found in protists, fungi and plants. Animal cells and some protists are normally naked. Fungi have a cell wall in which chitin is a major component. The cell walls of protists are very diverse in chemical composition and structure and many planktonic protists such as the diatoms, dinoflagellates and desmids have remarkably sculptured cell walls.

Plants and some kinds of protists have a type of cell wall where cellulose is a major structural component. The evolutionary origin of cellulose cell walls is obscure. The basic structure of cellulose seems relatively simple, essentially it is a polysaccharide with glucose as its basic unit, but it is a very large polymer with many possible variations in the degree of and kinds of bonding in its various parts. Cellulose is also produced by the acetic acid bacteria (*Acetobacter*), forming an outer coat or pellicle of cellulose that helps them to float at the surface where conditions are aerobic. Bacterial cellulose microfibrils are isolated from each other and do not form the strong material seen in plants. In plants and algae the microfibrils are closely associated with each other. Cellulose is synthesised as scales in the Golgi apparatus by some algae but more usually cytoplasmic membrane-bound cellulose synthase enzymes synthesise it. There are differences in the form of cellulose microfibrils produced among different algal groups and plants.

In Rhodophyta (red algae) the cell wall has two layers: the inner layer has cellulose or another polysaccharide and the outer layer is mucilaginous with a sulphated polymer of galactose. This gives the red algae their characteristic slipperiness. Red algal cell walls are harvested to provide agar or carrageen (carragheen). Agar is used not only as a culture medium but also in cosmetics and to produce capsules for drugs. Agarose is a purified form used in electrophoresis. Carrageen from *Euchema* is used as a stabiliser in dairy products, paints and cosmetics. In nature, continually sloughing this mucilaginous layer prevents other organism colonising the surface of red algae. Some coralline red algae (Corallinaceae) also deposit calcium carbonate in the cell wall and they may have a jointed or crustose form. Red algae are important components of coral reefs, and are also common on rocky shores.

1.3.2 The domains of life

The profound differences in membranes and cell wall types of the Eubacteria, Archaebacteria and Eukarya have encouraged some writers to speculate that cellular life has originated three times, once for each domain of life (Table 1.2).

Eukaryotes differ in two key respects from prokaryotes: the presence of membrane-bound organelles in the cytoplasm, such as mitochondria, and in photosynthetic organisms, plastids (chloroplasts and others); and the presence of a nucleus, itself a membrane-bound structure (a double membrane) containing the genetic material organised into chromosomes. The genetic material undergoes mitotic division controlled by the action of the cytoskeleton in eukaryotes.

One important advantage that eukaryotes have is that their cells are larger than prokaryotes. Their greater size is accompanied by greater internal structural complexity that compartmentalises different cell functions. The largest prokaryotic unicellular organisms are symbionts of surgeonfish called *Epulipiscum fithelsoni* that can be more than 0.5 mm long, but this is a very exceptional prokaryote. Most prokaryotes are in the range 1.0–4 μm long with a diameter of 0.25–1.5 μm. The Cyanobacteriaceae, on average, exceed this range

Table 1.2 The domains of life

Archaebacteria	Eubacteria	Eukarya
Prokaryote organisation	Prokaryote organisation	Eukaryote organisation
DNA-binding proteins HMf and HMt with homology to HU-1 and HU-2	DNA-binding proteins HU-1 and HU-2	Histones
1 RNA polymerase transcription factors not required	Several RNA polymerases transcription factors not required	3 RNA polymerases transcription factors required
Commonly inhabitants of extreme environments: high salt, low pH, or high temperature	Not normally inhabitants of extreme environments	Not normally inhabitants of extreme environments
Includes methanogens	Not methanogens	Not methanogens
No muramic acid in cell wall membrane	Muramic acid in cell wall	No muramic acid in cell wall
Membrane lipids ether-linked, some branched	Membrane lipids ester-linked, unbranched	Lipids ester-linked, unbranched
Ribosomes 70S	Ribosomes 70S	Ribosomes 80S
Initiator tRNA methionine	Initiator tRNA formylmethionine	Initiator tRNA methionine
Introns sometimes present	Introns mostly absent	Introns commonly present
Operons	Operons	Operons absent
No capping and poly-A tailing of mRNA	No capping and poly-A tailing of mRNA	Capping and poly-A tailing of mRNA

with a mean length of about 50 μm. In contrast most eukaryotic unicellular organisms have cell diameters 2–200 μm but some are much larger than this.

Geochemical evidence such as the presence of steranes, especially cholestane and its analogues, indicate the existence of 'eukaryotes' at least 500 million to 1 billion years before fossil eukaryotes are found. The earliest fossil evidence of probable eukaryotes is provided by the dark curl or spiral of *Grypania*, up to 0.5 m in length and 2 mm in diameter, in rock cores first observed in rocks dated at about 2100 Ma (Negaunee Iron Formation, Michigan, USA). However, the diversity of eukaryotes up to 1000 Ma in the early Phanerozoic was very limited.

1.3.3 The nucleus, the cytoskeleton and cell division

Cell division provides another trace of the presence of living organisms in rocks of a great age. The simplest kind, carried out by prokaryotic organisms is binary fission. The cell enlarges and then splits into two. There are many examples of fossils of Archaean age showing these stages.

A defining feature of eukaryotes is the presence of a nucleus and cytoskeleton, and with it a particular kind of organisation of the genetic material. The nucleus has a double membrane surrounding a matrix containing the chromosomes. The chromosomes are highly structured packages of the genetic material, DNA, complexed with proteins called histones, to form a material called chromatin. The DNA is wrapped around the histones forming bead-like structures called nucleosomes. Nucleosomes are linked like a string of beads by the chain of DNA running between them. This string is coiled and supercoiled into tightly condensed chromatin to form a chromosome. Each nucleus has several to many chromosomes, depending upon species and each chromosome carries different genes.

A peculiar chromosomal organisation is present in the dinoflagellates, planktonic algae with armour-like coats. They have 12–400 'chromosomes' attached to the nuclear membrane that unwind only slightly between cell divisions and they are the only group of eukaryotes that lack histones. Current thinking is that these peculiarities are highly derived features.

The nucleus divides with the aid of the cytoskeleton. The cytoskeleton is a network of protein filaments, called microtubules and actin filaments, extending through the cell. The cytoskeleton is involved in many aspects of cell movement and growth, for example directing vesicles towards the growing cell wall and aligning the growing cellulose microfibrils. Microtubules about 24 nm wide are built up from the protein tubulin in a helical structure at special places in the cytoplasm called microtubule organising centres. Sometimes microtubules are associated with contractile actin filaments 5–7 nm wide.

Perhaps the most important role of the cytoskeleton is in cell division. Microtubules arising from an area called the centrosome form the spindle or phragmoplast that controls the movement of chromosomes to daughter nuclei. Microtubules attach to chromosomes that have already replicated into two chromatids and are held together at their centromeres. By the action of the cytoskeleton the chromatids are separated, one to each pole of the spindle. In this way a regular and highly organised division of the genetic material occurs. Following nuclear division the cytoplasm divides by a process called cytokinesis, also controlled by the cytoskeleton. Either the cytoplasm furrows until the two cells are separated or a cell plate is formed across the cytoplasm.

1.3.4 Organelles

Organelles are intra-cellular structures that are either like mitochondria and chloroplasts, which are membrane bound, or centrioles, which are not. Mitochondria and chloroplasts have their own genome, DNA in circular chromosomes like those of bacteria. Almost all eukaryotes have mitochondria while plant cells also have plastids including chloroplasts. A few eukaryotic organisms, the Archezoa, lack mitochondria. This may be because they are truly primitive or that they have lost mitochondria because of their peculiar lifestyle as extra- or intra-cellular parasites. They also lack Golgi bodies or have peculiar kinds.

Endosymbiont = organism living symbiotically inside a host cell.

Endosymbiosis

The modern consensus among biologists is that the cells of eukaryotes have a fundamentally chimeric origin, from the fusion of two or more distinct organisms, and their organelles arose as endosymbionts.

Symbiosis between closely related bacteria enables bacteria to adapt rapidly to local conditions. The first stage of cooperation may have been the production of highly stratified bacterial mats where the physical conditions of light quality and oxygen concentration control the ecological transition from one dominating bacterial species to another, but with each relying on the transformation of conditions created by the species above. A more significant cooperation is seen in the so-called consortium species that consist of a symbiotic relationship between anaerobic heterotrophs and photosynthetic green sulphur bacteria: they cluster together in aggregates in anaerobic sulphide-rich mud. It is a short step from this to the formation of a chimera by endocytosis or horizontal gene transfer. One important example within the bacteria is the presence of two photosystems in the cyanobacteria; this is thought to indicate that they have evolved from a genetically chimeric prokaryote, something related to the Heliobacteriaceae fused with something related to the filamentous green non-sulphur bacteria.

Heterotroph = an organism requiring organic molecules to provide energy.

Eukaryotic cells may therefore be perceived as a special case of the general phenomenon of microbial associations, their plastids and other organelles such as mitochondria having arisen by a series of endosymbioses involving different lineages of prokaryotes. Originally, symbiosis may have resulted from endocytotic ingestion by the host cell. Endocytosis is the folding of the cell membrane around materials from the environment to make a small pocket lined by the plasma membrane, which is eventually sealed off to make a vesicle. Phagocytosis is endocytosis of a large solid particle. Mitochondria and plastids have a double membrane; one derived from the host cell and one from the ingested endosymbiont.

The emergence of partner species and their coevolution must have begun by at least 3500 million years ago. There is some evidence that a member of the prokaryotic Archaebacteria, an eocyte (a highly thermophilic and sulphur-metabolising archaebacterium), was the host cell in the endosymbiosis of two eubacterial species, which became the mitochondrion and plastid respectively. Archaebacteria are closer to eukaryotes in some respects than the Eubacteria. However, it is clear that even after the first eukaryotic lineage had arisen there was substantial horizontal gene transfer between different lineages so that the relationships and origins of the different components has become somewhat obscured.

Both mitochondria and chloroplasts contain circular DNA genomes and are capable of independent protein synthesis. In dinoflagellates the chloroplast genome is peculiar because each gene is on its own mini-circle chromosome. It is apparent that after endosymbiosis many of the previous functions of the prokaryote genome were subsequently lost or transferred to the nucleus of the host cell. For example, rubisco, the enzyme involved in carbon fixation, is a simple multimeric enzyme composed of small and large subunits: in plants the small-subunit of rubisco, has transferred from the chloroplast to the nucleus. Also there is some evidence that nuclear genes coding for mitochondrial proteins in higher plants are more similar in sequence to prokaryotic than to eukaryotic genes. The transfer of functions to the nucleus could be viewed as a move towards more efficiency. The dependency of the organelle upon the expression of nuclear genes, because of the loss or transfer of the majority of organellar genes to the nucleus, distinguishes organelles from obligate endosymbionts.

Nevertheless some genes have been retained within the organelle. These code mainly for proteins that maintain redox balance, which must be synthesised where they are needed to counteract the deadly side effects of ATP generating electron transport. Evolutionary divergence in mitochondria occurred very early in the evolution of eukaryotes. For example, plants and animals have flattened cristae compared to the tubular or discoid cristae found in many kinds of protists.

Mitochondria are thought to have arisen from formerly free-living purple non-sulphur eubacteria. These are the only Eubacteria apart from the cyanobacteria, which are both photosynthetic and not strictly anaerobic, although in the purple non-sulphur Eubacteria photosynthesis is inhibited by oxygen at relatively low concentrations. However, it was not their photosynthetic ability, which was important in the evolution of mitochondria, but their ability to utilise organic compounds in aerobic respiration.

The origin of chloroplasts

The origin of chloroplasts from something like free-living cyanobacteria is supported by evidence of the similar DNA sequences they contain. The Prochlorophyceae, which are a derived group from the cyanobacteria, seem to be strong contenders as ancestors because, like chloroplasts, they have chlorophyll a and b and carotenoids but do not have phycobilisomes. They also have stacked thylakoid membranes where the photosynthetic pigments are located. However, they are not direct ancestors of chloroplasts though they share common ancestry with them. A more direct cyanobacterial origin of chloroplasts in the red algae (Rhodophyta) is supported because they too have phycobilisomes like the cyanobacteria.

In eukaryotes endosymbiosis of a cyanobacteria-like organism is seen most clearly in a small group of fresh-water algae, the Glaucocystophyta, which contain a photosynthetic organelle called a cyanelle. Two features of the cyanelle show a direct link to cyanobacteria. There is a persistent peptidoglycan cell wall between its two plasma membranes and it has genes for both subunits of rubisco. Cyanelles were, therefore, thought to be the result of recent endosymbiosis and two have even been given names as species of

(a)

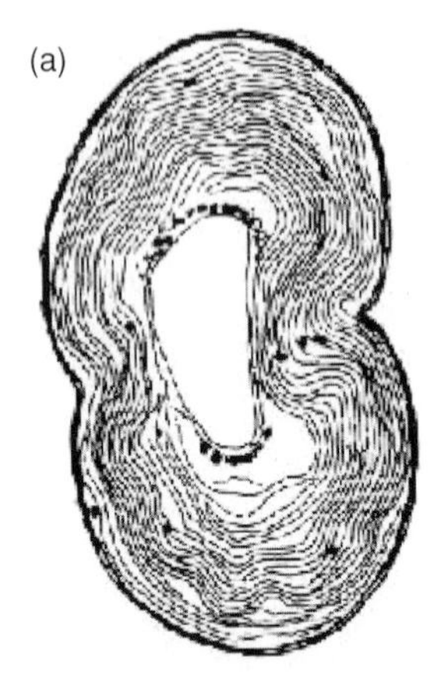

(b)

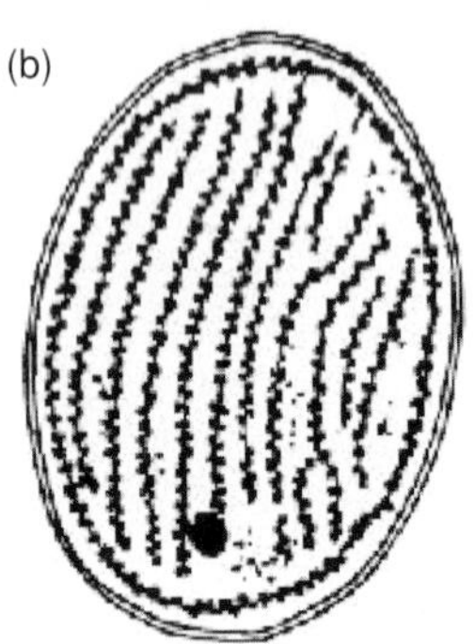

(c)

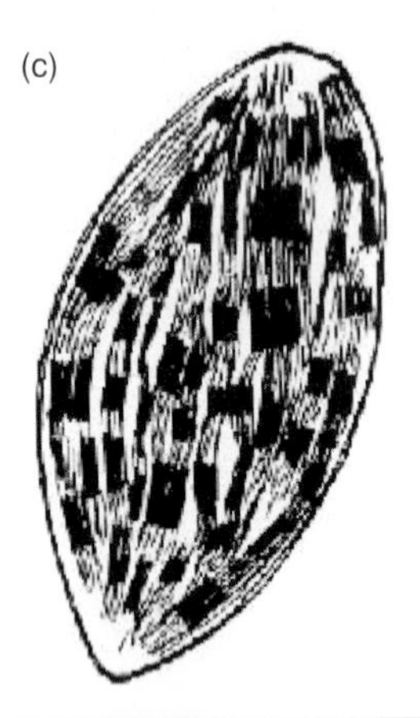

Figure 1.18. Photosynthetic apparatus of (a) a Cyanobacterium; (b) a red alga; (c) a chloroplast (from Lee, 1999).

cyanobacteria. Other groups of photosynthetic protists have acquired their plastids by secondary (or tertiary) endosymbiosis, with an endosymbiont eukaryote already equipped with a chloroplast. For example, in the photosynthetic euglenoids, three membranes surround the photosynthetic organelles.

Centrioles and flagella

There is a third organelle called the centriole, which may also have an origin as a highly modified endosymbiont. Centrioles are practically identical to the basal bodies of the characteristic flagella of eukaryotes, which are sometimes called undulipodia to distinguish them from the flagella of prokaryotes. Each flagellum has a characteristic structure of an axoneme, a ring of nine pairs of microtubules running as a core inside the flagellum membrane, and extending as nine triplet microtubules in the basal body where it is attached to the main part of the cell. Centrioles have an identical triplet microtubule structure as the flagella basal bodies. Centrioles, if present, are found in pairs perpendicular to each other in a region called the centrosome. There are two centrosomes near the nucleus. They function as microtubule organising centres and are associated with the production of the spindle in cell division. Centrioles are not present in conifers, flowering plants, and some other organisms that never produce motile cells.

There is considerable variation in the form and arrangement of flagella in algae. Flagella may be smooth (whiplash flagellum) or hairy (tinsel flagellum). One large and important group of algae the Ochrophyta (or Heterokontophyta), which include the brown algae and diatoms, have one of each kind. The hairs (mastigonemes) may be either non-tubular or tubular. The latter consist of a hollow shaft with terminal filaments. The tinsel flagella in the Heterokontophyta are tubular, a feature they share with 'fungal' oomycetes and several other non-algal groups that are placed with them in a group called the stramenopiles (= straw hair). The number, orientation and distribution of flagella differs among groups of algae.

The endosymbiont origin of flagella and centrioles is hypothesised to be from a spirochaete bacterium; a motile eubacterium with a flagellum that vibrates between the protoplasm and an outer flexible sheath, even though nothing like an undulipodium or centriole is seen in any prokaryotes, not even in spirochaetes. Because centrioles lack their own DNA there are few data on whether their evolutionary origin is through symbiosis. However, an endosymbiotic origin of undulipodia and centrioles is attractive for one very important reason. It may explain how mitotic cell division in eukaryotes arose. Symbiotic associations between motile and non-motile organisms are well known. To become well established, cell division between the symbiotic partners would have to be coordinated to ensure both partners divided at the same time. Perhaps the spindle first arose in the primeval eukaryote as a derived flagellum basal body as a means of coordinating cell division between a spirochaete and its archaebacterial partner. An alternative hypothesis is that the centrioles arose endogenously first to control mitosis and only later permitted production of undulipodia.

The basal bodies of undulipodia extend as various microtubule root systems. Different root systems delineate different algal groups. In some green alga, and in plants, the basal bodies are distinctive multi-layered structures that may function as microtubule organising centres.

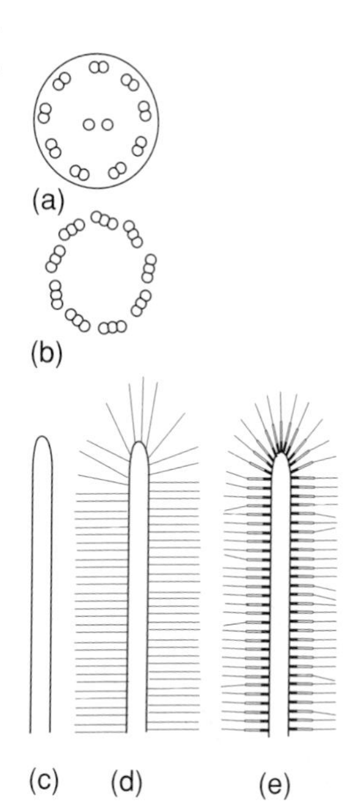

Figure 1.19. Flagellum structure: (a) 9 doublet + 2 singlet microtubule axoneme structure; (b) 9 triplet microtubule basal body and centriole structure; (c) whiplash flagellum; (d) non-tubular tinsel flagellum; (e) tubular tinsel flagellum.

1.3.5 Reproduction

Prokaryotic organisms are uniparental and partly because of this they potentially have very rapid rates of reproduction. Many are motile but most rely on passive means of dispersal. A few produce resistant dormant cells, endospores, with improved survivability while being dispersed. The cyanobacteria produce akinetes, enlarged, thick-walled resting cells that survive harsh conditions. Although prokaryotes are mainly asexual some kinds do have the ability to exchange genetic information from one parental lineage to another. Two cells come together and part of the DNA of one cell is transferred to the other, where it can become incorporated into its own genome. This process is called conjugation. Alternatively, in a process called transduction, DNA is carried from one cell to another by a bacterial virus, a bacteriophage, which first infects one cell and then the other. In laboratory experiments it has even been shown that DNA that has been released to the medium by the lysis of one cell, can be taken up, in a process called transformation, and incorporated by a recipient cell. Although all these processes have been demonstrated in laboratory cultures it is uncertain how important they are in nature. Conjugation does not seem to occur in the cyanobacteria but transduction might, since cyanobacterial viruses do exist. Prokaryotes potentially have a very fast rate of reproduction and new mutations are multiplied and rapidly propagated through the population.

Eukaryotes have a much greater size and complexity in their cells generally and in particular in their genetic material. They reproduce much more slowly and although, like prokaryotes, they rely on mutation as the primary source of variation, sexual reproduction is their main source of new genetically distinct individuals. The evolution of the cytoskeletal spindle and cytokinesis enabled regular highly organised mitotic cell divisions. Following division of the nuclear material the cell divides. Different groups of algae and plants differ in the form of spindle formation and cytokinesis (Figure 1.20).

Eukaryotes undergo asexual fission to multiply the individual, rather like the prokaryotes, but following a mitotic division. In most, including algae and plants, there is also sexual reproduction, which involves another kind of cell division called meiosis. Like mitosis, meiosis utilises the cytoskeletal spindle and cytokinesis, but whereas mitosis produces daughter cells with an identical duplicate genome, meiosis halves the number of chromosomes, producing haploid daughter cells (Figure 1.21).

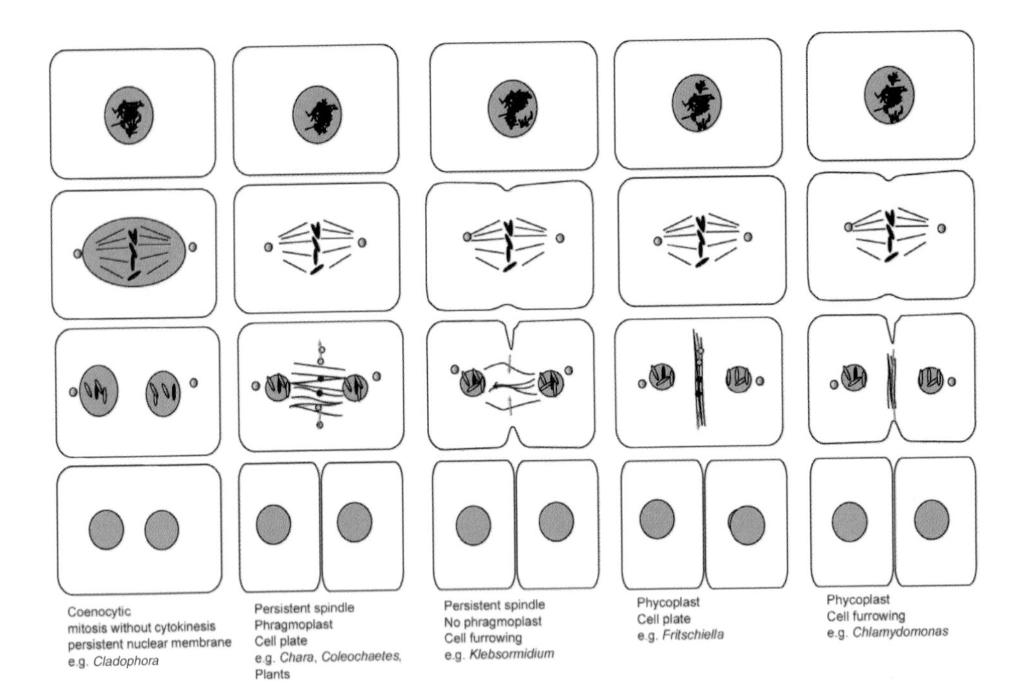

Figure 1.20. Spindle formation. There is considerable variation in spindle formation. In the 'green algae' group the Chlorobionta coenocytic organisms like *Cladophora* have nuclear division without cell division (cytokinesis). In other Chlorobionta the spindle soon collapses and a new system of microtubules called the phycoplast is formed perpendicular to the spindle. In the Streptobionta (Charophyceae and Land Plants) the spindle (phragmoplast) is persistent and survives until cytokinesis and a cell plate is formed.

Allele = version of a gene.

Sex

Eukaryotes alternate between haploid and diploid phases of the life cycle linked by the sexual cell division called meiosis. In meiosis haploid daughter cells are produced with half the number of chromosomes as the diploid meiotic mother cell. The sexual fusion (syngamy or fertilisation) of two haploid cells to produce a zygote restores the diploid condition. The zygote carries the genetic material from two different parents. It has pairs of homologous sister chromosomes, one from each parent, each homologue bearing the same genes, but perhaps bearing different alleles. In this way each zygote carries a different combination of different parts of the genetic material from different parents. When meiosis occurs, this pooled genetic variation is assorted randomly, so that each haploid cell produced, called a gamete, carries a different set of alleles. In these two ways, random mating and random assortment of genetic variation in meiosis, the range of genetic variation is greatly increased.

An important theoretical advantage sexual organisms have over asexual lineages arises from the way asexual organisms have a tendency to gain slightly deleterious mutant alleles by chance, by what is called genetic drift. Genetic drift is more likely in small populations or those that go through population bottlenecks. Like a cog that can only move forward because it is held in place by a ratchet, the number of slightly deleterious alleles, or genetic load, accumulates over time. This possibility was first pointed out by H. J. Muller. Sexual organisms have the ability to escape Muller's ratchet because, by sexual recombination, new lineages with novel combinations of non-deleterious

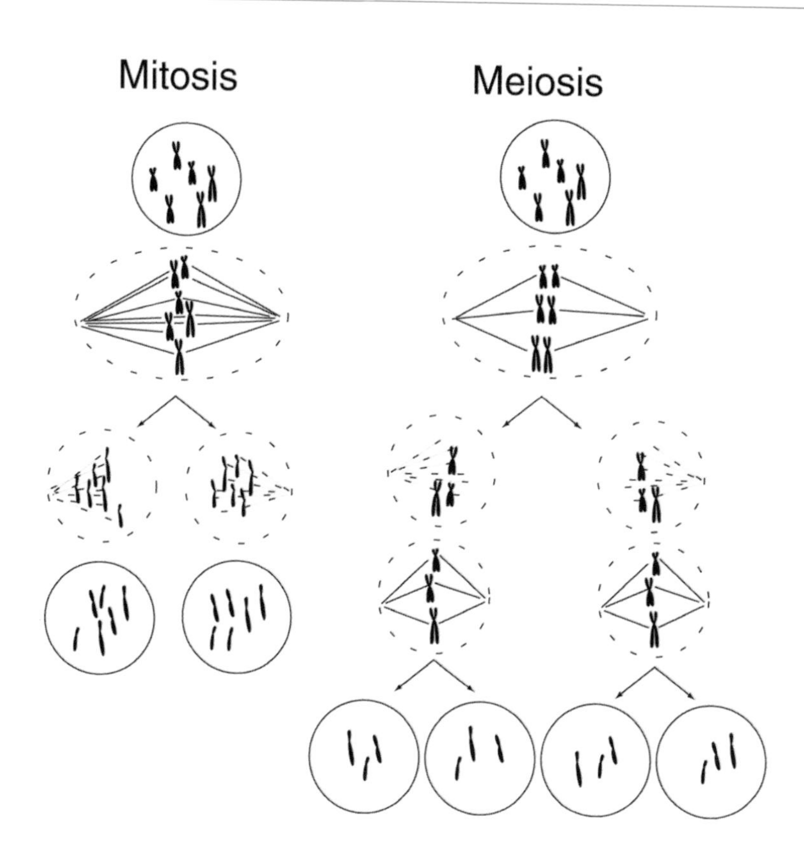

Figure 1.21. Mitosis and meiosis.

alleles are recreated in each generation of sexual reproduction. Another very important evolutionary aspect is that it is differences in sexual reproduction that reproductively isolate evolutionary lineages enabling them to diverge and become differently adapted.

There are various forms of sexual reproduction. In the vast majority of eukaryotes, including plants, gametes from two different parents fuse to form the zygote, a condition enforced by self-sterility. These organisms are called heterothallic. However, some organisms are self-fertile or homothallic. Note that this distinction between homo- and heterothallism is different from the distinction between male and female. In some heterothallic algae, for example, there is no distinction between male and female. The gametes look identical; they are isogamous (Figure 1.22). However, even in isogamous organisms sometimes there are different mating types and only some combinations of gametes will fuse together in fertilisation. Isogamy is quite widespread in the algae.

Alternatively, the two gametes that fertilise have the same form, but one, designated as the female, is larger than the male. This condition is called anisogamy. There is a third condition called oogamy. Oogamous species have a clear distinction between a small motile male gamete, called the sperm or spermatozoid, and a large immobile female gamete, called the egg or ovum. Some species of the green alga *Chlamydomonas* are isogamous, some are anisogamous and

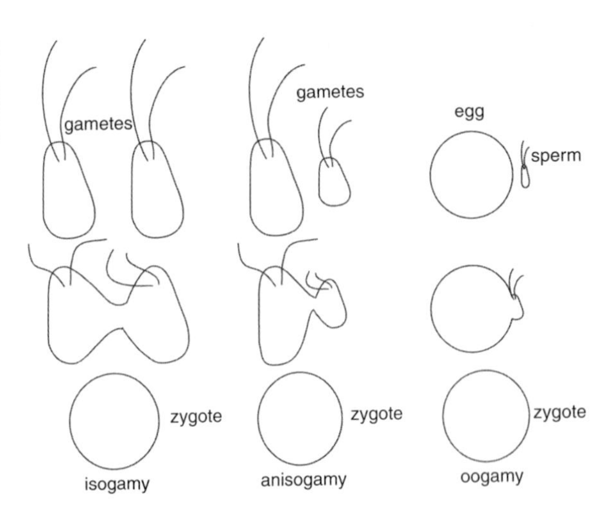

Figure 1.22. Syngamy, the fusion of gametes: isogamy, anisogamy and oogamy compared.

a few are oogamous. The evolution of differences from isogamy via anisogamy to oogamy represents a specialisation; between the male gamete, which is motile and produced in large numbers to maximise its chances of finding a female gamete, and a female gamete, which is large with food reserves for the zygote and produced in smaller numbers. This specialisation has evolved on many different occasions in different evolutionary lineages. The male gamete is normally flagellate and motile, although red algae, conifers and flowering plants have non-flagellate male gametes. In those species that have male and female gametes, these may be produced both by the same bisexual individual, a state called monoecy, or by separate male and female individuals, a state called dioecy. For example, in the brown algae, *Fucus* is dioecious and *Pelvetia* is monoecious.

It was sex that created huge potential advantages to eukaryotic organisms. As with the evolution of photosynthesis it was possibly the existence of a high-UV light environment, because of the lack of any significant ozone shield, which provided the spur for the evolution of sex. A high-UV environment is potentially highly mutagenic. Haploid organisms, which have a single copy of each gene, are disadvantaged if UV causes a deleterious mutation in an essential gene. Those organisms with multiple copies of genes could survive the destruction of one of the copies by mutation.

However, many aspects of cell life are determined by the relative concentration of gene products. A disordered, or unbalanced, multiplication of genes is potentially disadvantageous, because it could result in unfavorable proportions of gene products. Although nuclear division without cell division, endopolyploidy, does not unbalance the genome, the first time it happens it creates a diploid nucleus in which each chromosome has a homologous chromosome with exactly the same genes; however, if endopolyploidy is repeated several times, it results in a bloated genome and consequently a slow rate of cell division. Meiosis may have first evolved as a mechanism to prevent this by periodically halving the number of chromosomes. In meiosis homologous chromosomes are matched by becoming become aligned with each other before being separated to the two daughter nuclei. This matching is essential to ensure balanced daughters, but it also provides the possibility of some genetic repair: the DNA of one homologous chromosome can act as a template to the other damaged strand.

Multiplication and dispersal

The two distinct purposes of reproduction, multiplication with dispersal, and the generation of genetic diversity, are clearly differentiated in eukaryotes. Genetic diversity is maintained through sexual reproduction as new combinations of parental characteristics are created in the progeny. Multiplication and dispersal occur through the production of specialised cells called spores. A spore does do not undergo syngamy but is produced in large numbers to multiply and disperse the organism in space, or in time, as, for example, by allowing it to withstand a prolonged period of desiccation.

Motile spores have flagella and are called zoospores. Many organisms also produce resting spores, which survive dormant through a harsh season to germinate when conditions improve. Spores often have a thickened and protective outer cell wall. The resistant material, sporopollenin, is usually present. Spores may be produced by mitotic or meiotic cell divisions. If they are produced as a result of meiosis they are produced in fours called tetrads. Such spores are sometimes termed meiospores. Individuals growing from meiospores are genetically variable and different from their parents. Spores produced by mitotic cell division grow into individuals identical to each other and to their common parent.

In plants, spores are produced in sporangia arising from a diploid individual or tissue (the sporophyte) and sperm and eggs are produced in antheridia and archegonia, respectively, arising from a haploid individual or tissue (the gametophyte). A regular sexual cycle alternating between haploid and diploid phases became established in eukaryotes, but with many variations. There are two major categories of plants, bryophytes and tracheophytes, that differ in the relationship of the sporophyte and gametophyte. In bryophytes the sporophyte grows 'parasitically' out of the gametophyte. In the tracheophytes the sporophyte is an independent plant. This will be explored in more detail in Chapter 4.

Life cycles

Three major kinds of life cycles can be distinguished; haplobiontic, diplobiontic and haplodiplobiontic (Figure 1.23). Haplobiontic organisms do not produce a diploid multicellular individual and produce spores more or less directly from the zygote. Diplobiontic organisms do not produce a multicellular haploid individual and have a haploid stage restricted to the gametes. Haplodiplobiontic organisms alternate

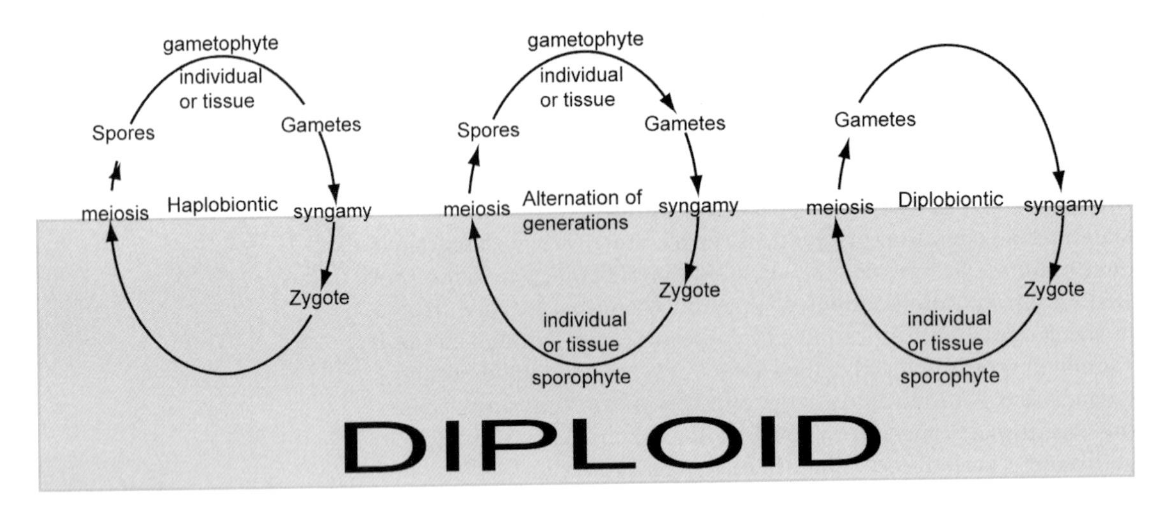

Figure 1.23. Life cycles.

between haploid and diploid multicellular individuals, often termed an alternation of generations. It is really a difference between when mitotic cell divisions take place, between meiosis and fertilisation only, fertilisation and meiosis only, or both. All these patterns are present in the algae. The haplobiontic life cycle is more common in the green algae and is exhibited by *Chara* and *Oedogonium* for example. The green alga *Ulva* is haplodiplobiontic. The brown algae such as *Laminaria* and *Desmarestia* also exhibit an alternation of generations. In *Fucus* there are separate male and female gametophytes.

In haplodiplobiontic organisms some, like *Ulva*, have an isomorphic alternation of generations; the diploid and haploid individuals are similar looking plants. In an aquatic habitat the two activities of sex and dispersal overlap to a considerable degree. Both zoospores and gametes can disperse by swimming or can be carried passively in water currents, and the sporophyte and gametophyte are similarly adapted for releasing them in water. Indeed in algae the spores are scarcely different from gametes except that they germinate directly into haploid plants whereas the gametes fuse in pairs (syngamy) to produce diploid zygotes that only then germinate into diploid plants.

However, in plants and haplodiplobiontic brown algae the haploid and diploid individuals are heteromorphic, differently adapted and startlingly different looking. The evolutionary explanation of heteromorphy lies in the way the haplodiplobiontic alternation of generations has permitted further evolutionary specialisation, separating the two different aspects of reproduction, sex from dispersal. More frequently the diploid spore-producing individual, the sporophyte, is larger and longer lived because successful dispersal is enabled if spores are produced in very large numbers. This is the case in the brown alga *Laminaria* and most plants. However, the gametophyte is large and the sporophyte is small in some organisms, such as the brown alga *Cutleria* and most bryophyte plants.

The red algae (Rhodophyta) have an alternation of generations but with three generations: two sporophytic ones and a gametophytic one! The first sporophytic generation established following syngamy produces diploid carpospores, which germinate into a second sporophyte, which produces haploid spores by meiosis.

The possession of an extended diploid phase in the life cycle, as exhibited in organisms with a diplobiontic life cycle or haplodiplobiontic life cycle, has several important theoretical advantages over the haploid condition. Syngamy between cells with homologous chromosomes, but carrying different alleles, each coding for a product with slightly different characteristics, creates a diploid that is heterozygous. In the diploid condition, if both alleles are expressed, the individual may be better adapted to a wider range of conditions because of its wider range of expression. In addition a range of different alleles may be present in the population, with different individuals having different combinations of alleles; the range of possible genotypes, fitting a range of circumstances, is much greater than the range of alleles. In each generation, random breeding and crossing-over give rise to new combinations of alleles. This kind of genetic variation between individuals provides the building blocks of evolutionary divergence through natural selection.

In trying to understand the importance of sexual reproduction and diploidy a reductionist approach focusing on individual genes and alleles is unhelpful. Alleles are expressed in the individual along with the alleles of many other genes, which can mutually alter each other's expression and selective advantage. It is the individual, not the allele that is selected. In addition, the individual is not just the sum of the activities of all its genes, its genotype. It is the external form of the individual, the phenotype, that is selected, and this is the product of an interaction between genotype and the environment in its broadest sense. The advantage individual alleles confer to the individual, in addition to being contingent on their genetic background, is also dependent upon the individual's history and present circumstance. The evolving species is a set of populations of individuals. Species, populations, and individuals are each a trajectory in space and time determined by interactions between their genotypes and the environments they experience.

The importance of diploidy may be that it may provide a mechanism to maintain high levels of genetic variation in sexual lineages. In haploid individuals all alleles are expressed, exposing the resultant characteristics to natural selection, and deleterious ones may be eliminated relatively rapidly. However, in diploid individuals, if an allele is recessive it is not expressed and is hidden from natural selection, even if it is disadvantageous. It is only expressed when it is present in the homozygous condition. A simple formula called the Hardy–Weinberg equilibrium allows us to estimate the frequency of homozygotes in a population: it is the square of the frequency of an allele, so that even if 1% of a population have a recessive allele only 0.01% of the population will express it. As a consequence even deleterious alleles may only be eliminated by natural selection very slowly because, for most of the time, they are hidden. Now, because the relative advantage or disadvantage of any allele depends both upon the particular genetic background and the particular environment the individual finds itself in, this delay may provide enough time for the allele to be expressed in a set of circumstances where it is advantageous and become selected for rather than against.

Eukaryotic sexual reproduction and an extended diploid phase in the life cycle enabled new pools of genetic variation to be created. Natural selection in these gene pools provided a mechanism by which organisms adapted to occupy more niches. There was an 'explosion' of organismal diversity from the Cambrian 590 million years ago that is associated with the origin of diploid eukaryotic cells and sexual reproduction.

1.3.6 Photosynthetic eukaryotes

Photosynthetic eukaryotes have traditionally been separated into two groups: the algae and land plants. All groups of algae are primarily aquatic but there are many terrestrial and subterranean species in addition to those that form part of lichen associations and those that live inside animals and plants. All land plants are primarily terrestrial and although some are free-floating or parasitic almost all are 'rooted' to the soil. However, this distinction between the algae and land plants is misleading. It is clear that while many algae, including both unicellular forms like the euglenoids (Euglenophyta) and complex multicellular forms like the brown algae (Phaeophyceae), are only distantly related to plants, others like the green algae are much more closely related.

The euglenoids originated as non-photosynthetic flagellates, which became photosynthetic, not by taking up a photosynthetic bacterium, but by taking up the chloroplast of a photosynthetic alga. Later some lost their photosynthetic ability but not their plastid. Even more remarkably there seems no end to this process of stealing the ability to photosynthesise by endosymbiosis; some kinds of algae are tertiary endosymbionts by taking up an alga. Secondary and tertiary symbiosis is widespread in the algae.

The algae are a heterogeneous mixture of groups and 'algae' is a term that should be used for a grade of organisation not a taxonomic group. It is not even clear how many major groups should be recognised. Between 7 and 11 divisions or phyla of algae have been recognised in different classification schemes, comprising approximately 70 000 living and extinct species. The seven most important divisions or phyla of algae can be placed in three main groups (Table 1.3). The divisions/phyla are classified mostly on the basis of their physiology,

Table 1.3 The algae

A. Algae with **chloroplasts surrounded by the two membranes of the chloroplast envelope only** – product of a primary endosymbiosis by the phagocytic uptake of a cyanobacterium by a protozoan.

Glaucophyta (Glaucocystophyta) (glaucophytes, important because of the possession of cyanelles) – 8 genera of mainly non-motile fresh-water of unicellular organisms.

Rhodophyta – (red algae, red seaweeds)(4000–6000 species): non-motile; some unicellular organisms but many are multicellular filamentous or pseudoparenchymatous seaweeds, mainly marine, and including calcified coralline forms.

Chlorophyta – (green algae) (17 000 species) marine and fresh-water, unicellular organisms or various multicellular forms.

B. Algae with **chloroplasts with one membrane of chloroplast endoplasmic reticulum outside the chloroplast envelope** – product of a secondary endosymbiosis by the phagocytic uptake of a chloroplast by a protozoan.

Euglenophyta – (euglenoids) about two-thirds are non-photosynthetic, and capture and ingest prey by phagocytosis – 40 genera (900+ species): motile by flagella or by wriggling (metaboly); unicellular, mainly fresh-water in many wetland habitats rich in decaying organic material.

Dinophyta – (dinoflagellates) very important in marine ecosystems, whose name comes from their spinning swimming motion) – 550 genera (2000–4000 species): motile; mainly marine; unicellular, planktonic organisms but including some that live symbiotically in reef corals and other animals.

C. Algae with **chloroplasts surrounded by two membranes of chloroplast endoplasmic reticulum either side of the chloroplast envelope** – product of a secondary endosymbiosis by the phagocytic uptake of a eukaryotic alga by a protozoan.

Cryptophyta – (named because of their inconspicuousness, they are small and burst readily, but they may be abundant in cold or deep water) 12–23 genera (200+ species): motile; fresh-water and marine, frequently incorporated in parts or as a whole as an endosymbiont in ciliates or dinoflagellates.

Prymnesiophyta – (also called Haptophyta from the presence of a haptonema, a thread-like external structure arising from a basal body that attaches to and pulls in prey): ~50 genera (~300 species); motile or non-motile; mainly unicellular organisms, some colonial; mainly marine; includes the coccolithophores with calcium carbonate rich scales whose fossils are largely responsible for the formation of chalk deposits.

Ochrophyta (Heterokontophyta or Chromophyta) – a very diverse group including both unicellular organisms like the diatoms, Bacillariophyceae, (10 000–12 000 species) and large complex seaweeds, the brown algae, Phaeophyceae (250 genera and 1500 species), multicellular seaweeds, along with several other important groups.

chemistry and reproductive behaviour, as well as the chlorophylls and accessory pigments present in the chromatophores.

There are three algal divisions that have chloroplasts derived by the primary endosymbiosis of a cyanobacterial-type of organism. These three divisions are the Glaucophytes (Glaucophyta or Glaucocystophyta), the red algae (Rhodophyta), and the green algae (Chlorophyta), and it is from the last of these that plants originated.

The Glaucophyta represent in some ways the earliest stages of the evolution of photosynthetic eukaryotes. They are small group

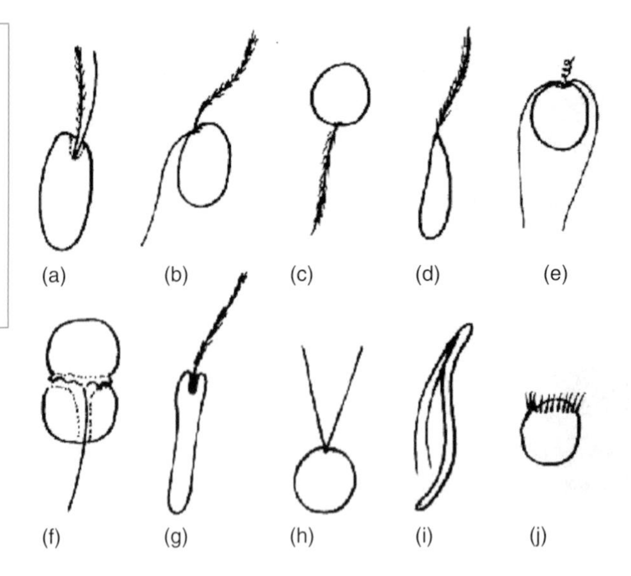

Figure 1.24. Motile unicellular algae: (a) Cryptophyta; (b)–(d) Ochrophyta; (b) Xanthophyceae, Raphidophyceae, or Chrysophyceae; (c) Bacillariophyceae; (d) Eustigmatophyceae; (e) Prymnesiophyta; (f) Dinophyta; (g) Euglenophyta; (h)–(j) Chlorophyta (from Fig. 1.6 of Lee, 1999).

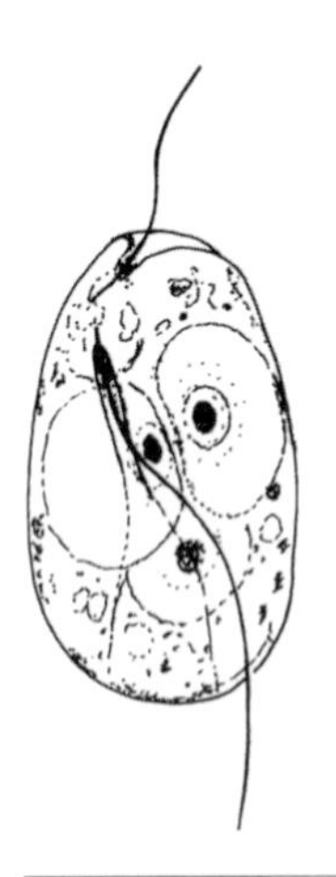

Figure 1.25. Member of the Glaucophyta (from Lee, 1999).

of three genera of fresh-water unicellular algae, which are important because of their unusual plastid, which retains a thin peptidoglycan cell wall located between the two outer membranes of the plastid. They are found as epiphytes in sphagnum bogs and other low-temperature aquatic habitats. They have chlorophyll a only, with secondary pigments including beta-carotene, zeaxanthin, and beta-cryptoxanthin. Phycobilisomes are present. Consequently the thylakoids are not stacked. *Glaucocystis* has a cell wall made of cellulose, *Gloeochaete* has a non-cellulosic cell wall and *Cyanophora* lacks a cell wall.

The Rhodophyta were certainly one of the earliest groups to evolve in the late Precambrian. The biochemical and structural similarities between red algae and the cyanobacteria are considerable. Both lack flagella and both possess phycobilisomes between the thylakoids. A fossil red alga of the group Bangiophycidae is known from silicified carbonate rocks dated at 1260–950 Ma from Somerset Island, northern Canada. Fossil coralline red algae from 700 million years ago have been described.

Plants (Embryophyta) are most closely related to the Chlorophyta: together they might be called the Viridiplantae (green photosynthetic organisms). Almost all algal phyla/divisions have at least a few unicellular species and several such as the Euglenophyta and Dinophyta are entirely unicellular. Most unicellular organisms are flagellate. The form and arrangement of flagella is an important character distinguishing the different groups (Figure 1.24). Some unicellular species have a resting vegetative phase when they lose motility. In *Chlamydomonas* this so-called palmelloid stage is short, but in some algae like *Phaeocystis* (Prymnesiophyta) it is permanent and motile cells are produced only for reproduction. The unicellular organism *Dinobryon* (Chrysophyta) has a dendroid habit and looks like a little tree. It is attached to the substrate by a stalk of mucilage. In an analogous way

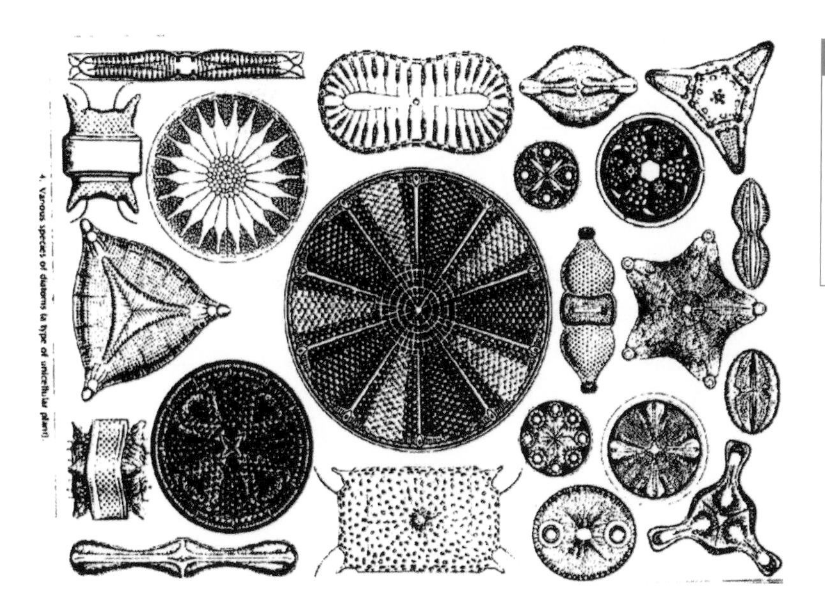

Figure 1.26. Diatoms (Bacillariophyceae) have cells surrounded by a two-part box made of highly sculptured silicon dioxide (silica). They lack flagella but some have the ability to glide over a surface (from Haeckel, 1904).

many brown and green seaweeds mimic flagellate unicellular organisms when they produce motile spores (zoospores) or male gametes.

Coccoid algae are rounded and non-motile for most of the life cycle. Although they may be free-living, they rely on altering their buoyancy to utilise water currents for dispersal. Some of these never produce flagellate cells. One example is *Cyanidium*, one of the few species of red algae (Rhodophyta) that is unicellular. It lives in acidic hot springs where it replaces cyanobacteria that prefer neutral or basic conditions. It can grow even where the acidity is one hundred times that of lemon juice and in temperatures up to 56 °C.

In the Ochrophyta most groups are either unicellular or have some unicellular species. By far the most important group is the diatoms (Figure 1.26), which is the main photosynthetic component of the plankton. No species of brown algae (Phaeophyceae) is unicellular but they do produce unicellular flagellate spores.

Unicellular green algae are quite diverse and include a range of motile forms, many of which are flagellate and planktonic. The primitive green algae, the Micromonadophyceae, have an unusual structure of flagellar attachment. There is an apical or lateral depression out of which two, four, or eight flagella arise. In other green algae, such as *Chlamydomonas* it is normal for the flagella to arise from an apical or sub-apical bump. *Chlorella* is the green alga that seems to appear everywhere in water or on soil that is exposed to the sunlight. Sexual reproduction has never been observed and the tiny highly simplified cells never produced flagella. It was the first alga to be grown in pure culture in the laboratory and it has been used extensively in photosynthesis research.

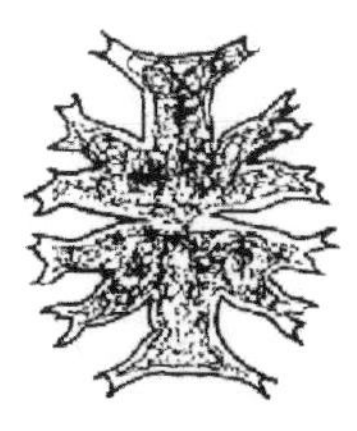

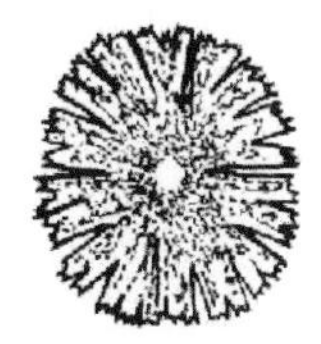

Figure 1.27. *Micrasterias* desmids.

The desmids are curious unicellular green algae. There are about 10 000 species in 40 genera, and they live as plankton or as benthic organisms. They have two nuclei. Some have a homogeneous cell wall

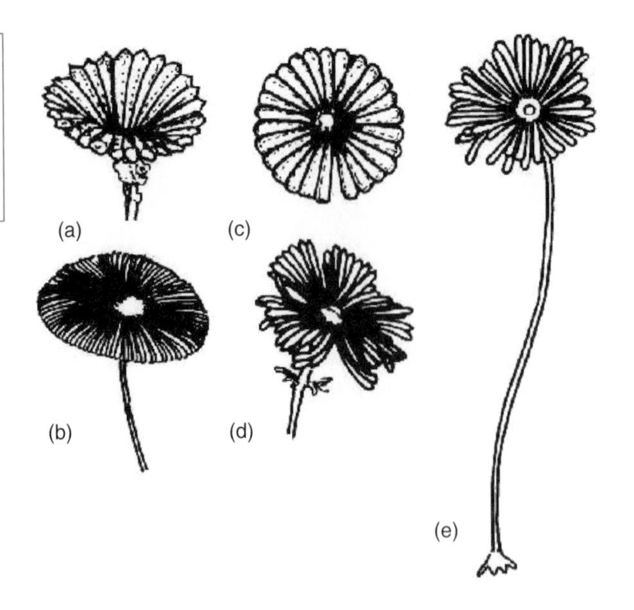

Figure 1.28. *Acetabularia*, a large unicellular green alga. Different species with distinct 'crowns': (a) *A. dentata*; (b) *A. acetabulum*; (c) *A. kilneri*; (d) *A. farlowii*; (e) *A. crenulata*.

Benthic organisms = attached marine organisms.

but many have a cell wall of two compartments connected by an isthmus. Like most other members of the Charophyceae they do not produce motile cells at any stage of their life cycle.

The green alga *Acetabularia* illustrates the extent and the limits to the size and complexity that a unicellular organism can achieve (Figure 1.28). It grows as a single cell up to several centimetres long with a distinct stalk and branched base. At maturity its nucleus divides meiotically to produce a coenocytic plant with many nuclei that migrate into a crown of rays at the top of the stalk.

It was through experiments grafting different species of *Acetabularia* with different shaped crowns that Hammerling showed that the nucleus contains the genetic information that directs cellular development of the different-shaped crowns in each species. However, experiments amputating the developing stalk have shown that the initiation of a crown can occur in the absence of a nucleus if a large enough stalk is present. The nucleus and cytoplasm interact in development and internal cues such as cell age and size are used to regulate reproductive onset. The genetic material AND the cytoplasm are part of the same evolving autopoietic system (see Section 1.1.2).

1.4 Pattern: multicellularity in the algae

Along with the evidence of living chemical processes and the form of cells, it is the presence of pattern that provides some of the earliest evidence for living organisms. Fossil microbes from cherts and shales, such as those from the Warrwoona Group from 3500–3300 million years ago, are filamentous or colonial. Over many millions of years greater complexity and diversity of organisms evolved. In the Gunflint

Formation of Ontario, Canada from 1900 million years ago, 12 different species including multicellular forms, have been detected.

Multicellularity originated separately in many different groups such as bacteria, fungi, different groups of algae, and in different groups of animals. Among the algae several groups have both unicellular and multicellular representatives. Unicellularity is unknown in the brown algae (Phaeophyceae) except as a short reproductive stage of the life cycle of the multicellular organism.

The advantages of multicellularity are not hard to understand. Multicellular individuals can be larger, thereby occupying space more effectively and competing for light or nutrients. But one of the difficulties of being large is the control of cell function; in a large organism proportionally a large amount of genetic material is required to produce enough gene products. By becoming multicellular multiple nuclei provide gene products throughout the individual.

Being large enough to occupy space is not the only advantage of multicellularity. Perhaps more important is the greater possibility of specialisation it permits. The organism can become patterned with different zones specialised for different functions. The simplest kinds of multicellularity are seen in some bacteria that form chains of cells but even here some kind of specialisation is often observed. In the bacteria, streptomycetes have branched chains in which some branches are specialised for reproduction by fragmenting into spores. A measure of the degree of specialisation that multicellularity permits is shown by the widely diverse forms exhibited by many of the algae. This may take the form of sheets, ribbons or nets, or it may be like sponge or jelly. Other algae are like fans or ears, disc-like or spiky, or they may be branched or unbranched. In branched forms, the branches may all be similar or different. In texture their tissues range from delicate films to fleshy, firm or hard structures like corals. What is remarkable is that similar patterns have evolved several times in the different main lineages of algae, especially in the Rhodophyta and Chlorophyta, and a single algal lineage may display several different forms.

1.4.1 Coenocytic or siphonous forms

Multicellularity is not the only way of overcoming the problems of largeness. Coenocytic or siphonous forms are unicellular but exhibit some of the specialisation seen in multicellular forms. *Grypania*, from 2100 million years ago, may have been a coenocytic alga. It has been likened to the unicellular green alga *Acetabularia*, which grows to a length of 3 cm or more, and becomes coenocytic or siphonous in its later stages of development; that is, the ‘cell’ contains multiple nuclei within a single continuous cell membrane. A multiplicity of nuclei can multiply the sources of gene products. Other examples of coenocytic green algae are *Protosiphon*, which consists of a single multinucleate sphere up to 0.3 mm in diameter, and *Ventricaria*, which may grow to the size of a chicken’s egg.

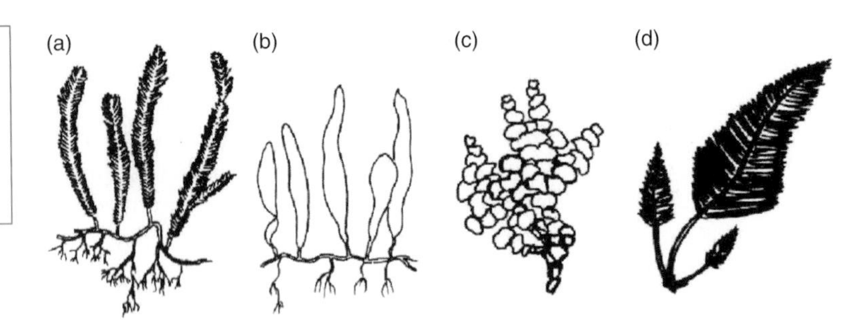

Figure 1.29. Coenocytic green algae: (a) *Caulerpa taxifolia*; (b) *Caulerpa prolifera*; (c) *Halimeda*; (d) *Bryopsis plumosa* frond (from Figures 5.35, 5.36, 5.37 in Lee, 1999).

Other coenocytes have a more complex form. *Halimeda* is a coenocytic marine green alga of coral reefs (Figure 1.29). It is subdivided into many flat segments each between 0.5 cm and 5 cm across. *Bryopsis* is similar but has a differentiated frond arising from a holdfast. The cytoplasm, with many nuclei and chloroplasts, lines the tube like stem and branches. *Caulerpa* also produces fronds but they are flattened. The large size of all these coenocytic algae places physical stress on the individual and they show various adaptations to overcome this. *Halimeda* produces a calcium carbonate skeleton. *Caulerpa* has stengthening ingrowths of the cell wall. The coenocytic form is vulnerable to loss of a large part of the cytoplasm if the cell wall is damaged. However, *Caulerpa* rapidly contracts the region around any wound so that it can be sealed off.

A different coenocytic pattern is seen in the green alga *Codium*. Here the body of the plant is a dense network of multinucleate filaments so that a pseudo-parenchymatous tissue is produced. Analogous forms are seen in both the fungi and the brown algae. Some coenocytes show remarkable differentiation but overall the possession of a coenocytic body does seem to limit specialisation. In all of them reproductive tissues differentiate after being cut off from the rest of the coenocytic body.

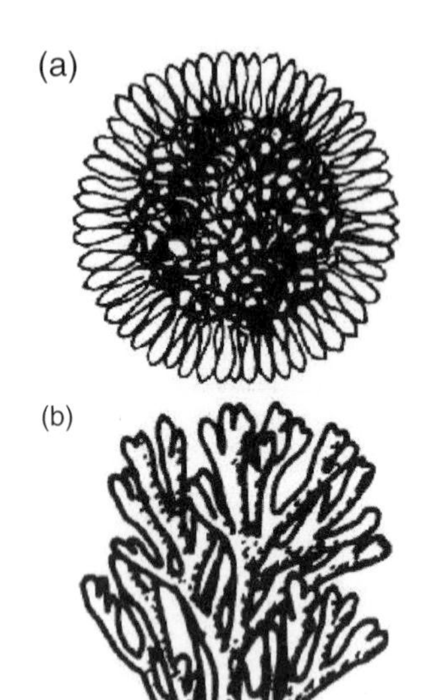

Figure 1.30. Green alga (Chlorophyta) *Codium*: (a) section through axis; (b) whole branched plant.

1.4.2 Colonial forms

There are two basic types of colonial forms. In one, the colony is indefinite and continues to grow by cell division and reproduces by fragmentation. *Hydrodictyon* is one beautiful example in the green algae. It produces net-like colonies up to 1 cm long. The cells of the colony are large and coenocytic.

In the second colonial form, which is usually free floating, there are a fixed number of cells at its origin, which do not vary much during its life-span. Examples from the green algae include *Gonium*, *Eudorina* (Figure 1.31) and *Volvox*, each with a different number of cells, differences in ornamentation or motility, and differences in degree of specialisation between cells.

In *Gonium*, which may have 4, 8, 16 or 32 cells depending on species, the concave colony moves by the cooperative action of the flagella of all its component cells, but each cell can divide to start a

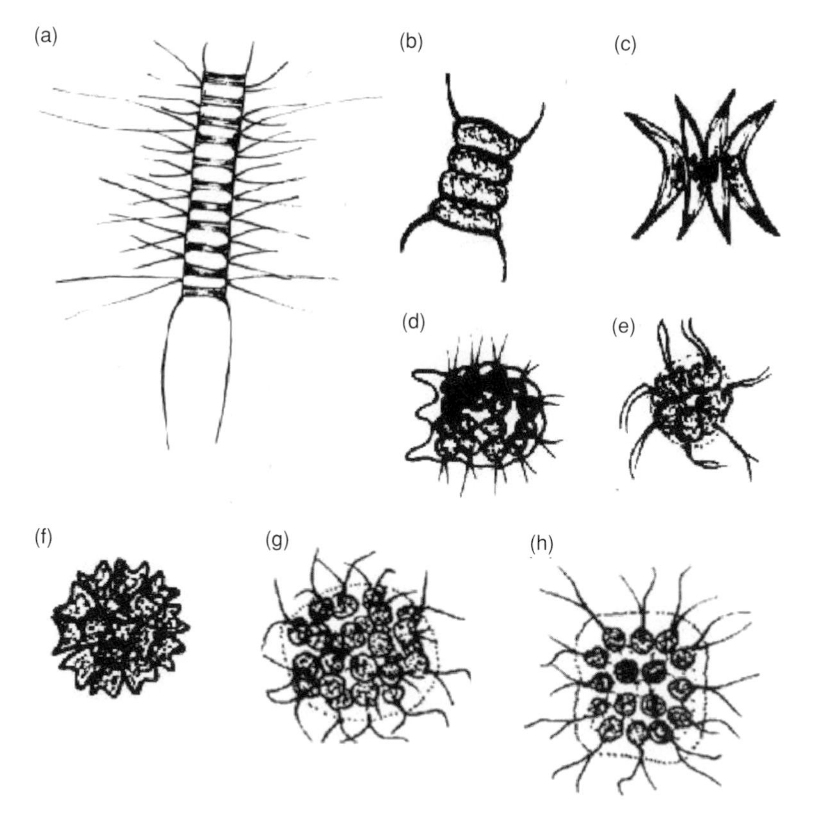

Figure 1.31. Colonial green algae: (a)–(c) *Scenedesmus*; (d) *Platydorina*; (e) *Pediastrum*; (f) *Pandorina*; (g) *Eudorina*; (h) *Gonium*.

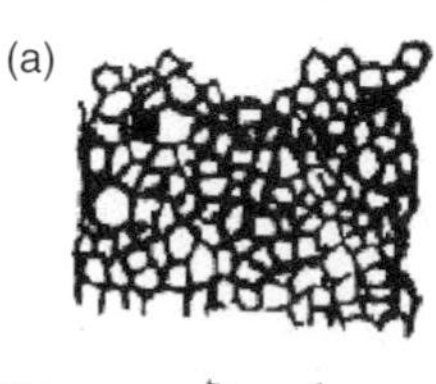

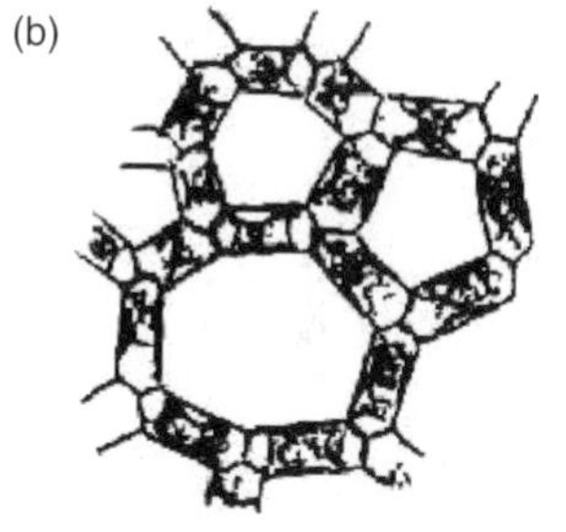

Figure 1.32. *Hydrodictyon*: (a) portion of net-like colony; (b) detail of net (a).

new colony. In *Scenedesmus* the four cells are ornamented with spines. In *Eudorina*, which can have 32, 64 or 128 cells, some posterior cells are small and incapable of cell division. In some forms like *Platydorina* there is a differentiation between the anterior and posterior part of the colony. *Volvox* is the most complex free-floating colonial alga. It has a single spherical layer of hundreds of cells, most of which are purely vegetative, and others that divide to give rise to juvenile spheroids, which are later released from the parent. These free-floating forms exist in a fairly homogeneous environment, the only polarity being the source of light, and consequently they show only weak differentiation, enabling them to orientate and swim towards light.

1.4.3 Filamentous forms

Filaments can consist either of a single line of cells (= uniseriate), or multiple lines of cells (= multiseriate). Because filaments are usually attached to a fixed substrate and have free-floating distal filaments they effectively exist in two environments and consequently show a greater degree of differentiation, specialisation of cells and patterning than free-floating colonial forms.

The filamentous type of multicellularity and specialisation is present in some cyanobacteria like *Anabaena*. It has, along with the normal photosynthetic cells, some large non-photosynthetic cells that

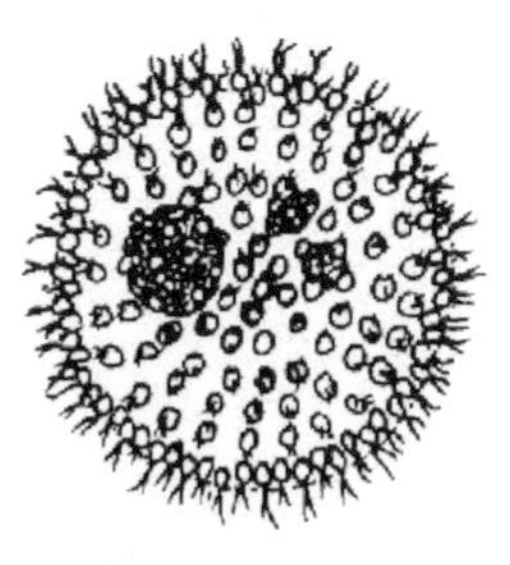

Figure 1.33. *Volvox*. Colonial green alga.

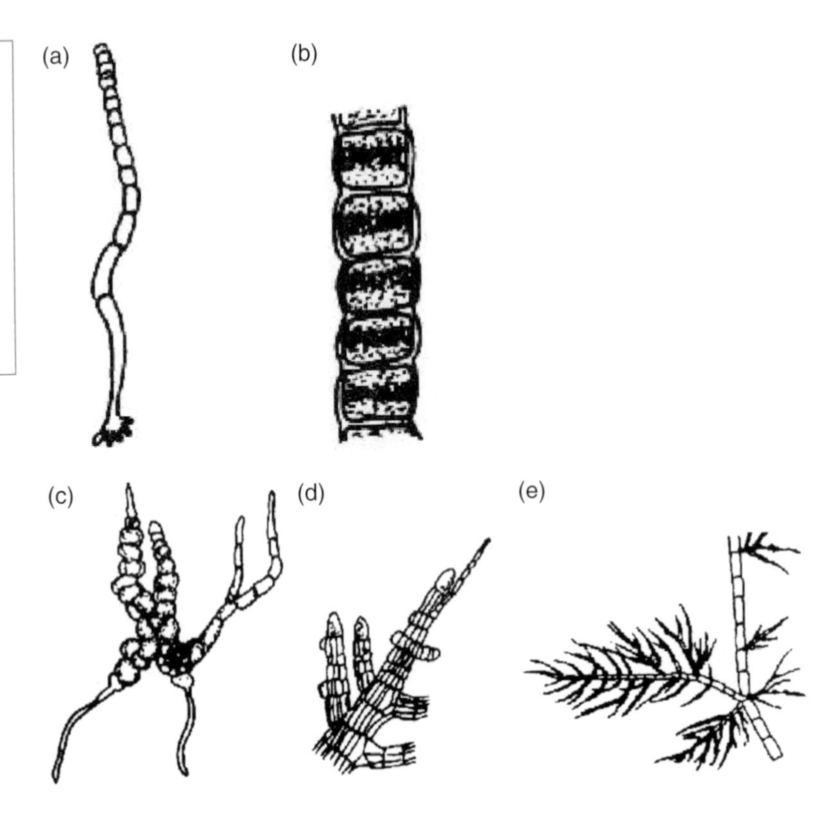

Figure 1.34. Filamentous algae: (a) *Chaetomorpha* (part of filament and holdfast) (b) *Ulothrix* (part of filament); (c) *Fritschiella* (filamentous terrestrial alga with differentiated aerial prostrate and soil penetrating filaments); (d) *Sphacelaria* (multiseriate filament); (e) *Draparnaldia* (part of filament showing main axis and branched laterals produced at nodes).

are specialised for nitrogen fixation. Some cyanobacteria also develop akinetes, enlarged cells with a thickened outer coat, that act as resistant spores permitting survival during periods of drought or heat.

The simplest filamentous forms in eukaryotes consist of a single chain of cells but, as in *Chaetomorpha* (Chlorophyta), the lowest cell is usually adapted as a holdfast to fix the filament to the substrate (Figure 1.34). *Bangia* (Rhodophyta) has filaments at first uniseriate, which later become multiseriate as the cells divide to produce a cylinder of up to eight wedge-shaped cells around a central core and surrounded by a sheath. A holdfast is made up of the elongated filaments of several basal cells. Fossils from 1200 million years ago from the Huntington Formation of Canada are very similar to living *Bangia* and represent the earliest record of the Rhodophyta.

There are many beautiful branched filamentous red and green algae. Growth in filamentous forms occurs by the activity of an apical cell which cuts off cells periclinally, dividing parallel to the surface or to the main axis. Each cell can then divide anticlinally to produce branches either singly or in whorls. There are diverse and complex structures, with different branching patterns and different degrees of differentiation between the main axis and branches. *Cladophora* is relatively simple, with lateral branches only weakly differentiated from the main axis. In *Draparnaldia* there is a clear differentiation between the main axis and branched laterals.

Fritschiella is a terrestrial green alga with multiseriate branches that are differentiated as either upright, prostrate, or positively geotropic, and it also has long colourless unicellular filaments or rhizoids that penetrate the soil.

1.4.4 Flat sheets (thalloid) and three-dimensional forms

Algae with sheets of cells display an alternative form of multicellularity. *Porphyra*, a red alga has sheets that are one or two cell layers thick (uni- or bistratose). The sheets are bistratose in the green alga *Ulva*. *Coloechaetes* produces disk-like forms. A three-dimensional tissue is also found in some red algae with filaments embedded in a mucilaginous matrix.

However, the most complex multicellularity comes when there are more than two-layers of cells because this permits specialisation and differentiation between surface layers and the interior of the organism. *Chara* has a thickened multiseriate axis. Although the axis in Charales has an apical cell that divides to elongate the axis, the thickened axis below is created, by changes in the orientation and asymmetrical divisions further down, producing files of cells of different dimensions. This gives the axis a 'corticated' structure, having a thickness of several cell layers. *Chara* has branched filaments (see Figure 1.40e). Branches arise at the nodes by 90° changes in the orientation of the spindle in the cell division of certain sub-apical cells. Long colourless unicellular filaments, rhizoids, are produced from the nodes near the base of the plant.

The interior of the organism may cease to have clear layers but forms a more mixed parenchymatous or pseudo-parenchymatous tissue that may have within it strands or islands of cells, which themselves are specialised for particular functions. In the brown algae the main tissue is a network of filamentous cells surrounded by a matrix. Cells may be differentiated with elongated cells in the middle for transport of mannitol and amino acids. The surface layer is specialised as a protective epidermis. An analogous kind of patterning has evolved in plants.

(a)
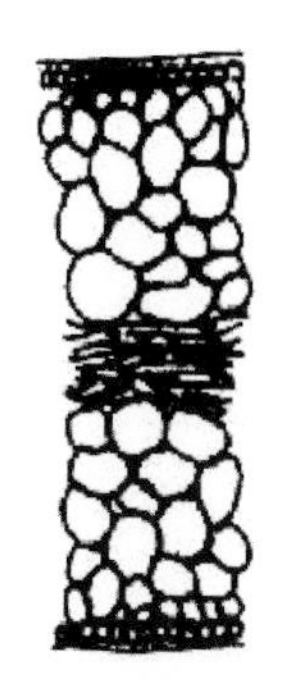

(b)

Figure 1.35. Anatomy of *Laminaria* (Phaeophyta) a stratose and parenchymatous algae (from Lee, 1999). (a) Section through the lamina; (b) stipe of showing trumpet connections between filaments and differentiated layers.

1.4.5 Inter-cellular connections and the differentiated body

A critical aspect of multicellularity is the control of inter-cellular communication. Inter-cellular connections have evolved separately in different lineages. For example, the cytoplasmic connections in red algae are quite different from the plasmodesmata of green algae and plants and plasmodesmata have multiple origins too. Other kinds of connection between cells are also formed; 'sieve-plates' between transport cells in brown algae, analogous to those of phloem sieve-tube cells of land plants, permit the transport of dissolved materials between cells. In the red algae there are distinct pit connections between adjacent cells in a filament. Such pits may arise *de novo* where the cells of adjacent filaments touch. The pit connections have a core of protein and a cap of polysaccharide.

(a)
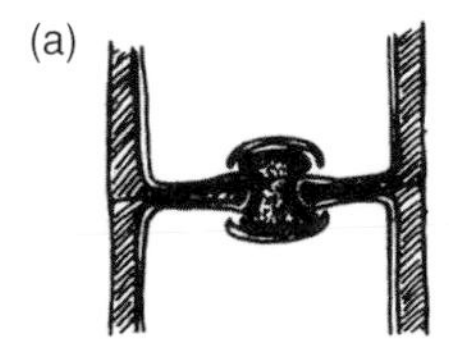

(b)

Figure 1.36. Diagrammatic representation of inter-cellular connections in algae and plants: (a) in red algae (Rhodophyta); (b) plasmodesmata in plants.

The presence of multicellular tissues with a network of connected cells permits a complex patterning of the organism. As the individual grows it differentiates in a highly organised pattern. Different parts of the body specialise for different functions. In photosynthetic sedentary organisms, both algae and plants, these parts and their functions are:

- to attach to the substrate as a holdfast;
- to penetrate the substrate as a 'root system';
- to provide an axis to which other parts can be attached;
- to provide a transport system connecting parts of the plant;
- to provide large exposed surfaces to capture light;
- to provide exposed regions for reproduction.

The way these parts are elaborated in plants is the subject of the following chapters. The brown seaweeds are described below for comparison.

1.4.6 The brown algae

The brown algae (Phaeophyceae) are the largest and most complex photosynthetic organisms that are not plants. Possible fossil brown algae have been found in North American and Asian rocks as old as 650–544 million years (the Vendian or latest Proterozoic), that is, long before the diversification of plants. Others have been described from the Ordovician, but only from much later, in the Tertiary, is there any degree of certainty about their presence.

Figure 1.37. Pelagic brown alga *Sargassum* (Phaeophyceae).

The brown algae are probably the most familiar of the algal groups since they comprise most of the large seaweeds of intertidal zones. There are numerous marine, but also a few fresh-water, forms. The genus *Sargassum* (Fucales) is unusual in having some large complex species that are free floating and have a pelagic existence in areas of the tropical Atlantic Ocean (the 'Sargasso Sea'). *Sargassum* is like an intertidal plant that has escaped to the open sea.

Intertidal algae are complex multicellular organisms and show a range of adaptations for life in the intertidal zone (Figure 1.38). A strong holdfast grips the substrate. A strong and flexible stipe and midrib absorbs the stresses of waves. The frond is flattened to capture light but divided to allow water to flow around it easily. Air vesicles buoy up the alga and help to prevent tangling.

The brown algae possess chlorophylls a and b and their food reserves are different from those of either red or green algae. The plant body of brown algae is made up of a pseudo-parenchyma of tightly packed filaments. The mucilage covering the whole alga aids in preventing desiccation at low tide, as well as helping to reduce physical damage.

They are usually divided into about 12 orders, the simplest of which is the Ectocarpales. No members of the brown algae are adapted to a life on land but frequently the complexity of the group is cited as being analogous to that of the higher land plants.

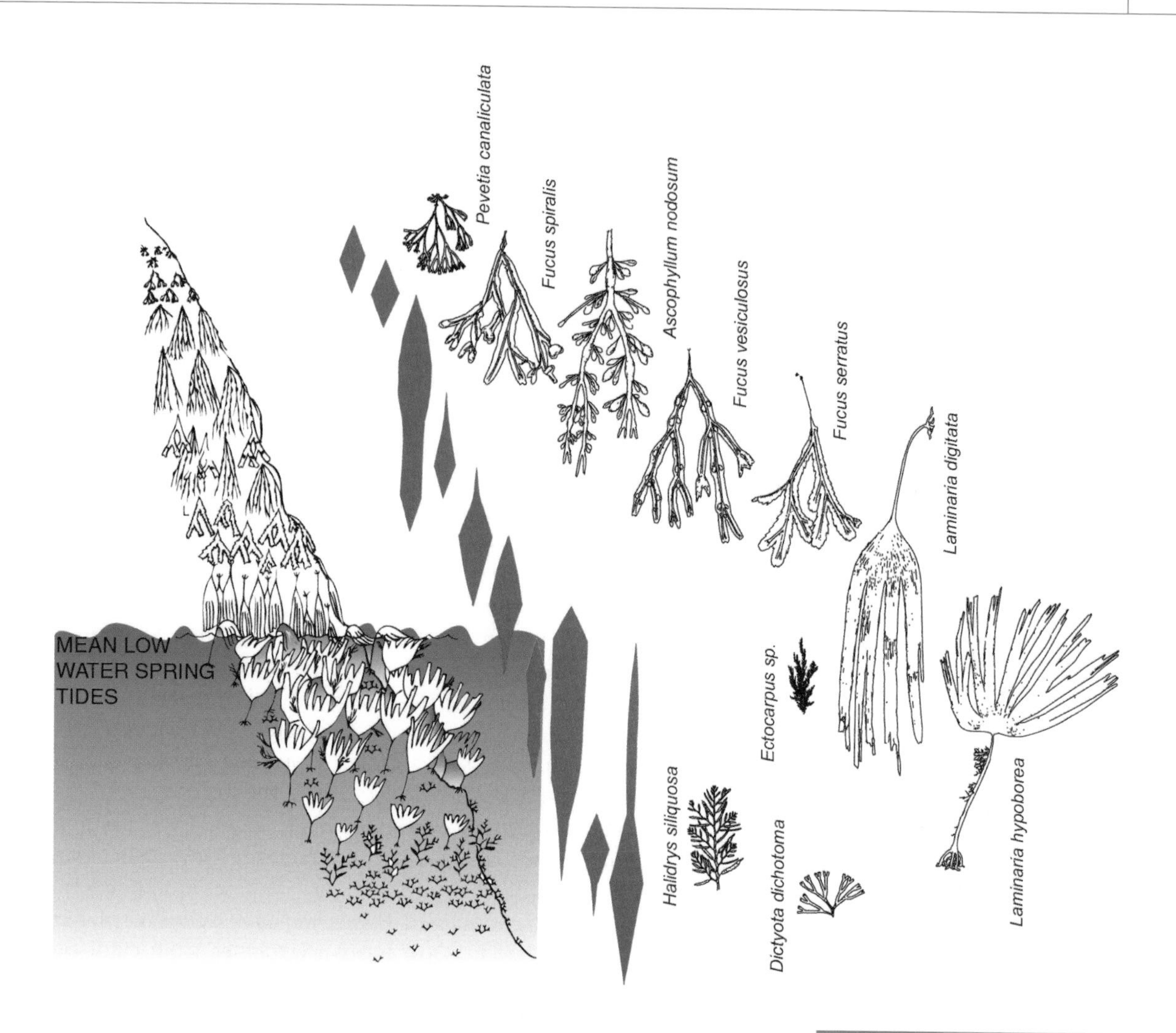

Figure 1.38. Intertidal zonation at an exposed and sheltered rocky site and on an unstable substrate from the NE Atlantic coasts showing the different distribution of seaweeds (modified from Hiscock, 1979).

The Laminariales, or kelps, are found in all the colder seas and oceans of the world and include the two largest known algae, *Macrocystis* and *Nereocystis*. The familiar 'tangle' of Atlantic shores is *Laminaria digitata*. They are most frequently found in the sublittoral zone just below the low-water mark. In *Laminaria*, the thallus is differentiated into a blade that, depending on the species, may be single or dissected into many segments. There is usually a strong, pliable stipe or stem connecting the blade to the holdfast that anchors the kelp to the rocks. In the stipe there is a specialised conducting tissue analogous to the phloem of land plants. It has sieve-tube elements, 'trumpet hyphae', which lack a nucleus at maturity. This conducting system probably evolved along with an increase in size. Growth of the thallus is rapid and constant abrasion by the pounding of waves against rocks is quickly made good by very rapid meristematic growth at the base.

1.4.7 Green algae (Chlorophyta)

Green algae were almost certainly the ancestors of plants. They are very diverse and include a range of organisms, approximately 500 genera and more than 16 000 species, from unicellular flagellates described above, to complex multicellular organisms. They have a very wide ecological range and though most live mainly in marine, brackish or fresh water, they are also include semi-terrestrial forms dispersed by wind, or kinds that grow in the soil or in association, even inside, plants and animals. *Paramecium* is a ciliate that engulfs unicellular green algae and keeps them in its vacuole. The euglenoids (Euglenophyta) have gained their chloroplast through a symbiotic association with a green alga. Green algae are also one of the partners in the lichen symbiosis.

The fossil record of the green algae begins in the Cambrian, predating by many millions of years the origin of the land plants. They may have arisen even earlier but the earliest putative fossil green algae come from the Neoproterozoic rocks 900 million years old in Australia and 700–800 million year old rocks in Spitzbergen. Some early fossils are referable to the Dasycladales, the order of unicellular stalked organisms that includes the living *Acetabularia*. About 120 fossil genera have been described. Some of the multicellular fossil algae from Spitzbergen resemble the living genera *Coelastrum* and *Cladophora*. Many of these fossils have been preserved because they secrete lime around the complex multicellular thallus.

The use of taxonomic names gets difficult here if we adopt a strictly phylogenetic approach to naming groups. The green algae that we have called the Chlorophyta, are not a monophyletic group, because, as we have defined them, we have excluded plants (Embryophyta). A more strict use of terms that arises from cladistic analysis of biological variation would include all green plants and algae in a single group (the Viridiplantae) and split this into the two distinct sister lineages, which have been called the Chlorobionta and the Streptobionta. Then again, if we recognise the plants as a distinct taxonomic group, the Embryobionta, the remaining Streptobionta cease to be a proper taxonomic group. The separation of the Chlorobionta (including only some green algae) from the Streptobionta (including some green algae and all plants) is based on what are regarded as fundamental features of cell division and flagella insertion, as well a molecular characters such as the DNA sequences of mitochondrial genes.

The complications are endless because there is a 'primitive' group of green algae called the 'Prasinophytes' by some, or the Micromonadophyta or Micromonadophyceae by others, and which includes several lineages related to either the Streptobionta or Chlorobionta or neither. For example, *Mesostigma viride* has the same pigments as the Ulvophyceae in the Chlorobionta but on the basis of DNA sequence data this prasinophyte is the earliest divergence at the base of the Streptophyte lineage that gave rise to plants. *Mamiella*, *Mantoniella*, *Micromonas* and *Pseudoscourfeldia* have a distinct light-harvesting

Table 1.4 The differences between the Chlorobionta and Streptobionta plus Embryobionta

Chlorobionta	Streptobionta and Embryobionta
Nuclear membrane persists throughout mitosis	Nuclear membrane disintegrates in mitosis
Phycoplast	No phycoplast
Phragmoplast transitory	Phragmoplast (spindle) persistent
Motile cells symmetrical	Motile cells flattened
Flagella apical and directed forward	Flagella sub-apical and at right angles to cell
Cytoskeleton not in flat broad bands	Cytoskeleton with flat broad bands
No glycolate oxidase or if present outside peroxisomes	Peroxisomes with glycolate oxidase (photorespiratory enzyme)

pigment–protein complex indicating a separate lineage from either the Chlorobionta or the Streptobionta.

Although the prasinophytes may not represent a monophyletic group they are an interesting grade of organisation and share several features. They are all flagellate unicellular organisms with a variable number of flagella arising from inside an apical notch. They can undergo encystment losing their flagella and becoming dormant inside a round resistant wall, which looks like a spore, though it does not contain sporopollenin. Most have an interesting scaly surface unlike that of the cell wall of plants or theca of other green algae. The scales arise inside the Golgi apparatus and are then transferred by a vesicle to the apex of the cell from where they join the several layers of scales that are already present.

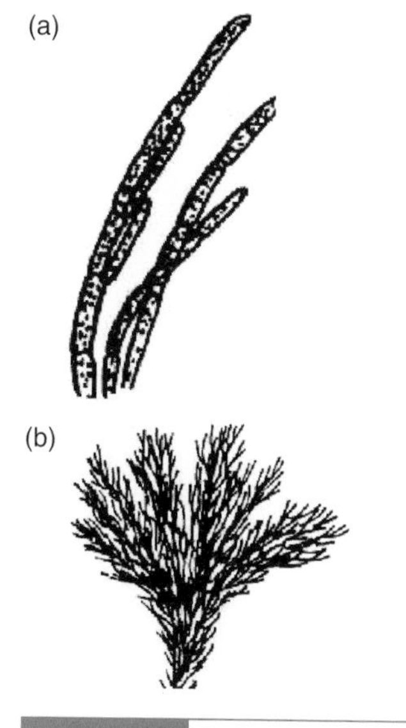

Figure 1.39. *Derrbesia* filamentous green alga with coenocytic cells.

The Chlorobionta include three groups of the green algae, the Chlorophyceae, the Pleurastrophyceae and the Ulvophyceae, and are represented by a range of unicellular, filamentous or stratose forms. One way in which these groups can be separated is by how the flagella in their motile cells are arranged. Like the hands of a clock, the Chlorophyceae have the flagella at either 1 and 7 o'clock or 12 and 6 o'clock, while the Pleurastrophyceae have them at 5 o'clock. The Ulvophyceae have them all at 12 o'clock. There is unlikely to be any adaptive significance to these arrangements. This is just one of those features that became fixed developmentally early in the evolution of a lineage.

Molecular data indicate that the charophytes in the Streptobionta are a monophyletic sister group to plants. Bryophyte plants share several features with them but these may have evolved in parallel. Charophyte fossils such as *Palaeonitella* are known from Silurian marginal marine sediments. They underwent their first major diversification in the Silurian and Devonian at about the same time as plants were originating and first diversifying. *Nitella*-like fossils are known from fresh-water Lower Devonian Rhynie chert. The first record of the living order Charales is from the Permian, and they underwent a major radiation in the Mesozoic.

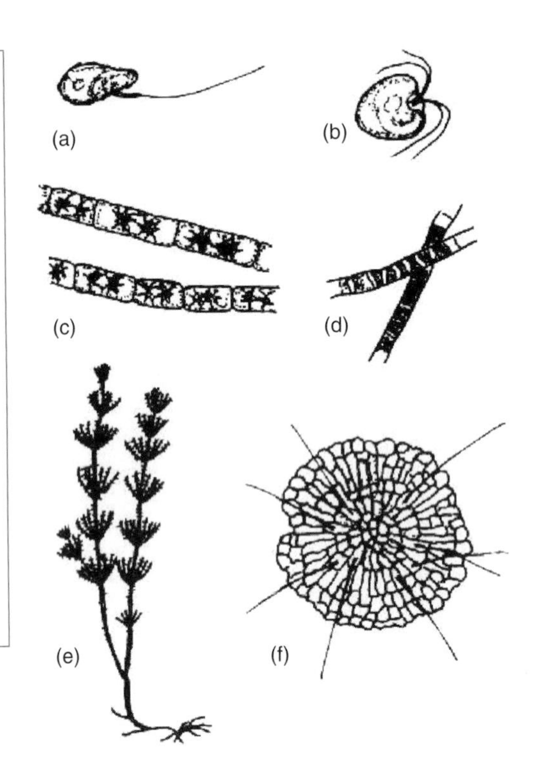

Figure 1.40. Charales, the green algal sister group to plants shows different levels of complexity of form: (a) *Mantoniella* (Micromonadophyceae) probably the most primitive charophyte is scaly with two flagellae, one very short; (b) *Pyramimonas* is also scaly – the scales are produced by the Golgi body; (c) *Klebsormidium* filaments do not have holdfasts and there are no plasmodesmata; (d) *Zygnematales* show a ladder-like (scalariform) conjugation in sexual reproduction; (e) *Chara* has a differentiated branched form with nodes and internodes; (f) *Coleochaete* – some species form a pseudo-parenchymatous disk, and have sheathed filaments.

There are four main groups of living charophytes (Figure 1.40), the Klebsormidiales, Zygnematales, Coleochaetales and Charales. The Klebsormidiales have undifferentiated uniseriate filaments and, unlike other charophytes, they are isogamous, releasing biflagellate gametes. The Zygnematales such as *Spirogyra* are multicellular and filamentous, or bi- or unicellular organisms. They include the very large group of planktonic organisms called the desmids. Very unusually for algae and plants the Zygnematales undergo conjugation, the filaments aligning to each other and fusing, to bring their non-flagellate gametes together. The Coleochaetales and Charales are the most complex filamentous and pseudo-parenchymatous multicellular charophytes. They are oogamous and have complex reproductive structures. The Charales are called the stoneworts because they are commonly heavily calcified.

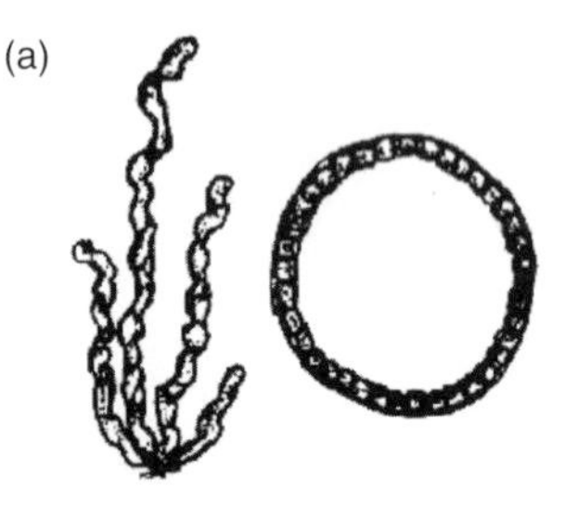

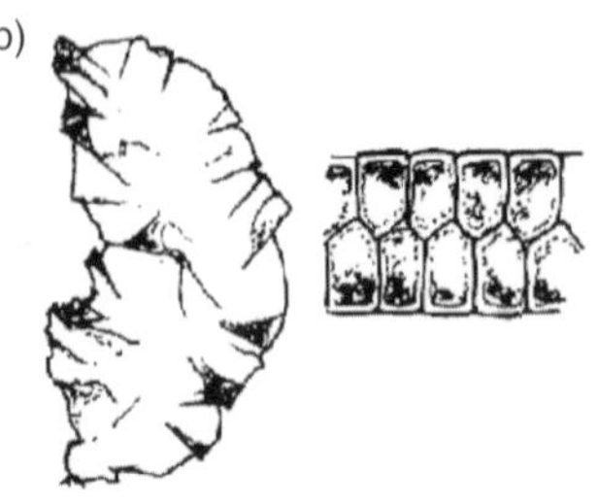

Figure 1.41. Thalloid chlorophyta in the Ulvophyceae. (a) *Enteromorpha* with a unistratose tubular lamina; (b) *Ulva* with a bistratose flat sheet.

1.5 What is a plant?

Plants are actually very strange living creatures indeed. Their life is alien to us. But this is their planet; they have made it and we live in their shadow. Plants seem static so that it is easy to forget that plants are living organisms. We associate movement with life. Though some plants look like pebbles (*Lithops*), they are not stones. If plants do not have life surging through them like animals, it trickles through them in a constant stream. Plants combine the stability of structure with

Table 1.5 The kingdoms of life

Prokaryotes	Bacteria	Archaebacteria	Archaebacteria
Eukaryotes	Plants	Eubacteria	Eubacteria
	Animals	Plants	Protoctista
		Fungi	Plants
		Animals	Fungi
			Animals

the fluidity of change. In the following chapters we explore this in detail.

Before the modern period of classification any organism not obviously animal was usually regarded as a plant. This applied to fungi and many marine animals such as anemones, bryozoans, corals, etc. A dichotomous view of the world, basically 'folk-taxonomy', was all-pervasive and influenced classification for two centuries. But there is a unity between plants and other organisms. It is now known that the similarities between animals and plants outweigh all their discrepancies. Differences in mode of life and structure are no more than differences of expression and economy. Nowadays, three to six kingdoms of organisms are usually recognised (or seven if viruses are included).

However, the most fundamental dichotomy of living organisms is at the cellular level between prokaryotes and eukaryotes (see Section 1.3). For most people, however, the differences between prokaryotes and eukaryotes are meaningless in comparison with the apparently real and culturally important distinction between plants and animals. The old dichotomous view still has relevance to our everyday lives. For this reason the differences between plants and animals are outlined in considerable detail below but this is not to be regarded as an endorsement of a two-kingdom classification.

Fungi and algae are still often regarded as plants but in most modern classification systems they are placed in separate kingdoms. The latter view is adopted in this book, so that when we write 'plants' this is equivalent to 'land plants' of other writers. Algae are placed in the protoctista or protists, which are mainly unicellular. They are exceedingly diverse in very fundamental ways. They have traditionally divided into heterotrophic (non-photosynthetic) and autotrophic organisms. The latter are 'the algae' of diverse sorts. Algae have traditionally been considered to be 'plants' in a broad sense and studied by botanists, and some at least of these algae have a direct place in the story of plants because they are related to the ancestors of plants. Botanists have also studied fungi, although it is now clear that they are more closely related to animals than they are to plants. However, they too have a place in this story because of their ancient and close ecological partnership with plants, in mycorrhizal plants and mycotrophic plants.

1.5.1 How do plants differ from other living organisms?

The most conspicuous differences between the majority of plants and animals are that plants are sessile organisms that remain in one spot throughout their lives. Apart from growth movements, they are mostly stationary, although they have a dispersal phase that has its counterpart in sessile marine animals. Unlike animals, plants have a very limited ability to choose their environment. The relatively complex growth patterns of higher plants are largely a substitute for motility and maximise potential responses to environmental fluxes.

However, plants do have remarkable abilities for dispersal mainly by the production of spores and seeds. Because of the phenomenon of dormancy, especially prevalent in seed plants, they can disperse in time, as well as space. They may be said to be localised in space but not in time, whereas animals are localised in time but not in space. These fundamental differences arise directly as a result of their respective modes of nutrition. Plants have a relatively passive mode of nutrition. Being photo-autotrophic (a minority are parasitic or mycotrophic) they derive their energy requirements from the breakdown of energy-rich carbohydrate molecules, which they elaborate themselves from the carbon dioxide of the air by the process of photosynthesis using sunlight. Plants concentrate water, minerals and carbon. Inorganic substances such as trace elements and mineral compounds are obtained in very dilute form from the air, soil and water and built into organic substances necessary for the maintenance of life. Rooted in the soil, and growing upwards to the light, green plants are searching organisms relying on very diffuse resources and do not obviously seek out and devour food (although some carnivorous plants ingest animals).

In contrast animals are mobile, most have bilateral symmetry and polarity, a head end and a tail end, and often have appendages such as limbs for movement. Food is obtained in already concentrated organic form. They have an active mode of nutrition, which requires that they direct their bodies to a source of food that must be searched for, detected and pursued. This presumably entails choice. Animals usually have some sort of nervous system and a brain for information processing, as well as organs of perception such as eyes, ears, chemo- and heat-receptors. Animals have evolved to a level of diversity far exceeding plants in terms of their adaptations for feeding and mobility. They have developed complex social systems, they can detect and flee from danger and they can feel pain.

The main exceptions are sessile marine animals such as corals, sea-anemones, fan-worms, barnacles, and feather-stars, etc., that, not surprisingly, resemble plants superficially in their overall form, modular construction and their defence systems. Many sessile animals have a radial symmetry or at least a symmetry more suitable for catching diffuse organic matter in the surrounding water. Like plants, they show an increase in the surface area:volume ratio in their parts that are directly concerned with obtaining nutrients. Unlike plants, however, these animals are not attached to the substrate by roots and do not require sunlight (although some such as corals actually have a symbiotic relationship with phototrophic algae). They are strictly heterotrophic organisms and the orthodox interpretation of their resemblance to plants is through their having to face similar environmental problems, that is how to obtain enough nutrients from a very dilute medium.

Fungi are heterotrophic like animals, but derive their energy requirements from the breakdown of dead complex organic matter or parasitism. Their energy source is diffuse and in some ways fungal mycelial growth is like the growth of fine roots in the soil.

The form of plants is such that there is a large surface area:volume ratio that is necessary to absorb material from extremely dilute solutions. They have a symmetry that is best described as radial or spherical. Mineral nutrition and light comes to the plant and the amount that is absorbed is proportional to the absorbing surface. This has necessitated that the plant body consists of a large number of ramifying components, each with a high surface area:volume ratio. This allows the plant to 'scavenge' the maximum amount of nutrients possible. Although plants are extremely sensitive to their environment, they are non-sentient, without any correspondence to a nervous system and as far as we know they feel no pain.

Being immobile, plants are rigidly ordained by the accident of their situations. Plants cannot run away. The structure of plants is both diffuse and vulnerable, both above and below ground. To counter this they have evolved a powerful armoury of physical and chemical defences such as spines, thorns, prickles, stinging hairs, noxious latex, resins and poisons. Size may be adjusted to the resources available. Plants grow themselves 'out of trouble' by processes best described under the broad term 'ontogenetic contingency'. This is the phenomenon whereby an individual plant's developmental programme and ground-plan can explain and constrain its capacity to respond phenotypically to environmental variation throughout its life. Plants persistently attain size-correlated variations in their forms and processes owing to the functional obligations imposed by their environment. Growth in plants is therefore somewhat analogous to behaviour in animals. Unless a plant enters a resting stage it must accept fluctuations in the environment by adjusting to them by some physiological or morphological means. Plants tend to respond to environmental stress by variation in reproductive rate without death. In other words, plant populations are very likely to have their numbers regulated by their environment.

1.5.2 The challenge of the land

Plants grow in a physically unsupporting and potentially damaging environment. The Ordovician and Silurian landscapes were hostile to colonisation by algae adapted to the aquatic environment. The thalloid and filamentous algal life forms that lay in damper indentations on the land and at the margins of pools and streams did not modify

Table 1.6 Major differences between plants and animals

Plants	Animals
Autotrophic, photosynthetic	Heterotrophic
Sessile	Mobile
Indeterminate growth	Determined growth
Modular construction	Unitary construction with head and tail ends
No nervous system but symplasm	Nervous system
Cellulose cell walls	Cell walls absent or various
Vacuoles present	Vacuoles absent
Sex cells arise late in development	Sex cells separated from body cells early in development

their environment in any profound way. The sun baked the land and the winds and rain scoured it. Soil development was minimal.

Plants have adapted to life on land by internalising the external atmosphere and exploring the soil in an intimate way. Plants required many adaptations in order to break away from the aquatic margins:

- to survive periods of desiccation – poikilohydry
- to restrict water loss – waxy cuticle
- to supply water to all parts of the plant – vascular system
- to enable gaseous exchange – stomata and internalised gas spaces
- to obtain mineral nutrients – rhizoid or root system
- to expose a large surface area for photosynthesis – branching system, secondary (thickening) growth and leaves
- to shield themselves from excess light or to dissipate excess light – pigments, physiological mechanisms
- to keep cool – transpiration
- to support themselves in the air – thickened cell walls (collenchyma) or lignified cell walls (sclerenchyma)
- to prevent being uprooted by wind or water – root system
- to reproduce sexually – archegonia
- to reproduce laterally – lateral spread, fragmentation
- to disperse – sporophyte, sporangia and air-dispersed spores.

The crucial difference between terrestrial and aquatic environments is that water is only intermittently available on land, but plants have adapted in many different ways to this. There is an important distinction between those organisms that have a very limited ability to remain hydrated in a dry environment, a condition called poikilohydry, and those that are able to remain hydrated even in a very dry atmosphere, a condition called homoiohydry. All algae, but relatively few plants, are poikilohydric. However, they include the very successful groups of the mosses and the liverworts. Nevertheless the evolution of homoiohydry was an important step in the colonisation of the land. Homoiohydric plants evolved at least 420 million years ago

The evolution of homoiohydry is associated with an increase in the maximum size attainable by plants. Poikilohydric photosynthetic organisms range in size from unicellular cyanobacteria 0.65 μm in diameter, and eukaryotic algal unicells 0.95 μm in diameter, to a maximum of up to 1 m tall in a few mosses. The size range of homoiohydric plants is much greater, ranging from 5 mm to 130 m in height. A consequence of this too is that, because they are larger, plants have a greater scope for increased complexity.

Geological periods from the Palaeozoic	Dates started (millions years ago)
Quaternary	1.64
Tertiary	65
Cretaceous	144.2
Jurassic	205.7
Triassic	248.2
Permian	290
Pennsylvanian	322.5
Mississippian	362.5
Devonian	408.5
Silurian	439
Ordovician	510
Cambrian	570

1.5.3 The characteristics of plants (Plantae, Embryobionta)

The name Embryobionta refers to the embryo, the multicellular structure produced at the earliest stage of growth at one stage of the life cycle, the first growth of the diploid sporophyte plant. However, another feature that distinguishes plants from almost all algae is the presence of complex reproductive organs like the female organ, the archegonium, and the male antheridium. The shared features of plants include the following:

- haplodiplobiontic life cycle and multicellular sporophytes
- archegonia (female reproductive organs)
- antheridia (male reproductive organs)
- sporangium (spore-producing organ)
- sporopollenin in the spore wall (also detected in some algae)
- cellulose cell wall
- cuticle (waxy outer layer of the epidermis)

1.5.4 The first plants

A particularly important fossil, because it is the earliest plant to show several adaptations for life on land, is *Cooksonia*. It had rounded or kidney-shaped sporangia, organs in which hundreds of air-dispersed spores were produced. The presence of these trilete spores, which have a characteristic shape from being produced in tetrads, is the earliest firm evidence of land plants in many fossil rocks. The earliest *Cooksonia* has been discovered in rocks older than 420 million years ago from the Devilsbit Mountain of Co. Tipperary in Ireland. Already several species existed, differing in the shape of their sporangia.

Cooksonia had a very simple branching pattern (Figure 1.42). The short stems forked at a wide angle and each stem terminated with a rounded knob of a sporangium. Plants probably grew adjacent to stream channels or on top of sand bars, rapidly maturing as the habitat dried so that there was a flush of sporangia. The outer layers of the stem and sporangia had cells with thick walls, sclerenchyma, which supported the plant and may have conferred some protection

Figure 1.42. Reconstruction of *Cooksonia* at the Royal Botanic Gardens Kew.

from drought. A cuticle and pigmentation protected against high-UV light.

The evolution of *Cooksonia* was a portent of a remarkable change in the world. For millions of years nothing much had changed on land. Now the plants were making the terrestrial environment potentially habitable for animals. For example at Ludford Lane near Ludlow, in Salop (Shropshire), 414 million years ago, along with *Cooksonia* there were two kinds of centipedes and a trigonotarbid arachnid that must have been preying on smaller arthropod detritus feeders. If the whole history of the Earth took place in one day, the first cells had evolved before 8 a.m. but the land was not properly colonised until 10 p.m. Then at the end of the Silurian Period and the beginning of the Devonian Period, about 400 million years ago, after all the waiting, like a kettle suddenly coming to the boil, a full terrestrial vegetation and an accompanying terrestrial arthropod fauna appeared in just a few million years.

1.6 Sub-aerial transmigration of plants

In many ways the subsequent evolution of plants merely plays out what the algae had already established (Figure 1.43). The most profound evolutionary steps had already been taken; multicellularity and an integrated differentiated form, and a multiplicity of variations in the reproductive cycle, in sex and dispersal, are all present in algae. This was the thesis of A. H. Church, championed by Corner, in his theory of the 'subaerial transmigration of plants' for the establishment of land plants. Perhaps they imagined a more direct ancestral

Figure 1.43. Green algae at the aquatic margin.

relationship between several groups of algae and land plants, an idea that has since proved wrong. However, they were very perceptive to emphasise in this way the shared genetic memory of plants and algae and the fundamental similarities between them.

Further reading for Chapter 1

Hiscock, S. *Field Key to the British Brown Seaweeds* (England: Field Studies Council, 1979).

Kauffman, S. A. *The Origins of Order. Self-organization and Selection in Evolution* (Oxford: Oxford University Press, 1993).

Lee, R. E. *Phycology*, 3rd edition (Cambridge: Cambridge University Press, 1999).

Niklas, K. J. *The Evolutionary Biology of Plants* (Oxford: Oxford University Press, 1997).

Taiz, L. and Ziegler, E. *Plant Physiology*, 2nd edition (Sunderland, MA: Sinauer Associates Inc., 1998).

Willis, K. J. and McElwain, J. C. *The Evolution of Plants* (Oxford: Oxford University Press, 2002).

Chapter 2

The genesis of form

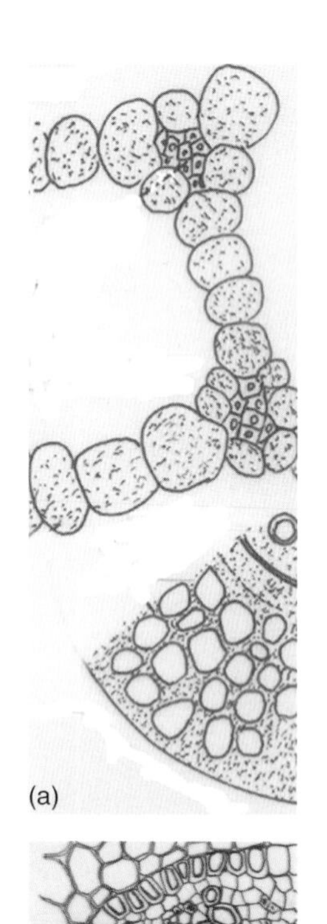

Figure 2.1. Aerenchyma: anatomical sectors through (a) *Potamogeton* stem and (b) *Iris* root. The large channels are formed by the death of cells.

The student of Nature wonders the more and is astonished the less, the more conversant he becomes with her operations; but of all the perennial miracles she offers to his inspection, perhaps the most worthy of admiration is the development of a plant or animal from its embryo.

Thomas Henry Huxley, 1825–1895

2.1 Plant development

Plants undergo an orderly succession of developmental changes (ontogeny) starting with the simple structure of the embryo and ending with the highly complex organisation of the mature plant, senescence and death. At least, this is the zoocentric view of the plant life cycle. In reality it is more complex than that because many plants, by shedding their parts, actually may be said to be constantly dying while they are living. Programmed cell death (PCD), called apoptosis in animals, occurs in the normal life cycle of all plants, for example, during maturation and senescence of leaves, flowers and fruits, and abscission in the regular seasonal cycle of temperate plants during leaf-fall, and as the result of stress. In aquatic and semi-aquatic plants cell death creates air channels that aerate the submerged tissues. In addition, the great propensity for vegetative reproduction gives plants almost immortality. Such serial changes contribute to the unfolding development of the plant as a whole, but they also occur at all levels of organisation, from cells and tissues to organs.

As a plant cell gets bigger, the surface area does not increase in proportion to the volume. It becomes less efficient at exchanging substances with its surroundings, and slower in its growth rate. Therefore, all plants get bigger primarily by increasing the number rather than the size of their cells. A sphere has the least possible surface area for a given volume, and any departure from this shape increases the surface area:volume ratio. Increasing the surface area of a plant increases photosynthetic potential but it also exposes the plant to greater water loss and heat stress. Plants must be able to achieve a compromise between the two conflicting components of the surface area:volume relationship. For plants that float freely in water, the best way to keep a high surface area:volume ratio is to remain small. This is not possible for land plants that must remain fixed in one spot, compete for light, and be able to obtain a supply of water. They must combine large size with a high surface area:volume ratio, and a large surface, relative to bulk, may be produced by elaboration of external form. Subdivision and flattening (for example, of branches and leaves) increases surface area enormously. A large surface area should be available to absorb carbon dioxide and oxygen as well as to maximise light capture. These processes are often carried to a maximum while simultaneously complying with the demands of robustness and mechanical strength. For ordinary plants the limit is about 30 cm^2 of surface area:1 cm^3 of volume. Internalised airspaces that ramify through spongy tissues permit gaseous exchange and minimise water loss, while elaborate external structures trap sunlight. Leaves have a tissue called spongy mesophyll connected to the exterior by closable pores called stomata.

Plant development involves two major processess: growth and differentiation. Growth and differentiation usually take place concurrently but not always. Growth is an increase in the amount of living material leading to an irreversible change in the size of a cell, organ or whole organism. At the cellular level this leads to increase in cell size and ultimately cell division. The external form or overall appearance of plants and their parts such as stems, leaves, and flowers, is the result of differential growth. This is ultimately determined by the quantity and arrangement of cells within different tissues and organs. Differentiation therefore refers to qualitative differences between cells, tissues and organs. Allometric growth is where there are differences in overall appearance of a plant, organs or tissues connected to the age due to variation in growth rate of the different parts. It is a kind of scaling effect. Parts that grow faster than the whole plant are positively allometric, whereas those that grow slower are negatively allometric.

2.1.1 Plant cells and tissues

Different types of plant cells can be distinguished by the shape, thickness and the constitution of the cell wall, as well as by the contents of the cell. Differentiation goes as far as programmed cell death so that some plant cell types are not fully differentiated and functional until their protoplasm disintegrates at maturity leaving the 'dead', but still functional, rigid skeleton of the cell wall. Large parts of the plant body may be composed of these dead cells. Plant cells are diverse but just a few basic types can be recognised in most land plants. In different combinations they form the tissues of the plant.

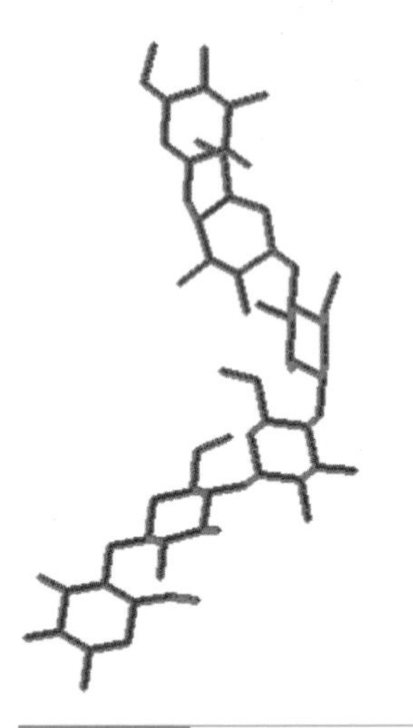

Figure 2.2. Cellulose: a small portion of the chain polymer.

2.1.2 The cell wall

The main component of the cell wall, but still only 20%–40%, is cellulose. This is a structural material. It is a fibrous polysaccharide that is 'spun' through a rosette of cellulase synthase enzymes embedded in the plasma membrane, plasmalemma. The cellulose is anchored to the inside of the pre-existing cell wall. Each molecule of cellulose is a ribbon-like polymer of between 1000 and 15 000 β-glucose units up to about 5.0 μm long. Each elementary microfibril of cellulose, which has a diameter of 3.5 nm, contains about 40–70 molecules spun off together in a highly H-bonded structure of great strength. The elementary microfibrils may be arranged in patterns of great complexity, the result of many 'spinning' rosettes acting together in an array to weave each layer of the cell wall like the warp and the woof of a complex fabric. Larger fibrils, 20–25 nm in diameter, form chains or rods, with alternating dense crystalline areas, called micelles and linked by unordered fibrils. Fibrils may be further combined to form macrofibrils up to 0.5 μm in diameter.

Even at this sub-cellular level it is differences in the polarity of the cell, in the orientation of the various fibrils and the composition of many other substances, such as the hemicelluloses, which fill the spaces between them that confer the mechanical characteristics of each cell. These substances, which include proteins, act either as glues or lubricants. Important constituents are the pectic substances that glue cells together at the middle lamella. Some cell walls, especially in wood, become impregnated with a matrix of lignin, a complex polymeric resin that helps to bind the fibrils together, and prevents them shearing apart. Others cell walls are impregnated or encrusted with the fatty substances, cutin or suberin, making them impermeable to water.

The composition of the cell wall is not fixed. Growth-promoting plant hormones make the wall extensible. In the maturation of fruits, cells become less firmly glued to each other. In abscission, cell walls break down. As a response to fungal infection or incompatible pollinations materials are laid down in cell walls to prevent invasion. Components of the cell called oligosaccharides, which are derived from partial breakdown of cell wall polysaccharides, can act as signalling molecules.

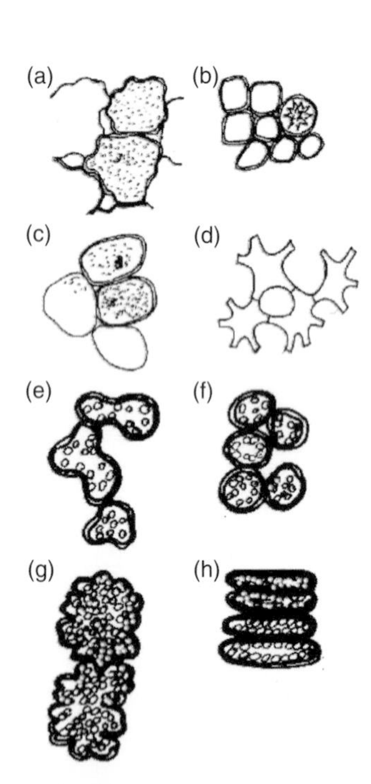

Figure 2.3. Parenchyma. (a)–(d) Non-photosynthetic parenchyma; (b) one cell, an idioblast, with a crystalline inclusion; (e)–(h) photosynthetic parenchyma (chlorenchyma); (e), (f) from mesophyll of flowering plants; (g) from pine leaf; (h) from palisade layer of flowering plant.

2.1.3 Parenchyma, collenchyma and sclerenchyma

The primary cell wall, which is formed by all plant cells, tends to have a rather disorganised net of cellulose fibrils, with many fibrils orientated transversely to the main axis of the cell. The cell can stretch and expand as it grows. A secondary wall may be produced. Those cells with only a primary cell wall are called parenchyma or collenchyma cells. In parenchyma cells, the cell wall is uniformly thin except in areas with dense plasmodesmata, the primary pit fields, where it is even thinner. Parenchyma cells can be many shapes, rounded, irregular or elongated.

Elongated cells packed with chloroplasts together form the palisade layer in the upper part of the leaf tissue. They are the most specialised photosynthetic cells in the plant. Below them, knobbly shaped parenchyma cells form the spongy mesophyll tissue with many intercellular spaces.

Collenchyma cells have a thickened primary cell wall. They may vary from short and isodiametric to long and fibre-like. The cell wall may be thickened at the angles of the cell or along one or more faces of the cell. The cell wall in collenchyma is rich in pectic substances. It is more organised than a parenchyma cell wall, with many layers of fibrils. The fibrils are arranged in alternating layers in parallel, either transversely or longitudinally. The cell wall is plastic, allowing the cell to deform without splitting or snapping and yet give strength to the plant as it grows.

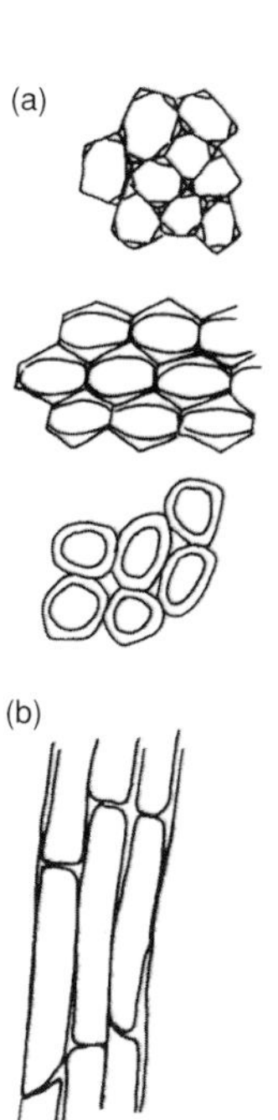

Figure 2.4. Collenchyma: (a) transverse; (b) longitudinal section.

Parenchyma cells and collenchyma cells may make up tissues called parenchyma or collenchyma, respectively. Parenchyma cells are found in many other kinds of tissues and may be specialised in various ways to carry out different functions such as photosynthesis or storage. Two important parenchymatous cells are the sieve tube elements and companion cells of the phloem that function in the translocation of nutrients. Collenchyma tends to be found only as a strengthening tissue in a peripheral location in the plant, but similar kinds of cells may be found elsewhere.

The other main cell type, sclerenchyma, has a secondary cell wall, which is produced inside the primary wall, after the cell has elongated or enlarged. It makes the cell elastic, allowing it to deform, but returning to its original shape after the stress is removed. The fibrils are regularly arranged: in parallel to each other, mainly longitudinal to the main axis, and in a weave with alternating layers at different angles. The more acute the mean angle of the fibrils to the main axis of the cell the greater the stiffness of the cell. However, if the fibrils are orientated more transversely, the cell is less likely to break, because the helically wound fibres can buckle inwards rather than snap when put under strain.

The secondary wall is often produced unevenly in bands, rings, helices or lamellae, conferring different characteristics to the cell; or more continuously, and then it is pitted. The secondary cell wall does not form over pits or primary pit fields but it sometimes arches over the pit as a dome with an aperture at its apex. Coming in pairs in adjacent cells these kinds of pits form a bordered pit pair. A tertiary non-cellulose cell wall may also be produced in some cells.

Many sclerenchyma cells lack a protoplast at maturity. They form a dead structural element in the plant and/or a conducting system in the xylem tissue.

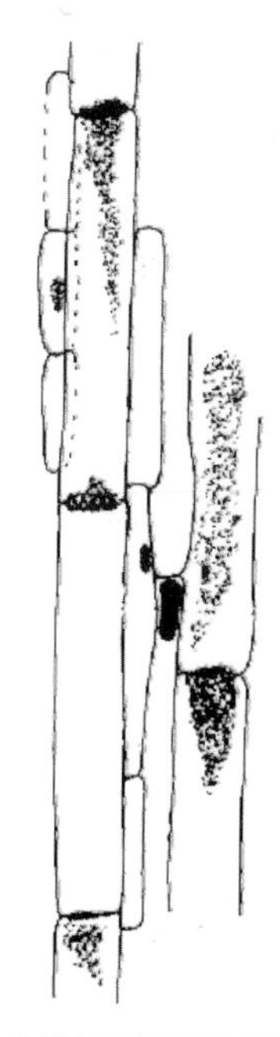

Figure 2.5. Phloem: sieve cells and companion cells.

Two main kinds of water conducting elements are characterised as tracheids and vessel elements, but, by looking at a range of plants, it is clear that there is a complete spectrum of types from fibres, fibre-tracheids, and tracheids to vessel elements (Figure 2.6). Non-flowering seed plants have only tracheids but flowering plants generally have a

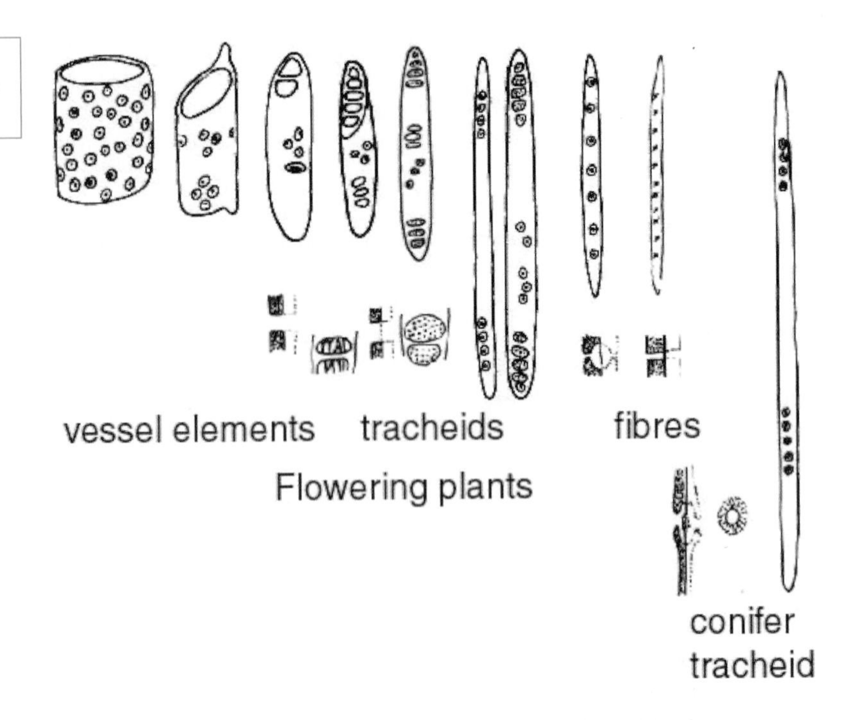

Figure 2.6. Xylem elements showing in detail the types of pits connecting adjacent elements.

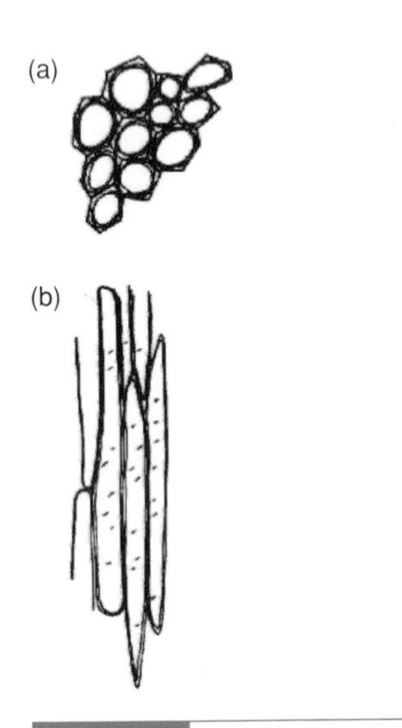

Figure 2.7. Sclerenchyma: (a) transverse; (b) longitudinal section.

mixture of fibres and vessel elements in their xylem. So-called 'non-vascular plants' have analogous cells called hydroids. However, even those sclerenchyma cells that are dead at maturity and have ceased to be conducting elements in the heartwood may have a prolonged life before dying by becoming the receptacles of plant metabolites.

Sclerenchyma cells may be found in a specialised sclerenchyma tissue but they are commonly found interspersed among other kinds of cells in other tissues, as idioblasts. They may make up a large part of the xylem. Sclerenchyma cells may be isodiametric ('stone cells' or sclereids) or elongated fibres. Fibres can be very long, up to 10 cm in hemp (*Cannabis*), and up to 55 cm in ramie (*Boehmeria*). They gain this length by an extended period of growth after cell division. The secondary wall is laid down after growth has ceased.

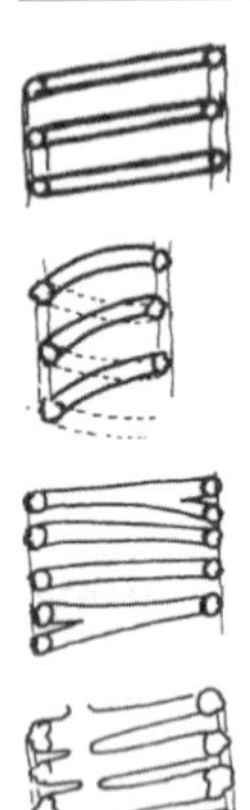

Figure 2.8. Secondary cell walls and pitting.

2.2 Plant growth and differentiation

Plants utilise soil nutrients and light in minute diffuse amounts, are rooted and relatively immobile organisms with rigid cells, and are compelled to grow by cell division in localised, more or less permanently embryonic regions called meristems. Those located distally in the root and shoot apices, from which permanent tissue (the primary plant body) is derived, are called apical meristems (Figure 2.11). There are other meristems at leaf bases and margins, known as intercalary, marginal and plate meristems that function in the maturation and shaping of mature leaves and reproductive organs, and in secondary thickening, increasing the girth of the plant. Secondary thickening occurs by cell divisions in the special intercalary meristem called the vascular cambium that lies between the xylem and the phloem, and in the phellogen (or cork cambium). The latter produces the protective bark. The tissue produced by these meristematic regions constitutes the secondary plant body, but not all plants have secondary thickening. The majority of herbaceous species do not. In woody plants, the tissues produced behind the apical meristem or primary tissues are like those produced in herbaceous plants, but are transformed later by secondary growth.

Although meristems are localised regions of cell division, all plant cells are theoretically totipotent. Meristematic potential can arise from almost any living cell, especially those that are relatively undifferentiated. The regeneration response is the response of already differentiated mature cells to revert to an embryonic state before becoming meristematic. This contrasts with the meristematic response to wounding, which is called the restoration response. In regeneration, the mature cells initially go through a process of dedifferentiation, become less vacuolate and more cytoplasmic before the first divisions occur. Both kinds of response may involve reiteration, the repeating of a phase or stage of growth. This is usually described as traumatic reiteration when the plant body is damaged and adaptive reiteration by the elaboration of the plant body in response to available resources. An extreme example is shown in the pollarded tree.

Examples of the regeneration response involving adaptive reiteration include the bulbils of *Oxalis*, *Allium*, bog orchids (*Hammarbya palludosa*), the turions of aquatic plants, the miniature plantlets of species of *Bulbophyllum* and *Kalanchoe* (Crassulaceae), and viviparous grasses that are produced in various parts of the plant and detach from the parent to multiply it (Figure 2.12).

Meristems produce two kinds of daughter cells; those that extend and differentiate into specialised cell types and tissues such as epidermis, cortex, vascular tissue, etc., and others which remain meristematic. Cells differ from one another as a result of unequal division, differences in positional information, or both. The primary meristems are the sites of almost all apical growth, and the shoot apex is carried upwards by differentiation and elongation of the primary meristematic daughter cells. The developmental potential of particular cells is determined within the primary meristem itself and not by cell signals from elsewhere, but the potential for growth of the whole meristem is influenced by environmental factors such as light or moisture, in addition to correlative inhibition by other organs.

Assimilation of water and minerals occurs primarily at the distal regions of root systems, while leaves, the principal organs of photosynthesis, are produced at the apices of shoot systems. The vascular system that connects these processes must therefore develop simultaneously, resulting in a complex system of communication throughout the plant. This communication system is largely by means of plant

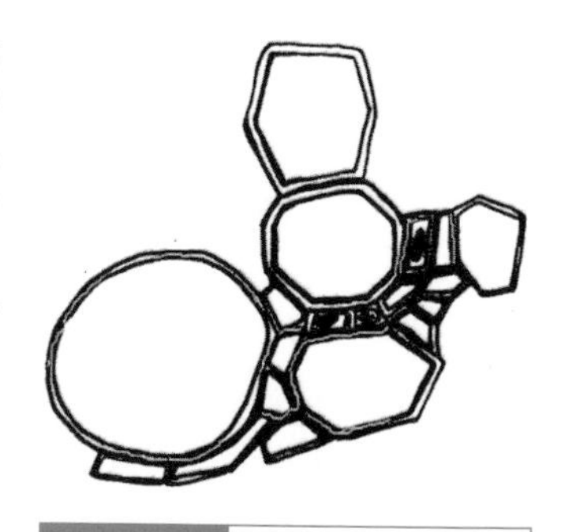

Figure 2.9. Xylem tissue in transverse section.

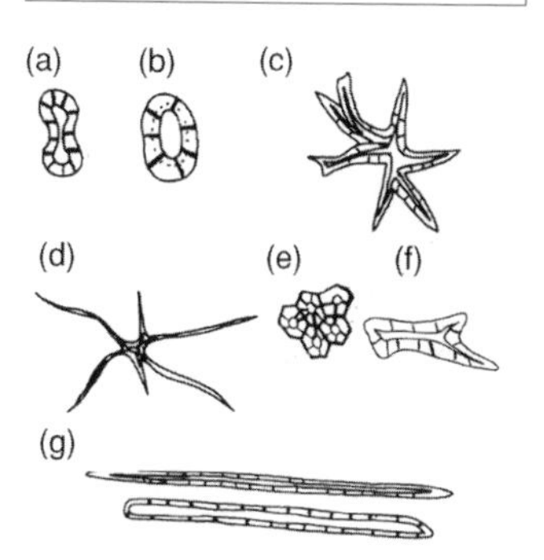

Figure 2.10. Idioblasts: variously different shaped cells generally with secondary walls.

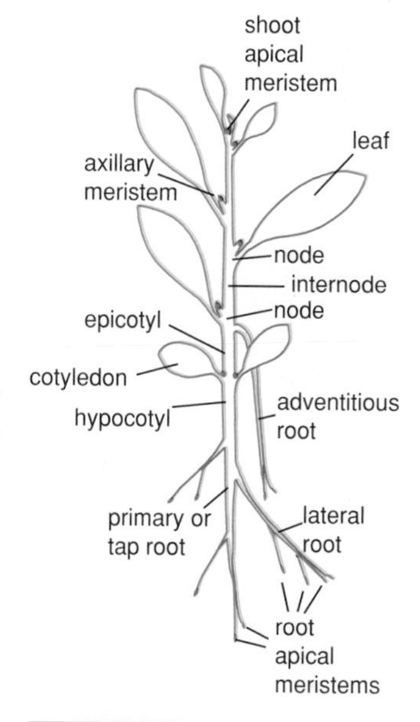

Figure 2.11. Plant showing main types of organ and location of plant apical meristems.

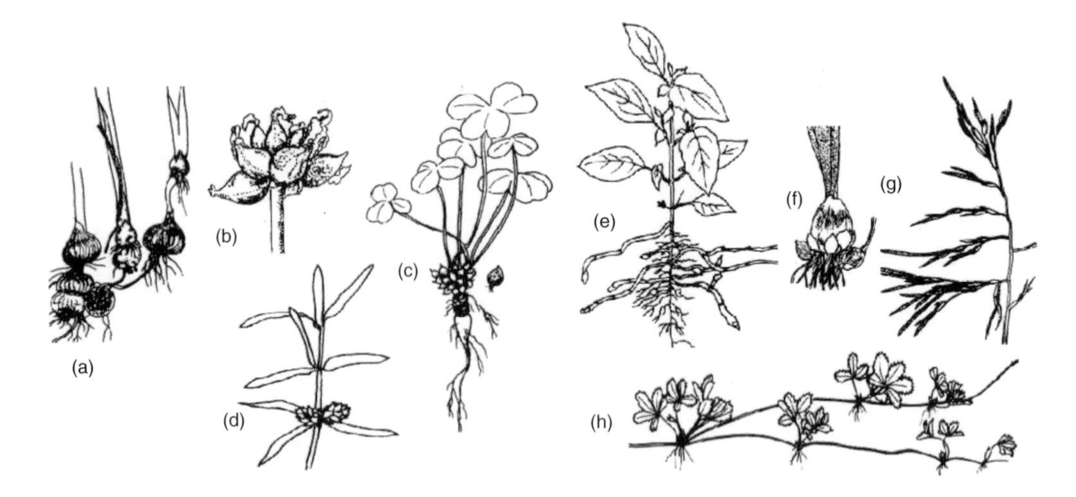

Figure 2.12. Vegetative regeneration: (a) corms (*Crocosmia*); (b) bulbils (*Allium*); (c) bulbils (*Oxalis*); (d) turions (*Hydrocharis*); (e) stolons (*Circaea*); (f) bulblets (*Hyacinthoides*); (g) viviparous plantlets (*Festuca*); (h) runners (*Potentilla*).

Figure 2.13. Pollarded trees: an example of the restoration response to wounding.

growth substances although electrical phenomena are also involved. These growth substances are sometimes referred to as 'hormones', but this is not strictly accurate, and again reflects the zoocentric bias inherited in botany. Plant hormones include auxins, cytokinins, abscissic acid and gibberellins. They are described in more detail in Section 2.3. below.

The earliest studies of plant development relied on direct observations of shoot and root apices, and of microtome sections. With the development of microscopy, especially scanning electron microscopy, these techniques were considerably enhanced, but the problem of integrated development and the actual genesis of form remained elusive.

An understanding of the role of meristems and the processes involved in tissue and organ formation is crucial for a deep appreciation of plant development and ultimately its place in plant evolution. It is important for the student to develop a dynamic view of the plant world, both in terms of the morphological potential that plants possess, and the way this potential expresses itself through ontogeny and phylogeny. The following account largely describes the processes that take place in the mature plant.

2.2.1 Shoot apical meristems (SAMs) and shoot systems

The apex should be regarded as part of a dynamic system of development and integration. Shoot apical meristems are found particularly in buds and produce shoot systems through the successive formation of leaf and bud primordia at nodes, separated by sections of stem called internodes.

The primordia are the small protuberances that enlarge around the margins of the apical dome of the meristem. As this meristem continues to grow, the new leaf primordia are usually protected by an outer layer of, as yet, unexpanded immature leaves that were produced earlier. As these primordia differentiate, cells in the axil of the leaf base remain meristematic and eventually develop into an axillary bud meristem that also has the potential to develop, either as

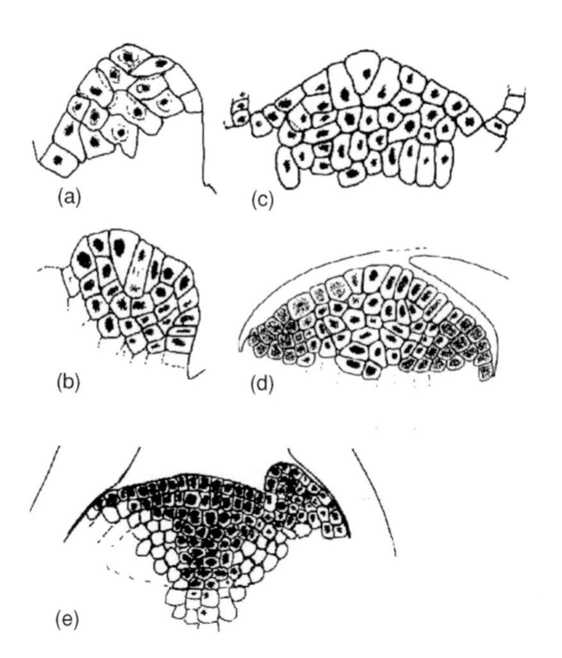

Figure 2.14. Apical meristems: (a) a moss (*Fontinalis*); (b) a whiskfern (*Psilotum*); (c) a fern (*Dryopteris*); (d) a conifer (*Pinus*); (e) a flowering plant (*Forsythia*).

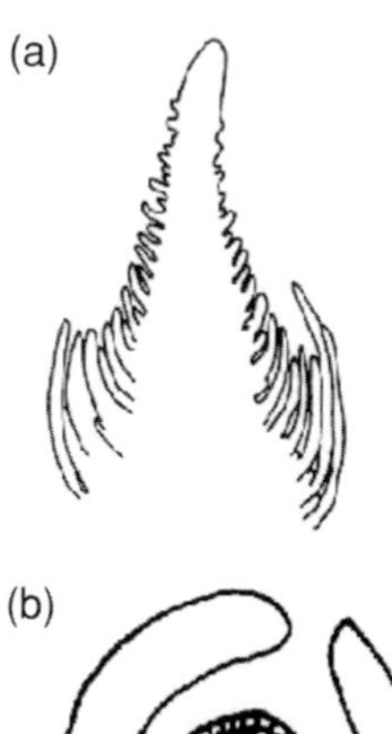

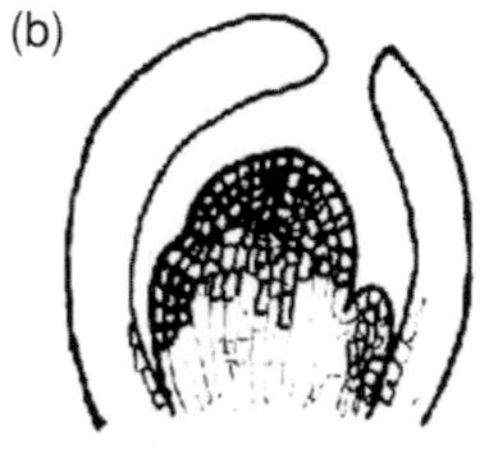

Figure 2.15. Apical meristems showing leaf primordia: (a) *Elodea*; (b) more typical flowering plant.

an axillary flower or as a lateral shoot. Each single leaf and its axillary bud, plus a portion of the shoot immediately below it, can be conceived of as a unit or metamer, and the timing between the production of each successive metamer in the meristem has been termed a plastochron.

Zonation of the meristematic region appears to be widespread in seed plants although the specific characteristics of the zones are varied. These zones function with a high degree of independence in the generation of shoot tissues although the capacity to regenerate entire apices is present in both central and peripheral zones. Based on cell structure and growth rate, two main zones with separate complementary functions may be delimited, the promeristem near the apex tip and the primary meristem below this.

Shoot apical meristems are radial or elliptical in shape and range greatly in size. The largest are found in pachycauls such as cycads and palms. Generally, the size of the promeristem and the apical meristem as a whole remains constant for a species, but there is usually some degree of relationship between meristem size and plant vigour. Shoot apical meristems have characteristic cells, which are almost isodiametric, small and thin-walled with dense cytoplasm and no conspicuous vacuole. The promeristem includes the apical initial cells and their most recent derivatives that have not yet undergone any of the changes associated with tissue differentiation. Apical initial cells are the source of meristematic cells, and retain their potential for continued division. At least one product of each division stays in the meristematic region.

In whiskferns (*Psilotum*), clubmosses (*Lycopodium*), horsetails (*Equisetum*), some ferns (e.g. *Dryopteris*), and in bryophytes, a single apical initial cell, shaped like an inverted tetrahedron or pyramid, may be recognised (Figure 2.14). It divides sequentially, producing daughter cells from each of its lower sides. There is however, considerable variation in the apical zonation of gymnosperms and ferns.

The patterns of cell divisions in apices of seed plants are less regular and do not define any unique locations for initial cells. All cell lineages can theoretically be traced back to single apical cells, the founder cells, but in most seed plants no specialised apical initial cells can be found. In experiments with chimeras, plants constructed by grafting two genetically distinct tissues together, there can be more than one initial cell. Relatively exact determination of the fate of developing cells is uncommon but there are some known examples. The flower primordium in *Arabidopsis* is probably initiated from four cells derived from the longitudinal division of two apical initials, but this may be a special case since *Arabidopsis* is a member of the Brassicaceae, a family also called the Cruciferae, a name that refers to its cross-shaped flowers with parts produced in twos and fours.

It is the relatively inactive promeristematic region that is the ultimate source of new tissues and overall patterning of organs even though the region itself forms few cells and pattern formation occurs before the primary meristem stage. In Buvat's theory the 'méristème d'attente', a central zone of hardly dividing cells is essentially passive, and it is the outer 'anneau initial' that generates new cells. However, during reproductive development, the cells of the méristème d'attente are activated and participate in the formation of inflorescences and flowers. The reproductive organs of lower plants do not conform to this pattern of development but may arise from any actively dividing meristematic tissue, often from marginal regions of the apex; for example, the leptosporangia of ferns may arise from superficial protodermal cells.

Promeristem cells are not clearly differentiated from one another. They are neither clearly polar nor do the microtubules at the tip of the promeristematic dome have consistent orientation. Nevertheless, differentiation occurs, in which at least four cellular types are recognisable by the time the primary meristem stage is reached. By then, cell divisions and cellular sizes increase, often rapidly. Cell divisions in the promeristems of shoots do have characteristic orientation, and, together with cell size, stainability, and internal architecture, have allowed the recognition of subzones, although these vary widely.

In ferns the outer layer is very obvious because the cells are elongated anticlinally, perpendicular to the surface. Commonly, in the flowering plants, an outer layer(s) of cells of the promeristem, called the tunica, and an inner mass of cells, called the corpus, are distinguished. The tunica comprises one to five (usually two) superficial layers of the apex above the level of the youngest leaf primordium, in which usually only anticlinal divisions occur. In the monocots and gymnosperms some periclinal divisions have been observed whereas, in the inner layers, cell divisions are in many different directions. In gymnosperms such as *Pinus* an apical zonation has been described where there is a group of slightly larger apical initials that have a low affinity for cell stains. They give rise to a group of central mother

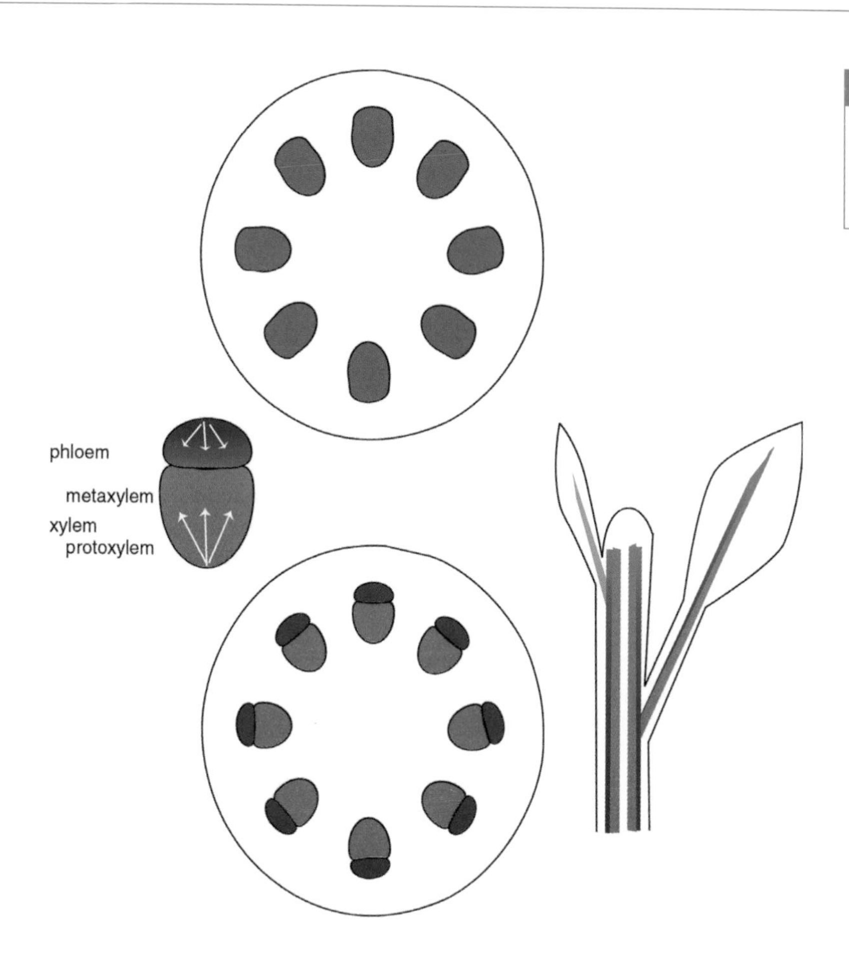

Figure 2.16. The development of the primary vascular tissue in the stem of a flowering plant: procambium (in purple) and (xylem in red, phloem in blue).

cells from which regular files of small vacuolated cells, called rib meristems, arise.

The tunica and corpus do show some biochemical differences between their component cells and the quantitative distribution of cellular organelles, both of which may be important for cell lineages but not for the organisation of the promeristem itself. Although the actual number of cell layers in the tunica is variable, usually two or more outermost layers of the apical dome are consistently at right angles to the surface and have the exclusive function of forming the protoderm and ultimately the epidermis. The other tissues formed by the promeristem are a ground meristem which matures primarily into various types of parenchyma such as cortex and pith and a procambium, normally produced in strands, from which the future primary vascular tissues are formed. In the stem each procambial strand becomes a vascular bundle.

Xylem cells are formed in the procambium in two phases (Figure 2.16). The protoxylem is produced while the organ is still elongating. Protoxylem cells are thickened by annular rings or helices of secondary cell walls, which allow the cells to stretch. There is a gradual transition to metaxylem cells that differentiate after the organ has elongated. Metaxylem cells have a wider diameter and have wide secondary wall banding in a scalariform pattern, or a more continuous secondary wall that is pitted. Two different kinds of phloem, protophloem and metaphloem are more difficult to distinguish.

The pattern of the vascular strands forming in the parenchyma reflects the growth of apical and lateral bud primordia, and is the result of the partial induction of the axial tissue. A very young leaf primordium, which is near the apex, influences the development of a wide section of stem tissue. Only below the apex does the inductive influence become canalised to the vascular system.

Each vascular tissue consists of a number of cell types. The close association of the components of the entire system results from an initial common specialisation perhaps mediated by auxin. Vascular strands induce and orientate differentiation by acting as sinks for any new flow of auxin. The development of the metaxylem radiates out from the pre-existing protoxylem, centrifugally in the stem (endarch development) or centripetally in the root (exarch development). New strands are orientated so that they form contacts with existing strands. The formation of such contacts is inhibited when the strands are connected to young leaves. A consequence of this is the pattern of gaps in the vascular system above and below leaves, called leaf gaps. Later, the different tissues each transport their own specific signal.

However, the complexity of the vascular system reflects a specialisation, a different response, of each of the different cell types in the xylem and phloem to different inductive signals. Gibberellins promote cambial activity and there have been indications that they specifically enhance phloem differentiation. Fibre differentiation is promoted by gibberellin in association with auxin. Thus, during the early stages of fibre differentiation, the cells are specialised transporters of gibberellins but this ceases when the cells mature. Parenchyma is often more elongated and more polarised the closer it is to the vascular tissue.

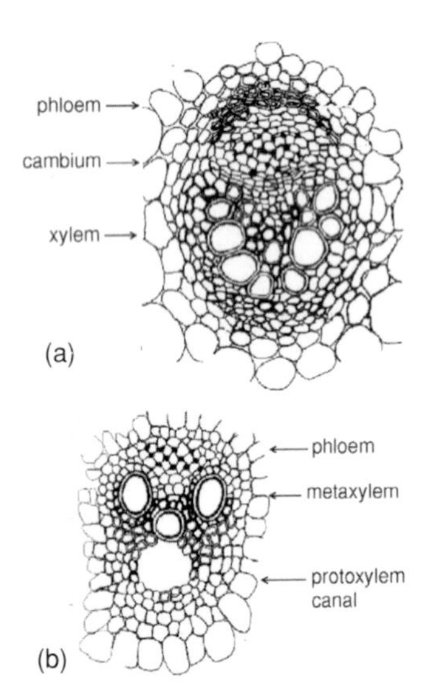

Figure 2.17. Vascular bundles: (a) open (*Ranunculus*); (b) closed (*Zea*).

Vascular strands consist of xylem and phloem. In the so-called 'open systems' an embryonic lateral meristem called the vascular cambium is present between xylem and phloem. Open systems have the potential for secondary growth by cell divisions of the vascular cambium, producing cells to the inside that differentiate into xylem, and to the outside that differentiate into phloem. Closed vascular systems, lacking a vascular cambium, are widespread but they are particularly associated with the monocots.

As well as vascular strand differentiation, other systems such as networks of lactiferous cells, laticifers or cells lining resin canals may arise in the cortex.

2.2.2 Differentiation in leaves

Leaves arise from leaf primordia produced in the shoot apex by the meristem. In higher plants leaves have a bud in the axil, in the angle between the lower part of the leaf and the stem. Commonly a stalk or petiole, a midrib and a blade (the lamina) can be recognised. The midrib is a continuation of the petiole into the blade. It is usually thickened and carries strengthening and vascular cells. Some leaves are sessile, lacking a petiole. The function of the petiole is to hold the leaf away from the branch to maximise reception of light. Petioles are often flexible so that the leaf shakes in the breeze. This may aid cooling and deter insects from landing. The petiole is the site of abscission.

All leaves are determinate structures that differentiate, mature, and lose their ability to grow any further. With few exceptions, leaves, even in evergreen species, have a limited life span, and of only one season in most deciduous species. In the very long-lived bristle-cone pine (*Pinus aristata*), leaves can live for up to 33 years. Unusually, these leaves, and those of some other evergreen species, have a capacity for secondary growth, though it is limited to the production of some secondary phloem. There are a few exceptional species with indeterminate leaves. For example, the leaves of *Welwitschia*, a peculiar gymnosperm from Namibia, grow continuously from a basal intercalary meristem, while the apex erodes away. The basal intercalary meristem of many monocot leaves is different, in that it does not provide indeterminate growth. Another example of indeterminate growth is found in the fronds of some climbing ferns such as *Lygodium*. These behave like shoots by growing continuously from an apical meristem, producing pinnae to each side as they grow forward. The difficulty of defining what a leaf is, the typology of leaves, is discussed in the next chapter.

The arrangement of leaves, the phyllotaxis of a plant may be regarded as the result of an interaction between two different space requirements; the requirements for arrangement of leaves to minimise shading, and the requirements of packing of the developing leaf primordia in the apical bud. How are the patterns actually initiated? Computer modelling shows that two controlling compounds, one diffusing down from the apical meristem and the other up from the differentiating vascular tissues could do it. Where both of these hypothetical compounds reach a critical concentration, a leaf primordium is initiated.

2.2.3 Branching

Much of the character of a plant body is determined by branching pattern. A stem may be thought of as a succession of internodes, while the combination of a leaf with its node and underlying internode represents a basic structural unit of plant architecture that allows the description of plant development regardless of the degree of complexity. These modules are sometimes referred to as 'phytons' or 'phytomers', and will be discussed more fully in Chapter 3.

During ontogeny these modules are repeated to reveal several levels of organisation that are really different stages of a common process of growth and transformation. The whole flowering plant may not only be regarded as a module, but also a branching axis or even a single branch may be a module. One way of defining modules is

Figure 2.18. Orthotropic axes grow in three dimensions.

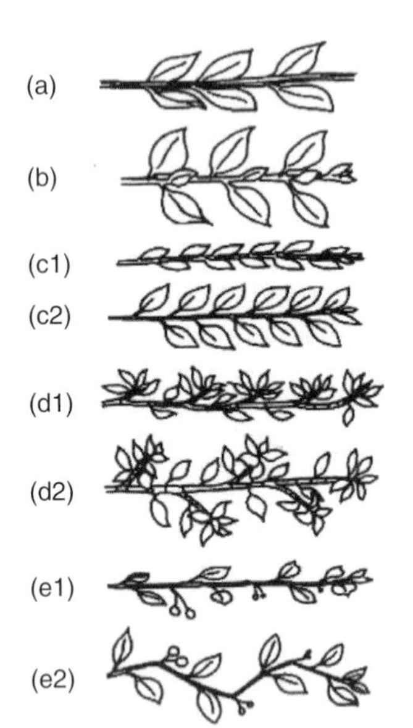

Figure 2.19. Plagiotropic axes often grow in two dimensions and can be of several types: (a) distichous arrangement with leaves in the same plane; (b) anisophyllous with leaves of different sizes; (c) simple plagiotropic growth; (d) plagiotropism by apposition; and (e) plagiotropism by substitution.

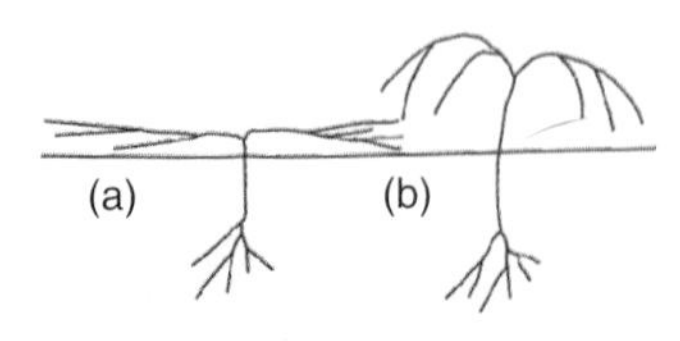

Figure 2.20. Orientation of branches: (a) procumbent; (b) decumbent.

that they each represent the growth of a plant ending in the production of a reproductive organ, either an inflorescence or another structure.

The concept of the 'module' was defined by Prévost as a 'leafy axis in which the entire sequence of differentiation is carried out from the initiation of the meristem that builds up the axis to the sexual differentiation of its apex'. In this terminology modules are then themselves made up of repeat units called metamers. Therefore, each metamer of the vegetative axis is the leaf, plus its place of attachment to the stem, and in the angle between the base of leaf and stem an axillary bud, together called a node, and the portion of stem between the node and the adjacent node called the internode.

One metamer is added to another, as the plant grows to form a stem or branch. It is possible to derive a great diversity of shoot systems from a number of simple rules for the timing of the production, development and rate of extension of the basic leaf–stem metamer, and the subsequent development of the axillary buds. Similar metamers may behave slightly differently becoming either a vertical or horizontal axes. A vertical axis, which is also called an orthotropic axis, has an essentially radial symmetry. The horizontal, or plagiotropic axis, is normally dorsiventrally flattened but can arise in other ways. A plant may have only one of these kinds of axes, or both. In a few species, a branch may start growing orthotropically and finish plagiotropically or vice versa. In addition branches may orientate themselves as they grow, either more actively, as in procumbent plants, or passively, as in decumbent plants.

Some plants have an unbranched axis. In branched species lateral branches may arise in different places, for example, either at the base or distally. Alternatively, they may be produced continuously along an axis, or rhythmically. A few plants have short-lived branches that behave like compound leaves. In the bryophytes branches arise below the leaves and not in the axils of the leaves.

The simplest kind of branching arises by the division of the apical meristem into two equal halves. This kind of branching, called dichotomous, was the earliest to evolve and can be observed in fossils such as *Cooksonia* of the Late Silurian, and in Early Devonian plants such as *Rhynia*. It can also be observed in living plants such as the whiskfern *Psilotum* and lycopods, an even in some relatively specialised plants such as palms, cycads and pandans, although frequently in these cases it has arisen not by true dichotomy of the apex but by an abortion of the apex and the growth of two lateral buds, and is hence called pseudo-dichotomous branching. In plants with sympodial growth each shoot apical meristem has a limited lifespan. The axillary bud nearest to the apex then takes the place of the apical bud, giving the shoot a characteristic zigzag appearance. The apical bud may be aborted or become determined as a reproductive organ.

Even in the Early Devonian some plants show a greater dominance of one of the dichotomising axes, as in *Asteroxylon*, illustrating a tendency towards the normal pattern most frequently seen today, i.e. axilliary branching in which a dominant main stem with its apical meristem gives rise to weaker lateral branches. The arrangement of lateral branches follows that of the leaves since branches arise from axillary buds at the base of each leaf. This evolutionary transition from dichotomous branching to axillary branching is the manifestation of the greater integration of growth of the plant, from one in which different apices of the plant behave entirely individually, to one in which there is coordinated growth of different parts of the plant.

Figure 2.21. Dichotomous branching in the Devonian plant *Rhynia*.

Branches may be produced immediately so that growth is continuous, a process called syllepsis. This occurs in herbaceous plants and many tropical trees. In most temperate trees, however, the bud meristems have a period of dormancy before producing a branch, so that growth is rhythmic, a process called prolepsis. The difference between a sylleptic and a proleptic branch can be recognised because, in the former, the basal internode is elongated and, in the latter, there may be many very short basal internodes sometimes marked by bud scales or scars. Rhythmic growth can also be recognised by regular changes in leaf size and by the formation of growth rings in the wood.

Figure 2.22. Sympodial growth and apical dominance in the Devonian plant *Asteroxylon*.

2.2.4 Root apical meristems (RAMs) and root systems

Despite some fundamental differences, the root apical meristem forms a root system similar in many ways to the shoot system. Although lacking the diversity of shoots, the root system is still a relatively diverse organ system, both morphologically and ecologically, and often exceeds the aerial shoot system in size. Some of the specialised root systems such as those in parasitic plants and mangrove plants, and the diverse mycorrhizal associations will be discussed more fully in Chapter 6.

Root meristems are clearly separated from the rest of the root by a well-defined elongation zone where the cells derived from the meristem elongate. However, cell divisions are not restricted to a highly localised apical zone but continue for a short distance down the root. The stele or root vascular tissues are often arranged in a distinctive star-shaped pattern in the centre of the root (Figure 2.24).

As with the shoot apex, there is a promeristem followed by a rapidly growing but determinate primary meristem but, unlike shoot meristems, root meristems do not produce any lateral appendages and there is no nodal/internodal organisation. Branching in the roots in almost all vascular plants originates endogenously in the pericycle (a layer of cells below the endodermis) of the central stele. The origin of lateral roots is determined in part by an interaction between the root apex and any existing laterals, so that in some highly homogeneous soils lateral roots are spaced regularly down the root. Laterals commonly arise opposite the xylem poles of the stele and may appear to be in ranks.

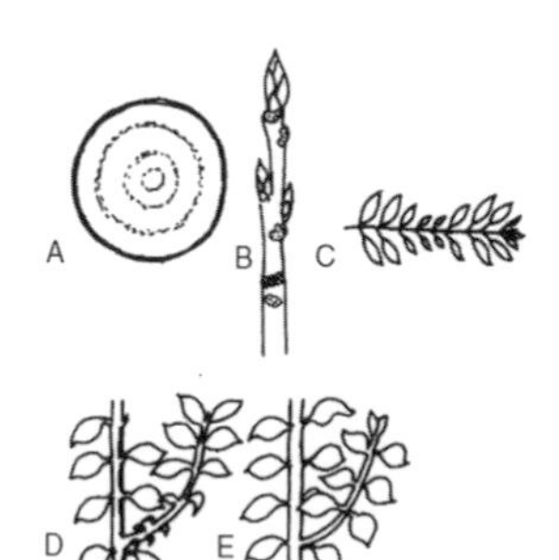

Figure 2.23. Rhythmic growth: (a) tree rings; (b) bud scar from overwintering bud scales; (c) in leaves along a twig; (d) prolepsis in side shoots; (e) sylleptic (continuous) growth in a twig.

The promeristems of roots are different from those of shoots. In roots, with the exception of families such as the Poaceae, and some species of the Brassicaceae, there are no longer discrete meristem

Figure 2.24. Development of stele in the root: phloem (blue); xylem (red). Xylem and phloem origins alternate, with xylem developing from the outside inwards (exarch).

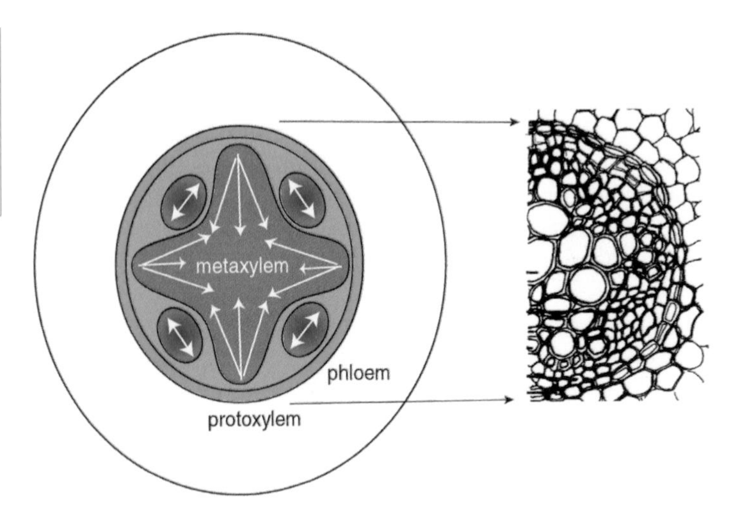

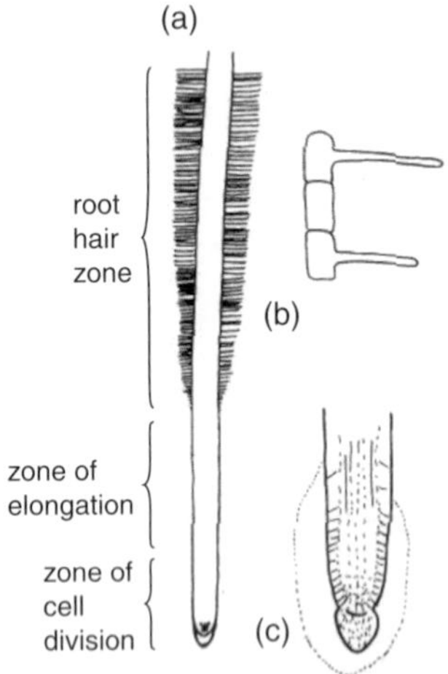

Figure 2.25. The primary root: (a) general view showing root cap, elongation zone and root hair zone; (b) detail of root hairs; (c) detail of root cap.

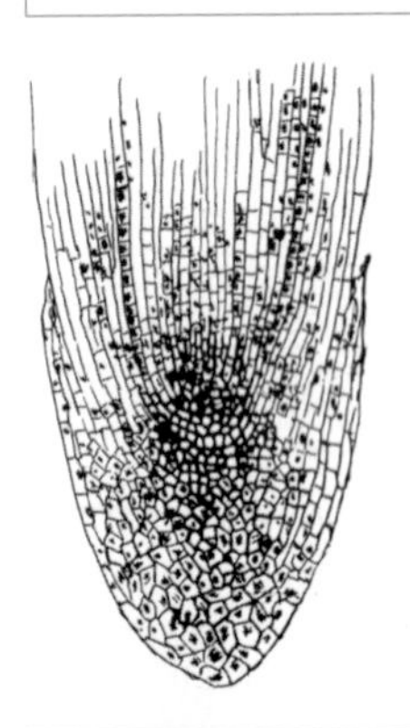

Figure 2.26. Anatomy of root meristem.

layers comparable to the tunica and corpus. Root meristems are distinguished by a quiescent centre at the tip that is characterised by little or no DNA synthesis or mitotic division, and which therefore may not grow at all. There is a root cap ahead of the root apex, with its own meristem, in the region known as the calyptrogen. The meristematic cells of the root apex therefore produce cells that mature both acropetally and basipetally. The root cap meristem may become isolated from the apical meristem by the quiescent centre. In this case the meristems are said to be closed. In other cases the root cap meristem may produce cells that become part of the epidermis or cortex of the root proper. It seems that one kind of root apical meristem may be converted into the other by small variations in their relative activity. The root cap protects the apical meristem proper as the root pushes through the soil. Cells from the root cap slough off as the root advances through the soil aided by lubricating mucilage that encourages a microflora in the rhizosphere. Much of the growth in roots occurs in the sub-apical region and the files of gradually maturing cells that make up this region can be traced to the sub-terminal region immediately below the root cap.

Initially roots are fine and delicate. The root apex is much smaller than that of the stem, generally about 0.2 mm in diameter, allowing it to penetrate most soil spaces. The small diameter of the root also increases the surface area available for uptake relative to the volume of the root but they do not absorb over their whole surface. Root hairs increase the surface area even further and absorption is concentrated in the root hair zone.

Some species may produce roots of two different sorts; long roots of unlimited growth and short determinate roots. Three categories of root systems have been identified based on the distribution of root diameter classes and the degree of development of root hairs: graminoid root systems are fine and delicate with profuse root hairs, resulting in a very high absorptive area to root volume ratio;

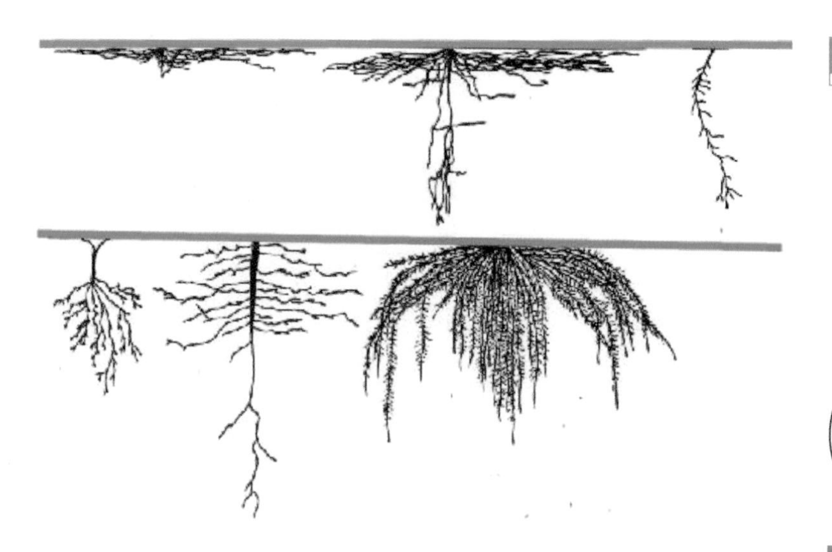

Figure 2.27. Rooting systems.

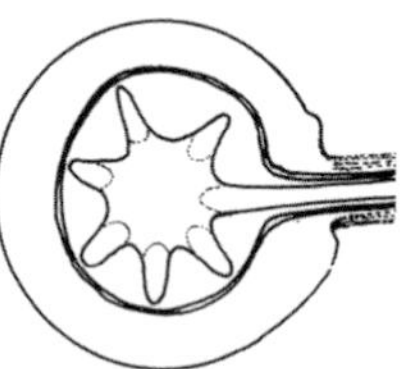

Figure 2.28. Endogenous origin of lateral roots.

magnolioid root systems have large diameter roots and fewer root hairs; and there are intermediate root systems. The functioning of these root systems cannot be viewed in isolation.

Roots generally exist in a mycorrhizal association with a fungus whose hyphae aid in absorption. The magnolioid root systems are compensated from having a low ratio of absorptive area to volume by having a rich mycorrhizal association. Graminoid roots tend to have a weak mycorrhizal association. The mycorrhizae are particularly important in low-nutrient soils; for example, the strongly developed mycorrhizae of the Ericaceae enable them to dominate acid heathland soils in many parts of the world (see Chapter 6).

Root development is so plastic that it is usually impossible to recognise the modular construction of roots (Figure 2.27). Only when some species are grown in a very homogeneous soil matrix or in liquid culture is the pattern of root growth regular enough to recognise a modular construction. Although nodes and internodes cannot be identified in the root, adventitious roots usually arise at stem nodes. Different orders of branching can be recognised but these are very different from the kind of modules described from shoots. Plasticity in root development has evolved in response to the extreme physical heterogeneity of soils with their mixture of particles of different penetrabilities and of many sizes: silt, sand, pebbles and solid bedrock. Lateral roots arise either thickly or thinly as required. There is also variation in root production over seasons as the availability of soil resources, moisture and nutrients changes very rapidly because of drought or rain.

For this reason, although the root system lacks the diversity of organs and branching patterns present in shoots, it is still a highly diverse and specialised organ system, both morphologically and ecologically. It often exceeds the aerial shoot system in size. A distinction can be made in seed plants between the primary root derived from the radicle and adventitious roots. In dicots the primary root may

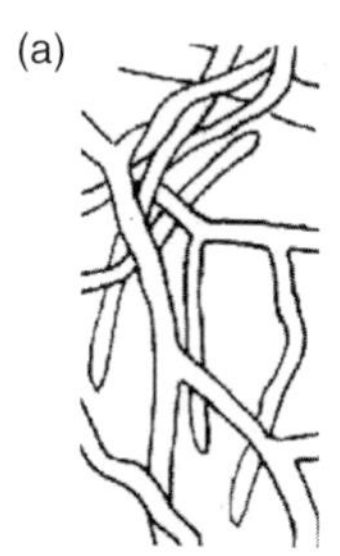

Figure 2.29. Diversity of root systems: (a) magnolioid; (b) intermediate; (c) fibrous.

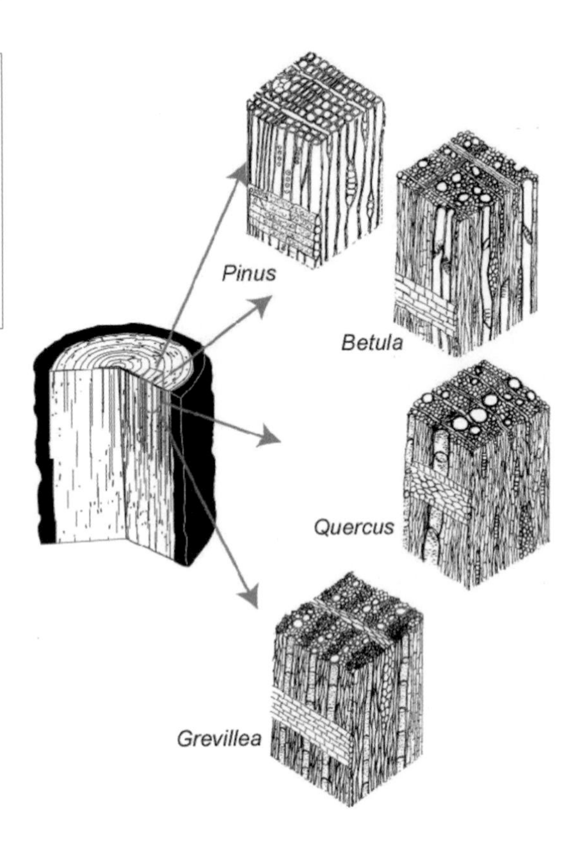

Figure 2.30. Three-dimensional diagrams of wood tissue illustrating contrasting anatomies: *Pinus* lacks vessel elements, present in the others, but has resin canals; *Betula* is diffuse porous and *Quercus* is ring porous; in *Grevillea* there are alternating bands of vessel elements and parenchyma.

become the taproot or become indistinguishable from the rest of the diffuse root system. In monocots the primary root is soon replaced by an adventitious root system.

2.2.5 Secondary thickening

After the development of the primary vascular tissue from the procambium, the secondary or wood tissue arises from the vascular cambium. Secondary growth allows the stem or root to increase in girth. In the shoot, a cambium arises in the procambium as a ring between the xylem and phloem of the vascular bundles (fascicular cambium). The arcs of fascicular cambium become connected into a continuous cylinder by an interfascicular cambium. In the root a cambium arises between the xylem and phloem, at first at the points of the xylem star, and then connects together to form a continuous ring. The cambium produces xylem cells towards the inside (centripetally) and phloem cells towards the outside (centrifugally) by tangential divisions of fusiform cambial initial cells. As more and more secondary tissue is produced the secondary stem and root grow more and more similar in appearance. The reason for this is that the primary root xylem star or the primary stem vascular bundles become insignificant in relation to the mass of secondary tissue. Each year more xylem is produced to form the woody tissue of trees and shrubs (Figure 2.30).

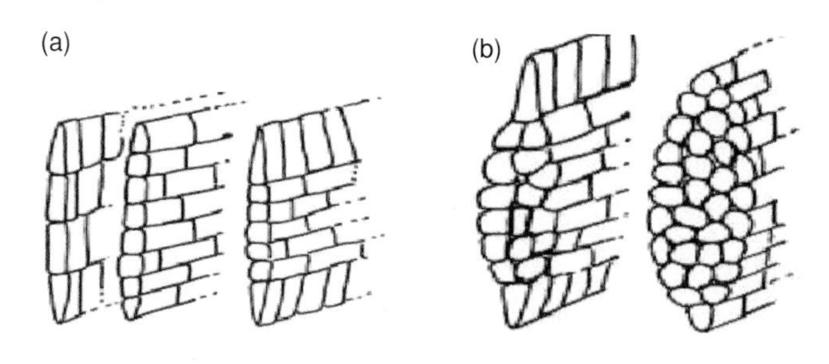

Figure 2.31. Rays: (a) uniseriate, the first two (from left) homogeneous, the third heterogeneous; (b) multiseriate, with heterogeneous (left) and homogeneous (right) types.

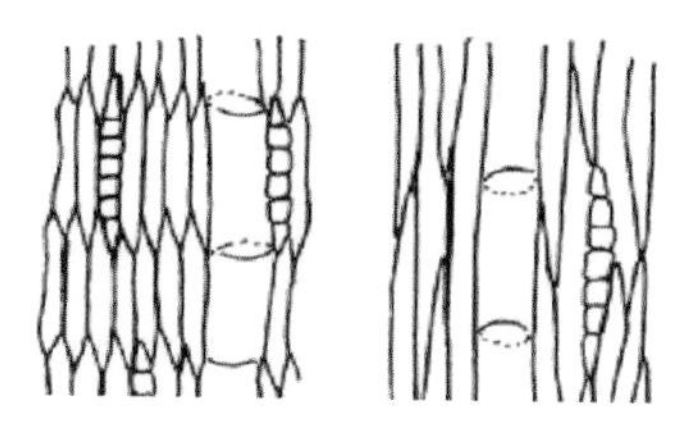

Figure 2.32. Storied (left) and non-storied (right) wood tissue; the latter arises from a non-storied cambium or by intrusive growth of differentiating elements into layers above and below.

The cambial intials are arranged either in horizontal tiers as a storied (stratified) cambium or, especially when the initials are long, with overlapping ends, as non-storied (non-stratified) cambium. The activity of the storied cambia may give rise to storied xylem tissues but there may also be some intrusive growth of xylem elements as they differentiate to produced a non-storied pattern.

These two kinds of cambia are the result of two different ways in which the cambium itself increases girth. In storied cambium the cambial initials increase in number owing to radial (anticlinal) division. In non-storied cambia the cambial initials divide by oblique cell walls and the daughter cells then elongate intrusively between pairs of adjacent initials above or below. This latter kind of cambial expansion is more primitive and seen more frequently in fossil and living pteridophytes and gymnosperms and primitive flowering plants. An even more primitive cambium was present in *Lepidodendron* that grew in the swamps of the Carboniferous Period. It had a unifacial cambium producing nothing to the outside. Fusiform (spindle-shaped) initials apparently only had a limited life span and new cambial initials were produced only occasionally as the diameter of the stem expanded. The cambial initials became wider to increase girth. Secondary growth was limited by the maximum size of cambial cell allowable, and by the absence of any secondary phloem. In addition some cambial initials divided across the long axis to produce ray initials that produce the horizontal files of parenchyma cells or rays across the girth of the xylem and phloem (Figure 2.31).

The activity of the vascular cambium is often seasonal, giving rise to marked growth rings in the xylem. The cambium ceases activity for a period and then later begins growth again. In the trees of seasonal environments this results in well-marked annual rings because the first-formed cells of the renewed 'spring' growth are larger than the last-formed cells of the previous late summer. In flowering plants, two kinds of rings are described, either as ring or diffuse porous, depending upon the relative size and position relative to the rings of vessel elements. Even tropical trees exhibit cycles of growth, although in the tropical forest they are not as well synchronised as they are in temperate latitudes.

Some living species have anomalous secondary growth. Members of the caryophyllid lineage of flowering plants (see Chapter 6), and many species of woody climbers, including the gymnosperm *Gnetum*,

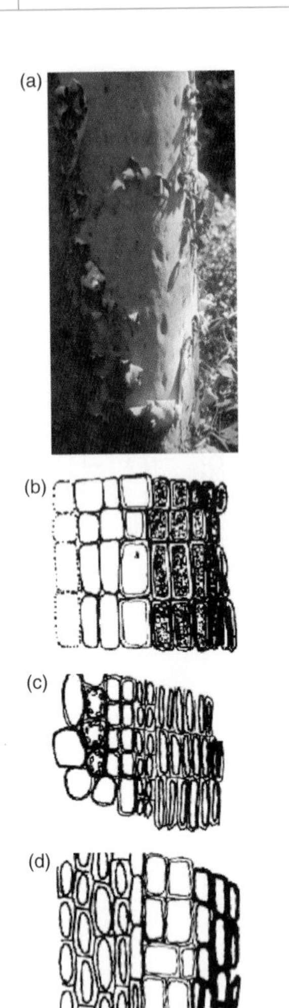

Figure 2.33. The bark exterior (left): (a) *Eucalyptus*; (b) *Quercus*; (c) *Pinus*; and (d) *Sambucus*.

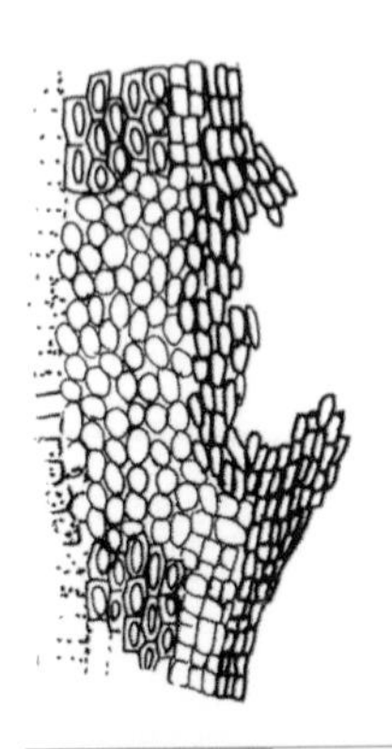

Figure 2.34. A lenticel from *Sambucus*.

produce successively new cambia in the cortex. Pachycaul trees of many different groups, such as palms, have what is called a primary thickening meristem, located beneath the leaf primordia, that carries out a kind of diffuse secondary growth.

2.2.6 The bark

The bark consists of the periderm and the other tissue layers, including the phloem exterior to the cambium. A periderm is produced, not only over the root and shoot systems of woody plants, but may also be produced elsewhere, as in the winter bud scales of some plants, where a protective layer is needed. The periderm arises from a lateral meristem, the phellogen (the cork cambium) that produces the phellem towards the outside. In some species the phellogen also produces a narrow band of tissue called the phelloderm towards the inside.

The phellogen arises initially in the primary tissues of the cortex. Subsequently, as the initial bark is pushed outwards by the production of phloem and is stretched, new phellogens arise in the phloem. A balance between the production of phloem and the rate of phellogens arising maintains the bark at a constant thickness. In parts of a plant that are wounded, a phellogen may arise in areas that were once deep inside the plant, including the secondary xylem, to protect and cover the wound. First, the outer wound layer becomes suberised and lignified forming a closing layer, and then a phellogen becomes active in the layer immediately below.

The phellem comprises tightly packed layers of cork cells and may be very thick, as in the 'cork' of *Quercus suber*, which is harvested for wine-bottle corks. The cork cells are suberised so that the periderm is impermeable to water, carbon dioxide and oxygen. The suberin occurs in layers alternating with wax on the inside of the primary cell wall. The cork cells are close-fitting 'bricks' forming a tight impermeable wall. Two types of phellem cells are commonly found; hollow air-filled thin-walled cells and thick-walled anticlinally flattened cells filled with dark resins or tannins. These two types can occur on the same plant. In *Betula* spp. they occur in alternating layers so that the bark peels in sheets like paper. In places, cork cells, called complementary cells, are produced loosely to form patches called lenticels which act as pores for gaseous and water exchange. In young roots, lenticels are produced in pairs on either side of the lateral rootlets. Elsewhere they may be arranged regularly in rows or irregularly.

The periderm of different species is very varied, and there are many different colours and textures. Some species shed their bark frequently, while others maintain the same phellogen for their whole lifetime of many years. The regular shedding of bark (exfoliation) is a means by which a plant can rid itself of lichens, bryophytes and other hitchhiker plants. It is a striking observation in tropical forests how some trees are covered with epiphytes of all sorts and others are clear. A heavy load of epiphytes potentially places a tree under severe mechanical stress. As a tree increases its girth the bark is liable to split providing many irregularities for the lodgement of the propagules of epiphytes. Regular shedding of bark maintains a smooth continuous surface. Good examples of this phenomenon can be found in the Myrtaceae (e.g. *Eucalyptus*, *Syzygium*, etc.) and the Ericaceae (e.g. *Arbutus*, *Rhododendron*, etc.).

2.3 The integration of developmental processes

Much of the organisation of meristems depends on internal processes of the meristems themselves, and is ultimately regulated by gene action. In the following sections we discuss the physical and chemical aspects of this organisation in a simplified way, but this is only part of the overall picture, and is a mechanistic model of plant development. The situation of course, in reality, is far more complicated and later in this chapter we will briefly explore some non-mechanistic ideas of plant development. Development in plants may vary owing to species age and many other correlative interactions. Biological form is specified in stages, each depending on stable or determined states reached during previous development. At any given time these may be influenced by intra-cellular factors that are, in turn, a function of the developmental history of the cells, the prevailing environmental conditions, and even the phylogenetic history of the organism. In Section 2.5 we deal with the epigenetics of tissue development and pattern formation, without which no picture of plant ontogeny is complete.

Although the apical region is, in a sense, autonomous in its ability to grow in the absence of developmental stimuli, it is dependent on interactions with the whole plant for normal development and functioning. This is particularly so for the initiation of reproductive growth and for the abortion of shoot tips, as seen for example in sympodial growth. Cells function together as an integrated system that suggests some kind of mechanisms for maintaining the organisation of the apex as a whole. For plant organs to function in an integrated way they require cellular as well as vascular connections. There is an invariable correlation between organ development and the differentiation and orientation of its vascular contacts. The induction of vascular differentiation requires integrative development through cellular communication and signalling. Developing shoot tissues actively induce vascular differentiation, though there is also a close correlation between the presence of an organ and the formation of non-vascular tissue of the axis.

Signals may be intrinsic or extrinsic. Intrinsic relations between organs depend on more than one simple control system, e.g. plant growth substances such as auxins and cytokinins exhibit major, although not unique, control of relations between organs. Responses to these intrinsic spatial signals are quantitative. Apices and other regions of rapid development are the major sources of signals but such signals originate in all tissues. The development of apices is limited by an exchange of signals with the rest of the plant and involves positive feedback. The more rapidly an apex develops, the

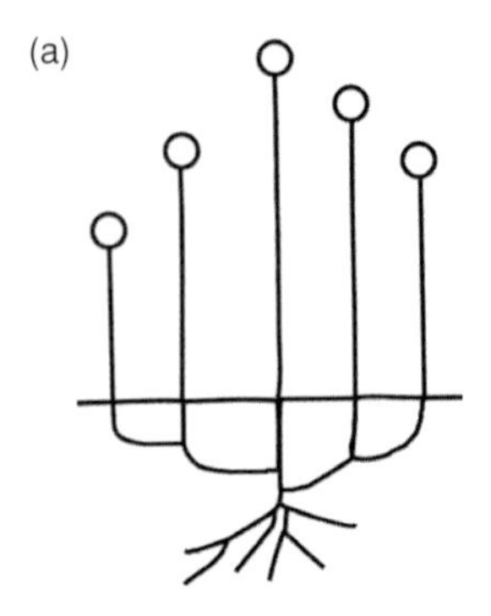

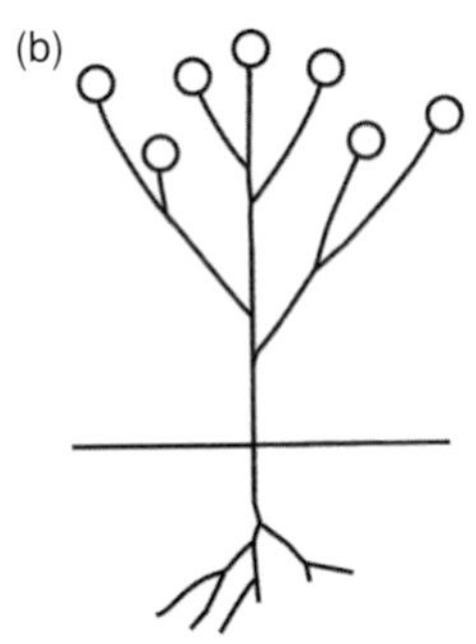

Figure 2.35. Branching pattern: (a) basal (basitony); (b) apical (acrotony) branching.

more rapidly it is likely to continue developing and the larger are its effects on the rest of the plant.

The initiation of new organs is due largely to the interaction of roots and shoots but these interact in different ways to produce basal or apical branching in different plants. These have complementary rather than identical effects on vascular differentiation. Developing shoots enhance root development, and vice versa. Although roots and shoots have continuous reciprocal effect on one another throughout development, they can compete with each other for metabolites. In contrast, signals received from the rest of the plant may merely limit the initiation of shoots and roots. The movement of substances towards the region in which they are used for growth serves as a signal of inhibition.

Expanding tissues, the ones with the largest influence, are the major sites of new organ synthesis. They may also be the sources of growth factor signals that are responsible both for inhibition, and for the development of the apex. The apex is the source of the original inductive signal, and the supply of limiting growth factors to an apex would be a function of its ability to enhance the development of complementary organs while simultaneously inhibiting competing ones. Thus, an apex polarises the axial tissues leading to a diversion of supplies towards the dominant apices. Such apical signals orientate differentiation in accordance with the axis of their movement through the plant thereby inducing complementary initiation where they accumulate.

Roots are the main sinks for auxins, resulting in the differentiation of vascular tissues along the axis from the developing tissues of the shoots to the roots. Roots, however, are also active sources of growth factor or other signals, and developing root tissues will orientate differentiation towards themselves. New organs dominate the plant and orientate tissue differentiation towards themselves. Apical parts of the shoot inhibit or dominate the growth of lateral buds, a phenomenon known as apical dominance. Such relations obviously influence the form of a plant, but there are variations in the sensitivities of subordinate tissues so that some plants branch mainly basally and others more apically. Laterals differentiate in the same way because of the presence of the same dominant organ, for example lateral shoots are frequently plagiotropic (e.g. *Araucaria*). The traits assumed by lateral branches (e.g. dorsiventrality) are often more extreme and less readily changed (more determined) than mere direction of growth.

There are a variety of structural gradients in plants owing to different developmental potentialities, such as differences in age along the axis. A gradient of the capacity to form roots could result in the first roots being initiated, preventing further root development and initiation. Inhibition and initiation of development are also exerted by other plant organs. Lateral buds that have started growing inhibit the development of the main shoot and rapidly growing leaves inhibit buds and induce root initiation.

Metabolites move preferentially towards dominant organs but, according to Sachs, a general role of metabolite distribution as a primary correlative signal is doubtful. Similarly, electrical signals are known in plants and could serve as signals, but the evidence for a wide role is not conclusive. Polyamines and oligosaccharides do influence development but are not known to have a role as signals. Other substances that are known to influence plant development include ethylene and abscissic acid (ABA), which may be signals of waterlogged roots and stressed roots respectively. Gibberellins have effects on apical differentiation and can replace root effects as well as having a specific effect on fibre differentiation.

2.3.1 Growth factors

Because of their importance in plant development, growth factors or 'plant hormones' are described here in more detail. Growth factors are intrinsic signalling compounds. There are two principal kinds involved in the integration of plant development: auxins and cytokinins, both of which influence plants at low concentrations. Interactions involving these growth factors control various relations between the parts of a plant.

Auxins are formed by and transported away from shoots but they are general signals and are not formed in any one type of cell in the shoot. They are based on the compound 3-indolacetic acid, which has a structural homology to the amino acid tryptophan. Auxins are the means by which growing shoots influence the development of the rest of the plant, primarily through the induction of vascular strands and the inhibition of lateral buds. It is their origin in developing shoot tissues that characterises them rather than any developmental process they elicit. The quantity of auxin could still indicate the size and rate of development of the shoot above, and the flow of auxin could specify direction of development.

Inhibition of lateral buds by auxin is not found in all plants under all possible conditions but exceptions are rare. Auxins have other effects, for example they prevent plagiotropic branches from turning upwards, and there is an inhibition of the development of abcission zones along the vascular connections between an auxin source and the roots. They also promote the growth in length and girth of young axial tissues that connect the source with the rest of the plant and, in addition, can induce the initiation of root apices.

Cytokinins are major signals of developing roots and generally have the opposite effect of auxins on shoot apices. Cytokinins include zeatin and several other related compounds that were first isolated from grain and coconut milk. Zeatin has a structure based on the purine adenine. Roots are the major source of cytokinins in plants but the role of cytokinins does not depend on their being formed exclusively in the roots. The enhancement of cell division has been regarded as the typical function of cytokinins but they have varied effects on plant tissues. Cytokinins inhibit the initiation of root apices in many plant tissues but promote adventitious initiation. They

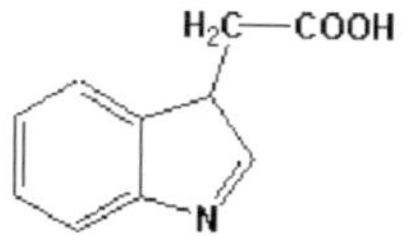

Auxin
(3-indolacetic acid)

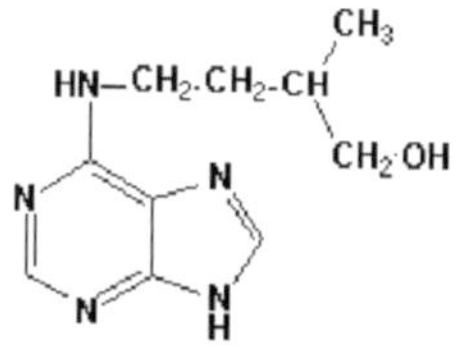

Cytokinin
(Zeatin)

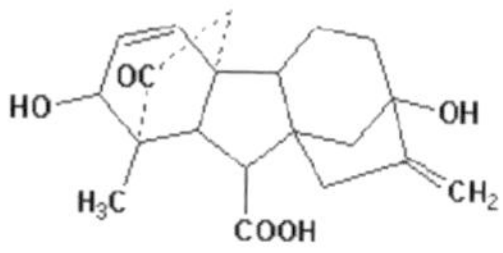

Gibberellin

$H_2C{=}CH_2$
Ethylene

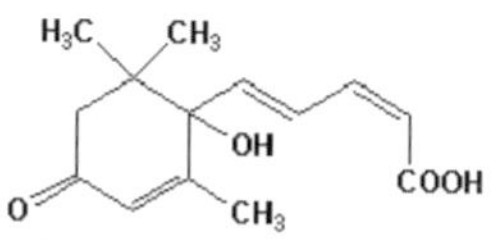

Abscissic acid
(ABA)

Figure 2.36. Plant growth compounds.

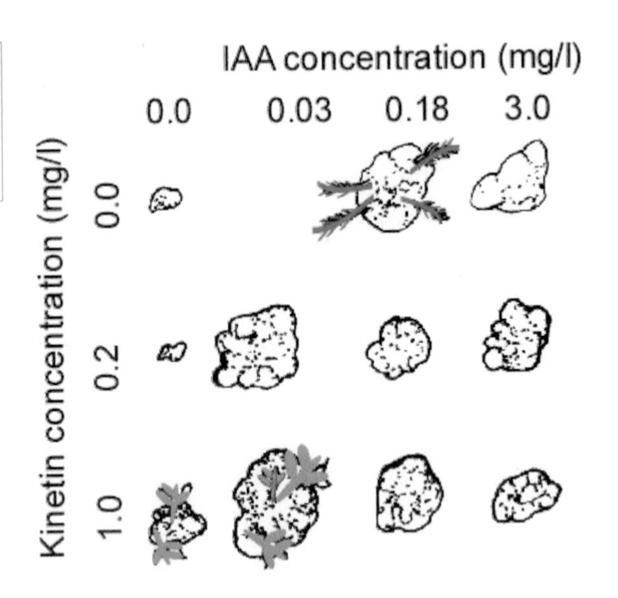

Figure 2.37. Interaction between plant growth compounds: cells grown as callus in tissue culture: shoots (green) and roots (red).

also counteract inductive effects of dominant shoots on lateral shoot apices and they are known to delay senescence.

Other important signalling compounds are gibberellins, ethylene and abscissic acid. Gibberellins are a family of about 70 polycyclic compounds and, like auxin, they promote cell elongation. They also stimulate the synthesis of enzymes that release stored nutrients in the seed. Ethylene accelerates senescence. Ethylene is a gas produced by ripening fruit and promotes the production of cellulases in abscission. It is important for inducing the death and autolysis of cortical root cells during lysigenous aerenchyma formation. It also seems to have a role in the determination of femaleness. Abscissic acid (ABA) acts as a stress messenger from roots to shoots at times of water stresss. It also induces winter dormancy by suppressing mRNA production. Brassinosteroids are a large and important class of steroidal signalling compounds that were first isolated in 1979. They are widespread in plants and algae and have an important role in cell division, enlargement and differentiation, as well as responses to stress and disease. There is a high degree of specificity between species and tissues at different stages of development.

There is a plethora of examples of causal relationships, synergy or antagonism, among the plant hormone signalling pathways. Auxin formation depends indirectly on the supply of cytokinins and vice versa. The ratio of auxin to cytokinin regulates morphogenesis in cells and tissues (Figure 2.37). Auxin–cytokinin polarity within a plant defines the architecture of that plant such as the number of lateral branches in the shoot and lateral roots below the ground. Gibberellins, auxin, and brassinosteroids have a stimulatory effect and ethylene, abscisic acid, and cytokinins have inhibitory effects on hypocotyl elongation. Cytokinins, brassinosteroids and auxin all participate in regulating the plant cell cycle. Auxin regulates ethylene biosynthesis. In contrast, cytokinins act antagonistically with brassinosteroids or ethylene to control leaf or fruit senescence, whereas abscissic acid acts antagonistically with ethylene and brassinosteroids. Brassinosteroids interact with gibberellins in the differentiation of xylem tracheids and vessel elements.

A growing apex influences the rest of the plant by being a source of its characteristic growth factor and by acting as a sink for the signals of the complementary apices. Its release is quantitatively related to the rate of apical development, while the development of an apex is limited by the supply of the growth factor it receives from the rest of the plant. This supply naturally depends on the current and previous vigour of the plant.

2.3.2 Inter-cell communication

Integrated development in plants and cellular patterning is crucially dependent on inter-cell communication. This is achieved principally by means of two main systems: a cytoplasmic network through the plasmodesmata in cell walls and the spaces between them, the symplasm; and an extra-cytoplasmic network the apoplasm. The early formation of plant cell walls is characterised by cytoplasmic strands spanning many neighbouring cells. These cytoplasmic strands are bound by a cell membrane. Plasmodesmata can also form *de novo* through existing cell walls. With immunological techniques, plasmodesmata have been shown to be extremely complex structures containing several unique proteins, including cytoskeletal elements, and that molecular trafficking in simple and branched plasmodesmata is different. Areas of dense cellular contact between cells with secondary walls are the pits and primary pit fields. In the phloem of vascular plants even the primary wall is absent and here there are relatively large areas called sieve plates where there are highly specialised plasmodesmata.

The ultimate development of cellular contact is to be found in the tracheids and vessels of the vascular system where openings between dead cells form a more or less continuous conducting system, part of the apoplasm that also includes the interstices of the cell wall itself. Regulation of apoplastic transport is effected at entry or exit at the cell membrane and its wide range of cross-membrane transport proteins that provide channels for various molecules including signals. For example aquaporins are highly regulated water channel proteins that play an important role in plant development and response to stress. Plants are not just some kind of elaborate plumbing network through which substances flow. Information flows through them too.

The condition of the cells at critical stages in their development determines the outcome of differentiation. Neighbouring cells can be radically different, even when they are the products of the same mother cell and are connected by plasmodesmata. However, the restriction of signals that integrate cells is normally required for differentiation. The cell wall is dynamic: plasmodesmata appear and rearrange, changing the cytoplasmic connectivity and the routes for signal molecules between cells. Erwee and Goodwin introduced the

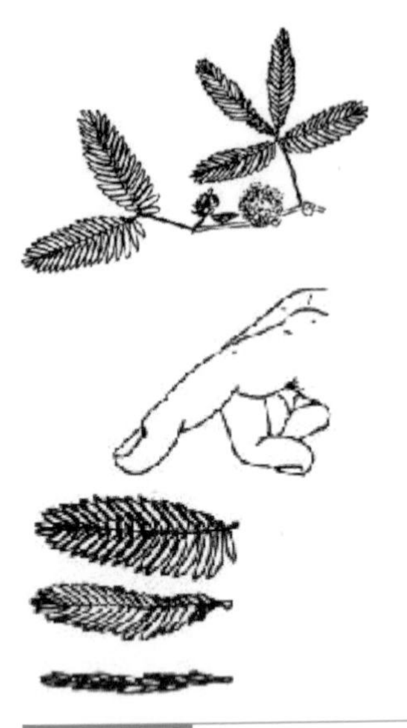

Figure 2.38. *Mimosa pudica* showing thigmotropic response.

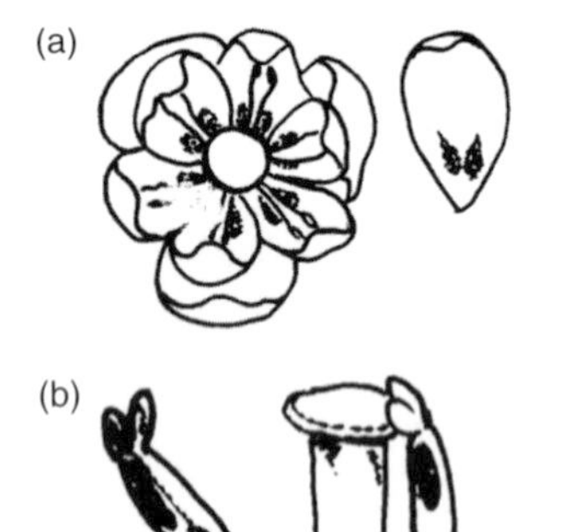

Figure 2.39. Sensitive stamens of *Berberis* react to the touch of a pollinator.

idea that plasmodesmatal conductance is lost or reduced between some tissues during differentiation, so that symplasm is segregated into domains, allowing cells within a domain to communicate freely with each other while communication between domains is restricted. For example, in guard cell differentiation, plasmodesmata connecting them to neighbouring epidermal cells become truncated and eventually non-functional. In contrast, in the phloem a small number of plasmodesmata are retained between the bundle sheath/phloem parenchyma cells and the sieve element-companion cells.

The relative performance and receptiveness to signals of different tissues varies depending on the genotype, environmental conditions and the stage of development. The trigger to different developmental processes depends on the competence and the previous developmental history of the responding tissue. Although cells are the units of division and of gene expression, it does not follow that they are necessarily the units of all aspects of development. At the whole-plant level there are important controls that may modify development but not necessarily individual stages at a cellular level. There are prepatterns or positional information, in the form of gradients or other patterned distributions of simple molecules.

2.3.3 Electrical signalling in plants

Plants do not have a nervous system but it is not quite true to say that only in science fiction do plants move. When touched, a wave of movement passes over the sensitive plant (*Mimosa pudica*) as it closes its leaves (Figure 2.38). *Cassia nictitans*, and species of the aquatic herb *Neptunia*, also have sensitive leaflets. Sensitive stamens are relatively widespread and are found in *Centaurea* and *Berberis* (Figure 2.39) for example, and have achieved a high level of development in the trigger-plants (Stylidiaceae). The carnivorous plants such as Venus' fly-trap (*Dionaea muscipula*) and the bladderworts (*Utricularia* spp.) respond very rapidly to the touch of their prey. In *Utricularia*, when the catch on the trapdoor is sprung, the lid of its bladder flips open and its prey is sucked inside to be digested. The suddenness of the response is brought about by an electrical signal, an action potential analogous to that of an animal nerve cell that passes along parenchymatous cells linked by plasmodesmata. The signal causes sudden changes in the permeability of the plasmalemma of the target cells. Water floods out and the hinge cells change shape in *Utricularia*, active excretion of water from the bladder resets the trap. In *Mimosa pudica*, special pulvinus cells at the base of the leaflets rapidly lose turgor, so making the leaf close.

There is also a broad range of slow plant movements, tropisms, which can be revealed by time-lapse photography. Some movements are more or less reversible. These are mediated by changes in turgor, although less explosively than the ones described above. There are heliotropic responses, orientating flowers or leaves to sunlight, or opening and closing of flowers, especially at dawn and dusk.

These reversible changes are not the main way in which plants 'behave'. From our zoocentric point of view it is easy to forget that

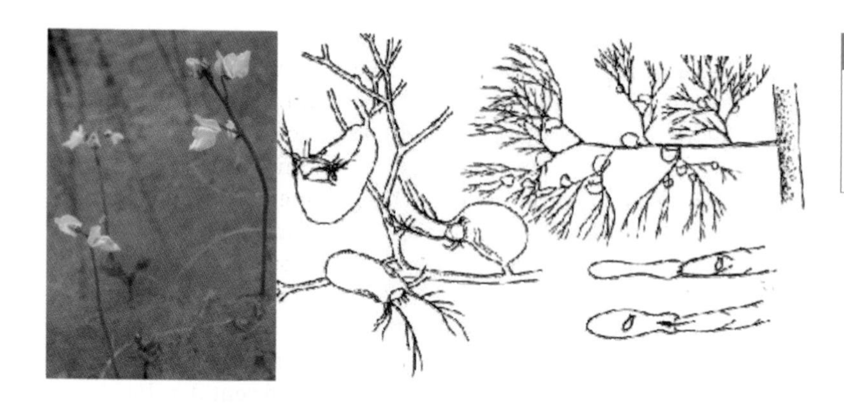

Figure 2.40. The bladderwort (*Utricularia*) has a thigmotropic resonse. The tiny bladders open on touch and suck in its prey, which is usually water-fleas (*Daphnia*).

plants do not really behave at all, but respond to stimuli by irreversibly changing the patterns of their growth. A single stimulus can bring about such a change. In an experimental situation, short periods of wind and fine rain, and even a gentle touch, brought about irreversible changes to the growth of *Arabidopsis* plants. Receptors in the cell wall or plasmalemma trigger a response by activating ion channels. Plants show some similarities to animals in this; each has long-distance signalling, by electrical signals and by growth factors. In the *Arabidopsis* experiments, calmodulin genes were switched on. Calmodulin is well known in animals as an important messenger that probably mediates responses by interacting with Ca^{2+} ions.

2.4 Cellular determination

So far our explanation of plant development has been merely descriptive of the sequential events that unfold during the ontogeny of apices, and of their integration through the activity of growth factor signals. From the time a cell is formed to the time it matures it can increase in volume by a factor of a thousand or more. In contrast to such quantitative change, a qualitative differentiation may prevail in those cells whose initial function depends on location, but the competence to differentiate is not limited to groups of special cells.

2.4.1 Polarity of organisation

Polarity of organisation is a characteristic of almost all multicellular plants. Differentiation follows existing polarity whenever possible and it occurs predominantly along the axes of vascular bundles. It is a general rule that roots are positively geotropic and regenerate on the basal parts of a plant while the positively phototropic shoots are initiated from dormant buds on the opposite, usually uppermost side. The localisation of new organs expresses a polarity of developmental processes that are present early on in the embryo; for example root and shoot meristems are established early in embryogenesis at opposite poles of the embryo.

When one thinks of polarisation one commonly thinks of electrical currents. Such currents are characteristic of polarised growth in plant systems but the extent of electrical phenomena in plant development is still largely unknown. Electrical polarisation has been detected in the algae, for example, in the zygotes of brown algae and may be partly caused by an orientated electrical current that is localised by, and further localises, specific channel proteins at one pole of the cell membrane. In the alga *Fucus*, asymmetry is established in the egg after it has been released from the mother plant and fertilised. In other organisms, the fertilised egg is primed with master proteins (conferring polarisation) from the mother plant.

The electrical contacts between cells may depend on the movement of ions through the plasmodesmata. However, the direction of movement of certain metabolites due to sources and sinks does not require innate polarity. Auxin transport is known to occur in new embryos and appears to play a role in the early stages of their organisation, but it is a late manifestation of an earlier, less specific polarity expressed by electrical currents. The transport of auxin therefore depends on the properties of the tissues through which it occurs and is independent of the location of sources and sinks.

Polarity of tissues may only be the sum of the polarities of their cells. From observations of organ regeneration, some early workers believed that this phenomenon is an expression of an innate tissue polarity based on each cell acting as a 'minute magnet' transporting morphogenetic signals in one preferred direction. Recent research has shown that the polarity of individual cells could neither be due to differential gene activity alone, nor have any orientation effect.

Polarisation involves many aspects of cell structure, the most likely initial events being differences in membrane characteristics such as the localisation of protein channels. Since the cells are polarised, something other than the genes or chromosomes must be orientated or localised within them. There are three possibilities:

- cytoplasmic gradients of substances or organelles;
- orientated aspects of the cytoskeleton;
- local differences in cell membrane.

Cytoplasmic gradients within cells have been recorded in many different systems and are expressed by localisation of the nucleus, concentration of cytoplasm, and by local densities of almost all organelles. Polarity is both induced and expressed by the orientated flow of auxin but there is no evidence that the polarity of auxin transport is caused by the unequal distribution of cytoplasm or of the organelles within the cytoplasm. Induction is a gradual process, the polarity of the tissues increasing as the flow of auxin continues, although what is actually observed in the plant is a wave of maturation, not one of induction.

The polar movement of auxin probably depends on a localisation of specific proteins at the basal side of the transporting cells that act as transportation channels. The largest components of the cytoskeleton are the microtubules, and their distribution is related to other aspects of polarity such as the plate of cell division and the orientation of the microfibrils in the cell walls. Microtubules are the skeletal structure of the orientated cytoplasm strands associated with tissue polarity but they cannot be the initial control of such orientation. There is evidence that microtubules can be re-orientated by auxin and influenced by ethylene, which may change the polarity and the polarising effects of auxin transport.

Young leaves and other sources of auxin induce both cambial activity and the differentiation of phloem and xylem. Cambial activity is regulated by functional needs and is induced by the excess auxin not transported by the mature phloem. Phloem activity reduces cambial activity by diverting inductive signals elsewhere. Radial polarity across stems is evident by anatomical structure, particularly that of the ray system that connects the phloem with the xylem across the cambium.

(a)

(b)

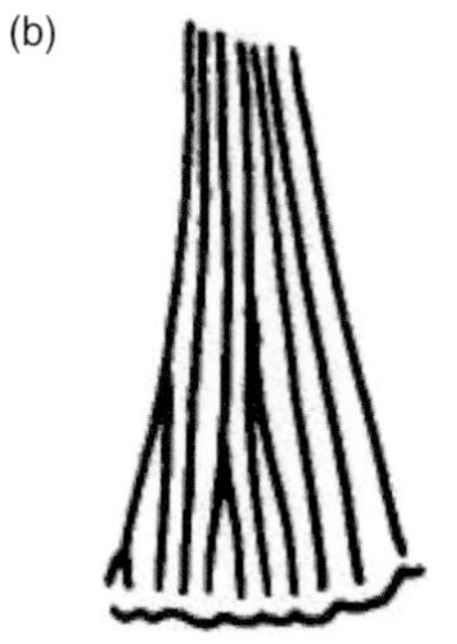

Figure 2.41. *Ginkgo* leaves: (a) detail; (b) showing dichotomous venation.

2.4.3 Meristemoids and local polarity in leaves

In most flowering plants, the development of the leaf is not synchronous and veins in leaf networks have no definable polarity. The correlation between complex vein networks and asynchronous leaf development favours the possibility that vascular differentiation is induced first along an axis and only later along a direction of polarity. Leaves, whose primordia develop along their margins, as in many ferns, do not have complex vein networks. Leaves may lack any vascularisation even if they have a thickened midrib or costa. The vascularisation, if present, may be simple and unbranched. For example, the needle-like leaves of conifers have one or two central unbranched vascular bundles. Leaves with a broad lamina have complex patterns of venation. A primitive pattern, which is seen in many groups including the ferns and gymnosperms, is a dichotomously branching, open venation. The veins do not interconnect and have blind endings. Occasionally adjacent veins may anastomose, as seen in *Ginkgo*. In some ferns there is a more regular pattern of cross-bridges between the veins giving a kind of reticulate pattern, although the underlying dichotomous venation is still obvious.

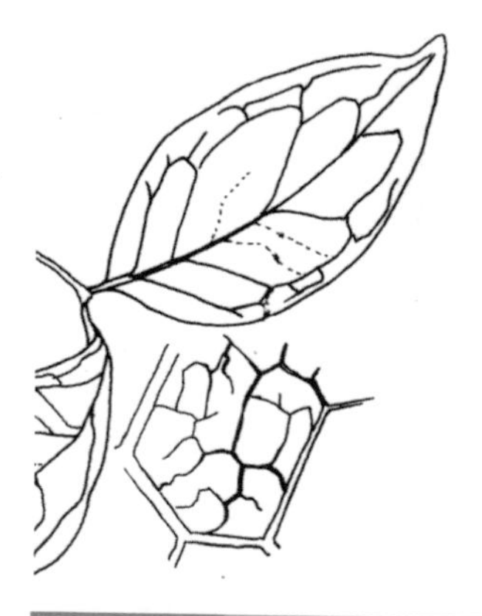

Figure 2.42. *Gingko* leaf showing detail of the reticulate venation.

The leaves of flowering plants, especially dicots but including some monocots, and the gymnosperm *Gnetum*, have a reticulate pattern of venation. First there is the vascular bundle or group of vascular bundles in the midrib that give rise to secondary veins, which in turn give rise to tertiary veins. The order or degree of lateral venation varies between different species. The finest veins surround areas called areoles. All cells of the mesophyll in an areole are very close to a vein. The system of venation is said to be 'closed' because, either the veins anastomose or, if they end blindly, the veins that give rise to them anastomose. In a reticulate closed venation, each part of the leaf can be served by many different routes. If any part of the venation is damaged, for example by herbivores, then no part of the leaf is isolated.

The finest veins make up by far the greatest part of total vein length, for example 95% in *Amaranthus*. Areoles may be very simple, consisting of only a few tracheids. The xylem is primary xylem because the finest veins are produced while the lamina is expanding, so the tracheids have to be able to stretch. Phloem may also be present but sometimes it is only present in the previous order of veins closer to the midrib. The vein may be surrounded by an obvious sheath of sclerenchyma or parenchyma cells. Transfer cells are found as a sheath surrounding the phloem. They are companion cells or specialised parenchyma cells.

With dichotomous venation the water diffusion pathway to the mesophyll and palisade cells is long because the veins are often quite far apart. In the cycads and conifers, however, this is compensated by the presence of a transfusion tissue surrounding the vascular bundles. Short, wide, nearly isodiametric tracheids are arranged radially around the bundle. In the cycads there are two transfusion tissues, one between the vascular bundle and the endodermis and another between the spongy mesophyll and the palisade parenchyma.

Most monocots and some dicots have parallel venation. The veins run longitudinally down the leaf connected by thin commissural bundles. Alternatively there is a midrib from which veins run out laterally in parallel. The arrangement is related to the way the leaf expands, either longitudinally or laterally. In those with longitudinal venation, there is a basal intercalary meristem. The potential for unlimited (i.e. indeterminate) growth by the activity of this meristem is limited because of the problem of conducting water into the lamina across this relatively immature region. In plants such as *Yucca*, the leaves are so slow growing that cells start to differentiate within parts of the meristematic region, thereby maintaining a vascular connection. Nevertheless even in these species the potential for leaf extension from the base is limited to very young leaves.

The differences between the parallel venation of monocots and the reticulate venation of dicot leaves has led some workers to suggest that monocot leaves are derived from the petiole, or petiole plus midrib, of the dicot leaf. Leaves of this sort are known from other groups, where they are called phyllodes.

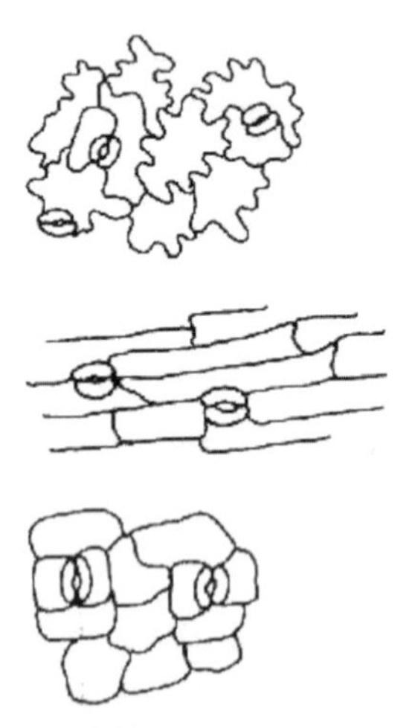

Figure 2.43. Epidermal cells: various stomatal subsidiary cell arrangements.

Internally the leaf is highly differentiated with an upper palisade layer and a lower spongy mesophyll (Figure 2.44). Support of leaf tissues, mainly conferred by the turgor of the cells, is enhanced by localised regions of collenchyma or sclerenchyma, especially as bundle sheaths around the vascular bundles. The bundle sheath may extend as a girder to reach either one or both epidermises. Nonvascular fibre bundles may also be present. Often the margins of the leaf are strengthened by the presence of sclerenchyma and a thick epidermis.

Usually the epidermis is different on the upper or adaxial surface from the lower or abaxial surface. The lower surface usually has stomata. Sometimes these are protected by being sunken in chambers,

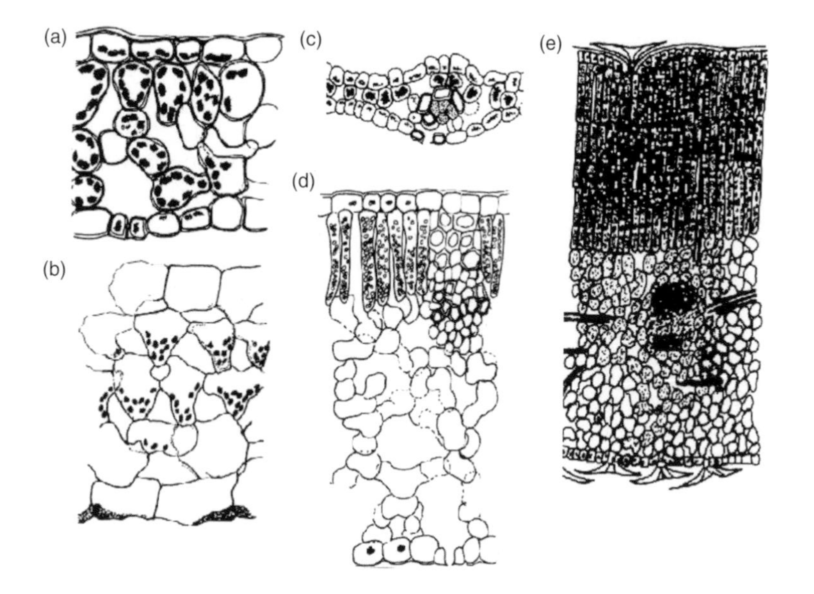

Figure 2.44. Leaves in transverse section showing variable dorsiventrality and differentiation. The upper palisade layer is most strongly differentiated in leaves of plants exposed to the highest light levels: (a) *Dryopteris*; (b) *Begonia*; (c) *Selaginella*; (d) *Helleborus*; (e) *Olea*.

furrows or by the presence of hairs. The adaxial surface is usually simpler, presenting a smooth mosaic of epidermal cells. Stomata are few or absent. Some plants, such as *Eucalyptus*, hold their leaves vertically, in which case the distinction externally between abaxial and adaxial epidermises, or internally between an upper palisade and lower spongy mesophyll layer, is not present. The dorsiventral differentiation (polarity) of leaf tissues is clearly highly canalised in some species of *Olea*, and relies little on cues (signals) from the environment. However, in others, such as some species of *Senna*, dorsiventrality is variable between closely related species, between populations of species, and even within individual plants.

The epidermis may contain specialised cells, many of which are modified from unicellular or multicellular trichomes. They are hair shaped, scale-like or globular. For example, there are the glands of insectivorous plants, the salt glands of halophytes, stinging hairs and nectaries. Some trichomes seem to function as antiherbivore devices making the leaf unpalatable. Alternatively, trichomes may protect the leaf from too much sun. Silica cells in the epidermis are also protective. Trichomes are often associated with stomata, helping to prevent too much water vapour loss from the open pore.

Local polarities are manifest in the development of hairs, stomata, glands and sclereids, and the early stages of their ontogeny are characterised by unequal cell division. The smaller of the two daughter cells forms the differentiated structure. The cells which exhibit this localised development are known as meristemoids. Meristemoids undergo special differentiation and depart from the axial organisation of the surrounding tissue. The differentiation and maturation of particular meristemoids and the timing of their development is related to their location within surrounding tissue.

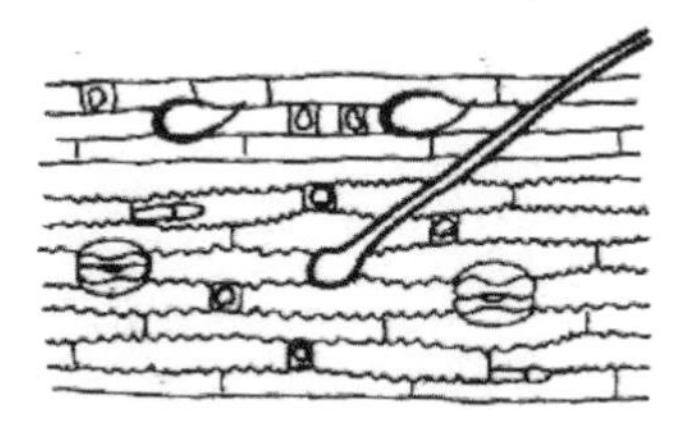

Figure 2.45. Epidermis of a grass showing long wavy cells and short cells containing silica bodies as well as microhairs, macrohairs and prickles.

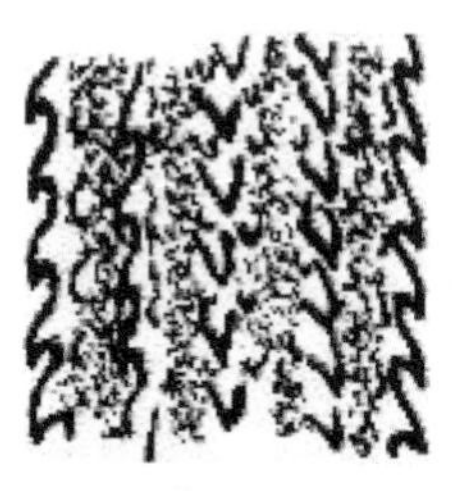

Figure 2.46. Surface of *Equisetum* showing strong development of siliceous cells.

Several different developmental processes result in different stomatal types in different taxonomic groups of plants. The patterns displayed by stomata appear to have ecological significance as evidenced by their distribution in many diverse plant groups such as succulents and gymnosperms. Groups of stomata tend to occur together and have the same orientation, although such complexes are polarised by the same long-distance signals that affect the rest of the plant body. The function of these stomatal complexes remains unclear, for example in the genus *Begonia*. Although the environment can control overall density of stomata, the control of patterning is located in the epidermis. The corollary of this reasoning is that such intra-cellular processes must also be implicated in the stomata-free regions. The distribution of stomata reflects a spacing pattern in which the inter-stomatal distances are actually greater than would occur in random spacing.

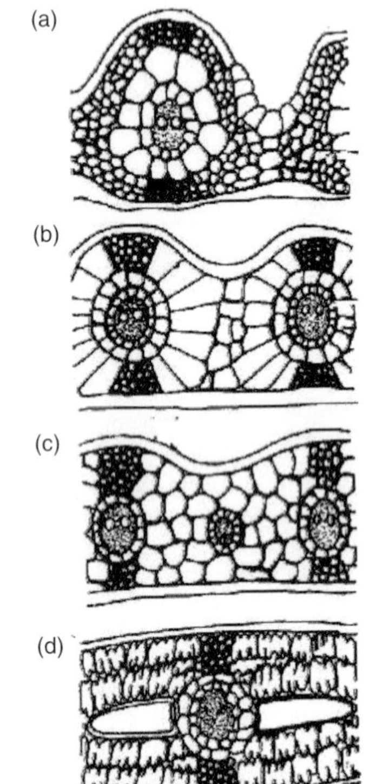

Figure 2.47. Leaf anatomies in grasses: (a) non-kranz; (b) kranz PS; (c) kranz MS; (d) bamboo-type.

2.4.4 The formation of cellular patterns

Induction by flow along established vascular axes, rather than by polar induction (which induces the initial vascular differentiation), is supported by a comparison of different stages of development in leaves. However, the pattern of specialised groups of cells must depend on information and events in the differentiating tissue itself rather than the mere exogenous source of an inductive molecule such as auxin. There is no doubt that information encoded in the plant genome is of prime importance in the sequence and expression of these unfolding events. However, the extent to which genes are implicated in developmental processes is still controversial.

The formation of cellular pattern requires two types of intercellular correlation: (a) an induction of the differentiation of similar cells along the future transporting channel; and (b) the inhibitory effects along the transverse axes, reducing similar differentiation. The pattern of the vascular system reflects the growth of primordia, the early stages of which have the greatest effects. Vascular strands consist of xylem, phloem and generally an embryonic cambium between them. Each of these tissues consists of a number of cell types. The close association of the components of the entire system results from an initial common specialisation and it is here that auxin could play a major role. Vascular strands induce and orientate differentiation by acting as sinks for any new flow of auxin. A fine example of the orientation of differentiated cells of different types around vascular bundles is shown in the contrasting leaf anatomies of grasses.

On a larger scale new strands are orientated so that they form contacts with existing strands but the formation of such contacts are inhibited when the strands are connected to young leaves, which could explain the formation of leaf gaps (Figure 2.48).

The inductive influence of a very young leaf primordium is on the development of a wide section of stem tissue. Only below the apex does the inductive influence become canalised to the vascular system. Later, the different tissues transport their own specific signal.

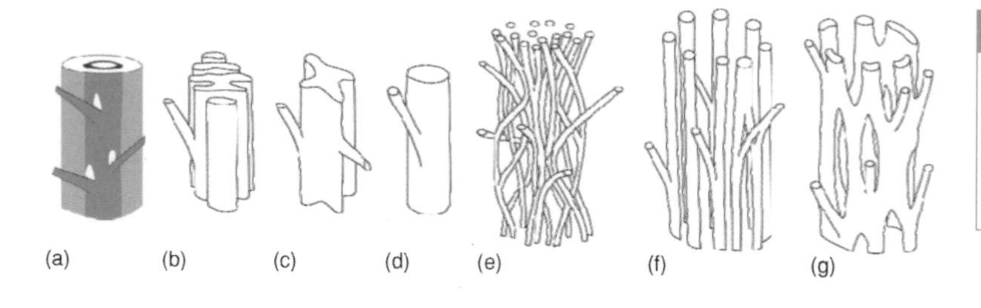

Figure 2.48. Leaf gaps and diversity of stele (dark green) types: (a) solenostele; (b) plectostele; (c) actinostele; (d) haplostele; (e) atactostele; (f) eustele; and (g) dictyostele.

Gibberellins promote cambial activity and there have been indications that they specifically enhance phloem differentiation. Fibre differentiation is promoted by gibberellins in association with auxins. Thus, during the early stages of fibre differentiation the cells are specialised transporters of gibberellins but this ceases when the cells mature. Parenchyma is often more elongated and more polarised the closer it is to the vascular tissue.

2.5 The epigenetics of plant development

So far in our discussion of plant development and differentiation we have described the development of the vegetative plant only. In this description we have not had to refer to any genetical control of the process but have explained what happens as a form of pattern development intrinsic to the plant of which genetical control is only part.

The progressive organization which becomes manifest during ontogenesis is the result of many interrelated serial, or sequential, processes; genetical, organismal and environmental stimuli being involved in the induction and regulation of the successive phases of development.

C. W. Wardlaw, *Organization and Evolution in Plants*, 1965

The reductionist, mechanistic approach to plant development can be taken further to embrace the role of genes in plant development, and there is no doubt that this approach is proving to be singularly successful in elucidating many of the mysteries of plant development that have puzzled plant morphologists through much of the twentieth century. In particular, the molecular analyses of gene expression in mutant phenotypes have provided some significant advances in recent years.

Plants have the paradoxical ability to make themselves by developmental processes from the zygote through various ontogenetic stages to maturity and senescence. How they achieve this is still the subject of considerable debate. From the late eighteenth century to the early twentieth century a lack of knowledge of gene action allowed vitalist theories to prevail as the only plausible explanation for sequential development. Nowadays, the dominant view is that development is guided sequentially by a chain of biochemical events that are ultimately initiated by information encoded in the genes. But how is this information transformed into development pattern formation, and what are the mechanisms for this transformation? Is gene action regulated through positional information, a view propagated by Lewis Wolpert, or by some means of self-organisation, or can it only be understood in a historical context, in the light of regulating and signalling systems of ancestral organisms, going back all the way to unicellular organisms?

We know very little about the interactions between genes, proteins and growth factors and the rates and direction of growth – the epigenetics of plant development. When we distinguish between the reciprocal communications between a plant and its developmental programme we are confronted with a paradox. It is analogous to the relationships between the components of an ecosystem that depend on a complex array of feedback systems. As Enrico Coen has so eloquently expressed it in his highly original book, *The Art of Genes*, 'Genes do not provide an instruction manual that is interpreted by a separate entity. They are part and parcel of the process of interpretation and elaboration'. Understanding the totality of a developing holistic system is virtually impossible. We have to take the system apart, reduce it to its components and analyse the bits separately and in a sequential fashion. Inevitably, our understanding will be incomplete, especially if we neglect to include historical processes.

All the cells of a plant, apart from the sex cells, carry the same complement of genes located on the chromosomes of the nucleus and those of the cytoplasmic genomes of the mitochondria and the chloroplasts. For simplicity, the genes can be thought of as belonging to two types, 'identity' genes and 'interpreting' genes, although, in reality, genes often perform both functions. The identity genes produce the frame of reference and encode for master proteins, while the interpreting genes are those genes that respond to the proteins produced by the identity genes.

Development in the embryo is initiated by master proteins contained in the cytoplasm of the mother cell and passed by maternal inheritance to the zygote. The interpreting genes in cells of the developing embryo respond to these prepatterns of proteins that were encoded for by the identity genes of the mother cell. The types of proteins contained in specific groups of cells determine the properties of those cells. For example, different proteins may determine the size, shape and chemical reactions of a particular cell. If each cell has the same complement of genes why do certain cells end up with different proteins and perform different functions?

The reason for this is that not all genes are 'switched on'. They may be 'off' or 'silent', or there may be gradients of activity and variation in binding-site affinity resulting in a graded response. Each of the genes carries coding regions of DNA, the base sequences of which encode for specific proteins, and a regulatory region that contains the binding site. The sequence of the regulatory region influences expression of the gene. There may be more than one binding site in each regulatory region. Each binding site is short (normally 6–10 bases long) and it is the presence or absence of proteins on the binding sites that determines whether a particular gene is silent or not. Proteins achieve their characteristics through their folding configurations and it is their unique shapes that allow them to recognise and bind to the binding sites. The regional pattern of gene activity is the expression pattern whether a specific gene is on or off.

Which state the interpreting gene is in depends on which master proteins are bound to its regulatory region, i.e. which facilitates transcription. The interpreting gene may be expressed in more than one site. The number of expression permutations depends on the number of identity genes and can theoretically number in billions. Any gene with a regulatory region can make its own interpretation of the master proteins. Because a master protein might bind to as many as one hundred different interpreting genes, an enormous constraint is imposed on the extent to which proteins can be modified. It is more likely that altered patterns of gene expression and interpretation, and consequently much of biological evolution, has involved changes in the binding sites within regulatory regions rather than in the master proteins themselves. Although only a subset of genes produces master proteins, essentially all genes are able to interpret them, including the genes for the master proteins themselves.

Mutations change the response to the prepattern of master proteins (i.e. their response to the common frame of reference). Mutations might start to couple the interpreting gene to a different master protein, modifying its expression, but they can also lead to a loss of binding sites. Genes may also act in combination to confer distinctions in identity. A different combination of binding sites leads to a different pattern of expression. The response depends on the biological and environmental relationships of the individual plant. A particular gene affects prepattern by initiating a particular response or outcome, one of which is selected out of a number of possible responses. Interpretation, therefore, is a highly selective process.

2.5.1 Differentiation in the axis

The control of development of tissue layers in the axis, in the shoot and root, is homologous, and even has some similarities to that of the development of floral whorls described below. For example, the root, endodermis and cortex, which derive from the same set of stem cells, are determined by the genes *SHORT-ROOT* (*SHR*) and *SCARECROW* (*SCR*), and mutations in these cause the endodermis and cortex to be replaced by a single tissue layer. These two genes are related in sequence, both encoding members of the GRAS family of transcription factors. *SHR* is required for the transcriptional activation of *SCR*. Importantly there is a flexibility of plant cell fate in meristems, as there is in tissue. Cell fate is position dependent. Anatomical studies of periclinal chimeras, which have genetically different ('mosaic') cell clones with easily scorable phenotypic traits such as albinism or ploidy levels, have revealed rare cell layer invasion events, demonstrating that stem and leaf cell fate is determined by position rather than by lineage, even at late stages of development.

2.5.2 Identity genes and floral development

The role of identity genes in development is seen most clearly perhaps in the process of sexual reproduction in plants. It has been especially elucidated in the development of the flower, but similar processes and related genes are involved in the switch to sexual reproduction, and

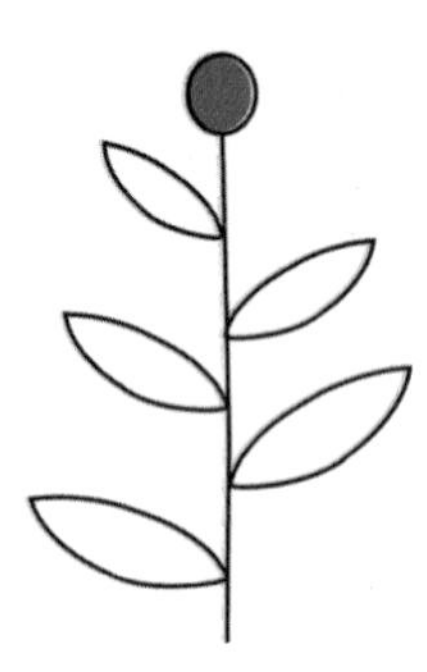

Determinate growth

Indeterminate growth

Figure 2.49. Two different kinds of growth separated by whether the apex of the main axis becomes differentiated as a reproductive axis.

the differentiation associated with it in non-flowering plants. At the onset of flowering, the shoot apical meristem switches from vegetative to inflorescence and/or floral development and usually from indeterminate to determinate growth. The inflorescence, in turn, switches development to floral meristems, which may be terminal or in the axis of a bract on the inflorescence.

A number of related genes have been identified in different plants whose activity is associated with the switch from the development of a vegetative shoot to that of a reproductive one. They have been given names such as 'Apetala1' (AP1), 'Leafy' (LFY) and 'Cauliflower' (CAL), which refer to the phenotype of the reproductive shoot when they are mutant. The same or closely related genes have been detected across a wide range of plants and, for example, are implicated in switching on the development of the reproductive cone in conifers and the flower in flowering plants. Once the switch has been turned on, the reproductive apex behaves as a promeristem, but one in which different types of lateral organ follow one another sequentially.

The determination of the reproductive shoot is not just the result of the autonomous activity of these genes, but arises from a complex web of extrinsic and intrinsic signals determining the age and timing of reproduction. The nature of these signals and the response to them is discussed below but it can give rise to many different kinds of behaviour:

- annual plants in which apices switch very rapidly to produce reproductive organs in the first year of growth
- biennials in which reproduction is delayed to the second year of growth
- perennials that delay reproduction but then switch some of their apices to reproductive growth either every year or rhythmically
- monocarpic perennials such as some species of agave and bamboo, that delay reproduction for many years but then convert all apices to reporoductive ones

Morphologically, the flower is a modified shoot, and reproductive apices share basic developmental processes with vegetative apices. A shoot apex may form various numbers of leaves and then change to form a flower or an inflorescence. Conditions external to the apices induce transition to reproductive development, but the transition can also be reversed. As transition proceeds, anatomical and cytochemical changes become more evident. General metabolic activation and enlargement of the promeristematic region are common, while zonation of the cells of the vegetative apex disappears.

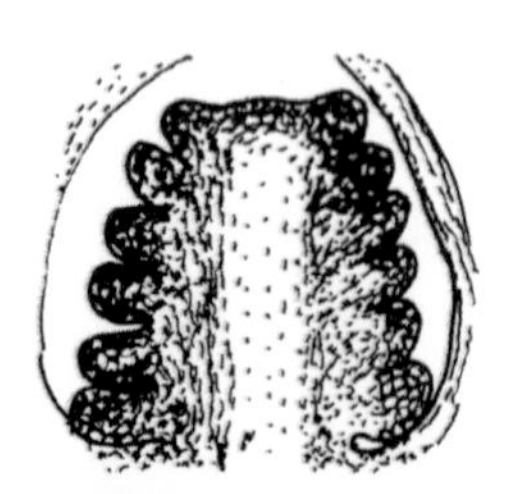

Figure 2.50. Floral meristem showing flattened apex and floral primordia.

An important trait of reproductive apices is that developmental stages, characterised by the nature of lateral organs, are not repeated once they have been achieved. The sequential changes in the reproductive apical meristem have been interpreted as meaning that each type of lateral organ induces the formation of the next lateral organ, i.e. sepals induce petals, and so on. The first steps in pattern formation involve the establishment of relative position of the organs. Floral organ primordia develop in concentric whorls, starting from the outer whorl. Sepal primordia arise first, in the outer whorl around the periphery of the meristem, then petals, then stamens and finally the gynoecia.

The idea that organisms develop epigenetically dates back to Caspar Friedrich Wolff in the late eighteenth century, although it was Johann Wolfgang Goethe who really elaborated the idea that floral organs and leaves are merely different manifestations of a common underlying theme. Through the study of developmental 'abnormalities' such as peloric flowers, Goethe was able to demonstrate that the organs of staminate whorls were capable of developing in a manner similar to the organs of the petaloid whorls. In many cultivated flowers, fuller double flowers have been produced by manipulating the expression of organ identity. This phenomenon was coined 'homeosis' by Bateson in 1894. There is a transition between organs in some species such as that between petals and stamens in *Nymphaea*.

(a) Normal monosymmetric flower

(b) Mutant peloric flower

Figure 2.51. *Antirrhinum* flowers.

The discovery of homeotic organ identity genes controlling the expression of repeating morphological units has accelerated the study of floral morphogenesis. Within plants these homeotic genes are called MADS-box genes. Changes in the regulation of these genes may have contributed to the establishment and structural evolution of flowers. The homeotic genes have arisen by duplication and have the same basic identity. The MADS-box is a highly conserved sequence motif found in a family of transcription factors. It codes for part of the master protein (the homeobox domain) that makes direct contact with the DNA at the binding site. The conserved domain was recognised after the first four members of the family, which were MCM1, AGAMOUS (AG), DEFICIENS (DEF) and SRF (serum response factor), and the name MADS (M) was constructed from the initials of these four 'founders'. Most MADS-box domain factors play important roles in developmental processes and have been called the 'molecular architects' of flower morphogenesis. A rough reconstruction of the history of MADS-box genes has already been made.

These monstrosities are so to speak, the experiments that nature made for the benefit of the observer: there we see what organs are, when they are not joined together; there we recognize what they really are, when an accidental case has prevented them from enlarging. A. P. de Candolle, *Organographie végétale* (Paris: A. Bellin, 1827).

Similarities in the base sequences of MADS-box genes suggests that these genes have evolved over millenia through the process of gene duplication, by the building of one set of events on another. This has allowed a reconstruction of land plant phylogeny based on such sequences. The last common ancestor of plants, animals and fungi that existed about one billion years ago had at least one MADS-box gene. In the lineage leading to green plants, the MADS-box (M) is followed by homologous genes which have been termed the IKC region.

The last common ancestor of ferns and seed plants that existed about 400 million years ago already had at least two different genes of the MIKC type. Large numbers of gene duplication events have produced a diversity of MADS-box genes in ferns (CRM type) that diverged very early on from seed plants. The common ancestor of gymnosperms, Gnetales and angiosperms, which existed about 300 million years ago, already possessed at least six different MADS-box genes that are present in living angiosperms (Figure 2.52). The ancestors of the angiosperms as we know them today possessed all the gene lineages from which the floral homeotic genes have evolved.

Figure 2.52. Evolution of MADS-box genes These are the AGAMOUS, AGL2-, AGL6-, DEF/GLO-, GGM13-, and TM3-like genes. The Gnetophytes form a different clade from the conifer lineage, which lacks the GM4, GM5 and GM66 genes. Among the angiosperms, the last common ancestor of monocots and eudicots also possessed AGL15-, AGL17-, SQUA-, as well as separate versions of the DEF- and GLO-like genes.

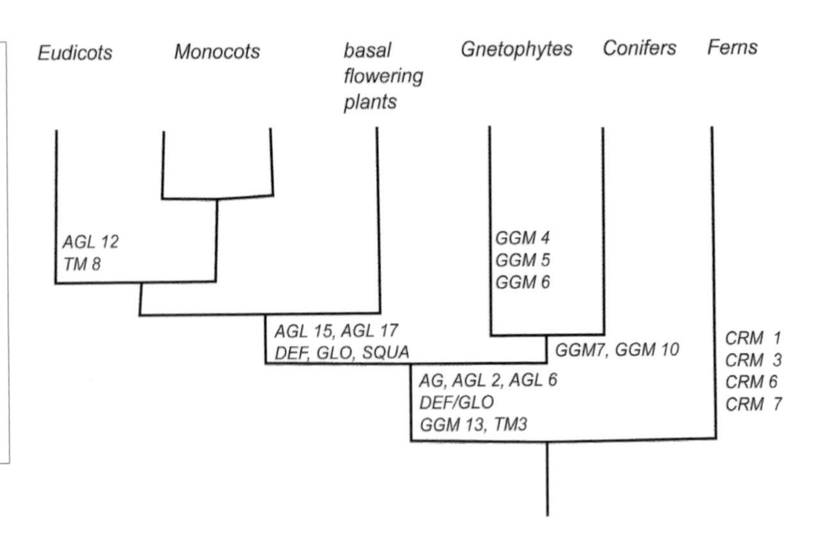

2.5.2 The ABC model of floral development

The diversity of angiosperm floral structure is based on more complex interactions between these floral homeotic genes. To illustrate this further we can use the two most studied angiosperms from which most of our current understanding of floral homeotic genes has been derived, *Arabidopsis* (Brassicaceae) and *Antirrhinum* (Scrophulariaceae).

The ABC model of floral development is a set of rules for predicting the morphology of flowers when individual genes are not expressed, usually because of mutation. The phylogeny reconstruction in Figure 2.52 shows that the MADS-box genes comprise several distinct lineages or clades of homeotic genes most of which, for each lineage, share related functions and similar expression patterns. The MADS-box genes providing homeotic floral functions, which have been termed A, B and C, each fall into separate lineages and determine the identity of various floral organs. Thus, SQUAMOSA provides for function A, DEFICIENS or GLOBOSA for B, and AGAMOUS for C.

The ABC genes involved collectively give several distinct possibilities of floral expression, usually by the absence of one or more whorls. In each whorl a different combination of one or more homeotic genes is expressed, and it is the particular combination of functions that determines organ identity in each whorl (Figure 2.54).

Flower development is a pattern formation process, the genes merely specifying the pattern of floral organisation of the apex, but not the details of how each floral organ will ultimately develop; for example, *Arabidopsis* and *Antirrhinum* both conform to the ABC model yet their flowers are radically different in appearance and in function. We know that certain other genes are involved in the final symmetry and shape of flowers, for example the '*cycloidea*' gene.

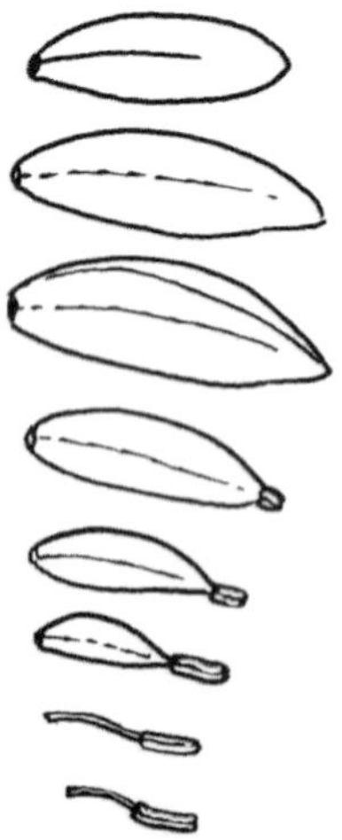

Figure 2.53. *Nymphaea* floral parts showing transition from bract to stamen.

2.5.3 The evolution of identity genes

The biochemical processes that occur in the development of living land plants have a history stretching back to the origins of life.

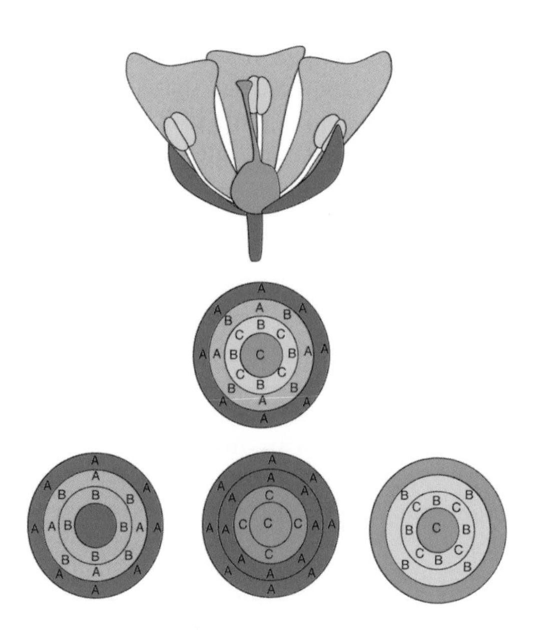

Figure 2.54. The ABC gene model of floral determination: class A genes function in whorls 1 and 2; class B in whorls 3 and 4; class C in whorls 2 and 3; In whorl 1, the formation of sepals is determined by class A genes alone; in whorl 2 petals are formed by the action of both class A and B genes; in whorl 3 stamens are formed by a combination of class B and C genes, while in whorl 4 the gynoecia are formed by class C genes alone.

According to Enrico Coen, the basic elements of gene expression evolved in unicellular organisms, primarily as a mechanism of temporal change as a means of adapting to changing environments. Coen believes that the spatial patterns of genes or master proteins did not play a major role in this process. Unicellular organisms achieve unique external shapes by the elaboration of other mechanisms such as those of the cell wall, and not by spatial patterns of genes.

With the evolution of multicellularity and the specialisation of groups of cells, the mechanisms that create diverse temporal patterns appear to have been conserved, but were co-opted to a new role, that of spatial patterns of genes. These genes are the 'identity genes'. Nevertheless, an understanding of the complexity of developmental processes may, in fact, also be aided by the examination of less simple forms. One of the revealing examples from among the 'primitive' unicellular algae is a rather unusual seaweed from the Mediterranean Sea called *Acetabularia*, the Mermaid's Cap. There are about 40 species in the genus, which belongs to the order Dasycladales. This group of green algae has been around for about 600 million years. The most striking fact about *A. acetabulum* is the size (several centimetres) and the complexity of cellular parts achieved by a single-celled organism. In its parasol, stalk and rhizoid-like holdfast it mimics multicellular algae (Figure 1.28).

Experiments by Goodwin and others have shown that it is the cell wall of *Acetabularia* that initiates the sequential changes along the developing stalk from juvenile whorls to the adult cap. The controlling mechanism is ultimately genetic, from proteins and mRNA molecules produced in the nucleus. There is active streaming of the cytoplasm, which distributes proteins and mRNA to all parts of the cell. However, differentiation and normal development of caps in *Acetabularia* does not occur until the concentration of Ca^{2+} ions in the surrounding seawater is 4 mM or greater. Interactions between the cytoskeleton and calcium produce spontaneous spatial patterns in the concentration of free calcium, involving positive feedback, and the mechanical state of the cytoplasm. Different rates of Ca^{2+} diffusion lead to alternating zones of higher and lower Ca^{2+} concentration, both the consequence and the cause of zones of a stiff and flexible cytoskeleton.

These findings from *Acetabularia* have profound implications for the morphogenetic processes involved in all organisms because, it should be recalled, all developing organisms have to develop their final form from initial stages involving few cells and which have simple symmetry. The processes that alter cell shape in *Acetabularia* demonstrate that the internal cell dynamics are involved in morphogenesis and act in concert with the genes and the environment rather than the cells being passively moulded by sequentially acting gene products. It supports the views of Wolpert that gene regulation may be via positional information. It also lends some credence to the ideas championed by early morphologists such as D'Arcy Thompson that many of the repetitious patterns or archetypes that we see in the biological world result from purely physical phenomena.

2.6 The theory of morphospace

According to Goodwin, the spherical zygote of *Acetabularia* has to break out of its simplicity into ordered complexity of form. This transition from higher symmetry (lower complexity) to lower symmetry (higher complexity) is a bifurcation event. In the life cycle of *Acetabularia*, the cytoplasm has a variety of possible dynamic states available to it (reaction norms). These range from a uniform steady state with everything in balance and no emergent patterns, to spontaneous bifurcation or symmetry-breaking from the uniform state to a stationary wave-pattern of calcium, and strain with a characteristic wavelength, determining alternative forms.

From the selectionist viewpoint these structures actually increase fitness; there is an adaptive landscape where only the peaks, certain types of form, have a selective advantage, and are evolutionarily fit. Of course the problem with this viewpoint is how to explain the origin of novelty because any route between adaptive peaks involves either a descent onto the plains before climbing to a different peak of fitness. In the journey, the individual has an intermediate form that is selected against. Alternatively, by some means, a lineage must leap across a valley, even a chasm, from one adaptive peak to another with a sudden transformation. Perhaps the idea of an adaptive landscape is just too simple. The landscape is defined not just by the physical universe, but by all other living organisms, all changing in time. It is more a rolling seascape than a fixed landscape, the crests and troughs ceaselessly forming and disappearing. Like corks bobbing on the waves, organisms dance restlessly across its surface. But then why do certain forms arise again and again?

Goodwin believes that the processes that alter the shape of *Acetabularia* during morphogenesis are those very processes that create generic forms in the organismic world. Despite the possibility for almost limitless forms, they explain the predominance of generic forms or archetypes over a limitless number of 'possible worlds'. This is because these structures are highly probable and arise from the dynamic processes of morphogenesis. Goodwin's theory is especially consistent with field models in which patterns are initially described by a set of harmonic functions (linear forms), but as pattern develops, the nonlinear features are expressed and distinctive wave-shapes emerge, as stable forms. A crucial aspect of this theory is that the parametric values in the equations describing the field are determined not just by environmental factors but also by genetic determinants that are the result of the previous evolutionary history of the organism. There may be some congruence between the Theory of Morphospace and the Morphogenetic Field Theory of Sheldrake, but more research is required to elucidate so many aspects of how organismal development proceeds and how the wondrous diversity of form in the living world is achieved.

Further reading for Chapter 2

Bell, A. D. *Plant Form: An Illustrated Guide to Flowering Plant Morphology* (Oxford: Oxford University Press, 1991).

Bonner, J. T. *On Development. The Biology of Form* (Cambridge, MA: Harvard University Press, 1974).

Coen, E. *The Art of Genes* (Oxford: Oxford University Press, 1999).

Coen, E. S. and Meyerowitz, E. M. The war of the whorls: genetic interactions controlling flower development. *Nature*, **353**, 31–37 (1991).

Goodwin, B. C. A structuralist research programme in developmental biology. In *Dynamic Structures in Biology*, ed. B. Goodwin, A. Sibatani and G. Webster (Edinburgh: Edinburgh University Press, 1989).

How the Leopard Changed its Spots (London: Weidenfeld and Nicolson, 1994).

Harrison, L. G. and Hillier, N. A. Quantitative control of *Acetabularia* morphogenesis by extracellular calcium: a test of kinetic theory. *Journal of Theoretical Biology*, **114**, 177–192 (1985).

Harrison, L. G., Graham, K. T. and Lakowski, B. C. Calcium localization during *Acetabularia* whorl formation: evidence supporting a two-stage hierarchical mechanism. *Development*, **104**, 255–262 (1988).

Howell, S. H. *Molecular Genetics of Plant Development* (Cambridge: Cambridge University Press, 1998).

Jablonka, E. and Lamb, M. J. *Epigenetic Inheritance and Evolution. The Lamarckian Dimension* (Oxford: Oxford University Press, 1995).

Lewontin, R. *The Triple Helix. Gene, Organism, and Environment* (Cambridge, MA: Harvard University Press, 2000).

Müller, G. B. and Newman, S. A. (eds.) *Origination of Organismal Form: The Forgotten Cause in Evolutionary Theory* (Cambridge, MA: MIT Press, 2003), pp. 3–10.

Sachs, T. *Pattern Formation in Plant Tissues* (Cambridge: Cambridge University Press, 1991).

Sheldrake, R. *A New Science of Life. The Hypothesis of Formative Causation.* (London: Blond and Briggs, 1981).

Steeves, T. A. and Sussex, I. M. *Patterns in Plant Development* (Cambridge: Cambridge University Press, 1989).

Thompson, D. W. *On Growth and Form: A New Edition* (Cambridge: Cambridge University Press, 1942). (Republished as *On Growth and Form: The Complete Revised Edition.* (New York: Dover Publ. Inc., 1992.))

Wareing, P. F. and Phillips, I. D. J. *Growth and Differentiation in Plants*, 3rd edn (New York: Pergamon Press, 1981).

Webster, G. and Goodwin, B. C. *Form and Transformation. Generative and Relational Principles in Biology* (Cambridge: Cambridge University Press, 1996).

Weigel, D. and Meyerowitz, E. M. The ABCs of floral homeotic genes. *Cell*, **78**, 203–209 (1994).

Wolpert, L., Beddington, R., Brockes, J., Jessell, T., Lawrence, P. and Meyerowitz, E. *Principles of Development. Current Biology* (London and Oxford: Oxford University Press, 1998).

Chapter 3

Endless forms?

. . . there is a grandeur in this view of life with its several powers, having been originally breathed into a few forms or into one; and that, whilst this planet has gone cycling on according to the fixed law of gravity, from so simple a beginning endless forms most beautiful and most wonderful have been and are being evolved.

Charles Darwin, *The Origin of Species*, 1859

3.1 The living response

3.1.1 The plant in its world: macrocosm and microcosm

The environment of plants exists on vastly different scales. Plants are the primary producers and are basal to almost all food chains except marine ones where they are replaced by the algae, and a few others such as some deep-sea hydrothermal vents where chemoautotrophic organisms live. They play a vital role in the flow of energy through all ecological cycles. The whole system of life rests solidly on their industry without which the evolution of many other organisms could not have occurred. Vegetation forms the macrocosm of life on Earth, yet the relationships of individual plants to their environment operate on a microcosmic scale. Many of the adaptations of plants to life on land have involved internalising the external, creating their own atmosphere in the spaces between their cells in their leaves and stems or garnering moisture and nutrients by colonising the soil in the finest possible way. It is impossible to understand the spatial and temporal distribution of plant species if their environment is characterised only as a property of the physical region in which they grow, because their space is defined by their own activities. Plants have a marked effect on soil pH, salt concentration, water-table, etc., as well as influencing aspects of plant communities such as boundary layer, light levels, humidity, turbulence, etc.

Figure 3.1. *Dryas octopetala*, like many flowers, shows a solar tracking response.

The relationships of a plant to its environment are therefore extremely complex, changing over a single day, over the seasons and over the lifetime of the plant. It is a dynamic relationship that also involves microclimate, succession, seedling recruitment, epiphytism, parasitism and pathogens. In the soil, the interaction of plants with other organisms, symbiotic relationships of plant and fungi in mycorrhizae, and plant and bacterium in root nodules, is brought into prominence. The formation of soils is dependent on these interactions. Transpiration by the plant transports water from the soil to the air. The organic material from decomposing vegetation is an essential component of the soil matrix. It stabilises soil aggregates and modifies the chemical behaviour and water retention characteristics of the soil. The rate of colonisation of purely mineral soils such as dune sands, volcanic ash or glacial sediments by plants can be studied by recording the degree of incorporation of organic material.

Plants are not just the objects, the victims, of evolutionary forces, but, in a real sense, they participate both in creating their own environment and in creating a potential for their own evolution. Plants are not passive acceptors of their environment. They can react to it, for example, by changing the orientation of their leaves, or by growing. They absorb carbon dioxide and release oxygen. They create shade and modify air currents. By opening their stomata and increasing transpiration, plants can markedly reduce the temperature of their leaves. Their presence modifies the flow of the wind. In hot conditions, by transpiration, plants humidify and cool the air, while in humid conditions they trap moisture from the air. Both the aerial and subaerial zones vary over time as well as spatially. The vegetation is an important determinant of climate, especially the microclimate of the aerial zone. With the destruction of world vegetation almost everywhere we are beginning to realise the importance of plant life for global climates.

Figure 3.2. *Cymbalaria muralis*: after fertilisiation the peduncle changes behaviour, elongates and seeks out cracks in a rock wall to plants the seed.

3.1.2 Responding to the environment

Plant senses are extraordinary. We have no words to describe them. We must resort to anthropocentric words like 'see', 'smell', 'taste' and 'feel', but they seem woefully inadequate. Behaviour in animals has no real counterpart in plants. The responses of plants to environmental variables are mostly growth responses or tropisms, but also include nastic responses. Shoots grow towards the light (phototropic) and roots normally grow downwards (geotropic). Flowers like *Dryas* seek the Sun (heliotropic) and are rotated on their pedicels to face it. More subtle tropisms exist so that the elongating pedicel of *Cymbalaria muralis* seeks out dark cracks in the rock to plant the ripening seed it bears. Climbers seek the feel of the bark of a host tree to support them. Although we can think of growth as analogous to behaviour, it is with difficulty because of the way plants track time. It is only when we see rapid movements in plants, for example the collapsing leaflets of the sensitive plant (*Mimosa pudica*), or the sudden snap of a

Venus's fly-trap (*Dionaea muscipula*) that we somehow connect this with behaviour, with being alive.

Wind and other environmental stresses can trigger responses such as wind-pruning, and drought and frost can seriously disrupt the transpiration stream, causing severe wilting. Osmotic effects can alter the appearance of many succulents, while mineral deficiencies can drastically alter the appearance of most plants. Buffeted by the wind, trampled by animals, or trimmed by lawnmowers, plants respond in various ways, either to compensate for traumatic injury or to adapt to changing conditions. They do this by altering the distribution of their body parts, as well as their overall mass, a phenomenon remarked upon several hundred years ago by Goethe, and subsequently by Geoffroy Saint-Hilaire, and Darwin.

We are just beginning to unravel many fascinating relationships in the plant world involving chemistry. The detection of herbivores by plants is also extraordinary. For example, maize plants respond to a compound called volicitin in the saliva of the beet army-worm caterpillar, and immediately produce mixtures of volatile compounds that attract the wasp *Cotesia margini-ventris*, a parasite of the caterpillar. Wounded tomatoes produce the volatile odour methyl-jasmonate that is detected by neighbouring plants, and immediately start to produce chemical defences. One of the best known defensive strategies of plants to the competition of other plants is allelopathy, which is quite a common phenomenon in Mediterranean and semi-desert vegetation, and may be more widespread than is generally thought. Chemical substances such as alkaloids, terpenoids and phenolics are released into the soil by plant roots and act as germination inhibitors. Numerous plants of high latitudes and elevations, and where the nutrients are 'locked up' and unavailable to the plant, on acid podsols, for example, have small needle-like or imbricate evergreen leaves that are retained on the plants for long periods. In turn the leaf litter that builds up takes a long time to decay in cold anaerobic conditions, and further contributes to the acid conditions. Most plants growing in such soils have a very close mycorrhizal association in their roots, and are prime examples of the dynamic interaction between a plant and its environment. This is adaptation in action, a 'chicken and egg' situation where it is difficult to separate cause and effect.

The main external signals that plants respond to are moisture and light. The control of morphogenesis by light is more accurately called photomorphogenesis, and is a process that is relatively independent of photosynthesis. There are two important stages of photomorphogenesis: pattern specification in which cells and tissues develop and become competent to react to light; and pattern realisation during which time the light-dependent process occurs. These reactions need to be amplified, normally by gene activation to initiate morphogenesis.

The part of the plant that 'perceives' the stimulus is the receptor. Many plant organs contain photo-sensitive compounds, each reacting very specifically to certain wavelengths of light. These light-sensors

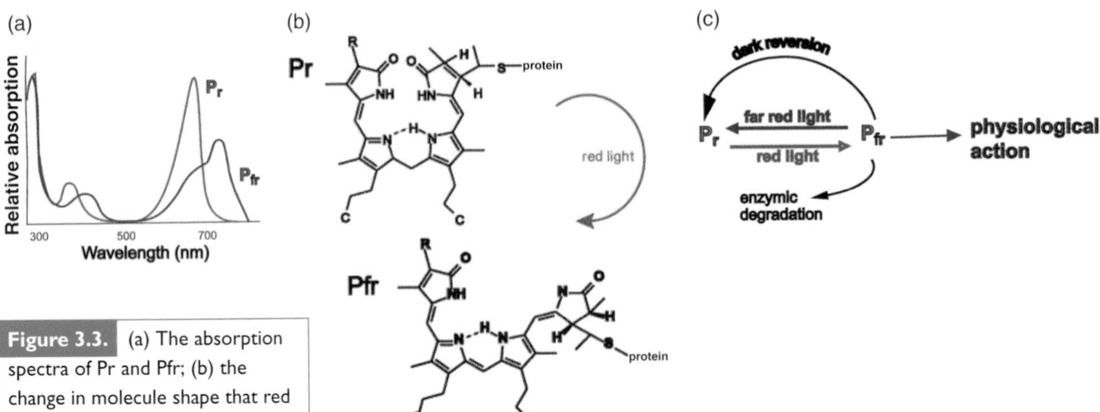

Figure 3.3. (a) The absorption spectra of Pr and Pfr; (b) the change in molecule shape that red light causes to Pr (*cis* form of phytochrome) to Pfr (*trans* form of phytochrome); (c) the relationship of the phytochromes enables day length/night length to be measured and trigger different physiological responses.

tell the plant if it's day or night, how long the day is, how much light is available and the direction from which the light comes. Plants also detect harmful ultra-violet rays and start producing pigments that filter out these rays. The main photoreceptor that responds to light is a pigment called phytochrome. There are two basic kinds of phytochrome (Pr and Pfr) that absorb mostly strongly red and far-red light as well as blue (Figure 3.3). The other photoreceptors are cryptochrome, which absorbs blue light and long-wavelength ultra-violet between 320 and 400 nm; UV-B photoreceptor, which absorbs ultra-violet light between 280 and 320 nm, and protochlrorophyllide a, which absorbs red and blue light and becomes reduced to chlorophyll a.

Phytochrome and other receptors control seed germination, and in the developing seedling it promotes chlorophyll production, leaf expansion and root development. Each stage in the development of plants is affected by the quantity and quality of light, culminating in the development of flowers and seeds. Individual stages of the developmental processes are highly specific for each species. For example, in hypogeal germinators, energy reserves in the cotyledons or endosperm are used primarily to extend the stem in darkness (i.e. through the soil), in contrast to leaf expansion in epigeal germinators. Seedlings deprived of light become etiolated.

3.1.3 Responding to the environment

Plants respond immediately to the environment by tropisms and nastic movements. Both tropisms and nastic movements are often the result of directional differential growth, but also by the reversible uptake of water into motor cells that collectively form a pulvinus. Tropisms are responses in which the direction of the environmental stimulus determines the direction of movement. With nastic movements, which include daily leaf movements, stomatal opening/closing and trap mechanisms in carnivorous plants, the external stimulus does not determine the direction of movement.

Temperature-induced movement of flowers is a nastic movement (thermonasty).

The perception and response to light is mostly in the apical meristems of the shoot. Shoots are usually positively phototropic, whereas roots seldom exhibit phototropisms. Light acts firstly as a trigger for the bending response, and secondly by decreasing the sensitivity of the organ to subsequent light (the tonic effect). In shoots it stimulates auxin to migrate from the irradiated side to the shaded side, with a corresponding inhibition (the inhibitor is indole acetic acid, or IAA) of the irradiated side. As a result, the shoot tip bends towards the direction of the stimulus. By the use of time-lapse photography we can see how a stem tip appears to trace a more or less regular ellipse as the stem grows. This movement was called circumnutation by Darwin, who suggested that all plant movements were modifications of this basic phenomenon. In this way, plants can adjust the positions of their stems and leaves to maximise light interception and minimise shading of neighbouring leaves. This is most clearly demonstrated in shade plants (sciophytes) where leaf mosaics are a common strategy.

In many plants, particularly those of high latitudes, solar tracking is a frequent phenomenon. The leaf blades remain nearly at right angles to the Sun throughout the day, return to a resting position at night, and resume tracking the following morning before sunrise. Solar tracking takes a variety of forms, for example in some desert plants 'negative' solar tracking (facing away from the Sun) occurs as a means of reducing heat and water stress. Some plants of high latitudes track the Sun with their flowers, thereby exploiting the slight temperature rise in the receptacle to lure pollinating insects attracted by the warmth. Leaf orientation is controlled by motor cells in a pulvinus where the blade joins the petiole, and movement of water in and out of these cells, which is regulated by osmotic solutes such as potassium. The cells in the major veins of the lamina detect the direction of the Sun's rays, and send a message to the motor cells of the pulvinus where the response may be finely tuned to signals from many different veins. These responses may involve auxin.

In tropical forests there is another growth phenomenon called skototropism, but this time the stimulus is shade. When the seeds of vines (e.g. *Monstera*) in tropical forest germinate, they have to find a supporting tree. The seedlings do not search randomly, but instead seek out the darkest sectors of the horizon, which are usually caused by tree trunks. The maximum distance the seedling can be away from the tree depends on the size of the trunk, but it is usually not more than about one metre. When it touches the tree, gravitropism, phototropism or thigmotropism take over. Thigmotropisms are the responses to touch, for example by the tendrils of a climbing plant.

Gravitropisms are growth movements towards or away from the Earth's gravitational pull. Primary roots are generally orientated more vertically than higher-order roots, which may be hardly gravitropic

Tropism	Growth towards
Phototropism	Light
Hydrotropism	Water
Chemotropism	Chemicals
Electrotropism	Electricity
Hygrotropism	Humidity
Skototropism	Shade
Gravitropism	Up or down

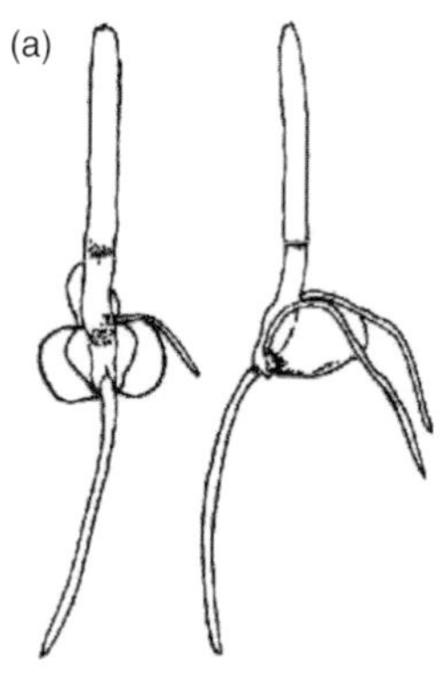

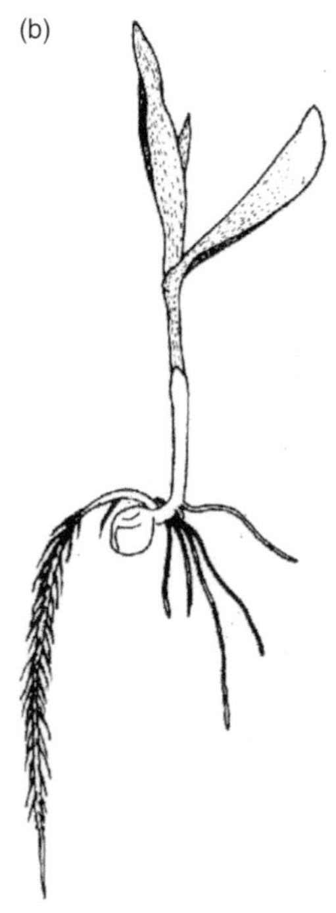

Figure 3.4. Phototropism and gravitropism exhibited by *Zea* seedlings.

Figure 3.5. Growth response to light: different shape of *Morus* leaves in canopy shade to full daylight.

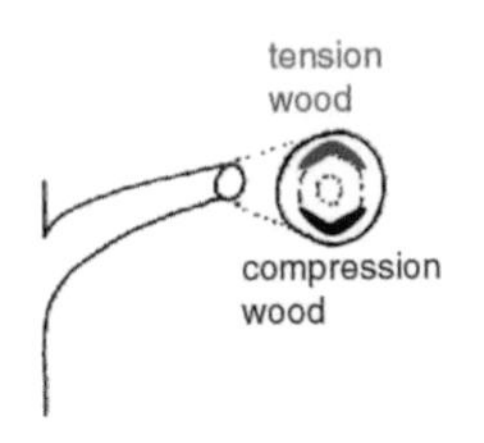

Figure 3.6. Reaction and tension wood (gelatinous fibres detailed).

at all. These differences allows the root system to explore the soil more thoroughly. Roots grow downwards (positive gravitropism) and stems upwards (negative gravitropism) in response to gravity. Vertical growth is called orthogravitropism, whereas, branches, petioles, and rhizomes are more horizontal and display diagravitropism. Plant organs with no response to gravity are called agravitropic. Gravity is perceived by plastids, especially amyloplasts, in the cells of the root cap.

3.1.4 Physiological responses

Leaf shape in different parts of the canopy may change in response to the quality of light the leaves receive.

A growing tree undergoes different mechanical stresses, for example, horizontal branches suffer two kinds of stress because of gravity: tension on their upper sides and compression on the underside. Stressed trees produce reaction wood and gravity seems to be the main trigger for its development. This is an increase in xylem and may be produced either on the upper or lower side of a branch by more rapid division of the vascular cambium on that side.

There are two main types: in conifers, reaction wood occurs on the lower side and pushes limbs upright by expansion, and is called compression wood. It is characterised by rounded, thick-walled tracheids with intercellular spaces, and cell walls with higher lignin content. About 60% of angiosperms produce reaction wood that forms on the top side, and contracts to pull the branch towards the trunk. This is tension wood, which is characterised by gelatinous fibres with cell walls in which the innermost layer has little or no lignin, but is rich in hemicelluloses. In old wood neighbouring parenchyma cells may bulge into tracheary cells through the pits to produce tyloses.

Many of these responses are physiological and promote changes in the form of plants as part of their normal life cycle. For example, the tracking of time and the onset of flowering. These photoperiodic responses can be quite complex and diverse, depending on the climate and the species in question. De Mairan (1729) showed that the daily leaf movements of *Mimosa pudica* persisted for several days after he placed them in darkness. Both Linnaeus and Darwin were intrigued by plant circadian rhythms. In the 1920s Garner and Allard studied the induction of flowering in tobacco and soybean and coined the term photoperiodism. They classified plants as being either long-day plants, short-day plants or day-neutral plants (Figure 3.7) but circadian rhythms are present in the normal daily activity of plants, no more obviously than in CAM (crassulacean acid metabolism) plants where circadian activity of assimilatory phosphoenolpyruvate carboxylase (PEPc) enzyme controls the uptake of carbon dioxide. More recently, attention has focused on the photoreceptors, the cryptochromes and phytochromes that entrain the circadian clock to the day/night cycle.

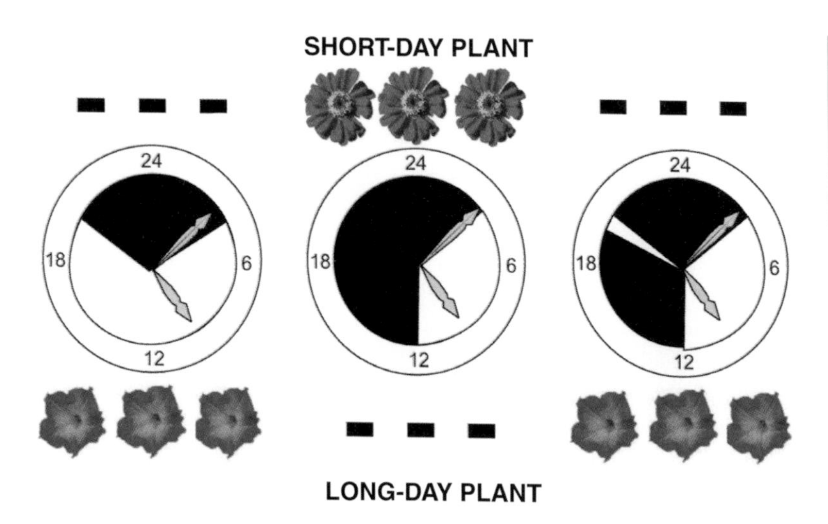

Figure 3.7. Long-day and short-day plants. Plants measure the length of the night with phytochrome. A flash of light converts a long night into two short nights.

Plant meristems are exceedingly sensitive to temperature and most usually have highly specific temperature requirements. Temperature changes, usually in combination with other factors such as light levels, initiate many critical stages in the development of a plant, for example, seed germination, flowering, and the onset or breaking of dormancy. In the process known as vernalisation, the onset of flowering in spring is promoted by low temperatures encountered during the winter, but there are several different kinds of responses among plants. Winter annuals respond to low temperatures as seeds, germinate in winter and flower the following spring. Biennials are dormant during the winter but germinate during the spring. During the first summer they do not flower, and it is the cold period of the second winter that induces flowering in the second spring. Some biennials require the additional onset of long days to induce flowering.

Other intrinsic processes are also involved in plant morphogenesis, which may greatly alter the morphology of plants, for example, controlled cell death, which occurs in the normal life cycle of all plants during maturation and senescence of flowers and fruits. The regular seasonal cycle in temperate plants involving leaf abscission and leaf fall is perhaps the most dramatic of such processes (Figure 3.8).

This is a regular phenomenon in seasonally arid climates. Some trees and shrubs cope with seasonal aridity by being drought-deciduous; for example, many African *Acacia* spp. lose most of their leaves at the start of the dry season. Such abortion may even include branch pruning as in species of *Aloe* from southern Africa or in *Cecropia* from Central America. Others have seasonally dimorphic leaves, the more drought-resistant ones being grown in summer.

The physiological response to drought is an important factor which may radically alter the form of plants, and is best seen in succulents such as the Cactaceae. A primitive genus in the Cactaceae,

Figure 3.8. Deciduous trees in the Mexican seasonally arid zone.

Pereskia, has leaves. In the ancestral cacti, leaves became reduced in size, and fleshier, eventually becoming ephemeral and minute, or reduced to spines called glochidia. Loss of leaf auxins may have promoted a number of changes in the stem: a delay in the completion of the vascular cylinder; the production of large primary rays; and the development of a long/short shoot arrangement to give the typical areole pattern of cacti. Each areole is a short outgrowth bearing spines. All these changes encouraged a shift to succulence in the stem. Finally CAM photosynthesis arose.

Succulence has also evolved in halophytes because of the high chloride concentration of cell sap (see Chapter 6).

3.2 The nature of evolutionary processes

Gravity and wind constrain the height to which plants can grow, but they cannot make plants that are genetically 'programmed' to be small any taller. Other environmental factors such as daylength and temperature, and biotic influences may have direct or indirect effects on development, flowering times, etc., but adaptations to such processes alone do not establish the overall construction of plants.

Before the development of genetics and modern microscopy, the dominant view of development was that of the preformationists who believed that a tiny version of a human being was encapsulated in a sperm cell, and developed in the mother's womb until birth, the mother providing nutrition only for the developing foetus. According to Lewontin, modern developmental biology is a version of the preformationist idea in disguise, i.e. genes determine the final form, while the environment is merely the platform for gene expression and organismic sustenance.

3.2.1 The limitations of orthodox Neo-Darwinian views

The orthodox Neo-Darwinian view of evolution regards plant form as the result of natural selection acting on random variations arising from mutations, and that similarity of form in unrelated groups is viewed as the result of convergent evolution. From this perspective, form is ultimately determined by genetic makeup as well as accidents of history, and the environment is regarded as the agency responsible for moulding form. Form is thus contingent upon fitness, and anything is possible. The curious fact is that plant diversity is characterised by much novelty, so the orthodox view is apparently borne out by the evidence from nature. Is form therefore the mere caprice of nature?

The Neo-Darwinian view gives a primacy to the genes in determining form and being the items through which natural selection acts. This view has been given remarkable support in recent years with the dramatic switch to the study of the genetics of development, particularly that of floral organs. We now know much about the genes which control development, the homeotic MADS-box genes. There is an epigenetic transition from genotype to phenotype, but the processes by which the forms of tissues, organs and the organisms themselves unfold during development still remains as one of the greatest challenges for current botanical research. As Futuyma has stated: 'Our ignorance of how genotypes produce phenotypes is, I believe, the greatest gap in our understanding of the evolutionary process . . .'

However, form in plants is the expression of a harmonious organisation that reflects the nature and relative arrangement of their parts. Development does not occur in isolation but is the result of equilibrium between endogenous growth processes and the constraints exerted by a particular environment, the physical and chemical parameters of which have set limits to the expression of plant form. These endogenous growth processes and their integration are ultimately under genetic control.

3.2.2 The integration and harmonisation of plant form

Although the continued survival of innumerable plant species in a diversity of habitats and climatic regimes throughout the world is adequate testimony to their fitness and adaptation, there are repeated common morphological patterns among them. The angiosperms, gymnosperms, and earlier-evolved groups such as ferns, lycopods, bryophytes, and even algae, all show intrinsic developmental constraints that limit their morphological expression. The '*Hippuris*' Syndrome among water plants is a prime example (Figure 3.9).

It is gradually emerging from the study of convergences in evolutionary form and the conservatism of morphological patterns that form in plants involves other processes that have hitherto been given little weight in studies of plant adaptation and evolution. The various

Figure 3.9. *Hippuris* Syndrome in (a) *Hippuris* and (b) *Equisetum*.

Figure 3.10. The serial development (meristic series) of flowers in *Heliconia*.

so-called 'parts' of plants display repeated relations to one another to form discernible patterns, not only at the level of the external form, but also at the level of the arrangements of cells and tissues. In serial organs, for example, often neighbouring parts are more similar than distant parts. Frequently one member of such a series assumes the form and characteristics of another, either of the same species or a different one, suggesting, in addition to constraints, some kind of equivalence or correspondence between the different forms. This is an example of homeosis.

No longer can we be satisfied that plant form arises through the simple process of natural selection acting alone on random mutations. Epigenetic processes involving integration and harmonisation of form at the cellular, tissue, organ, organismal, and environmental levels must all be considered. We have to examine the complex and constructive effects of certain genes, and the effects of selection on the correlation of single characters in the whole system of the organism. Genetic pleiotropism, allometric growth, and the effects of compensation, provide considerable explanation of morphological transformation even at the level of systemic changes. Compensation is when part of a plant becomes altered as a result of changes in its environment or as the result of damage to the plant itself. It is another way of expressing the harmonisation of the plant, i.e. its ontogenetic contingency.

There are no known laws of growth involved in these explanations, yet Darwin himself recognised that these processes must be accounted for if we are adequately to explain evolution. In the absence of known processes that could explain the epigenesis of form, many biologists, particularly in the late nineteenth century, were attracted to persuasive theories such as orthogenesis and vitalism.

These theories have remained marginal to orthodox Neo-Darwinism, which is essentially a mechanistic explanation for organic evolution. What is frequently overlooked is that this only constitutes part of Darwin's theory of evolution. Many biologists rightly feel that there is something incomplete about simple mechanistic explanations.

A frame of reference for developmental biology of only genes and cell organelles, while environment is merely a background factor, is clearly inadequate for the total developmental epigenesis of plants. Development is a transformational theory of change whereas current evolutionary theory is a variational theory of change. What is long overdue is an integration of both theories in a holistic or systems approach that emphasises the way organisms function as interconnected wholes. In recent decades there has been a re-awakening of generative or structuralist aspects of plant development.

Two quotations from chapter five of *On the Origin of Species* (1859):

I know of no case better adapted to show the importance of the laws of correlation in modifying important structures, independently of utility and, therefore, of natural selection, than that of the difference between the outer and inner flowers in some Compositous and Umbelliferous plants.

. . . we see that modifications of structure, viewed by systematists as of high value, may be wholly due to unknown laws of correlated growth, and without being, as far as we can see, of the slightest service to the species.

3.2.3 The dance through morphospace

The historical emergence of different forms may be more fully understood in the context of a 'theory of morphospace' such as that described by Goodwin. As Stephen Jay Gould says of evolution, 'it has no purpose, no progress, no sense of direction. It's a dance through morphospace, the space of the forms of organisms'. This approach provides botany with a logical, dynamic foundation and somewhat reduces the role of history as an explanation of plant form. The 'new' biology is a science of complex systems concerned with dynamics and emergent order. Viewed in this way, all the old biology changes. Instead of the tired metaphors of conflict, competition, selfish genes, climbing peaks in fitness landscapes, what you get is the metaphor of evolution as a dance, and it is the rules of the dance that is the subject of evolutionary biology.

Individuals of a plant species can be thought of as constituting a dynamic evolving system interacting with an environment that is simultaneously heterogeneous and dynamic. The net result is that the course of evolution can proceed along highly diverse pathways. Without a dynamic changing Earth there would be no major organismic diversity. Therefore, the successful study of plant form ideally should be within a four-dimensional ecological framework involving the interaction of plant, environment and time. This is a systems theory of plant evolution, i.e. the evolutionary dynamics of plants are determined by the structural organisation of the genome and the epigenetic system, and embracing mechanisms of change such as mutation, recombination, random drift and selection. It must also stress the cohesion of the genotype and the integration of the developmental system. However, it goes beyond orthodox interpretations, which regard organism–environmental interactions as theoretically reversible. Under a systems model of plant evolution, change is irreversible since it is essentially correlative, and the outcome destroys the conditions that made such evolutionary changes possible in the first place.

3.3 Order, transformation and emergence

As we described in Chapter 1, in the language of complexity theory we can think of a living organism such as a plant as a dissipative structure. To stay alive and maintain order, it needs to continually exchange gases, water and chemical compounds with the environment. This continual exchange involves thousands of chemical reactions, while the vast network of metabolic processes keeps the system in a state far from equilibrium. Only when the plant dies and all these processes come to a halt, does chemical and thermal equilibrium exist. Feedback loops are a crucial property of the system, and act as self-balancing and self-amplifying processes that may push the system further away from equilibrium until it reaches a threshold of stability. This point is called a 'bifurcation point'. It is a point of instability at which new forms of order may emerge spontaneously, resulting in development and evolution.

There is a high degree of order in plants. If not, their integrity would break down under environmental impacts. Through all levels of evolution, plants remain well-balanced and harmonious systems. The level of order and organisation (morphological harmony) may be adjusted to suit the ecological constraints, through processes of differential growth, growth inhibition, integration and compensation. These enable the primary resources of the environment to be constantly tapped, and an optimal form of organisation and order to be maintained. The development of the individual plant is through progressive differentiation of originally undifferentiated cells, and that its complexity is gradually built up until the final form is reached. As a result new systemic qualities of the organism arise from correlative processes. What are those correlative processes? How are such integrated processes or 'synorganisations' brought about?

Each adjustment or transformation is governed by specific rules affecting the organism as a whole. Such progressive transformation is largely epigenetic. It is a genetically mediated response to internal environmental changes, but also involves mechanisms of cellular communication. An analogy is to the BIOS (basic input/output system) in computer systems that integrates the hardware (genotype) with the operating system (phenotype) that provides an internal environment for the software (the plants response to the environment) (Figure 3.11). These internal integrative processes are dominant over the effects of the environment.

Similarly in a biological system there is mutual influence and induction of tissue, but that is not all. Obviously control by genes is crucial. Chromosomal linkage, position effects, pleiotropism, mutual reactions between the processes brought about by different genes, signals and enzymes all contribute to the systemic differentiation of the plant. There are volatile effects, such as the methylation that modulates gene expression.

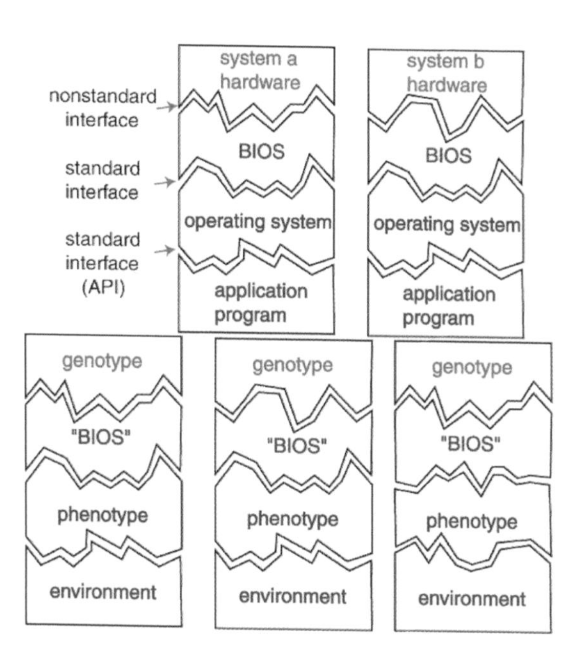

Figure 3.11. The integration of the genotype and the phenotype is analogous to the BIOS in computers. The BIOS integrates a computer's hardware and its software. The BIOS is part burned or flashed into a ROM and read-only part included on ROM chips installed on adapter cards, and part on additional drivers loaded when the system boots up. Variations in the biological BIOS can result in similar phenotypes in plants with different genotypes or different phenotypes in plants with similar genotypes.

During the development of plants the growth ratios of single organs and the structures in relation to the whole body usually remain constant for certain periods, but there can be changes in developmental timing, such as at the onset of flowering. It was Goldschmidt who first developed a clear idea of accelerations and retardations of certain gene-controlled developmental processes that are due to quantitative differences, and their mutual interaction during development. Thus, gene-control is implicated in the establishment of growth gradients that are correlated to the size of the whole plant body. During certain periods, an organ or a structure can grow more quickly than the body as a whole (positive allometry), more slowly (negative allometry), or with the same speed (isometry). Similarly, one can refer to growth of a certain part of an organ in relation to the whole (Figure 3.12).

Figure 3.12. Allometric changes in leaf shape in *Phyteuma* (Campanulaceae) from the base of the plant (far left) towards the inflorescence.

Allometric growth can thus occur simultaneously at different levels within an individual, but this phenomenon can also be seen

Figure 3.13. Heteroblastic development during ontogeny may occasionally suggest the recapitulation of phylogeny. Here in *Chamecyparis* the juvenile foliage seems to exhibit a kind of adult foliage found in earlier stages of cypress evolution.

between unrelated species. Large and small specimens of the same species, or closely related large and small species will hardly ever show the same type of proportions. By studying the different stages of allometric growth in comparative studies of different species, rules can be established that govern the correlation of body size and organs in the course of phylogenetic differentiation, and the relative influence of selection processes. Such gradients of differentiation in body proportions were ably demonstrated by the Cartesian transformations of D'Arcy Thompson.

The idea that stages in the ontogeny of an individual may somehow afford clues to phylogenetic relationships was not as clearly expressed in botany as in zoology, where the so-called biogenetic law 'ontogeny is the short and rapid recapitulation of phylogeny' was given much currency by Haeckel in the late nineteenth century. It was applied to botanical problems by Takhtajan who considered that it was a theory that 'penetrates botany with difficulty'. Alterations that may affect the evolution of plants may, theoretically, occur at any stage in the ontogeny of individuals, and, according to Takhtajan, the nature and extent of these alterations may be conceptualised as four modes of change, i.e. prolongations, abbreviations, deviations, and neoteny.

3.3.1 The developmental sequence

The developmental process even at its simplest level is very complex and involves the communication between different elements, multiple hypercycles, and feedback loops. Consequently, there are relatively few ways in which novel forms arise in development without a complete loss of integration. The idea of a developmental sequence of a number of stages A → B → C is too simplistic. It is more like a web of relationships that together fix a developmental process in space, but nevertheless, this sequence provides a framework for the discussion of developmental changes.

Addition of a stage in a sequence of development, what Takhtajan called prolongations, is extremely common in plants, for example pollen grains, seed coats, pericarps, and various parts of flowers, especially all types of outgrowths, such as the development of wings on seeds and fruits. An addition is much more likely at the end of an existing sequence of development, a terminal prolongation, than at its beginning, or in the intercalation of a new stage because such a pathway is less disrupted.

Addition is more common than subtraction or abbreviation. Indeed, the complexity of developmental relationships, the network of hypercycles, feedback loops and multi-dimensional influences, make subtraction of part of a developmental process particularly problematic. Abbreviations or subtractions are regarded as the omission of certain stages of development, the opposite of prolongations. Vestigial structures are regarded as cases of terminal abbreviation. Reduction in floral and vegetative parts, are examples of terminal abbreviations.

Expansion (A →B →C) or reduction (A →B →C) of a stage in a sequence of development is much more frequent than a complete deviation since it is less likely to disrupt the network of developmental relationships. Expansion or reduction can occur in space, expanding or reducing the size of an organ or part of an organ relative to others. In many cases, loss of a developmental stage (subtraction) can be shown to be no more than the reduction of that stage. Extension (A→BB →C) or shortening are related to expansion or reduction but occur in time. Extension of a particular stage of development often results in expansion, whereas shortening often results in reduction. If there are finite resources or limited time, extension of an earlier developmental stage might result in a neotonous organ, resembling an earlier stage of development (A →BC).

Takhtajan considered neoteny to be of prime importance, and referred to it as 'Peter Pan' evolution. The term is used in a phylogenetic sense when it is considered that the ontogeny of a plant is truncated, leading to a premature completion of development of the whole plant or part of it. It is often considered synonymous with 'paedomorphosis' or 'juvenilisation'. Neotenic changes depend on a simplification or despecialisation of the phenotype. Takhtajan was apparently swayed by the arguments of Koltzoff who pointed out that such simplification does not affect the genotype. Subsequent mutation could then lead to an evolutionary radiation of forms with 'new', juvenilised phenotypes.

Expansion/reduction or extension/shortening of development in part of an organ results in a change in shape or orientation, what Takhatajan called deviations. The profundity of the deviation depends upon its timing, earlier deviations more profoundly influence later stages.

Multiplication or combination of developmental sequences is particularly common in plants because of their repetitive (iterative) construction. Multiplication increases the number of organs or parts of organs, normally by the division of a meristem, whereas combination results from the fusion of two or more neighbouring meristems. Stebbins called this intercalary concrescence. Multiplication and combination/fusion in some cases are the consequence of expansion/reduction or extension/shortening of developmental stages. Extension of a developmental stage prior to meristem division may not prevent multiple organs developing eventually, but often results in them being fused together, whereas reduction or shortening of a developmental stage may lead to the early multiplication of an organ.

Multiplication and combination are the two most profound means by which a lineage can escape the developmental boundaries of an autopoietic system, creating new evolutionary potential. It is fascinating that this occurs at different levels in the hierarchy of organisation. For example, the evolution of genes has occurred by exon duplication (multiplication) and shuffling (combination). We describe below several examples of multiplication and/or combination in the

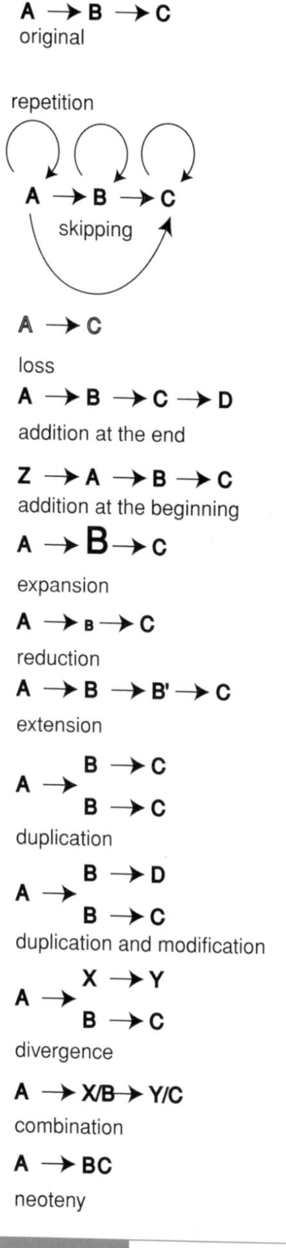

Figure 3.14. Possible different modes of developmental changes in the evolution of plants.

evolution of reproductive organs. Is it stretching the point too far to see the same process in speciation by geographic isolation (multiplication) and by hybridisation (combination), the two most profound influences on plant speciation?

It probably is stretching the point too far to see the same processes in the development of ecosystems but perhaps not. Imagine the multiplication of an ecosystem by the colonisation of an island and the combination of different elements by this process and look at the consequences in the development of novel island ecosystems with changed relationships.

Multiplication and combination are such profound triggers for evolutionary transformation because they can free an organ to function in a different way or adopt a novel function. Novel adapted organs do not arise out of the blue, as fully integrated functioning systems but normally arise by a transfer or change of function of an existing organ. Describing this in terms of autopoiesis, it is this change of function that so disturbs the network of developmental hypercycles that a bifurcation point arises, releasing the organ from its earlier developmental constraints. In this new developmental landscape evolution can be particularly rapid.

A word of caution is needed here. The classification of ontogenetic development into several different modes of change may not be the most useful way to proceed in developmental studies. There is every gradation between each mode, and obvious overlap, so that confusion may result. In addition, it cannot be assumed that certain stages ever existed. The logic behind such a scheme is basically typological and is the traditional thinking behind conventional ideas of homology. By dividing phenomena into defined units in space and time, for example leaf, shoot and root, it is too easy to makes hypothetical assumptions about presences and absences. Because the ontogeny of some plants may be conceptualised as serial steps, it should not necessarily be applicable to all, or ad hoc hypothesis used to explain discrepancies. It may be more useful to think of plant development and evolution in a dynamic way, for example the process evolution of Sattler in which plants are examined throughout their entire life cycle and with reference to a wider spectrum of plant form.

(a)

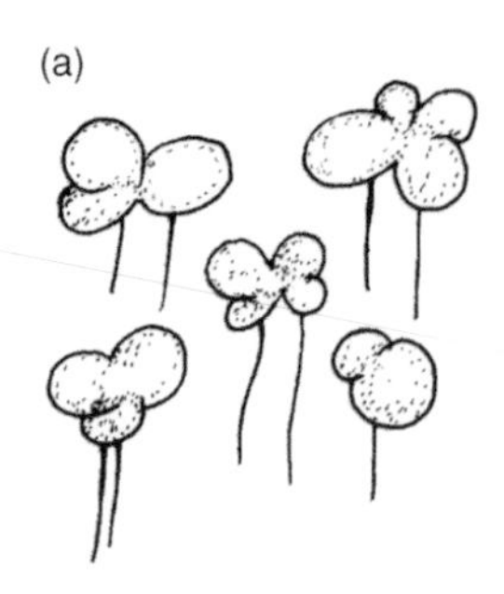

(b)

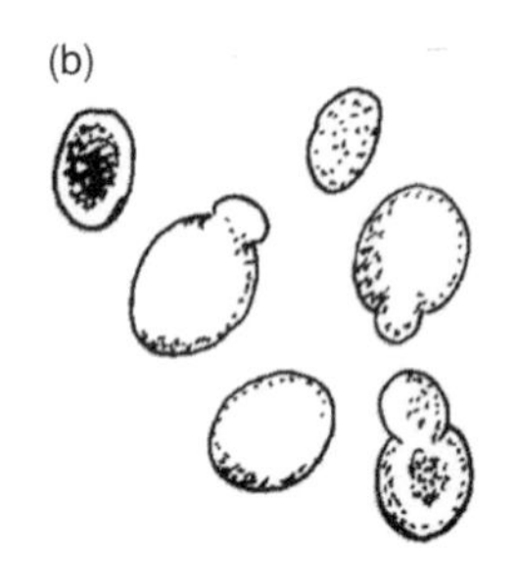

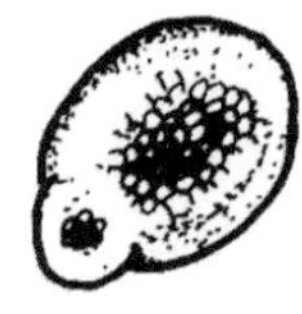

Figure 3.15. Two highly reduced forms in the Lemnaceae: (a) *Lemna*; (b) *Wolffia*.

3.4 Macromutation and evolutionary novelty

Changes in developmental timing are frequently considered an important macroevolutionary process. It increases new possibilities in evolutionary terms. As Huxley pointed out 'It is this possibility of escaping from the blind alleys of specialisation into a new period of plasticity and adaptive radiation which makes the idea of paedomorphosis so attractive in evolutionary theory'.

Stebbins' discussion of the role of what he calls intercalary concrescence fits in here. Abbreviation of a developmental stage in the apical meristem or a prolongation of a previous stage can result in the growing together of meristem primordia to result in union or fusion of parts. This has been of profound evolutionary significance in the evolution of reproductive structures, from the fused sporangia of some pteridophytes to the astonishing diversity of floral structures. Fusion of primordia that give rise to the same organ, connation, frequently results in tubular structures. Fusion of primordia that give rise to distinct organs, adnation, gives rise to novel compound organs. Both connation and adnation release evolutionary potential.

However, the acquisition of evolutionary novelty usually requires a shift in ontogenetic development and, since this represents a saltatory step, it poses serious problems for defenders of a gradualistic theory of evolution. The solution to the problem was perhaps provided by Darwin himself who pointed out that a change in structure must simultaneously involve a shift of function. Severtsov was one of the first to point out that an 'intensification' of function is all that is needed for the adoption of a new function. In the course of evolutionary change a morphological structure may have certain additional characteristics that are, initially, selectively neutral, but become increasing co-opted to perform new functions. Mayr stated somewhat ambiguously that, 'in most cases, no major mutation is necessary in order to initiate the acquisition of the new evolutionary novelty; sometimes, however, a phenotypically drastic mutation seems to be the first step'. Is evolutionary novelty or macroevolution the outcome of macromutations or not? As a preliminary to an answer, it may be constructive to consider the studies of evolutionary novelty by Jong and Burtt in *Streptocarpus*.

3.4.1 Growth forms in the Gesneriaceae

The Old-World genera of the Gesneriaceae (Cyrtandroideae) are well known for the unequal growth of their cotyledons (anisocotyly), while many have rather atypical growth patterns, such as continuous growth of one cotyledon, and epiphyllous inflorescences. *Saintpaulia* is well known for its ability to regenerate from single leaves. Atypical growth is found in a range of genera but species of *Streptocarpus* subg. *Streptocarpus* are the best known, and are all characterised by the continuous growth of one cotyledon. This enlarged cotyledon functions as a foliar organ but in many respects, it differs from a true leaf. The term 'phyllomorph' was coined by Jong and Burtt to distinguish the peculiar leaf-like structures of *Streptocarpus* from true leaves and cotyledons (Figure 3.15). Jong further differentiated the phyllomorph structure into a foliose component called the lamina and its rooting petiole-like stalk called the petiolode. The inflorescence of most species usually arises from the base of the midrib of the phyllomorph.

There are three groups based largely on the number of phyllomorphs they possess, and are either unifoliate or rosulate to varying degrees. In the unifoliate species of *Streptocarpus*, there is only

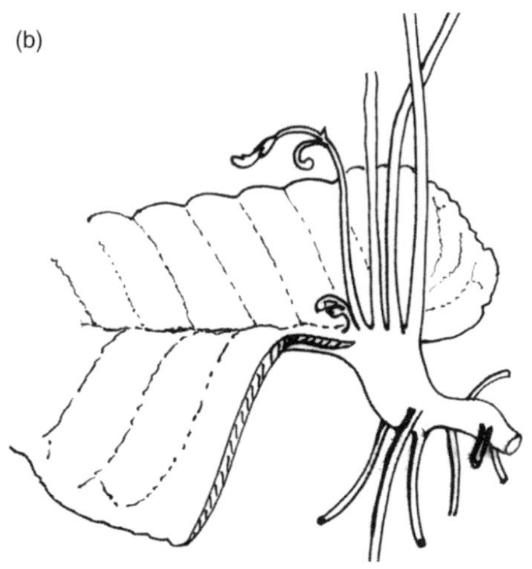

Figure 3.16. A *Streptocarpus* species with a phyllomorph. Jong differentiated the phyllomorph into a foliose component called the lamina and its rooting petiole-like stalk called the petiolode.

one phyllomorph (the cotyledonary phyllomorph), and the plants are monocarpic, whereas the rosulate species, which are perennial, have been termed 'colonial unifoliates', and comprise few to numerous repeating phyllomorph units.

In a study of *S. fanniniae*, Jong and Burtt found that there is unequal growth of the cotyledons in the seedling. The larger cotyledon continues to grow from a basal meristem, and eventually is raised above the level of the smaller cotyledon by intercalated tissue called the mesocotyl that eventually differentiates as the cotyledonary petiolode. In *S. fanniniae* there is no plumule and the meristematic tissue is sunk in an adaxial groove in the cotyledonary petiolode. It is from this groove meristem of the petiolode that new phyllomorphs and the inflorescence subsequently develop, while roots grow from the abaxial surface. This groove meristem is functionally the equivalent of the conventional apical meristem. In the regions at the base of each lamina lobe, and where the petiolode merges with the lamina there remain intercalary meristematic tissues called the basal and petiolode meristems, respectively. The basal meristem is responsible for continued growth of the lamina, whereas the petiolode meristem is responsible for growth of the midrib, and the elongation of the petiolode. In some species, part of the lamina dies back during an unfavourable dry season, and a zone of abcision is clearly recognisable. Growth of the lamina recommences in favourable conditions with activity in the basal meristem.

The petiolode has, by its possession of gaps in the vascular cylinder, and roots, a shoot-like nature. In contrast, its dorsiventrality and terminal lamina suggests a petiole. Although the phyllomorph is interpreted as a basic unit of structure combining features of both leaf and shoot, and applicable to the Gesneriaceae, similar ideas have been expressed by Sattler, and Arber. Jong and Burtt state that the coordinated activity 'gives the phyllomorph the stamp of distinction, a morphogenetic innovation that has provided new possibilities in the evolution of form'. '*Steptocarpus* might, indeed, seem to exhibit the acme of neoteny in flowering plants'.

In some species of *Chirita* and *Streptocarpus* subg. *Streptocarpella* flowers are occasionally produced by the cotyledon, suggesting that the evolution of the unifoliate condition evolved from ancestral plants that could develop facultatively in this way. In *Streptocarpus nobilis*, unifoliate growth is also facultative. Hilliard and Burtt speculated on the affect of genetic change on the development of the plumular bud in a caulescent plant that already possessed an accrescent cotyledon, and suggested that the unifoliate habit could develop in this way. Central to this argument is the role played by the environment. The unifoliate habit is well suited to relatively unoccupied habitats such as steep banks in forest or sheltered cliff faces, or on mossy tree trunks. If such novelty can arise from facultative ability (i.e. the ability to be flexible or plastic) then perhaps we should focus our attention of the genetics of phenotypic plasticity rather than macromutations. It should be recalled that small genotypic changes might produce massive phenotypic effects. Perhaps, by making a contrast between micromutations and macromutations we are indulging in semantics, with the result that we ask the wrong questions.

3.5 Unity and diversity; constraint and relaxation

One of the most compelling characteristics of organisms, and one that we often take for granted, is that no two individuals are alike. Each one is uniquely different yet, at the same time recognisable as belonging to a species. Individuality is the hallmark of all species though the distinction between individuals is obscured in clonal organisms such as plants. With their remarkable ability to reproduce asexually, plants can produce separate individuals that are genetically identical.

Organisms are biologically constrained to conform to relatively standard physical form and behaviour, and yet have an almost limitless diversity of individual appearances and behavioural preferences (novelty). These two processes are the stuff of evolution. One provides continuity over time, while the other provides the basis for evolutionary change. This continuity, this faithfulness to already-existing form is often considered deterministic, i.e. the organism develops according to a 'blueprint' encoded in the genes. Individual variation is then regarded as the outcome mainly of minor genetic variation passed on in the genes from parent to offspring and chance (stochastic) processes operating within the growing body and in the environment. Minute random variations in the internal or external environment may have a profound effect on the growing individual. However, novelty may not necessarily be the outcome of stochastic processes alone. It may be just one possible result of a dynamic spectrum of possibilities involving deterministic processes both at the individual and organismic levels of evolution.

Figure 3.17. Predictable architectural forms in non-flowering plants that follow a determinate pattern of growth: (a) lycopod; (b) *Araucaria heterophylla*.

The expression of a particular genotype in a particular environment is called the phenotype. Deterministic factors ultimately have a genetic basis, and are responsible for the conformation to the physiological and morphological parameters of the species. Nevertheless, environmental or opportunistic factors impinge on plant form and the ecological requirements of a species. Both factors combine to give a historical trajectory to a species (heredity). It is from the relative contributions of these contrasted processes (genotype and phenotype), and their interaction with the environment that plant form emerges. The way plants develop varies widely and, in the majority of species, there is a norm of reaction, i.e. there is a spectrum of possible outcomes. Ability to respond to chance environmental variables is ultimately determined by genotype, and differs widely in different groups of plants, in closely related species, and even between individuals of the same species. Such responses also vary at different times in the life cycle. In some species with more precise growth, development appears to be more deterministic; for example certain palms and arborescent monocots usually cannot respond to crown damage because they lack lateral meristems. Predictability of vegetative form can be shown to be widespread in non-flowering plants from lycopods to conifers. In other groups of plants, predictability of final form is more elusive. This stochastic response is usually expressed by reiterative vegetative growth and is one of its most important adaptive features of plants, but it can only be assessed after the initial deterministic component of growth has been recognised. Thus, it may be impossible to predict how any individual plant will develop in a given environment.

In marked contrast to vegetative parts, floral and fruit morphology are highly deterministic, with little scope for opportunistic development, which is one reason why these structures have lent themselves to a typological classification, and are so important for plant systematics. Naturally, there are differing views as to the extent to which plant form is determined by their genes or by chance events produced by the interaction of the genotype with the environment.

3.6 The phenotype

3.6.1 Developmental reaction norms

The phenotype of a plant is the sum total of its observable characteristics, and is the outcome of a complex relationship between the genetic coding of the individual (the genotype) and the environment. The relationship between phenotype and environment is referred to as the norm of reaction. How this harmonisation process works on the 'laws of growth' is still largely obscure, but certainly it is manifest through the normal developmental processes of the plant. Therefore, it is more appropriate to talk about a 'developmental reaction norm' or DRN. This is the set of ontogenies that can be produced by a single genotype when it is exposed to internal or external environmental variation. Natural selection operates on the DRN instead of on individual traits. The potential to respond to environmental cues is ultimately linked to the evolutionary history of the plant in question and, in the majority of species, there is a spectrum of possible outcomes.

3.6.2 Ontogenetic contingency

This ability to respond ontogenetically to environmental variables is called phenotypic plasticity. It is often regarded as adaptation by growth within the lifetime of the plant, i.e. a physiological response, but this is a narrow interpretation. It can apply also to plant populations that persistently differ over time, yet are practically identical from a genetic point of view. If the phenotype of the individual can respond and change in phase with the environment, then the plastic response is labile, for example in the lifetime of a single plant, when it also has been called 'ontogenetic contingency'. This harmonisation process allows the plant to survive relatively rapid vicissitudes of different environments encountered during ontogeny.

Figure 3.18. Plasticity in *Gentianella*: the cliff-top plant exposed to high winds grows as a stunted dwarf while the meadow plant is tall and flowers profusely. Both of these are mature plants.

3.6.3 Evolution of the plastic phenotype

Phenotypic plasticity has to be distinguished from situations where related plants differ as a result of having inherited genetic differences. The classical Neo-Darwinist interpretation of these phenomena is to regard the latter example as adaptation by the accumulation of genetic mutations acted upon by natural selection. In reality the differences between these two phenomena are seldom clear cut. In both cases the plant is adapted to its situation. Most variants occupy a range of environments, and if they are able to remain constant phenotypically, there can be some confidence that it is at least partly determined genetically.

We simply don't know what the effect the genotype has on the ability of a plant to respond in plastic way. The action of genes is contextual and it is misleading to think that it is the genes and their expression (traits) that are being selected. The genes, like other parts of a plant, do not exist in isolation but are an integral part of the whole organism. The orthodox view is that genetic variation and plastic response represent alternative and mutually exclusive means of dealing with environmental variability but plasticity responds significantly to selection and has been shown to potentially evolve independently of the mean of a particular trait.

Sometimes plants may be under such directional selection that eventually environmental parameters that are very different from the ancestral populations become the norm for these plant and they are then said to have reached a new adaptive peak by a process of canalisation. In such circumstances the modality of expression in the new

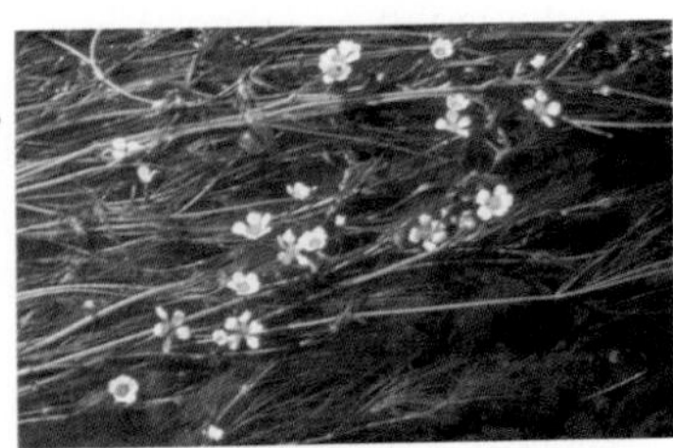

Figure 3.19. Water buttercups (*Ranunculus* subg. *Batrachium*) have some species with finely dissected submerged leaves, some with flat floating leaves and some species that can produce either or both types of leaves depending upon the situation.

population may display extreme values of the ancestral population or even may have gone beyond their bounds. In some instances the descendants of phenotypically plastic plants show a reduction in the amplitude of plasticity.

Marked plasticity is one of the features that distinguishes flowering plants from other land plants, and perhaps partly explains their evolutionary success. Form and development are variable between different groups, and even between members of the same species. In contrast, many non-flowering plants, are remarkably uniform, and conform to simple architectural models, i.e. they do not vary their pattern of growth, either within the plant, or between different individuals.

Species may differ in the extent of their plasticity, or may respond plastically in different ways, to the same stimulus. For example, different species of grasses do not differ in their overall level of plasticity, but they respond to different nitrogen levels, in different ways, by adopting unique phenotypes. Some garden plants are well known to be extremely adaptable to a wide range of conditions and able to respond by growing differently. Others, such as some alpines, are very precise in their requirements and are to be recommended for expert horticulturalists only.

Phenotypic plasticity in plants is mostly expressed in the number, size, and form of the vegetative parts, but it can also be physiological and biochemical and it can also be conspicuous in reproductive structures. The most spectacular examples of a plastic response at different stages in the life cycle of an individual are those associated with heterophylly in aquatic plants. Some species of water buttercup (*Ranunculus* subg. *Batrachium*) readily produce dissected or entire leaves as a plastic response to growing in water or above it, respectively (Figure 3.18). Yet, other species of water buttercup do not, producing only dissected or entire leaves.

Plasticity is very apparent in leaf shape. In many plants no two leaves are identical in shape. Plasticity differences of several kinds have been recorded between populations and species from shaded and open habitats, and even between different parts of the same plant. Leaves produced in the shade are different from those in the light. There are differences in leaf area and thickness, petiole and internode length, but plastic changes occur also in the internal anatomy. In bryophytes, morphogenetic responses to light intensity exert a sensitive control over the angle of inclination of the shoots. The effect of shade brings about the progressive elevation of the shoot in some bryophytes, whereas it induces a more horizontal growth in carpet-forming species. Many experiments have been done on plants removed from their original home and transplanted in a different environment, and these have been used to determine the relative contribution of genes and environment towards the developmental reaction norm.

Phenotypic plasticity has been thought to supplement the role of genetic variation in heterogeneous environments, and thus to act as a buffer against natural selection, allowing genetic variability to be maintained. The 'Baldwin Effect' is said to operate when a plant responds to the bottleneck effect of natural selection by wide amplitude of natural plastic variation that has already been 'pre-selected' or harmonised by a mosaic of microenvironments. Contrast this with the Darwinian/Neo-Darwinian model that offers variation due to random mutation in individuals conferring selectively advantageous traits. By concentrating on the trait, it effectively ignores the integration process with the whole plant.

Phenotypic plasticity is also a buffer against spatial or temporal variability in habitat conditions. From a reproductive point of view, it could also be considered as a mechanism that allows plants to control their reproductive effort. It is a mechanism that compensates for a lack of mobility in plants. Certainly, it is of vital importance in resource acquisition by plants. In productive habitats, high morphological plasticity is part of the foraging strategy adopted by competing plants, and highly plastic species therefore seem to have wide ecological amplitude.

3.6.4 Adaptive landscapes

Fisher's conception of the genetic architecture of traits was based on the idea of genes as independent factors – their effects could be added together to produce the phenotype. Fisher believed that only additive effects were important in determining responses to natural selection and that much of continuous variation is caused by multiple Mendelian factors rather than by environmental influences. Furthermore, Fisher believed that adaptive evolution proceeded on a global landscape with a single fitness peak as the 'ultimate objective'. This classical Darwin/Fisher model for adaptive phenotypic change through many genes with small effects is challenged by recent work on molecular and developmental genetics.

Current opinion favours the ideas proposed by Haldane, Wright, and Waddington, i.e. there is a major role for random drift and epistasis, with the potential to move a population across a 'valley' separating two 'adaptive peaks'. Wright believed that phenotypic expression was strongly dependent on both pleiotropic and epistatic effects and incorporated them in his Shifting Balance Theory. He

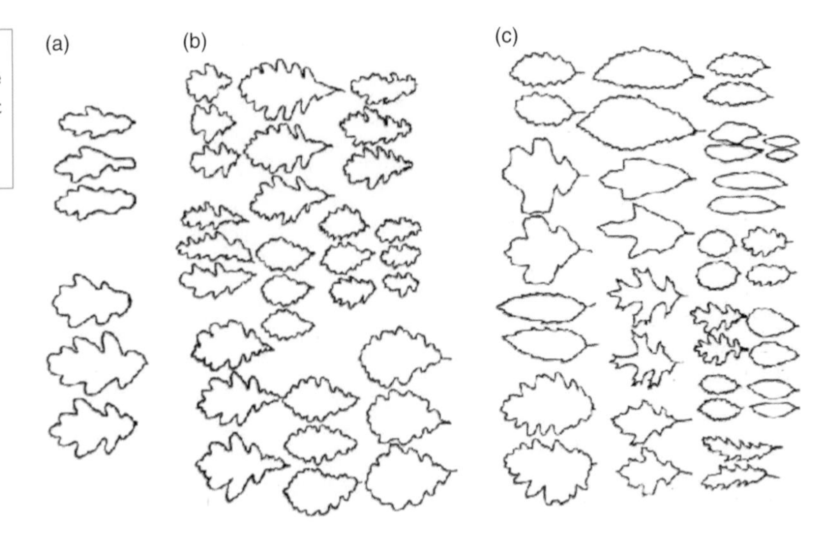

Figure 3.20. Oak leaves: (a) spring and summer leaves from the same tree; (b) leaves from different trees in a single woodland; (c) leaves from different species.

was convinced that not only were there manifold effects of single genes (pleiotropy), but that systems of interacting genes (epistasis) were pervasive. He believed these interactions were context dependent. Unfortunately, it was within the context of other genes present rather than these plus an ecological context as well. Adaptive evolution in Wright's view is local, rather than global, and that the fitness landscape permits multiple peaks of various heights. It was his Shifting Balance Theory that allowed phenotypes to move from local optima.

3.7 Variation and isolation

Genetic variation is shared with other individuals in a given population because they may share parents and are able to breed with each other (panmixis). However, even within populations individuals vary. Average differences between populations may be greater. Sets of populations may be distinct from other sets of populations. It is at this level of distinction, correlated with different ecological and geographical populations, that taxonomic species are recognised and described. For example the scale and extent of variation only differs in degree within an individual from that between species.

Leaves of a single oak tree vary in shape (Figure 3.20a), different trees of a single species also have on average a different-shaped leaf (b), while different species of oak have a different-shaped leaf (c). This last difference is used to help identify the species. The variation is of the same kind in each case, in the degree of lobing, the size of leaf and many other characters, but the source of variation is said to be different in each case. It is found within an individual, between individuals, and between species. The trees in (b) all come from the same wood. They are part of a single variable population. The mean leaf shape of different populations of a species may also differ.

It is the variation between individuals within and between populations that is the raw material of speciation. Natural selection occurs for or against particular individuals, but these are only temporary holders of part of the genetic variation of the population. It may result in changes in the population means of certain characters. Reproductive barriers between populations usually result in speciation, although such populations may remain interfertile, especially when the barriers are geographical.

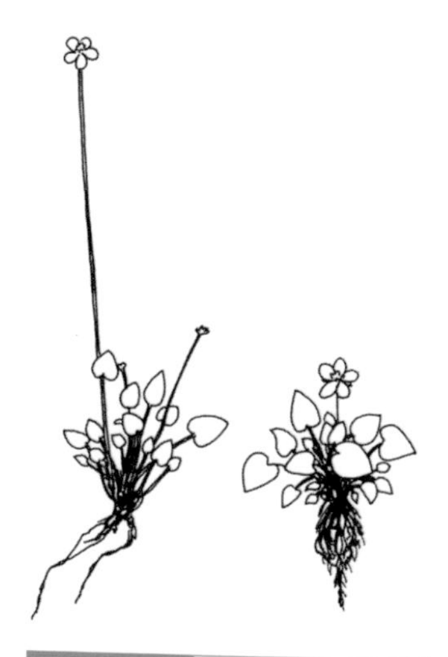

Figure 3.21. *Parnassia palustris* has different genotypes: the meadow variant that normally grows in tall grass produces its flowers on a long peduncle even if it is growing in short turf.

3.7.1 Naming diversity

The measurement of patterns of variation within and between populations gives a picture of the first stages of speciation. One variant, or race, may abruptly replace another. When these races are associated with ecologically distinct areas, they are called ecotypes or ecodemes. Sometimes these ecological variants are given formal taxonomic status. For example, Grass of Parnassus (*Parnassia palustris*), has two varieties/ecotypes in Britain: var. *palustris*, which grows in the tall grass of wet meadows, has a long stem, small flowers and leaves and a weakly branched rhizome; and var. *condensata*, which grows in the turf of coastal habitats, has a short stem, large flowers and leaves and an extensively branched rhizome.

Some of the complexities of trying to understand plant variation are illustrated by *Melampyrum pratense*, which displays many different kinds of variation. At least three kinds of variation have been detected in this species; geographical, ecological and local. In addition to leaf shape, there is geographical variation in a range of characters. Within a geographical region there is some clear ecological variation; for example, populations on calcareous soils have broader leaves, but there is also variation in flower colour, which is more difficult to understand. In Britain it has been possible to record some of these morphological patterns in a formal taxonomic hierarchy:

Melampyrum pratense
- subsp. *pratense* (on acid soils)
  - var. *pratense* (white or pale-yellow flower)
  - var. *hians* (golden-yellow flower)
- subsp. *commutatum* (on calcareous soils)

A range of characters distinguishes the two subspecies. A purple-lipped variant is only given the lesser rank as form *purpurea* because it is of sporadic occurrence. Most of the clinal variation is not recognised because of the overlap between regional populations and ecotypes. Much of the pattern of variation in flower colour in *Melampyrum pratense* is likely to have arisen by chance. Certainly the haphazard distribution of form *purpurea* suggests this. The more constant varieties, var. *pratense* and var. *hians*, may record the accident of a pattern of colonization of Britain shortly after the Ice Age, rather than being particularly adaptive to present-day conditions. If a population was founded by a golden-yellow individual all the plants in that population were golden yellow.

Species exist as series of isolated and semi-isolated populations. The degree of isolation between populations and the size of the populations are important factors determining the patterns of variation that arise. In the process of sexual reproduction, there is a chance sampling of the total variation in each generation. Thousands of male and female gametes are produced but only a few take part in fertilisation. Thousands of zygotes are produced but only a few grow into mature reproductive adults. Because of this sampling process gene frequencies can change by accident. This process is called genetic drift. In a small isolated population, chance effects are magnified because they are not swamped. A similar process can occur in populations that become very small for only a temporary period. In addition, marginal populations founded by a few individuals may be different from central ones. As a result of these accidental processes, populations that are very different from each other can arise without natural selection.

The pattern of variation in *Melampyrum pratense* is complicated because the patterns in Britain run counter to those in the rest of Europe. In some parts of Europe the golden-yellow var. *hians*, which is only taxonomically recognised in Britain because it occurs in pure populations, turns up sporadically, mixed with the normal pale-flowered plants. In addition the characters used to distinguish the British subspecies, subsp. *pratense* and subsp. *commutatum*, partly overlap with those used to distinguish other recurrent ecotypical variants called 'autumnal' (found in woods and scrub), 'aestival' (found in meadows), 'montane' (found in mountain pastures) and 'segetal' (found in cornfields). These ecological variants flower at different times in the year and recur in different species of *Melampyrum*, a parallelism that is a strong indication of natural selection acting in a similar way in different species.

The functional significance of character variation, even if there is a simple cline, is not easy to understand. It may not be possible to interpret the observed differences as functional ones. Are the *Melampyrum* plants growing on calcareous soils adapted to the soils or are they there by accident? How does having a broad leaf adapt them to calcareous soils? Why is a golden-yellow flower found in pure populations in Britain but turns up only sporadically in Europe? Only a beginning has been made to try to understand character variation. For example, in *Melampyrum arvense*, a difference in branching pattern results in a greater seed set for the more branched plants.

For some continuous characters the mean value of the character changes regularly along a geographical distance, a pattern which is called a cline. When the cline is expressed over a large geographical distance it is called a topocline. Alternatively, where the differences between plants are qualitative, clinal variation can still occur as the proportion of plants of each kind change. In some cases a clear correlation has been demonstrated between the changing character and an environmental variable. Commonly a climatic factor is identified as the correlating variable in a topocline. On a smaller scale the correlating environmental variable may be altitude, shade, aspect, slope or some other determinant of the microclimate or a character of the soil. These small-scale clines are often called ecological clines, or ecoclines. Such clinal patterns and other environmental correlations suggest the first stages in evolution, an adaptive shift in a character due to natural selection. However, it is an extra step from demonstrating a clinal relationship to demonstrating that the environment is actually the selective agent determining the plant characteristics.

Sometimes a rather haphazard pattern of variation is observed with neighbouring populations containing different variants or where a population appears to be polymorphic. These variations may be the result of autogamy (selfing) or agamospermy (asexual seed-setting). Agamic complexes or agamospecies are not strictly part of the same population even though they may coexist with each other. In the *Limonium binervosum* aggregate in the British Isles almost every population on different cliff headlands is statistically different from the others. The situation with *Taraxacum* reveals the impossibility of trying to give names to each of these agamospecies. In the UK, up to thirty agamospecies may occur at the same location. In these cases the adaptive significance of the variation, if any, is even harder to understand. There are circumstances in which a range of variants is selected for at a single location. There may be frequency dependent selection where the most abundant variant is selected against. It is also possible that these patterns of diversity are the result of chance events; mutation, migration and changes in population size.

3.8 Conceptualising plant form

3.8.1 What is an individual plant?

Although plants are complex organisms, we can see, by the process of abstraction, that their various so-called 'parts' display repeated relations to one another to form discernible patterns, and these patterns are central to our formation of concepts. This patterning process applies not only at the level of the external form as mentioned above, but also at the level of the arrangements of cells and tissues. We see the world of plants as discrete objects rather than as processes. Individuals too become further divided into discrete parts or characters, which are, in essence, symbolic rather than biological. These categories form the basis of all homology concepts. This has powerful implications for the way we classify plants and study their evolution and phylogeny, for we are liable to confuse the metaphor with the real thing. The questions we ask often reinforce the metaphor, with the result that we overlook those aspects of phenomena which do not fit the metaphor.

3.8.2 The classical shoot model

To describe and understand plant form one subdivides the plant body in accordance with rules that may only be partly consistent with nature. Nevertheless, the precision of the definitions we use for various abstract components of plant form must be rigorous, and there must be a consensus as to their use. The framework of developmental study in the higher plants has largely been provided by postembryonic stages. This is usually broken down for convenience into several components or approaches, none of which is totally adequate. The 'classical model', a stem with leaves arising at nodes and anatomically different roots, is the most frequently used and most popular model, perhaps because, for practical reasons, the main properties of vascular plant construction are sufficiently covered by the terms stem and leaf and their positional relations. The classical model embraces categorical or typological concepts since its basic units of construction such as stem or leaf, modules or metamers, represent organs with mutually exclusive sets of characters with seemingly precise positional relationships.

Looking at the evolution of stems in early land plants the classical model breaks down. In the earliest plants such as *Cooksonia* there were no roots and the aerial stem was only differentiated from the terrestrial axis by its orientation. However, even by the Early Mid-Devonian, plants that can be described in terms of the classical model were emerging. Lateral branches and leaves arising from a main stem and specialised terrestrial horizontal axes were present for example in *Asteroxylon*, although even at this stage terrestrial axes lacked some of the features we now associate with roots. The terrestrial axis in *Asteroxylon* arose and multiplied by branching and not by endogenous origin.

Root systems are very difficult to dissect into modules, metamers, or parts that can be named. Alternative ways of conceptualising plant construction might be justified because they present other perspectives of plant form, or they may have heuristic value.

Figure 3.22. Fern crozier.

3.8.3 Leaf form

Classic morphology still provides the conceptual framework for most phytomorphological investigations and for systematics. The logic is Aristotelian; any structure must be either one kind or another, and diversity of form is reduced to mutually exclusive categories. One only has to consider the striking diversity of leaves (Figure 3.23). Some have a clasping or sheathing leaf base or petiole. In the family Polygonaceae, the leaf sheath is extended into a kind of collar called an ochrea. Grass leaves have two auricles and a ligule at the junction between the leaf sheath and lamina. Many leaves have a stipule or pair of stipules arising below the leaf. Stipules are flaps or projections, sometimes leaf-like, which may serve to protect the developing bud and leaf primordium. Many aspects of a leaf vary, including the margin, the texture, thickness, hairiness, glandulosity and colour. Leaves may be modified to form scales, bracts, sepals, petals, thorns, spines, tendrils, flasks, storage organs, and root-like structures. The last occur in the water fern *Salvinia*. The leaves of carnivorous plants

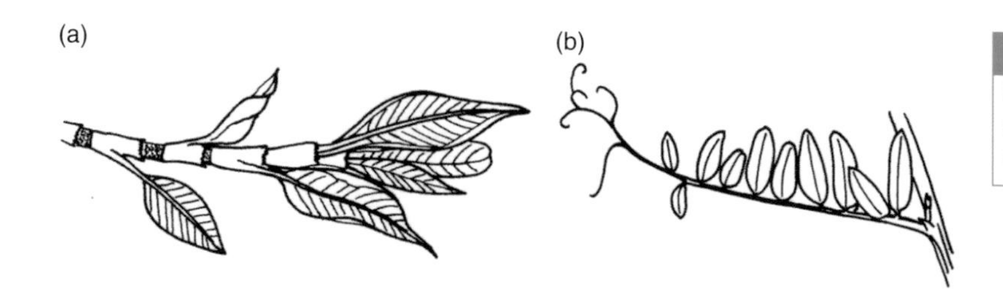

Figure 3.23. Diverse leaves: (a) leaves in the Polygonaceae have a sheathing base; (b) pinnate and tendrillate leaves in the Fabaceae.

are particularly diverse. Leaves vary in size from tiny scales a few millimetres in size to the huge compound leaves of some palms and ferns.

Compound leaves are divided pinnately or palmately into leaflets or segments. The segments of a pinnate leaf (pinnae) may be further divided (into pinnules) and subdivided up to five times. The midrib of a compound leaf is called the rachis. Each segment of a compound leaf arises from a separate primordium. This can be seen very clearly in the unfolding fern frond with its tightly packed primordia, in the fern crozier. Some plants have pseudo-compound leaves where leaf segments arise by the splitting along predetermined lines of weakness of a single lamina. Pseudo-compound leaves are found in banana, palms, and the ubiquitous houseplant *Monstera* for example. Compound leaves maximise the photosynthetic area while releasing the potential mechanical strains a single surface would suffer. Anisophylly may be present with the leaves held in a lateral position are larger than those that are either above or below the shoot.

Plant morphology, including morphogenesis, remains relevant to practically all disciplines of plant biology such as molecular genetics, physiology, ecology, evolutionary biology and systematics . . . Most commonly, morphology is equated with classical morphology and its conceptual framework . . . plants . . . are reduced to the mutually exclusive categories of root, stem (caulome) and leaf (phyllome). This ignores the fact that plant morphology has undergone fundamental conceptual, theoretical and philosophical innovation in recent times . . . If, for example, plant diversity and evolution are seen as a dynamic continuum, then compound leaves can be seen as intermediate between simple leaves and whole shoots. Recent results in molecular genetics support this view.
From R. Sattler and R. Rutishauser, 1997

As with the classical model of shoot architecture, study of early plants in the fossil record challenges our preconceptions of the nature of leaves. The earliest plants were leafless. It is clear that leaves have originated in several ways. Bower suggested that they evolved as enations from spines that became vascularised.

The telome theory of Zimmermann suggests microphylls might have evolved by reduction of a non-fertile lateral branch system, and megaphylls by the planation and webbing of such a lateral branch system.

The foliage of a late-Devonian tree such as *Archaeopteris* is a multi-branched shoot system clothed in small wedge shaped microphylls. The development of the fern crozier harks back to the last of these origins but some fern-like Devonian plants such as *Eospermopteris* did not have the finest division of webbed pinnules to be seen in modern ferns (Figure 3.25).

Almost all living land plants either have leaves or are leafless, and have evolved from leafy plants by secondary loss. The latter have a photosynthetic stem or, in the case of some epiphytic orchids, photosynthetic roots. Leaf-like structures, specialised appendages for the reception of light have evolved from different kinds of organs. In some species there are structures called cladodes, which look and function

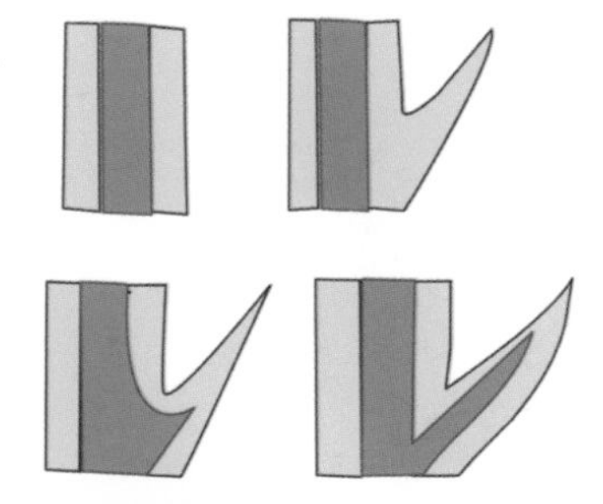

Figure 3.24. The origin of leaves as enations of the stem – microphylls.

Figure 3.25. Leaves of plants from the Devonian Period.

Figure 3.26. The origin of leaves as flattened lateral branch systems – megaphylls.

Figure 3.27. Phyllodes in *Acacia*.

like leaves but are actually modified stems or branch system. The degree of resemblance to true leaves varies, and can be remarkably close.

In some cases, as in *Phyllonoma* and *Ruscus*, it is difficult to be sure about the nature of the 'leaf' until it produces a flower or fruit on its margin or surface, something a true leaf never does. But what is a true leaf? Leaves are very diverse in structure and development. Fern fronds are leaves that bear the reproductive sporangia. They have been described as megaphylls and are not considered homologous to the leaves of bryophytes or many pteridophytes such as lycopods and selaginellas.

The leaves of mosses and liverworts are also often called microphylls but they are not homologous either to the microphylls of pteridophytes or even to each other. They share the characteristic of typically arising in three ranks from a tetrahedral apical meristem cell. The existence of a group of thalloid liverworts (Fossombroniaceae) with the thallus lobed and folded into leaf-like structures has suggested a mode of origin of liverwort leaves from a thallus or vice versa. The leaves of liverworts lack the central thickened midrib (costa) that is observable in many moss leaves, although only in some does this appear to have a vascular function. The leaf trace arising from the midrib of the leaf may only just penetrate the cortex of the stem. Even in *Polytrichum*, a moss with a relatively well-developed water conducting system, some of the leaf traces do not connect with the conducting tissue in the centre of the stem. The leaf trace may act as a kind of wick or be simply for anchorage and support of the leaf. The lamina of moss and liverwort leaves is typically very thin, usually comprising only one cell layer, but several mosses show considerable complexity in leaf structure.

'True' leaves of higher plants such as angiosperms are thought to be megaphylls, but, as we saw with the phyllomorphs of the

Gesneriaceae, there are exceptions. In *Utricularia*, it can be demonstrated that stem and leaf are not two mutually exclusive categories. The Podostemaceae also demonstrate the futility of rigid adherence to mutually exclusive categories. How plants have adapted to the challenge of energy capture, each in their unique ways by the use of planar surfaces such as microphylls, megaphylls, phyllomorphs, cladodes or phyllodes, is perhaps one of the greatest manifestations of their dance through morphospace. Nevertheless, the diversity of leaf shape exhibited by plants also provides an illustration of the limitations of the dance. There is a bewildering range of forms but particular shapes have arisen again and again. The forms are not endless. There are rules controlling the dance, all of which we do not yet understand, but some involve extrinsic factors, adaptation to the physical and biotic environment, and also, probably, to intrinsic factors, laws of development arising from the relatively few ways they can be arranged on a stem.

Figure 3.28. Cladodes of (a) *Ruscus*; (b) *Phyllonoma*.

3.8.4 Phyllotaxis

Phyllotaxis is more useful as a means of describing axial development than as an explanation of organogenesis. There is nothing inherently mystical or mathematical about this process. Phyllotaxis, by the constraints of geometry, symmetry and space-filling is by default often a spiral or helix. This patterning is particularly apparent in plants with compressed internodes (i.e. high-order phyllotaxis) such as *Crassula* spp., pineapples, pine cones, etc., and in the arrangement of other parts of plant such as the tubercles of cacti (Figure 3.30). It is erroneous to think of this pattern as the result of an intrinsic genetic or generative spiralling, although in some plants there is a secondary twisting of the stem, but these later movements are normal in all plants.

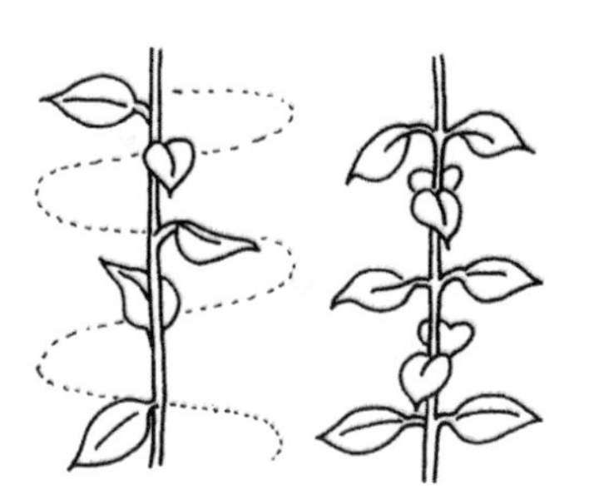

Figure 3.29. Arrangement of leaves on a stem: spiral (left) opposite and decussate (right).

Phyllotaxis is described by several rather esoteric terms that reflect the pattern produced by the regular production of leaf primordia at the shoot apex. The rate of initiation and growth of the leaf primordia, and the timing of the plastochrons, determines the phyllotaxis or arrangement of leaves on a stem. The patterns that result often form beautiful logarithimic spirals, and are called parastichies. They are, in reality optical 'illusions' produced by the juxtaposition of appendages produced at regular intervals round the perimeter of the elongating stem, a direct result of the apical processes.

There are usually two sets of intersecting parastichies. The secondary connecting parastichies between plastochrons in different turns of the spiral are usually more obvious than the primary sequence. Plants produce both right- and left-handed primary parastichies. The angle between primordia in the parastichy tends towards 137.5°, the so-called golden section or Fibonacci angle, especially where the primordia are packed closely together. The golden section produces the most even and gradual division of the apical dome,

Figure 3.30. Phyllotaxis in cacti: count the number of full turns round the genetic spiral to obtain a perfect overlap between tubercles and divide by the number of tubercles to get the phyllotaxy.

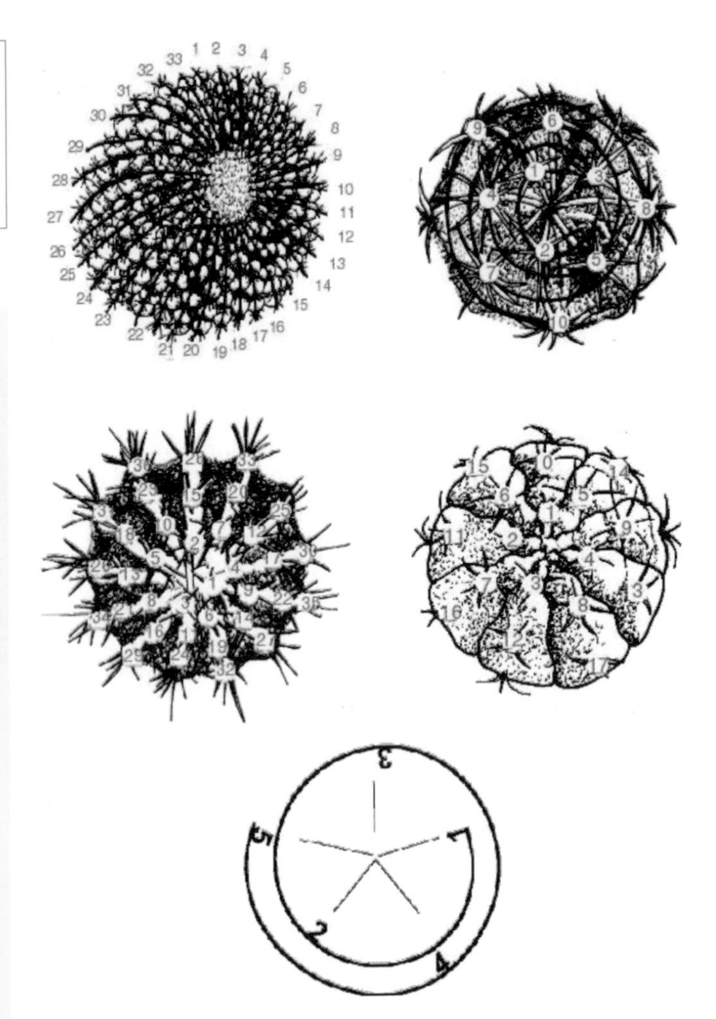

Pick up a pinecome and count the spiral rows of scales. You may find eight spirals winding up to the left and 13 spirals winding up to the right, or 13 left and 21 right spirals, or other pairs of numbers. The striking fact is that these pairs of numbers are adjacent numbers in the famous Fibonacci series: 1, 1, 2, 3, 5, 8, 13, 21 ... Here, each term is the sum of the previous two terms. The phenomenon is well known and called phyllotaxis. Many are the efforts of biologists to understand why pinecones, sunflowers, and many other plants exhibit this remarkable pattern. Organisms do the strangest things, but all these odd things need not reflect selection or historical accident. Some of the best efforts to understand phyllotaxis appeal to a form of self-organization. Paul Green, at Stanford, has argued persuasively that the Fibonacci series is just what one would expect as the simplest self-repeating pattern that can be generated by the particular growth processes in the growing tips of the tissues that form sunflowers, pinecones, and so forth. Like a snowflake and its sixfold symmetry, the pinecone and its phyllotaxis may be part of order for free.

Stuart Kauffman, *At Home in the Universe*, 1995

maximising the packing of primordia of different sizes and at different stages of development. The numbers form the Fibonacci sequence

$$1, 1, 2, 3, 5, 8, 13, 21, 34 \ldots$$

Each number is obtained by the addition of the two previous numbers, and is the basis for the description of phyllotaxis.

Lines connecting primordia lying on the same radius are called orthostichies. If there are five orthostichies and two full turns round the spiral before the first overlap then the phyllotaxis is 2/5. If there are eight orthostichies and three full turns round the spiral before an overlap then the phyllotaxis is 3/8. Other kinds are 5/13, 8/21, 13/34 and so on. In fact, the number of orthostichies relates to the size of the apical dome and rate of production of plastochrons. In a single stem, therefore, the phyllotaxis can change from 2/5 to 3/8 as growth speeds up. Note how all these kinds of phyllotaxis give approximations of the Fibonacci angle.

$$2/5 \times 360^\circ = 144^\circ \quad 3/8 \times 360^\circ = 135^\circ \quad 5/13 \times 360^\circ = 138.5^\circ$$

The significance of this is that, in a plant with leaves, it limits the overlap of leaves on the same stem: the orthostichies are lines of leaves directly overlapping and shading each other on the stem. This is an adaptationist explanation of phyllotaxy but, more fundamentally, phyllotaxis is one of the few examples where differences between mature plants can be described in simple terms of a developmental formula, in terms of morphospace.

In some plants, the axis is very short so that the leaves are crowded together in a rosette. The leaves are arranged in a complex mosaic filling light space. Orthotropic shoots combine the functions of gaining height with catching light, whereas plagiotropic shoots act primarily to occupy light space. The phyllotaxis in orthotropic shoots is usually either spiral or in pairs opposite each other but with each successive pair at 90° to the former, called a decussate arrangement. In plagiotropic shoots, the leaves may either be arranged distichously, strictly on two sides of the shoot, or have a spiral phyllotaxis, but in both cases they are arranged in one plane. If the phyllotaxis is spiral, this involves re-orientation of the leaves or the twisting of the shoot further back from its apex.

3.8.5 The architecture of plants

Goethe called architecture frozen music. Our attempts to notate the dance of plants through evolutionary morphospace are doomed to failure, because it is impossible to delineate their architecture by any simple system of terms. One system of notation that may be useful is to dissect their form by the concept of modules, and then to describe their architecture by the way the modules are arranged, for example by their methods of branching. In plants with monopodial growth the apical meristem is long lived, though it may have periods of dormancy over a harsh season. The distinction between sympodial and monopodial growth may be hard to detect because in sympodial growth the superseded apical bud may be obscured by subsequent growth. In both cases, the meristems of axillary buds can give rise to lateral branches. A stem may be thought of as a succession of internodes and the combination of a leaf with its node and underlying internode represents a basic structural unit of plant architecture that allows the description of plant development regardless of the degree of complexity. These modules are sometimes referred to as 'phytons' or 'phytomers'.

Organisms with modular construction may be further subcategorised as fragmentary (when isolated ramets become isolated from each other) or colonial (where the parts retain organic continuity). The modular construction of plants that do not fragment is usually complex (as in trees) because a high degree of physiological interaction between different modules is possible.

The modular construction of plants breaks down to a degree in those trees where there is a clear distinction between trunk and branches. The trunk is a set of metamers so modified it is impossible to identify them separately. However, the modular construction is still obvious in the terminal parts of the branch system. The modular construction is also difficult to observe in some highly modified forms and in thalloid plants. Nevertheless, the value of describing plant form and growth in terms of a modular construction is emphasised by physiological experiments, which have shown that most of the energy that goes into producing a reproductive structure comes from the close vicinity of its own module, rather than translocated from elsewhere in the plant.

There is a distinction between modules that have terminal reproductive structures (hapaxanthic) or lateral reproductive structures (pleonanthic). The former have determinate growth. They cease to grow after a while, often after producing a reproductive structure. The others have indeterminate growth. The axis is potentially immortal. A whole plant may be determinate or indeterminate or an indeterminate plant may have a mixture of determinate and indeterminate axes.

Figure 3.31. *Larix decidua* foliage showing long shoots and needles clustered on the short shoots.

A difference in branch systems that is often observed is that between long and short shoots. The short shoots have short internodes and limited growth, whereas the long shoots have longer internodes. Some conifers such as *Larix* and *Cedrus* have shoots that can switch between long- or short-shoot growth. Differently shaped leaves may be produced by each kind of shoot. Short shoots may be produced as laterals on a long shoot or, in sympodial systems, by the replacement of the forward growth of a shoot by a lateral. The terminal part of the original shoot then has very short internodes: it is the short shoot, and the branch system is therefore said to be plagiotropic by apposition.

Plants are often described, particularly by ecologists, by differences in the way they are constructed. These differences reflect differences between unitary and modular construction, but in reality there is every gradation between the two extreme types. In those that possess unitary construction there is a closed growth because of a determinate number of parts, the kinds of parts (organs) usually being numerous and diverse with a marked division of labour between them. In contrast, in those plants with modular construction there is open growth because of an indeterminate number of parts, the kinds of parts are usually few and one part may have a diversity of functions because of organisational plasticity.

Branching can take different patterns along the plant axis, and form different patterns of overall form (Figure 3.32). Hallé and Oldeman, in their studies of plant architecture, have most clearly developed the idea that plant architecture is under genetic control. According to these authors the architecture of a plant is the momentary expression of a growth plan, and the architectural model is the result of the 'genetic blueprint' that ultimately determines that growth plan. They believe that, if the form of a plant is regarded as the visible expression of its genes, it will surely be most clearly expressed in a 'standard' environment. However, it is difficult to determine just exactly what a 'standard' environment is for a given species. Can it

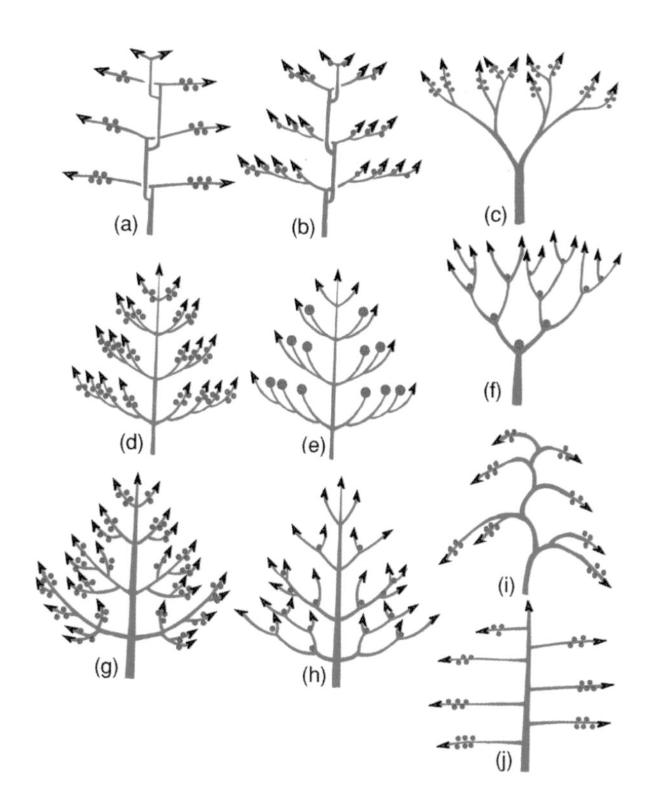

Figure 3.32. Tree architectural models: most models differ in whether the reproductive axes are lateral or terminal and whether the main axis is dichotomous (c), pseudo-dichotomous (f), sympodial (a, b, i) or monopodial (d, e, g, h, j). Names of models: (a) Nozeran's; (b) Prevost's; (c) Schoute's; (d) Aubréville's; (e) Petit's (Fagerlind's); (f) Leeuwenberg's; (g) Attim's (Rauh's); (h) Stone's (Scarrone's); (i) Troll's; and (j) Roux's (Massart's). A further distinction can be made depending whether growth is continuous or rhythmic (in parentheses).

be exactly defined and is it the same for all members of the same species?

So far, we have been discussing the architecture of the above-ground parts of the plant. There are great difficulties in classifying root systems because of their plastic development. The endogenous origin of roots allows a very flexible pattern of branching, which is part of the great plasticity of the root system as a whole. The earliest plants lacked roots. Early root systems were rather shoot-like, more tightly constrained, more modular, in their form. The freedom from constraint of modern root systems, their adaptability, is an aspect of their specialisation and it defies a simple categorisation.

3.8.7 The whole plant

Developments in ecology have imposed more sophisticated modifications on the typological approach, for example the numerous sub-units or plant construction (long shoots, short shoots, plagiotropic branch complexes, rhizome, inflorescence, cone, flower, etc.) are now interpreted in an integrated approach in which it is recognised that meristem interaction may be mediated by complex hormonal balances.

Fragmentation is implicit in empirical analysis and, in order for humans to transmit knowledge of morphology to each other, it is necessary to use a language whose vocabulary is based on the same general principles of fragmentation and reduction. Botany today remains reductionist and analytical. The various genes, cells, tissues and organs are usually studied in isolation and organisms have all but disappeared as fundamental entities in modern biology. While it cannot be denied that this methodology has been highly successful, it has limited application when we try to understand the dynamics of cybernetic living systems such as individual plants, populations and ecosystems, and it may not be sufficient for a more complete understanding of reality. The world in which we find ourselves is not an inert or mechanical object but a living, open dynamic landscape.

Perhaps the most logical way to conceptualise plants within their environment is to view them within a holarchy, starting with ecosystems and descending through a series of further holarchies, each of which contains a nested reticulum of complexity levels or holons. At the level of the individual plant (which is not always recognisable as such) the network embraces, in descending order, the organs, tissues, cells, organelles, molecules, etc. We stress the integration of plant parts, such as cells, tissues or organs, rather than treat them in isolation. Plant morphology therefore cannot be divorced from plant physiology, but are two dynamic interacting components of the same living phenomenon. The whole plant, in turn, cannot be divorced from its ecology. We need to be able to think in four dimensions to understand the multi-layers of complexities within natural systems, whether it be at the level of the community or at the level of the cell. The central concept of a holarchic organisation is the set of relationships that determine or characterise the system. For some botanists, organic form is synonymous with process.

Organisms are better viewed as a continuous flowing process of unfolding which encompasses the entire unbroken movement from fertilisation to death, a movement which initiates similar movements such as reproduction.

Rolf Sattler

The organism has no static qualities or properties. It is a complex process of flow, not a thing. A categorisation of processes may be as important as a categorisation of parts. Continuum or process morphology acknowledges gradations between typical structures. Homology then becomes a matter of degree. By 'form' we mean the totality of morphological expression found in plants, not just external appearance but anatomy as well. Such expression surely has greater meaning within a general theory of biological evolution that embodies change over time.

Plant morphology can be seen as a discipline that can be carried in the direction of population biology to the extent to which the plant can be considered a 'metapopulation' of either meristems or some other developmental or structural unit, and in the direction of ecology to the extent to which shapes are adaptive. Because process morphology may be approached at various nested complexity levels in a continuum, from molecules to ecological levels of the plant community, it is a more appropriate method for the study of evolution and phylogeny than typology, but applying it may be more difficult.

Despite the efforts of Hay, Mabberley, Jeune, Sattler and Rutishauser towards a more dynamic concept of 'continuum' or process morphology with degrees of homology, much of descriptive plant morphology is still based on a limited set of empirically based rules. Although their ideas and efforts are highly imaginative and provide a better way of seeing plant morphology as a dynamic flowing process, the greatest strength of their arguments is that it highlights a more plausible ontological theory of plant form. Such a theory also comes very close to being metaphysical and is in danger of parting company with science altogether, which may or may not be desirable, depending on one's point of view.

Some biologists see a partial validity in both a mechanistic (reductionist) and an organistic (holistic) approach, a reciprocal illumination between different hierarchical levels of evidence, between pattern and process. Both approaches become facets of the same problem when we study evolution. In descriptive studies of inflorescences, the typological approach has achieved a very high degree of sophistication. Sattler has suggested that different views of the universe are not necessarily in conflict with one another and that they may even be complementary. The acceptance of a complementary relationship between two (or more) apparently contradictory views might be the first step towards the synthesis of a new model.

For practical investigation of morphology and phylogeny reconstruction, most botanists still work with metaphors or concepts that view morphological characters as 'frozen in time', as static 'slices' of a continuous process. Development in plants is best studied using time-lapse photography. With continued refinement of computer programs such as 'Virtual Plant' it will eventually be possible to gain a better perspective of process morphology. The whole question of the ontological status of organisms and how we deal with it in our developmental, phylogenetic and systematic studies, is ripe for further investigation and debate. In the following chapters we will examine morphospace in three distinct ways: the adaptation of plants in sex, multiplication and dispersal; the relationship of plants to their environment; and the phylogeny of plants, tracing the path of plant lineages and ordering them.

Further reading for Chapter 3

Arber, A. *The Natural Philosophy of Plant Form* (Cambridge: Cambridge University Press, 1950).

Bateson, W. *Materials for the Study of Variation* (London: Macmillan, 1894; reprinted Baltimore: Johns Hopkins University Press, 1992).

Briggs, B. and Walters S. M. *Plant Variation and Evolution*, 3rd edition (Cambridge: Cambridge University Press, 1997).

Goodwin, B. C. A structuralist research programme in developmental biology. In *Dynamic Structures in Biology*, ed. B. Goodwin, A. Sibatani and G. Webster (Edinburgh: Edinburgh University Press, 1989).

How the Leopard Changed its Spots (London: Weidenfeld and Nicolson, 1994).

Gould, S. J. *Ontogeny and Phylogeny* (Cambridge, MA: Belknap Press, 1977).

Grant, V. *Organismic Evolution* (San Francisco: Freeman, 1977).

Hallé, F. Oldeman, R. A. A. and Tomlinson, P. B. *Tropical Trees and Forests: An Architectural Analysis* (Berlin: Springer-Verlag, 1978).

Hay, A. and Mabberley, D. J. On perception of plant morphology: some implications for phylogeny. In *Shape and Form in Plants and Fungi*, ed. D. S. Ingram and A. Hudson, Linnean Society Symposium No. 16a, London: Academic Press, 1994), pp. 101–117.

Jean, R. V. *Phyllotaxis: A Systematic Study of Plant Morphogenesis* (Cambridge: Cambridge University Press, 1994).

Jong, K. and Burtt, B. L. The evolution of morphological novelty exemplified in the growth patterns of some Gesneriaceae. *New Phytol.*, **75**, 297–311 (1975).

Kauffman, S. *At Home In The Universe* (Oxford: Oxford University Press, 1995), p. 151.

Levin, D. A. *The Origin, Expansion and Demise of Plant Species* (Oxford: Oxford University Press, 2000).

Lewontin, R. *The Triple Helix. Gene, Organism, and Environment* (Cambridge, MA: Harvard University Press, 2000).

Müller, G. B. and Newman, S. A. (eds) *Origination of Organismal Form: The Forgotten Cause in Evolutionary Theory* (Cambridge, MA: MIT Press, 2003), pp. 3–10.

Niklas, K. J. *Plant Allometry: The Scaling of Form and Process* (Chicago: The University of Chicago Press, 1994).

Sattler, R. *Biophilosophy – Analytic and Holistic Perspectives* (Berlin: Springer, 1986).

Homeosis in plants. *Am. J. Botany*, **75**, 1606–1617 (1988).

Towards a more dynamic plant morphology. *Acta Biotheor.*, **38**, 303–315 (1990).

Process morphology: structural dynamics in development and evolution. *Can. J. Botany*, **70**, 708–714 (1992).

Homology, homeosis and process morphology in plants. In *Homology: The Hierarchical Basis of Comparative Biology*, ed. B. K. Hall (New York: Academic Press, 1994), pp. 423–475.

Sattler, R. and Rutishauser, R. The fundamental relevance of morphology and morphogenesis to plant research. *Ann. Botany*, **80**, 571–582 (1997).

Schlichting, C. D. and Pigliucci, M. *Phenotypic Evolution: A Reaction Norm Perspective* (Sunderland, MA: Sinauer Associates, Inc., 1998).

Stebbins, G. L. *Flowering Plants: Evolution above the Species Level* (Cambridge, MA: The Belknap Press of Harvard University Press, 1974).

Takhtajan, A. L. Patterns of ontogenetic alterations in the evolution of higher plants. *Phytomorphology*, **22**, 164 (1973).

Webster, G. William Bateson and the Science of Form. Essay (pp. xxix–lix). In Bateson, W. *Materials for the Study of Variation* (London: Macmillan, 1894; reprinted Baltimore: Johns Hopkins University Press, 1992).

Webster, G. and Goodwin, B. C. *Form and Transformation Generative and Relational Principles in Biology* (Cambridge: Cambridge University Press, 1996).

White, J. The plant as a metapopulation. *Ann. Rev. Ecol. Syst.*, **10**, 109–145 (1979).

Chapter 4

Sex, multiplication and dispersal

flowers are . . . constrained by history: they have a phylogenetic burden. Many levels of their evolutionary history have imprinted their marks on them. They cannot escape them. One can often see traces of earlier phylogenetic (sub)strata in the structure of flowers. One should not forget that each flower, however harmoniously functioning at any time, is a mixture of features that are of different evolutionary ages. Different historical levels are incorporated and work together. The notion of 'evolutionary tinkering' is especially apt for flowers.

P. K. Endress, 1994

4.1 The yin and yang of reproduction

The trio of sex, multiplication and dispersal are the pillars of the evolution of life on Earth. Multiplication comes through the processes of sex and dispersal but requires neither. Many organisms reproduce only, or mainly, asexually and have no special mechanisms for dispersal. For organisms living in water, dispersal was never much of a problem. Water currents dispersed them haphazardly. But dispersal on the land is a formidable challenge, a cliff that the first land colonists had to scale. The landscape was only patchily friendly and even if a foothold could be established the leap from one damp patch to another was a huge challenge. The plants that first met this challenge, and the structures they used to do it, their spores, provide the first record of complex terrestrial life. In colonising the land, plants transformed it making it an easier place for other colonists.

Evolution has occurred without sex but sex was the triumph of eukaryotic organisms, vastly increasing their evolutionary potential. Sex and dispersal are yin and yang, passive and active, incomplete without each other. Each has the potential to multiply the organism. Sex creates and maintains genetic variability among individuals, but without dispersal it is pointless as the organisms mate with themselves or near relatives limiting the expression of their genetic differences. While sex creates different individuals, dispersal tries them out in different situations where they may be adapted. In this way the rondo of sex and dispersal has been the engine of evolution, an engine that has shifted gear with every new specialisation.

Plants have an alternation of generations, a heteromorphic haplodiplobiontic life cycle (see Figure 1.23), where the different phases of the reproductive cycle are adapted for sex or for dispersal. Different lineages of plants each have a distinct pattern of reproduction that represents a particular solution of how to maintain genetic diversity and how to multiply and establish the plant. Key events have been the evolution of spores, heterospory, seeds, flowers and fruits. It was the evolution of flowers, that soft machine for sex and dispersal, that unleashed a bewildering potential for diversity, a rococo of seemingly endless forms.

The structures described are clearly fitted for their purpose, adaptive, promoting sex, multiplication and dispersal; the physical rules of nature do sometimes channel adaptation in certain directions. A good example of this is how spore size can be understood in terms of Reynold's Law of buoyancy described below. Plants having a similar pollination mechanism may become adapted in a similar way and evolve convergently, but this does not answer why they share any *one* particular form. There are many different ways a flower might be adapted for pollination by a bee, so why are there particular norms or archetypes that are repeated again and again in different lineages?

It is not sensible only to dissect the complex form of plants, to reduce the integrated form of a plant to a set of unitary adaptations, as if a living organism was a machine in which each part can be labelled with a particular function. An equally valid part of the story of plant reproductive diversity is the limitations of form that development imposes, and how, when those boundaries are broken, a new landscape of form becomes available for evolutionary exploration, permitting new levels of reproductive complexity and sophistication.

4.2 Sex

4.2.1 Sex organs and cells

The gametophyte bears the sex organs, the female archegonia and male antheridia. Each archegonium contains a single egg in a multicellular bottle-shaped container. The egg and the archegonium arise from the same cell but thereafter their development is separate.

Plants have sometimes been called the 'Archegoniatae' because the female organs in algae, called oogonia, are remarkably different from archegonia. Oogonia differ from archegonia in having no protective layer of cells and multiple eggs are produced by subdivision of the oogonium. Although normally the oogonium is naked in some algae they may be produced in cavities in the thallus. In some charophyte algae a kind of surrounding sterile coat is present, but it is produced by the growth of surrounding filaments with a different anatomical origin to the oogonium. This seemingly obscure technical difference is actually highly significant because it is one of the bits of evidence that indicates that plants may have had a single common ancestor.

Archegonia are strikingly similar throughout plants. Each is made up of two parts; the lower chamber, called the venter, and the neck. Archegonia differ mainly in the number of cells of which they are composed, the relative size of the neck and venter and in whether they develop from a superficial or deeper cell. An important stage in the development of the archegonium is the separation of the wall from the interior cells, freeing its differentiation. The lowest of the interior cells becomes the single egg. Above it, there is a ventral canal cell and several neck canal cells. The canal cells disintegrate at maturity to create the canal via which the sperm enters the archegonium. Archegonia produce a chemical sperm attractant that in ferns can be experimentally replaced by malic acid.

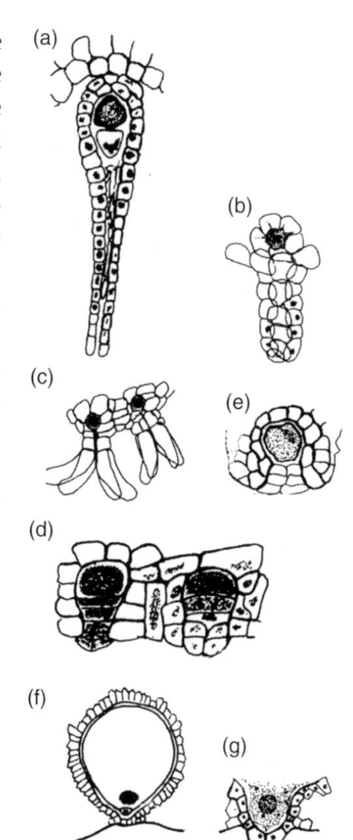

Figure 4.1. Archegonia: (a) liverwort – *Pellia*; (b) fern – *Dryopteris*; (c) horsetail – *Equisetum*; (d) clubmoss – *Selaginella*; (e) *Isoetes*; (f) *Ginko*; (g) *Dioon* with detail of neck at mature stage.

In seed plants there is an evolutionary trend in the reduction of the archegonium. A jacket layer of cells surrounds the central cells. The number of central cells varies. In *Ginkgo* a large egg cell, a ventral canal cell and four neck cells but no neck canal cells are present. In cycads the pattern is similar but a cell wall is not produced between the egg nucleus and a ventral canal nucleus that is short lived. Conifers may follow either pattern but frequently have fewer neck cells than *Ginkgo*. In contrast, in *Ephedra* there are 40 or more neck cells. In *Gnetum*, *Welwitschia* and the flowering plants no readily identifiable archegonia are recognisable but the egg nucleus, or egg, is buried in, and surrounded by, other nuclei or cells that cannot be directly related to archegonial wall cells.

The male organ is called an antheridium. It consists of a jacket of sterile cells surrounding the developing sperm. Rather similar looking multicellular gametangia are produced by the alga *Chara*. The antheridium usually consists of a spherical or ovoid sac in which sporogenous cells give rise to many sperm. Antheridia vary greatly in size between different groups, and the number of sperm produced by each antheridium varies from hundreds, even thousands, in some bryophytes and ferns, to only four in *Isoetes*. In the seed plants there is no antheridial sac. The antheridium is reduced to only one or two cells and only two sperm or sperm nuclei are produced.

Several different kinds of sperm are found in different groups. In whiskferns, horsetails, *Isoetes*, ferns, cycads and *Ginkgo*, a multiflagellate sperm is produced with the flagella arranged in a spiral around the pointed end of the rounded cone-shaped body. In the mosses, liverworts, hornworts, clubmosses and *Selaginella* there is a biflagellate sperm. The possession of a biflagellate sperm may be a primitive feature because it is shared with many algae. In the conifers, the

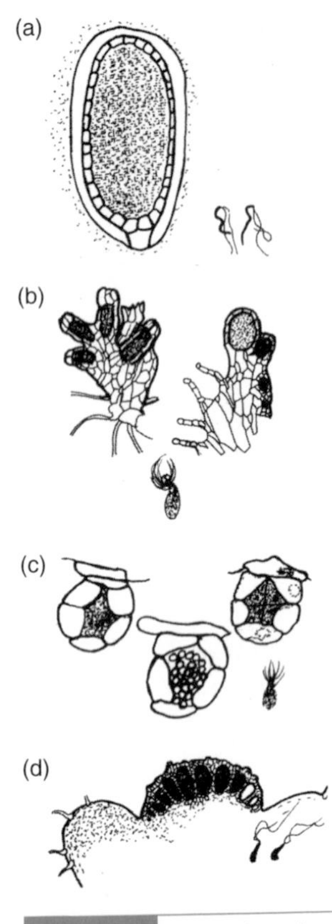

Figure 4.2. Antheridia and sperm: (a) liverwort – *Marchantia*; (b) horsetail – *Equisetum*; (c) fern – *Dryopteris*; (d) clubmoss – *Lycopodium*.

gnetophytes, and the flowering plants the sperm is non-flagellate but instead the male cells or nuclei are conveyed towards the egg in a pollen tube in a process called siphonogamy. A pollen tube is also produced in cycads and *Ginkgo* but in these groups it has a haustorial nutritive function and is not directly involved in fertilisation. Instead flagellate sperm are produced and released into a chamber above the archegonium. Flagellate sperm have very limited ability to swim but they have been observed to be splashed distances up to 60 cm by rain.

4.2.2 Sexy plants and tissues – the gametophyte

There is some evidence that the first land plants had an isomorphic haplodiplobiontic life cycle with the structure of the gametophyte and sporophyte plants identical except for the reproductive organs. Some of their proposed algal ancestors in the Charales have an isomorphic haplodiplobiontic life cycle and some primitive living plants like the whiskfern *Psilotum* have a gametophyte that is very similar to the rhizome of the sporophyte. Several gametophyte fossils have been identified: *Langiophyton mackiei*, *Lyonophyton rhyniensis*, and *Kidstonophyton discoides* from the Rhynie chert of the Devonian Period and elsewhere. These are relatively large and complex plants but different from their taller sporophyte partners. So it seems that one of the earliest adaptations of land plants was the greater specialisation of each reproductive generation. These fossil gametophytes bear their sex organs in cup-like apices such as the splash cups of mosses.

Gametophytes of all living plants are small plants or tissues. This is even true of the mosses, liverworts and hornworts where the sporophyte is dependent on the gametophyte and is normally even smaller than the small gametophyte. Gametophytes are small because of the necessity for water for sexual reproduction, as if they are amphibious, relying on a rain shower to have sex. Its no accident that the largest bryophyte gametophytes, procumbent or pendulous forms up to a metre in length, are aquatic.

The life of the moss gametophyte begins when the spore germinates to produce a juvenile filamentous stage called the protonema. This is particularly interesting because of its similarity to some kinds of filamentous green algae such as the terrestrial alga *Fritschiella*. It is possible that this is how plants originated and everything else is a baroque elaboration upon it. The filaments are usually uniseriate but the protonema can form an extensive branched axis with several kinds of filaments.

Prostrate creeping filaments are green and have either transverse cell walls (chloronema) or oblique cell walls (caulonema). The latter are identical in structure to the non-green rhizoids that are also produced. Upright filaments, in which the cells are rich in chloroplasts, have transverse cell walls. In the lantern moss, *Andreaea*, the protonema is multiseriate. Normally the adult gametophyte arises when cells in the caulonema round up and, by oblique cell divisions, produce the apical bud cell. However, in the bog moss, *Sphagnum*, the filamentous protonema soon enters a thalloid phase from which the mature gametophyte then arises.

The reduction of the juvenile protonemal phase is an evolutionary advanced feature in bryophytes. In the moss *Dicnemon* and the thalloid liverwort *Conocephalum*, and some hornworts, development of the adult gametophyte is direct from the multicellular spore. In most liverworts there is a juvenile filamentous phase but it consists of only two or three cells. In thalloid liverworts the apical cell of the protonema divides to produce a quadrant of cells, giving rise to a plate of cells in which the apical cell of a growing thallus appears. In leafy liverworts the protonema produces a clump of cells, sometimes called the sporeling, from which the mature liverwort arises.

The diverse leafy or thalloid adult gametophytes of bryophytes are described in Chapter 5. They bear archegonia and antheridia either separately or together, sometimes with organs such as sterile hairs called paraphyses or a cup-shaped ring of leaves (perichaetium – female apex, perigonium – male apex) that help trap water. Rain drops splash sperm from one plant to the next. In some complex thalloid liverworts sperm transfer is aided because antheridia and archegonia are raised up on mushroom-like stalks.

Smallness in mosses, liverworts and hornworts is a specialisation enabling a rapid life cycle to take advantage of transient but abundant water. Many can then dry out to rehydrate rapidly when water becomes available again. Many bryophyte habitats have been created by the evolution of other large plants that have provided shade or places for the bryophyte to grow as an epiphyte or epiphyll. Bryophytes dominate these niches that require smallness. It is intriguing that evolutionary reduction in the three bryophyte groups, liverworts, hornworts and mosses, which have long separate evolutionary lineages, has led to evolutionary convergence, so there is a resemblance between the leafy gametophytes of mosses and liverworts, and between the thalli of the hornworts and thalloid liverworts respectively. One subclass of mosses, the Buxbaumiidae, is very interesting because it shows a great reduction of the gametophyte.

It is this same evolutionary path (where the sporophyte is large and the gametophyte inconspicuous) that the non-bryophytes have taken. There are three main growth forms of gametophyte in lycopods, and in ferns and their allies. The gametophyte may be green and grow at the soil surface like a very simple thalloid liverwort or hornwort; indeed it is often called a prothallus. Most ferns have a gametophyte like this (Figure 4.8). Alternatively, as in many lycopods, the whiskfern and adder's-tongue fern, it is a subterranean tuber-like structure that is usually associated with a fungus (mycotrophic) and lives saprophytically. The distinction between these two types is not sharp because small amounts of chlorophyll may be found at the apex of these mainly saprophytic gametophytes if they are exposed to the light, and green thalloid gametophytes often also have a fungal symbiont. A third kind, found in all 'higher' plants, the gymnosperms and flowering plants, is so small that it develops only

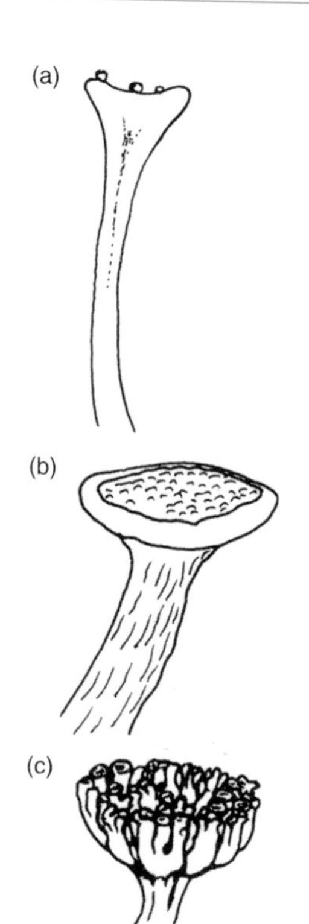

Figure 4.3. Reconstructions of gametophytes from Devonian fossils: (a) *Lyonophyton*; (b) *Kidstonophyton*; (c) *Langiophyton*.

Figure 4.4. Leafy moss gametophytes (*Polytrichum*) with the sporophyte growing up from them.

Figure 4.5. Protonema in mosses showing caulonema and multicellular chloronema with budding leafy plants.

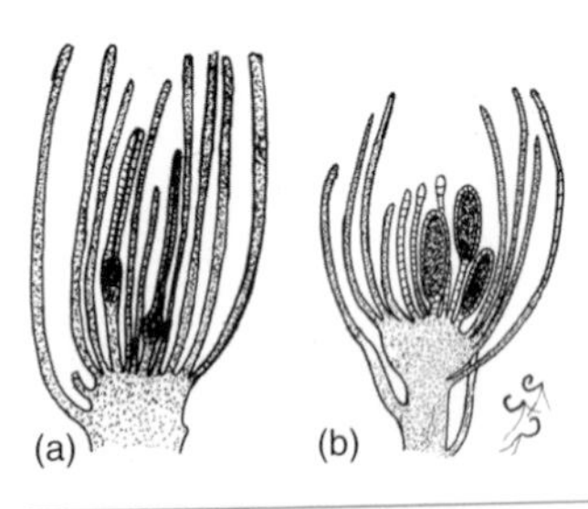

Figure 4.6. Sections through fertile apex in *Mnium*, a leafy moss, with perichaetium, paraphyses and sex organs: (a) female with archegonia; (b) male with antheridia and biflagellate sperm detailed (sperm not to the same scale).

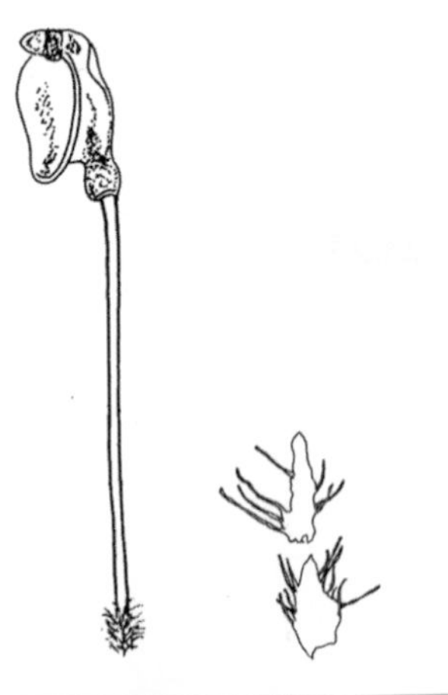

Figure 4.7. *Buxbaumia* is only obvious when the sporophyte is present. The gametophyte is tiny. The protonema produces side branches that bear perigonial or perichaetial leaves (detailed).

endosporically, i.e. within the spore wall. A tiny endosporic gametophyte is part of the story of the evolution of seeds that is told below.

Most free-living gametophytes in lycopods, ferns, and their allies bear or are capable of bearing both male and female organs. *Equisetum* is the exception in being heterothallic, producing male and bisexual prothalli that look different. The green prothalli grow on the soil surface, have rhizoids and are very small (1–8 mm). Male gametophytes are lobed with a few sterile lamellae. The other kind of gametophyte is protogynous: at first it has many sterile lobes with the archegonia situated between; later, after several weeks, female development ends and antheridial lobes are produced.

Sex in these non-seed plants is an epic carried out on a tiny scale. The sperm are tiny unicells liberated into an environment that is only temporarily wet. Every raindrop lands and explodes, throwing the sperm towards or away from their target. They can swim a little, wriggling in the film of water that coats the grains of soil or fragments of leaf litter. They scent their prize and orientate towards it but soon become exhausted. By rare chance one finds its way to the neck of an archegonium and forces its way down the neck pushing aside the debris of the neck canal cells. Finally it reaches the egg, fusing with it to form the zygote thereby initiating the next stage of the life cycle, the growth of the sporophyte.

4.3 Dispersal

One of the most important adaptations for plants colonising the land was the ability to disperse propagules in the air. Plants produce air-dispersed spores in organs called sporangia produced by the sporophyte. Sporangia differ from the spore-producing organs of algae because, early in their development, two distinct cell lines are produced; one develops to provide the sterile outer wall of the sporangium and the other, after a meiotic cell division, develops to become the haploid spores. The haplodiplobiontic heteromorphic alternation of generations can be regarded as an adaptation permitting the greater specialisation of spore-producing structures, because a large, multicellular sporophyte is potentially more complex. In most plants, the sporophyte generation is dominant. It is a relatively tall phase, and tallness helps the aerial dispersal of spores. Dispersal of spores across the land is unpredictable, and large plants can produce more spores. Sporophytes are much more varied than gametophytes. Dipoloid individuals potentially have a greater range of expression and evolutionary potential than haploid organisms.

The spores are produced in fours called tetrads that normally then break apart so that they are released individually. As a consequence of being produced in a tetrad each spore may have a characteristic trilete scar where it was attached to the others in development. The trilete scar is a useful marker for detecting the presence of fossil plant spores. Fossils from the Late Silurian and Early Devonian show a remarkable diversity of sporangia.

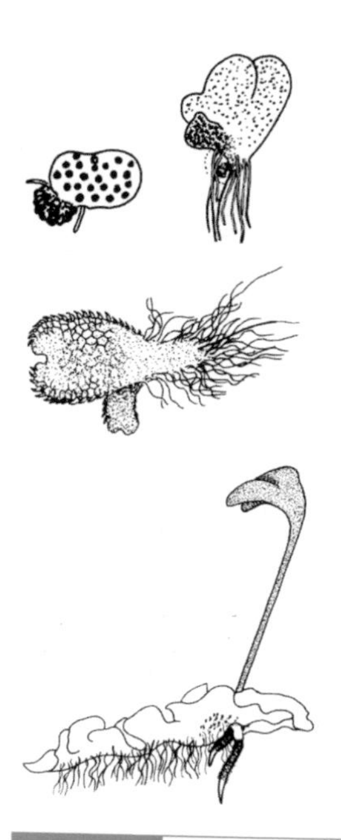

Figure 4.8. Gametophyte of the fern *Dryopteris* from germination to mature prothallus after fertilisation and growth of young sporophyte.

4.3.1 Spores, elaters and sporangia

Spores are dispersal units, types of disseminules or diaspores. They are produced in large numbers and propagate the plant, so they are also called propagules. They are protected by a spore wall that is resistant to desiccation. Spores arise directly from meiotic divisions within the sporangium, but they may undergo different degrees of development before spore release, and may be unicellular or multicellular at the time of release. Multicellular spores are released in some mosses. A defining feature of all plants is that a compound called sporopollenin protects the spores. Sporopollenin is a carotenoid-like polymer in the outer wall of each spore that protects it from desiccation and is highly resistant to deterioration except through oxidation. Consequently spores are also protected from degradation in anoxic environments like lake sediments and waterlogged peats, aiding the process of fossilisation. Sporopollenin is found independently in dinoflagellates and acritarchs.

The sporangium of vascular land plants, sometimes called polysporangiophytes because the sporangia are produced on a branched stem, is usually a simple sac like organ. The earliest were on some species of *Cooksonia* from the Late Silurian. They are multistratose, i.e. they have have a thick wall with several cell layers. This kind of sporangium is called a eusporangium, in contrast to the kind with a thin, unistratose wall, called a leptosporangium. Developmentally, eusporangia have a multicellular origin from a group of cells on the surface of the plant that divide periclinally (parallel to the surface) to give rise to an inner sporogenous layer and an outer wall layer. The wall develops by subsequent periclinal and anticlinal (at right angles to the surface) divisions. It includes an inner layer (or layers) called the tapetum that nourishes the developing spores and disintegrates as the spores mature. Eusporangia are present in all 'higher' plants except the modern ferns. Hundreds or even thousands of spores are produced in each eusporangium. They have no special adaptations aiding spore discharge but split open along a line of weakness called the stomium. *Equisetum* has a strobilus of closely packed peltate sporangiophores and the wall of the spore itself partly peels back to form a strip-like elater that remains attached to the spore and helps fluff up the spore mass.

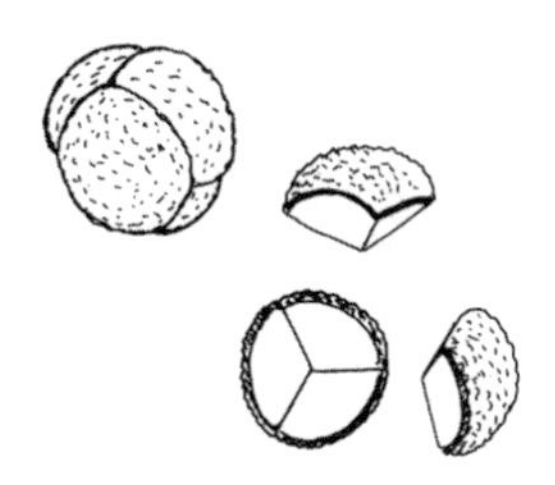

Figure 4.9. Spore tetrad and broken into monads.

Leptosporangia are present in modern ferns. A leptosporangium arises from a single superficial cell. Periclinal and oblique divisions occur successively producing an apical cell and stalk cell, a jacket cell and internal cell among others. From these a unistratose sporangium wall, a two-layered tapetum, a sporogenous tissue and a stalk arise. Leptosporangia differ at maturity from eusporangia in being small, having a single-layered wall, in being long stalked, and in containing usually 64 spores or less. *Osmunda* is a primitive leptosporangiate fern in which a large number of spores, 256 or 512 per sporangium, are

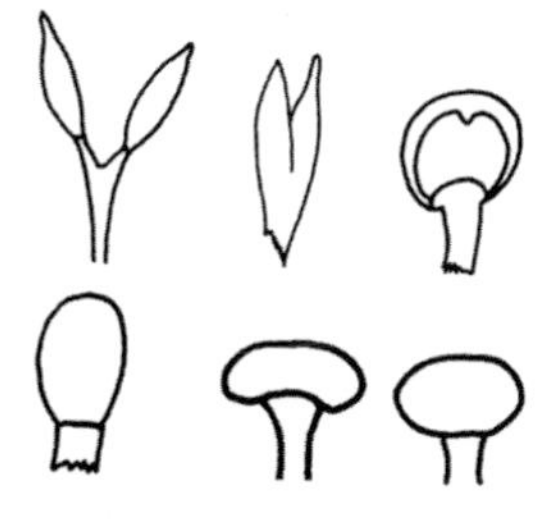

Figure 4.10. Sporangia of fossils from the Silurian and Early Devonian.

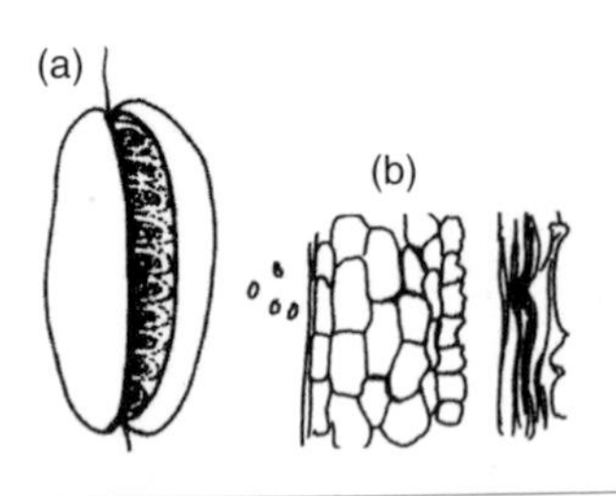

Figure 4.11. Eusporangium in the fern *Marattia*: (a) external view; (b) detail of sporangium wall showing multistratose structure before and after dehiscence.

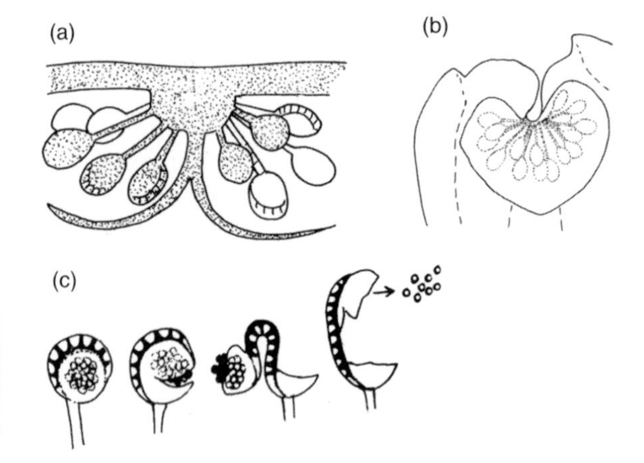

Figure 4.12. Sori showing protective flap called the indusium: (a) section through a *Dryopteris* sorus; (b) 'false indusium' of *Adiantum*; (c) Fern leptosporangium in spore release.

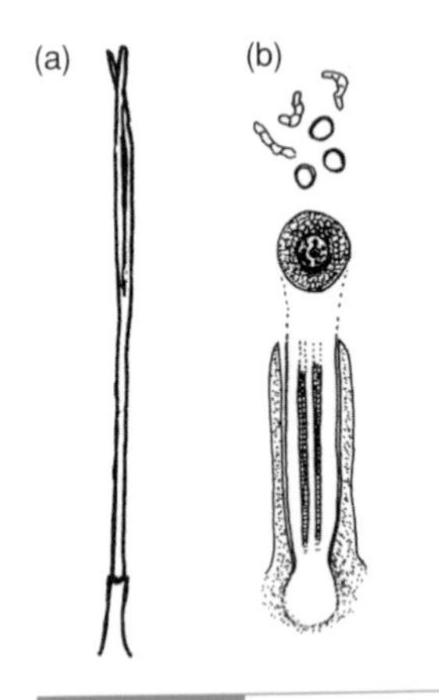

Figure 4.13. Hornwort (*Anthoceros*) sporangium: (a) external view; (b) sections showing columella, spores and pseudoelaters.

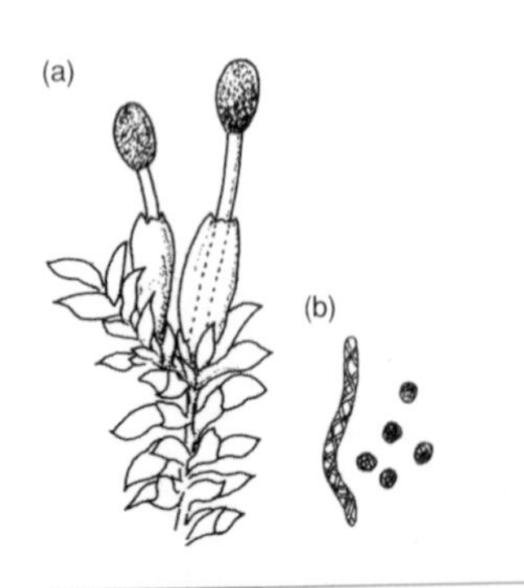

Figure 4.14. Liverwort capsule: (a) *Diplophyllum*; (b) elater and multicellular spores from *Pellia*.

produced. The leptosporangium has a ring of cells with thickened cell walls on three sides that force the sporangium to split open, flinging out the spores (Figure 4.12).

Sporangia may be stalked or sessile. Frequently they are associated with leaves called sporophylls. The sporophylls may be concentrated in a particular part of a shoot or more commonly they are aggregated into a terminal compound structure, a cone or strobilus. In many ferns the sporangia are grouped in sori (singular = sorus) that are epiphyllous on the under-surface or at the margin of a frond (Figure 4.12).

Capsules in bryophytes

The bryophytes have very different sporangia from those of the tracheophytes, in which the whole sporophyte consists of a haustorial foot buried in the gametophyte, a stalk called the seta, and a single sporangium, called the capsule.

The hornwort sporophyte at first enlarges inside the thallus protected, beneath a projection of the thallus. As it elongates to become mature the sporangium breaks out and the calyptra is left as a kind of collar or involucre. A basal meristem elongates the sporangium over an extended period. Inside the columnar sporangium there is a central sterile columella surrounded by sporogenous tissue. Spores and multicellular sterile elaters mature from the apex downward as the sporangium elongates. The sporangium splits at its apex, in the region where the spores and elaters are mature, to release the spores. The sporangium of the hornworts is similar in some respects to the columnar sporangium of the fossil *Horneophyton* of the Devonian Rhynie chert, though this has a lobed structure.

In liverworts the archegonium swells to accommodate the developing gametophyte. Eventually the seta elongates breaking the archegonium and pushing the capsule out. The remains of the archegonium are left as a tattered collar around the base of the seta. In some liverworts like *Marchantia* the capsules are elevated not just on the seta but, because the whole sporophyte arises from a stalked platform that is part of the gametophyte and is called the archegoniophore. In most liverworts the capsule is spherical or ovoid. The wall of the capsule may be unistratose, or multistratose. Accompanying the spores in the liverworts there are sterile cells called elaters that promote dispersal (see Section 4.5.1).

The capsule splits longitudinally, either regularly or irregularly, along lines of weakness, into valves, although in some an apical opening is produced. In *Asterella* a cap called the operculum falls off. In *Cyathodium*, tooth-like structures are present around an apical opening. In some liverworts like *Monoclea* the capsule is elongated and the capsule splits by a single longitudinal slit. In *Riccia* there is no seta and no elaters, and the spores are released by the capsule disintegrating.

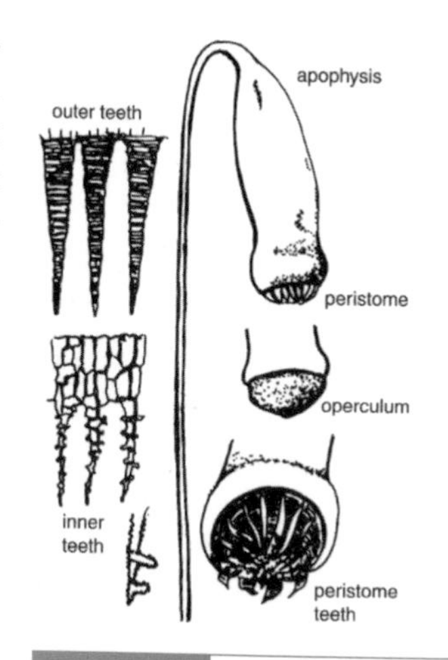

Figure 4.15. Moss capsule.

In mosses the seta elongates before the capsule has matured. The archegonium splits open in such a way that the top part, called the calyptra, is carried up as a kind of a hat over the apex of the capsule. Here it continues to influence the development of the capsule until it shrivels and falls off. Moss capsules are complex (Figure 4.4). Often they have a basal sterile photosynthetic region called the apophysis. Here there are stomata connecting with intercellular air spaces between the photosynthetic cells. The apophysis can make a considerable contribution to the energy budget of the sporophyte. The sporogenous part of the capsule has a central core of non-sporogenous tissue called the columella, which runs through the centre of the spore mass. At the apex of the capsule a ring called the annulus is present forming an obvious junction with the operculum. Below the operculum, in the class Bryopsida, there is a simple or complex peristome. *Sphagnum* lacks a peristome but has an operculum that is exploded off to release spores.

Figure 4.16. Spores with elaters from *Equisetum*.

Pollen, microsporangia and stamens

The dispersed spores of seed plants are called microspores or pollen. Pollen is produced in sporangia (microsporangia) called pollen sacs. In most gymnosperms the pollen sacs are grouped together on the abaxial surface of special scale leaves called microsporophylls. In some cycads such as *Cycas* there are over a thousand pollen sacs on each microsporophyll. In the cycad *Zamia* there are only five or six. Similarly, in the conifers each sporophyll may bear many pollen sacs, as in the Araucariaceae, but it is normal to have between two and nine pollen sacs to each scale: two in *Pinus*, four in *Cephalotaxus*. In both cycads and most conifers, the microsporophylls are arranged spirally in a male cone or strobilus (*Pinus*) (Figure 4.19) or on a pendulous catkin as in *Podocarpus*. In *Taxodium* the male cones are arranged in a pendulous compound microstrobilus. A number of other arrangements are found. The Taxaceae (*Taxus*) are peculiar among the conifers in having a stalked structure, a peltate microsporangiophore, bearing

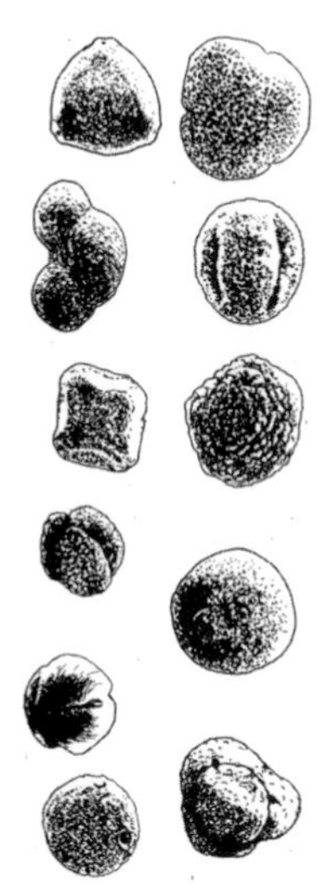

Figure 4.17. Spores and pollen.

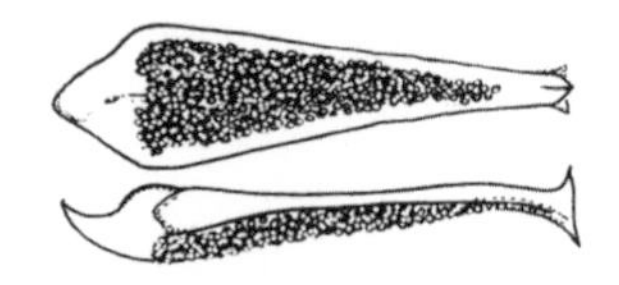

Figure 4.18. *Cycas* microsporophyll.

Figure 4.19. *Pinus* male cones: (a) clusters of male cones (notice arrangement of needles on short shoots); (b) section showing microsporangia.

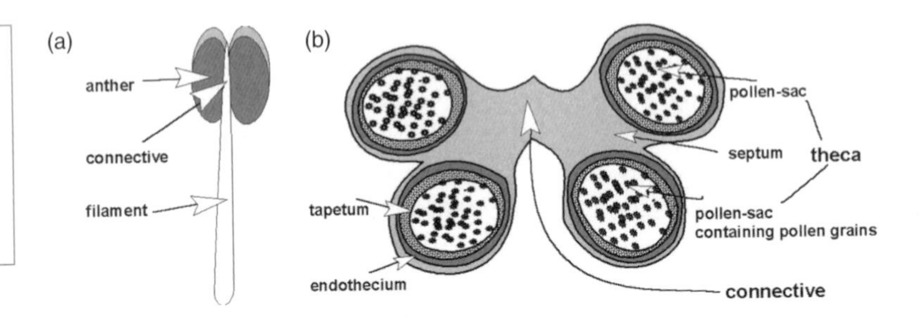

Figure 4.20. Stamen in a flowering plant. At maturity the wall between the pollen sacs may break down. Pollen sacs are joined in pairs by the connective to the filament. (a) whole stamen; (b) section through the anthers.

6–8 pollen sacs. The microsporophylls of the Taxodiaceae (including *Taxodium* and *Sequoia*) are somewhat similar to those of *Taxus* although their derivation from the more usual microsporophyll is more obvious. In *Ginkgo* the microsporangia are also borne on microsporangiophores arranged in a catkin. The microsporophyll of flowering plants is called a stamen (Figure 4.20).

4.3.2 Cross-fertilisation and establishment

Spores are units of dispersal and the gametophytes that arise from them carry out sexual reproduction. Arising from a single spore, a self-fertilising bisexual gametophyte can found a new colony. The early land plants probably were all homosporous. Although by no means certain, this probably meant that these spores germinated into bisexual gametophytes. In the evolution of plant reproduction there was a functional shift from spores as the units of dispersal to being units that promote sexual reproduction, with the function of dispersal carried out by seeds instead. The evolution of heterospory and endosporic development were intermediate stages in this revolution in reproduction, stages that can be observed in living land plants and which have also been discovered in fossils.

Heterothally

The horsetail *Equisetum* is homosporous but heterothallic, partially herkogamous and dichogamous. It produces spores of only one size but some spores germinate to produce a large gametophyte that is bisexual but protogynous. Other spores produce small and entirely male gametophytes. This spatial and temporal separation of archegonia and antheridia promotes cross-fertilisation.

Anisospory

Small spores are more readily dispersed but larger spores carry more energy reserves and more effectively establish the gametophyte. Sporangia containing spores of different types have been found in fossil plants such as *Barinophyton* but the smaller spores have sometimes been interpreted as abortive spores. However, ultrastructural comparisons of the spores of the fossil *Barinophyton*, which fall in two size classes, 30–50 μm and 650–900 μm, indicate that they were both functional. Anisospory has been likened to the variation in seed size within some species of flowering plants where it results from competition between seeds for the same resources in development within the ovary. Perhaps it has adaptive value where the environment is very heterogeneous and two different reproductive strategies, dispersal (small spores) and establishment (large spores) are favoured.

Anisospory gradually increased in the Devonian but it is very uncommon in living land plants, where it is confined to a few kinds of mosses. In these it is associated with heterothallism. Although a large proportion of bryophytes are dioecious (over 60% of the Bryidae) most are homothallic. In a few, however, dioecy is accompanied by striking anisospory and heterothallism. In *Macromitrium* spores of two sizes are produced. Two of the spores of each tetrad remain small giving rise to male plants, while the larger spores produce female plants. The male plants are dwarf and grow epiphytically on the female plants. Under the chemical influence of the female host the males consist only of an involucre (perigonium) surrounding the antheridia, and a few rhizoids. In this peculiar genus, males, when growing separately, may be either dwarf or identical vegetatively to the female.

Heterospory

By about in the Middle Devonian, 386 million years ago, plants such as *Chauleria* clearly had two sizes of spore produced in different sporangia: megaspores produced in megasporangia and microspores in microsporangia. Heterospory evolved several times in distinct lineages in Devonian plants. This can be traced from homosporous to strongly heterosporous forms in the calamites (fossil relatives of *Equisetum*). For example, *Calamostachys binneyana* is apparently homosporous, although this fossil may be the microsporangiate cone of a heterosporous species where no megasporangiate cones have yet been discovered. *C. americana* is certainly heterosporous with megasporangia containing megaspores three times as large as the microspores in the microsporangia borne on the same cone. In *Calamocarpon* the megaspores are relatively larger again and each megasporangium contains only one spore.

A short glossary of reproductive terms

Homospory = spores of the same size.
Anisospory = spores of different sizes in the same sporangium.
Heterospory = spores of different sizes, and normally different sexes, produced in different sporangia.
Heterothally = gametophytes of different sizes, and normally different sexes.
Endosporic development = development of the gametophyte inside the spore.
Herkogamy = spatial separation of the male and female organs of reproduction.
Dioecy = separate male and female plants.
Monoecy = bisexual individuals.
Dicliny = separate male and female reproductive axes on a single plant.
Dichogamy = temporal separation of the male and female organs of reproduction.
Protogyny = female organs maturing first.
Protandry = male organs maturing first.

Figure 4.21. *Platyzoma*, a heteromorphic heterosporous fern.

Figure 4.22. Megaspore and micropore to same scale from *Pilularia*, a heterosporous aquatic fern.

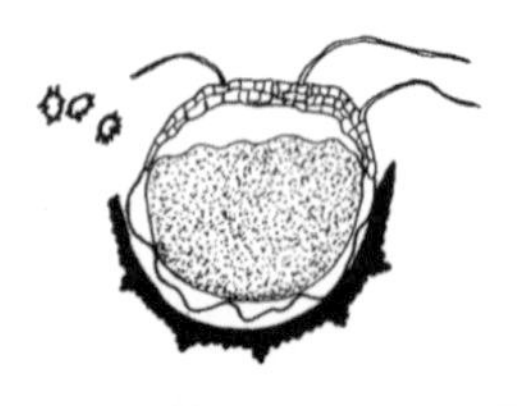

Figure 4.23. Endosporic gametophyte (section) in *Selaginella* with rhizoids arising from the surface (cell wall shown in black). Three microspores are shown to the same scale.

Megaspores normally give rise to female gametophytes and microspores to male. Megaspores enhance the survivability of the female gametophyte and microspores enhance the potential for cross-fertilisation since they are small, produced in large numbers and are easily dispersed. The living fern *Platyzoma*, which grows in north-eastern Australia, is not very markedly heterosporous, but is quite strongly heterothallic. It produces 32 microspores (diameter 71–101 μm) in each of its microsporangia and 16 megaspores (diameter 163–183 μm) in each of its megasporangia. The microspores produce a filamentous male gametophyte that produces only antheridia. The megaspores produce a spatulate bisexual gametophyte which initially produces archegonia and later antheridia.

There is a paradox here because heterospory, with the dioecy that normally accompanies it, places limitations on dispersal. Larger megaspores are less easily dispersed in the air, and microspores and megaspores have to settle in close proximity to each other for fertilisation to be effected. Perhaps for this reason it is in aquatic and semi-aquatic environments that heterospory first evolved. Megaspores are as easily dispersed in water as the microspores. Heterospory achieved its greatest development among the lycophyte trees of the Carboniferous swamps. For example, the tree *Lepidophloios* had separate male and female cones. Each megasporangium in the female cone produced a single megaspore protected by a sporangium wall and integuments, called an aquacarp.

Among living plants, there is a preponderance of heterospory in aquatic ferns such as *Pilularia*. All seed plants (Spermatophytina) are technically ‘heterosporous’ although the megaspores are highly altered and not dispersed. It was the evolution of seeds as dispersal units, diaspores, that permitted the seed plants to overcome the poor dispersability that is the consequence of heterospory on dry land.

Endosporic development

Megaspores are produced in smaller numbers than microspores but their greater size enhances the survivability of the (female) gametophyte, maximises the production of eggs and enhances the early life of the sporophyte after fertilisation. A trend in reduction of the number of megaspores produced in a sporangium can be observed in the fossil record. Heterospory in living plants is invariably associated with the phenomenon of endospory whereby the micro- and megagametophytes germinate within the spore and hardly extend their growth outsde the spore wall. Endosporic gametophytes are largely non-photosynthetic and depend on a readily available supply of nutrients from within the spore itself.

This is the case in *Selaginella* and *Isoetes* where the female gametophyte scarcely breaks free of the spore wall, merely sending some rhizoids out to absorb moisture. There is a period at first of free nuclear divisions and then apical nuclei become separated from each other by cell walls. At the tri-radiate apical scar the spore wall splits open and rhizoids grow out. The female gametophyte protrudes from the megaspore but usually it remains non-photosynthetic, relying on the nutrient reserves in the undifferentiated mass below. Archegonia develop in the surface layer of the exposed region. The male gametophyte also has greatly reduced requirements. The microspore is tiny and produces a male gametophyte that consists of only a single antheridium. When the antheridium is mature it disintegrates and the spore wall ruptures to release the sperm. One consequence of endospory is an accelerated gametophytic phase, thereby shortening the vulnerable sexual phase of the life cycle.

The nourishment of the female gametophyte may be more prolonged if the megaspore is not released and remains within the megasporangium so it can be fed by the sporophyte.

Figure 4.24. *Isoetes* after fertilisation showing the gametophytic tissue surrounding the embryo sporophyte. The spore wall is still present at the base.

4.3.3 Ovules and seeds

The evolution of seeds was a very important event in the adaptation of plants to life on land. A seed develops from a fertilised ovule. An ovule is derived from a megasporangium containing a single megaspore that is never released, and in which the megasporangium is protected by a layer called the integument. The process of evolution of seeds from a megasporangium surrounded by a ring of sterile appendages is traced in the fossil record by a series: from a 'pre-ovule' surrounded by sterile branches, through the fusion of these branches to form a fully integumented ovule surrounding a megasporangium/nucellus. *Hydrasperma* had four ovules surrounded by an outer cup of sterile branches. In an ovule the tissue derived from the megasporangium wall is called the nucellus. The integument is open at the apex where it is called a micropyle.

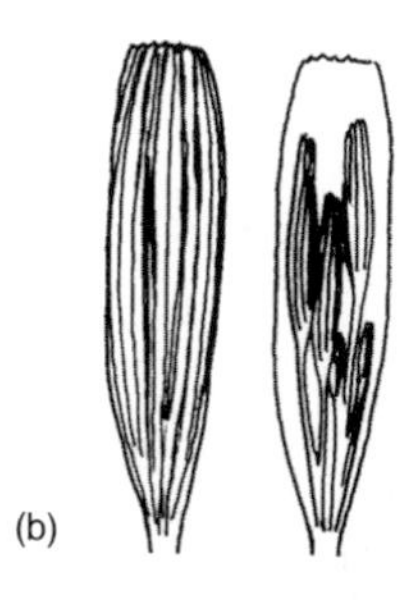

Figure 4.25. Stages in the evolution of seeds; (a) a tress of branches surrounding ovuliferous branches; (b) ovules protected in cupulate structures.

Accompanying the evolution of the ovule there was the evolution of a reproductive system whereby pollen was deposited at the micropyle of the ovule in a process called pollination. By this means male and female gametophyte tissues were brought into close proximity. A pollen grain is the microspore of a seed plant. Generally some development of the male gametophyte has taken place inside it before it is released. In primitive seed plants the male gametophyte released a motile sperm which then fertilised the egg in the ovule, as it still does in living *Ginkgo* and cycads. In more advanced seed plants the male nucleus is conveyed to the egg by a tube in a process called siphonogamy. The integument may have evolved for several reasons:

- to promote more efficient pollination
- as a means of protection from predators or drought/exposure
- to promote more effective dispersal by providing a wing or floats

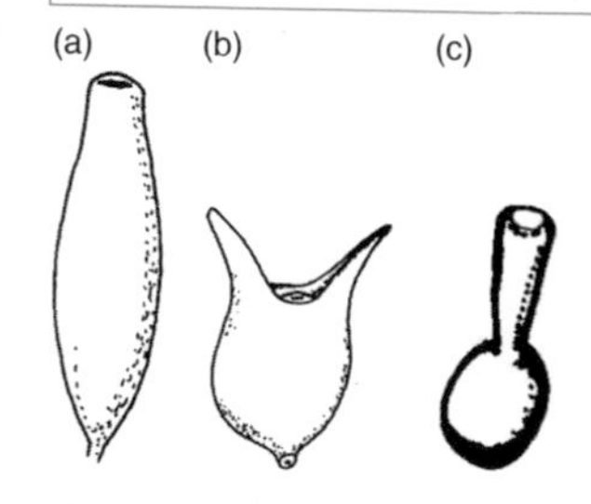

Figure 4.26. Devonian ovules with the integument adapted for pollination and/or dispersal of the seed. Note that (c) has a projecting salpinx.

The earliest seed plants had an ovule that terminated axes rather like the living *Ginkgo*. The ovule of gymnosperms is not normally hidden, although it may be protected in a cupule or by the tightly closed scales of a cone. They are said to have 'naked ovules' in contrast to the

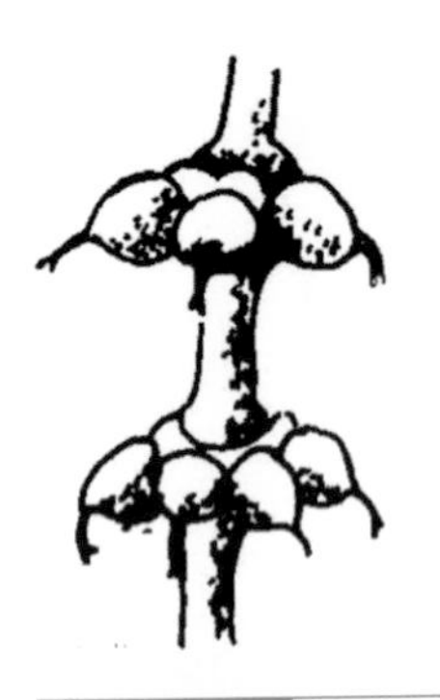

Figure 4.27. Part of a female catkin in *Gnetum* with tiers of ovules.

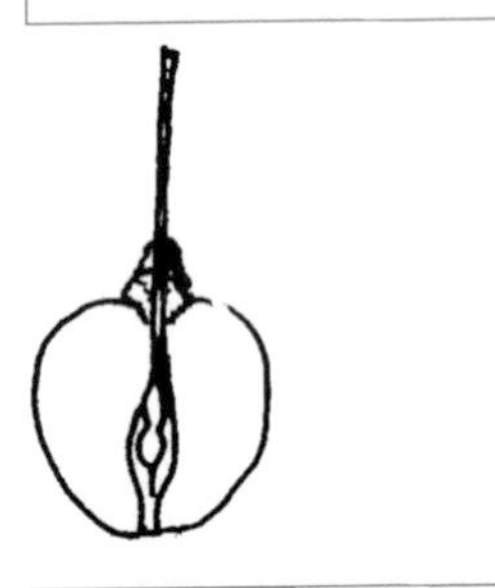

Figure 4.28. Projecting integuments in the ovule of *Welwitschia*.

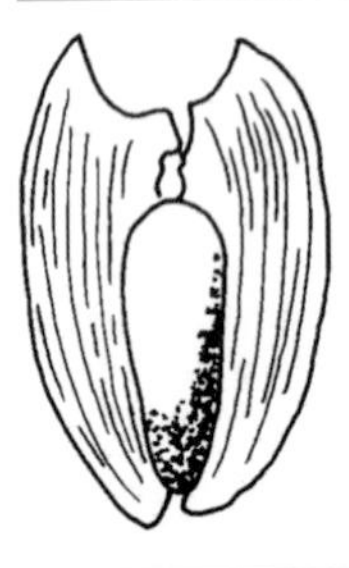

Figure 4.29. Platyspermic fossil seed.

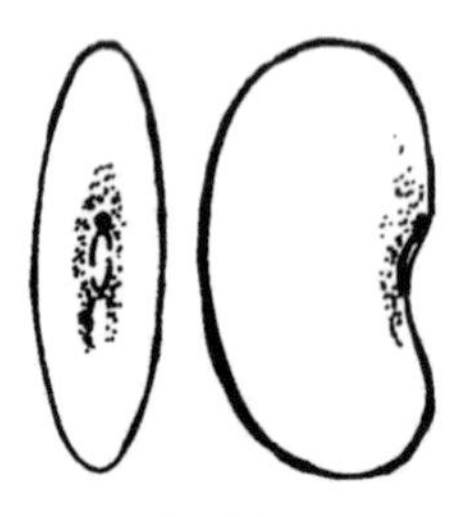

Figure 4.30. Seed (*Vicia* Fabaceae) showing the hilum, the scar where it was attached by the funiculus to the ovary placenta.

flowering plants (Magnoliopsida) in which the ovules are protected by a structure called the ovary. Before fertilisation the unreleased megaspore (nucellus) forms a protective and nutritive layer around the female gametophyte, and the integument provides an additional protective layer. In flowering plants the ovary wall provides a further layer of protection.

The ovule functions in pollination by the provision of a pollination droplet to trap pollen. Some Devonian fossils like *Hydrasperma* and *Genomosperma* from the Lower Carboniferous have an integument drawn out into an elongated funnel called the salpinx, which caught the air-dispersed pollen and directed it to a pollen chamber above the archegonia. Projecting integuments are found in the living gnetophytes. The lobed structure of early integuments modified patterns of airflow to allow pollen to settle on the nucellus. Alternatively the integuments may have prevented rain interfering with pollination in pendulous ovules.

The ovule/seed represents a concentrated, highly nutritious energy source that is very attractive to predators, and even with a protective integument, in living plants there is a very high loss of ovules/seed both prior to and after dispersal. Glands on the cupule of the fossil *Lagenstoma* may have further discouraged predators. Another possibility is that the integument provided protection from the environment, perhaps drought. This is especially important in large plants where the ovules/seeds are produced high up and in an exposed position. After fertilisation the ovule develops into the seed and the integument is transformed into the tough seed coat (testa) as an embryo seedling develops from the zygote. The mature testa is a complex of several or many cell layers. Hardness is conferred by the presence of sclereids. Cell walls may be impregnated with cutin, phenols and other compounds as well as lignin. Hairs may be present, as in cotton seeds. On the surface of the seed is the hilum, the scar where the ovule was connected by the funiculus, either to the ovuliferous scale or to the pistil.

One way to understand the evolution of seeds is that there has been a transfer of function so that the primary function of spores is no longer dispersal but as a part of sexual reproduction. Now dispersal occurs AFTER fertilisation by the release of seeds.

The rounded seeds of *Hydrasperma* are said to be radiospermic but the evolution of the integument as an aid to dispersal is most clearly indicated by the existence of another kind of seed called platyspermic because of its flattened bilateral symmetry. The flattened winged form maximised wind dispersal after the abscission of the seed. The importance of seed as an aid to colonisation is indicated in the conditions of fossilisation of one of the earliest seed plants *Elkinsia* from Elkins in West Virginia. It appears to have colonised open habitats near shorelines where there was little competition. Similar situations are indicated for other late Devonian seed plants. By the Carboniferous seed plants were also taking advantage of open disturbed upland habitats. Early evidence for seed dispersal is seen in the abundant isolated platyspermic seeds of the fossil *Caytonia*.

Seeds have the advantage that they are very well protected so that they have high survivability. They can be dispersed in both space and in time. Seeds are time capsules because a tough resistant testa allows some seeds to survive dormant in the soil for many years before germinating.

However, it is strange that there are very few fossil seeds in comparison to fossil ovules. This shortage could mean that following fertilisation abscission occurred quickly and germination of the seedling was direct with no intervening dormancy. Nevertheless the provision of energy reserves in the seed greatly aids the rapid establishment of the seedling. The rapidly dividing triploid endosperm tissue in flowering plants lays down a nutritive tissue after fertilisation. The endosperm may be cellular from inception, or free nuclear at first, or for the length of its life. The endosperm surrounds the embryo in the seed at the time of seed dispersal except in exalbuminous flowering plants where it disappears as its nutritive reserves are absorbed into the embryo. In this case food reserves are present in seed leaves, the cotyledons, and directly available to the growing apices of the seedling. Seeds with an endosperm generally require a longer period of incubation in warm moist conditions before they germinate. Large-seeded flowering plant families tend to have an endosperm. In gymnosperms, the surviving tissues of the female gametophyte and nucellus provide a nutritive tissue called the perisperm, and also in a few flowering plants the nucellus survives to provide a perisperm surrounding the embryo.

Figure 4.31. Albuminous seed of *Zea* (Poaceae) sectioned: the largest part is the endosperm tissue overlying the absorptive scutellum attached to the embryo.

Some of the ovules of some gymnosperms, especially in the cycads and *Ginkgo*, reach mature size after pollination but before fertilisation. This prior provisioning of the seed may be wasted if fertilisation is not achieved, and unpollinated ovules are aborted. Conifers are more efficient than cycads or *Ginkgo* by either reallocating the food reserves of unpollinated ovules or by increasing the size or quality of the food reserves, as measured by dry weight, only after fertilisation.

Figure 4.32. Primitive ascidiate carpel from *Thalictrum* (Ranunculaceae) does not have a well-defined style but a stigma in which the lobes of the carpel are pressed together.

4.3.4 Carpels, pistils and fruits

Flowering plants (angiosperms) differ from gymnosperms in having a structure called a carpel or pistil within which an ovule or ovules are protected. The carpel or pistil comprises a chamber called the ovary that encloses the ovule(s), a specialised area called the stigma that, in pollination, receives the pollen, and the style that connects them and contains a tissue or canal that conducts the pollen tubes towards the ovule(s). A carpel is a primitive kind of pistil. It is like a folded or peltate fertile leaf, with ovules arising inside. Normally the carpel is highly modified and/or fused with other carpels to form a compound pistil.

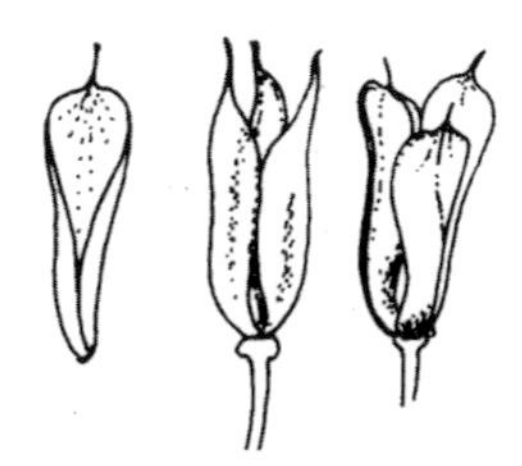

Figure 4.33. Follicles in *Aquilegia* (Ranunculaceae) after seed dispersal.

After fertilisation the ovary wall develops into a fruit wall, called the pericarp. The pericarp is adapted in many different ways to promote the survival and dispersal of the seed and the successful

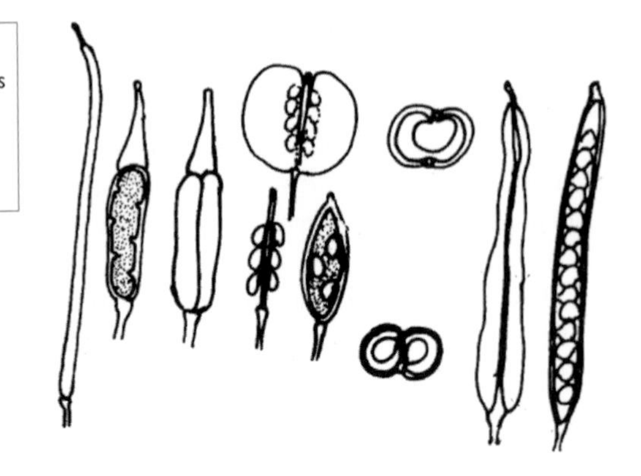

Figure 4.34. Fruit variation in the Brassicaceae. The fruit consists of two fused carpels separated by a septum. Siliques are elongated and siliculas are broad.

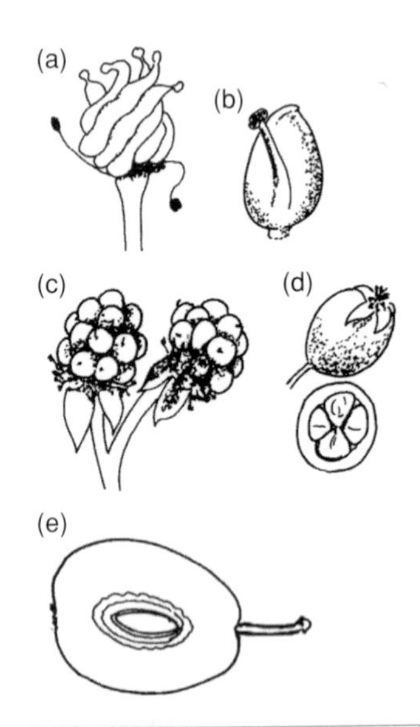

Figure 4.35. Some different ovary and fruit types in the Rosaceae: (a) follicles in *Filipendula;* (b) perigynous achene in *Alchemilla;* (c) drupecetum in *Rubus;* (d) berry in *Crataegus;* (e) drupe in *Prunus* cut in half to show hard endocarp.

establishment of the seedling. The simplest pistil types, most like a folded leaf, become fruits called follicles that split open to release their seeds. The silique or silicula present in the Brassicaceae (Figure 4.34) arises from a pistil of two fused carpels.

The rose family Rosaceae exhibits a wide range of fruit types. The subfamily Spiroideae has one to many dehiscent follicles. Some species have winged seeds. The cherries, peaches, almonds and plums (subfamily Prunoideae) have a single superior carpel with one ovule producing a fleshy drupe. The inner part of the pericarp is sclerenchymatous giving the familiar stone of these fruits. The subfamily Rosoideae is very variable. They have single-seeded indehiscent fruits. In *Acaena* these are achenes with the calyx modified as long hook-tipped spines. In *Alchemilla* the style is basal. The fruits are often combined with other fleshy tissue derived in different ways. In the rose, *Rosa*, and the strawberry, *Fragaria*, the fruits are hard achenes, although in the former they are contained within the fleshy swollen hypanthium forming the hep, and in the latter the achenes are situated on the exterior of a swollen torus. In *Rubus*, the blackberry and raspberry, the fruits are fleshy themselves, drupelets (little drupes) grouped together on the torus, to give the 'berry' (drupecetum). In both *Fragaria* and *Rubus* there are a large number of carpels that are arranged spirally on the extended torus. In *Rosa* and *Fragaria* there is one ovule per carpel, in *Rubus* there are two but only one develops. Subfamily Maloideae has an ovary produced below the petals (inferior) with 2–5 carpels, producing a fleshy pome. They exhibit a condition called pseudosyncarpy. The carpels are free on their ventral (inner) surface but held together by the fleshy fruit tissue.

4.4 From sex to establishment

4.4.1 Sexual development

Sexual development in the seed plants exhibits two evolutionary trends. One is for the greater reduction of sexual organs and their associated tissues. Both the size and length of development of the tissues is reduced leading to a kind of paedogenesis. This is exemplified by the reduction of the megasporangium/nucellus and of the female and male gametophyte. The other trend is the recruitment of more and more layers of tissues exterior to the sex cells in the development. This is exemplified in the evolution of ovuliferous cone scales and of the carpel. The whole process of sexual reproduction has been compressed back, so gradually earlier developing vegetative organs are incorporated and transmogrified into parts of the sexual apparatus as integuments or the ovary.

In the gymnosperms the megasporangium/nucellus is relatively well developed with several cell layers (Figure 4.36). Female gametophyte development usually begins with a free nuclear phase and then cell walls are formed. The size and number of nuclei that are produced varies between groups. In some cycads the ovules are up to 6 cm in length and between 1000 and 3000 nuclei have been counted. In *Ginkgo* there are up to 8000 nuclei, and 2000 in the conifer *Pinus*. This free-nuclear phase may last for a considerable time. In *Pinus* there is a halt at the end of one season, at the 32 nuclei stage, and nuclear division resumes in spring. Cell-wall formation begins at the periphery of the coenocytic gametophyte and proceeds centripetally. In the cycads and *Ginkgo* regular hexagonal shaped cells called alveoli are produced. After cellularisation some of the cells at the micropylar end become archegonial initials. The number of archegonia produced varies considerably. Only two are produced in *Ginkgo*, usually between 2 and 6 in cycads, and sometimes many in conifers.

The flowering plants show several reductions on this pattern. Although a similarly multistratose nucellus, is present in many, especially primitive flowering plants that are termed crassinucellate, more advanced flowering plants are tenuinucellate and have a nucellus with a single layer (unistratose). This distinction provides one of the most important characters in flowering plants separating the two major eudicot groups, the Asterids (tenuinucellate) from the Rosids (crassinucellate).

In some ways it is the parallel of the evolution of the distinction between eusporangia and leptosporangia that separates primitive and modern ferns. The tenuinucellate condition is very strongly associated with another kind of reduction because, whereas most crassinucellate flowering plants have two integuments surrounding the ovule, most tenuinucellate flowering plants have only one.

As the ovule matures in most flowering plants, the nucellus disintegrates, except in a few mainly primitive families like the Piperaceae and the more advanced Caryophyllidae where it is retained as a 'perisperm'.

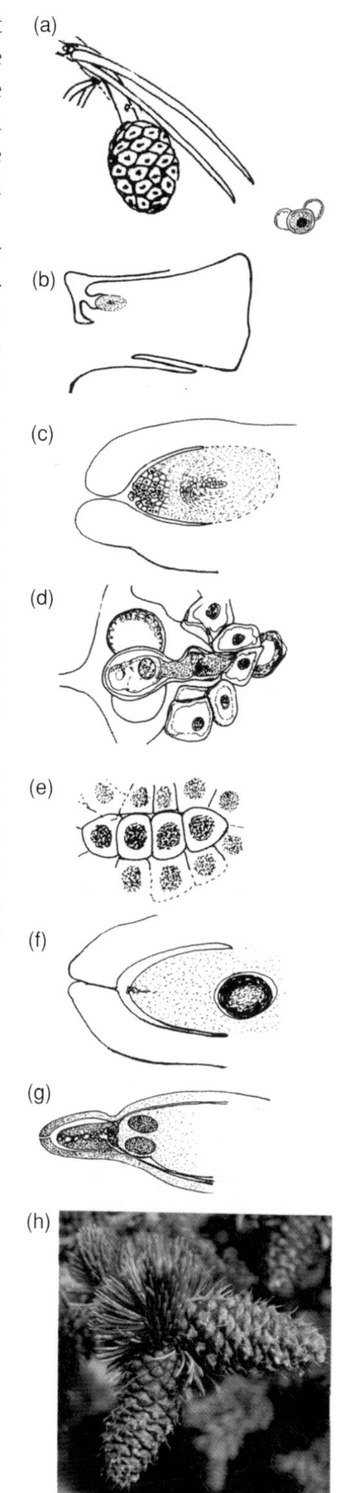

Figure 4.36. Sexual development of the female tissues of the conifer *Pinus* is shown in: (a) young cone with pollen grain inset (not to same scale); (b) section through megasporophyll; (c) developing female gametophyte at end of meiosis showing linear tertad of megaspores (detailed); (d) pollination, germinated pollen on nucellus with pollen tube; (e) linear megaspore tetrad; (f) multicellular female gametophyte before fertilisation; (g) at fertilisation showing large two large archegonia; (h) female cones in *Pinus aristata*.

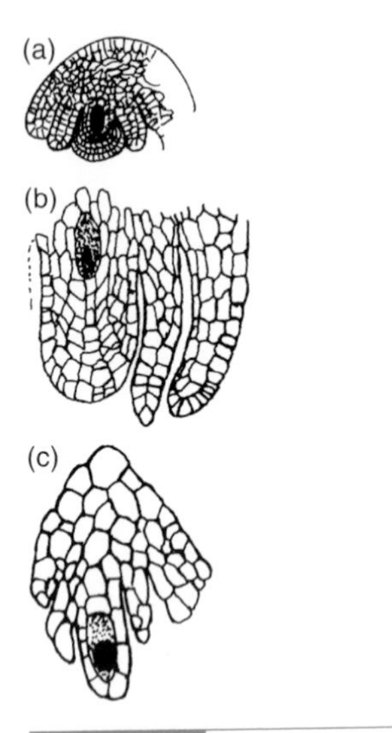

Figure 4.37. Ovules: (a,b) crassinucellate (several layers of cells surround the megaspore mother cell); (c) tenuinucellate (a single layer of cells surrounds the megaspore mother cell).

Another example of evolutionary reduction is in megaspore wall formation. The megaspore wall is particularly conspicuous between the four spores of the linear tetrad in cycads, not or only scarcely visible in some conifers, and is absent in flowering plants. Flowering plants and gymnosperms are further evolutionarily reduced, having monosporic development; producing a tetrad of which three nuclei disintegrate. Alternatively some flowering plants are tetrasporic, and no cell wall of any sort is formed between the four nuclei of the megaspore tetrad. This represents a co-option of all three megaspore nuclei into the development of the single female gametophyte, thereby shortening the developmental pathway. This evolutionary specialisation avoids the waste of three quarters of the tetrad. Tetrasporic development has arisen several times in as diverse flowering plants as *Adoxa*, *Plumbago*, *Peperomia* and *Fritillaria*. In other flowering plants such as the onion *Allium*, development is bisporic with two megaspore nuclei together producing a single female gametophyte.

In flowering plants the female gametophyte, called the embryosac, is very small. There is considerable variation both within and between different taxa. At least 12 different patterns have been named. However, in all the embryosac consists of only between four and sixteen nuclei and is only partly cellularised. In one common pattern three nuclei at the chalazal end (away from the micropyle) develop cell walls and are called antipodal cells. Two polar nuclei that migrate to the centre from opposite poles of the embryosac may fuse before fertilisation to form a secondary nucleus or central cell. The remaining cells are two synergids and an egg.

In the pollen sac, cell divisions give rise to several cell layers in its wall. The innermost layer gives rise to the pollen mother cells. The layer outside this is called the tapetum and has a special role in the formation of pollen. It nourishes the developing pollen grains and may determine the morphology of the outer wall of the grain. Cells here may be multinucleate and polyploid. One kind of tapetum, called amoeboid, has cells with protoplasts that penetrate between the developing pollen grains. Another kind, called glandular or secretory, eventually breaks down and is absorbed into the pollen grains. Below the epidermis of the pollen sac is an endothecium consisting of cells thickened on the anticlinal and inner periclinal wall. When the anther is mature the endothecium loses water and its outer wall shrinks pulling the pollen sac open at a line of weakness called the stomium.

Meiosis occurs in the pollen mother cells to give rise to a tetrad of haploid cells which eventually break apart and develop into the pollen grains without increasing in size. Development is either simultaneous or successive. In some groups, such as the Ericaceae, the four grains in the pollen tetrads do not break apart before dispersal. In *Acacia* the pollen is released in pollinia of 16 or 32 grains. In the Asclepiadaceae and Orchidaceae the pollen is further aggregated into pollinia which can combine all the pollen of a single anther locule into a single mass.

Male development in seed plants

Gymnosperms

microspore → prothallial cell
→ antheridial cell → tube cell
→ generative cell → stalk cell
→ body cell → sperm

Flowering plants

microspore → tube cell
→ generative cell → 2 sperm

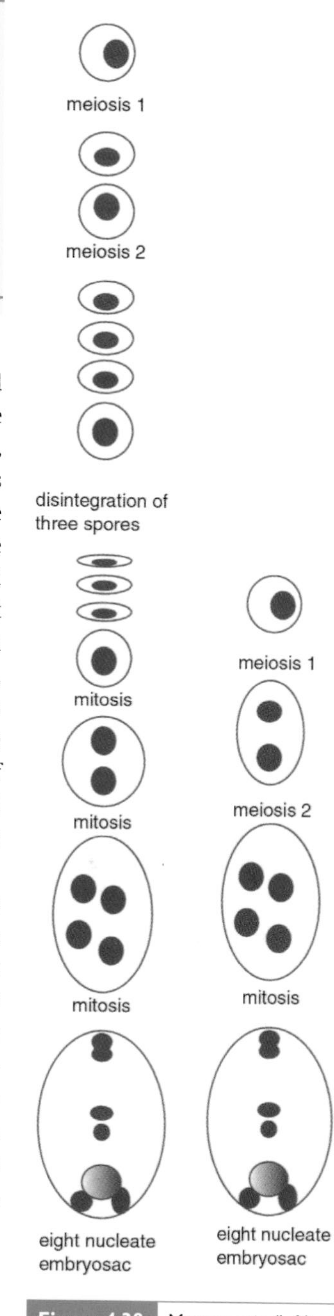

Figure 4.38. Monosporic (left) and tetrasporic (right) development of the female gametophyte (embryosac) in flowering plants.

The male gametophyte is very reduced. No clearly differentiated antheridium is formed and no jacket cells are present. In cycads the male gametophyte is a single vegetative cell called the prothallial cell, and an antheridial cell. The latter divides to produce a tube nucleus and a generative cell. In its turn the generative cell divides to produce a stalk and body cell. The body cell then divides at pollination to give two sperm cells i.e. a total of five nuclei are produced. In *Ginkgo* and the conifers, there are two divisions at the prothallial cell stage but the first prothallial cell has disintegrated at time of pollen release. In some conifers, from the families Araucariaceae and Podocarpaceae, many more prothallial cells are produced. In the flowering plants there is no prothallial cell stage and the generative cell gives rise directly to the two sperm nuclei. In flowering plants the division of the generative cell may occur either before or after the pollen is shed and so the dispersed pollen grain may contain either two or three nuclei.

The wall of the pollen grain is a complex structure with two main layers, the exine and intine. The outer exine may be sculptured in many different ways, with spines and ridges. In flowering plants its structure is tectate: a roof-like tectum is supported by little columns (columellae). Signalling molecules that mediate the process of pollen germination, either from the anther tapetum or from the male gametophyte, are carried on or within the exine. The exine is not continuous but broken by various slits, furrows (colpi) and pores. The pollen germinates through these apertures. The sculpturing and shape of the pollen grain is so varied that many different species can be identified from their pollen grains alone.

4.4.2 Pollination

In many gymnosperms, including *Pinus*, a viscous drop first catches the pollen grains and then, as it is reabsorbed, they are drawn through the micropyle into the pollen chamber above the nucellus. Only in a very few flowering plants pollen is caught by a liquid trap although this arises from the stigma. Normally the stigma is dry although the pollen may be 'sticky'. In gymnosperms, once the pollen has been drawn into the pollen chamber, the pollen grain germinates and a pollen tube is produced, which grows into the nucellus. In

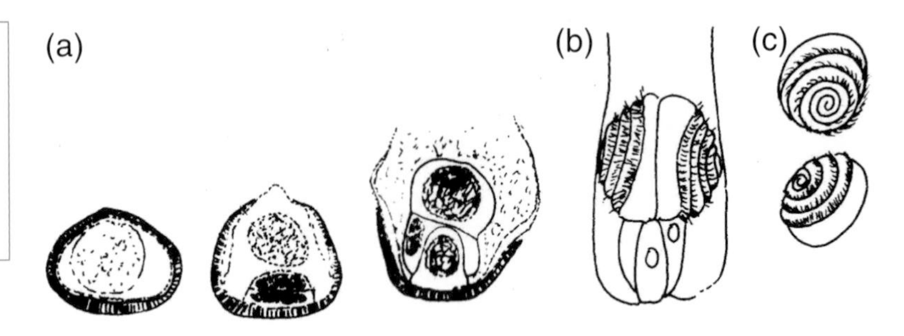

Figure 4.39. Male development in cycads: (a) *Dioon*, development of the male gametophyte in the spore up to germination; (b) *Zamia* formation of the two sperm; (c) *Cycas* liberated sperm with spiral arrangement of flagellae.

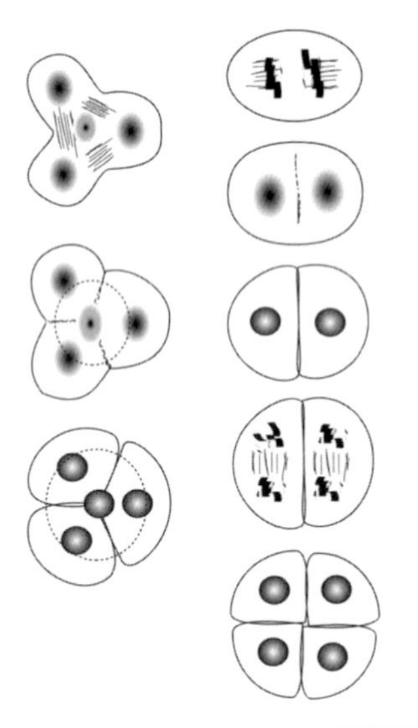

Figure 4.40. Pollen development (microsporogenesis): simultaneous on the left, successive on the right.

Siphonogamous = passage of male gametes to the ovule by a pollen tube.

cycads and *Ginkgo* the pollen tube is branched and purely haustorial, providing the male gametophyte with nutrients. In *Ginkgo* and the cycads the pollen tube gradually digests the tissue of the nucellus, finally breaks into the archegonial chamber, a space lying above the archegonia, and releases the motile sperm (Figure 4.39).

In conifers the pollen tube is initially branched and haustorial but develops eventually to transfer the sperm nuclei to the archegonium. In some species of *Pinus* there is a period of over a year between pollination and fertilisation. The development of a pollen tube may have allowed the nucellus to act as an anti-pathogen screen, recognising the pollen tube with its sperm nuclei and allowing it to grow but screening out the hyphae of pathogenic fungi. Perhaps it is in this self-recognition process that sexual-incompatibility between different species and self-incompatibility within a species has its origins. In *Pseudotsuga*, pollination is dry and the pollen tube conveys the male gametes from the integument lobes all the way to the archegonium.

Flowering plants are siphonogamous. The pollen grain germinates directly on the stigma, usually producing only a single pollen tube, but in most of them the pollen tube is not haustorial; instead, it grows down the style. In the angiosperm family Malvaceae both haustorial and absorptive pollen tubes are produced. Usually only 12–48 h may elapse between pollination and fertilisation. However, in some trees such as *Quercus* and *Corylus* from seasonal habitats, pollen tube development may be arrested for weeks or longer after a period of initial growth.

Two types of stigma have been described in angiosperms, called 'wet' and 'dry'. A 'wet' stigma produces an exudate (cf. pollen drop of gymnosperms) that helps to attach the pollen grain to the stigma but seems primarily to hydrate the pollen grain, providing a medium for pollen tube growth and prevents the stigmatic papillae from drying out. Pollen tubes penetrate the middle lamellae at the base of the stigmatic papillae and enter the stylar tissue. Pollen attaches to a 'dry' stigma with pollenkitt, a glue-like coating of lipoprotein derived from the tapetum. 'Wet' and 'dry' stigmas are associated with gametophytic and sporophytic self-incompatibility respectively (see below).

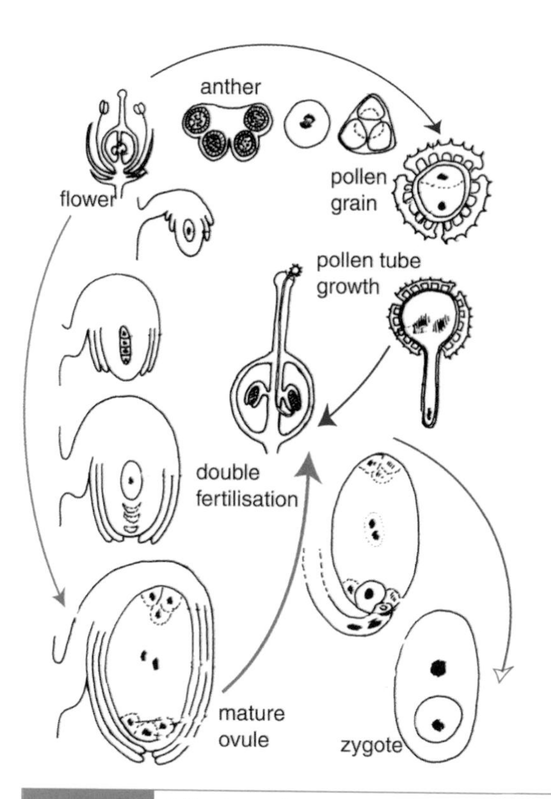

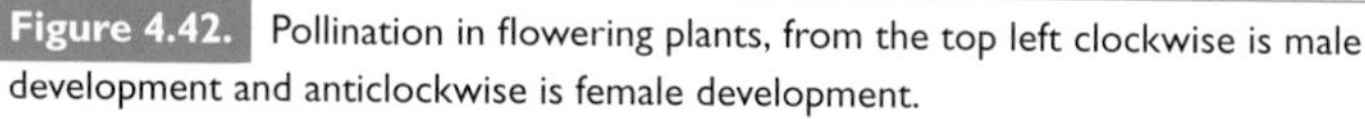
Figure 4.42. Pollination in flowering plants, from the top left clockwise is male development and anticlockwise is female development.

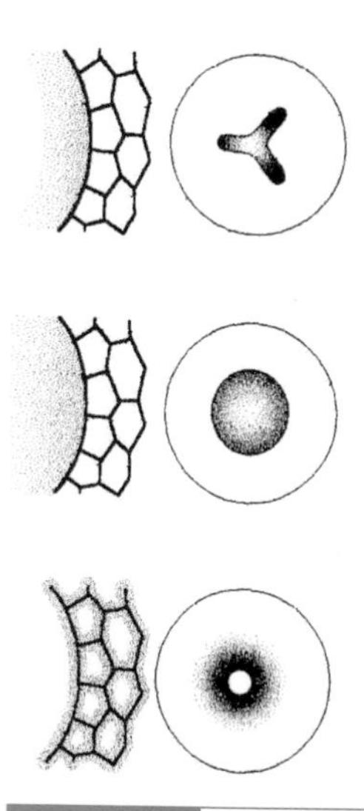

Figure 4.41. The pollen tube transmitting tract lies down the centre of the style. It takes various forms and involves both mucilages exterior to the cells and the cells themselves.

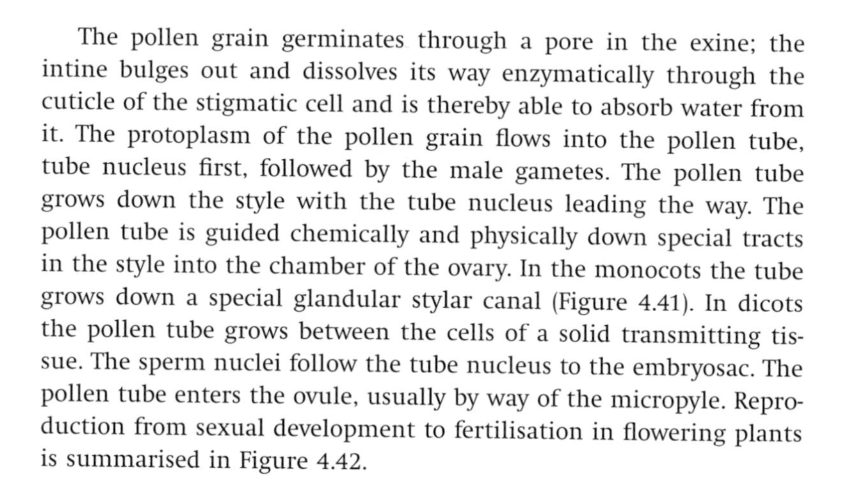

The pollen grain germinates through a pore in the exine; the intine bulges out and dissolves its way enzymatically through the cuticle of the stigmatic cell and is thereby able to absorb water from it. The protoplasm of the pollen grain flows into the pollen tube, tube nucleus first, followed by the male gametes. The pollen tube grows down the style with the tube nucleus leading the way. The pollen tube is guided chemically and physically down special tracts in the style into the chamber of the ovary. In the monocots the tube grows down a special glandular stylar canal (Figure 4.41). In dicots the pollen tube grows between the cells of a solid transmitting tissue. The sperm nuclei follow the tube nucleus to the embryosac. The pollen tube enters the ovule, usually by way of the micropyle. Reproduction from sexual development to fertilisation in flowering plants is summarised in Figure 4.42.

4.4.3 Fertilisation and embryogenesis

In gymnosperms each of the several archegonia may be fertilised so that several embryos may arise in each ovule (polyembryony). A single sperm nucleus fuses with the egg nucleus to form the zygote. In cycads and *Ginkgo* part of the egg protrudes into the archegonial chamber and engulfs a sperm. In *Welwitschia* tubular processes of the

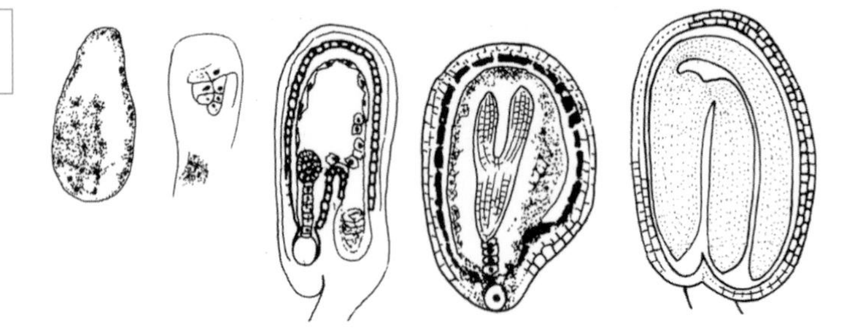

Figure 4.43. Embryogenesis in *Capsella*.

egg grow up to meet the pollen tube. In conifers, *Ephedra*, and *Gnetum* the pollen tube grows right up to the egg (siphonogamy). In flowering plants the pollen tube enters the embryo sac via one of the cells beside the egg (a synergid) that has a special receptive filiform apparatus. Then both sperm are released by a pore in the tip of the pollen tube. They migrate, and fuse, in a double fertilisation, one to the egg, the other to the polar nuclei, with the formation of a diploid zygote (sperm nucleus plus egg) and a triploid endosperm nucleus (male gamete plus two polar nuclei).

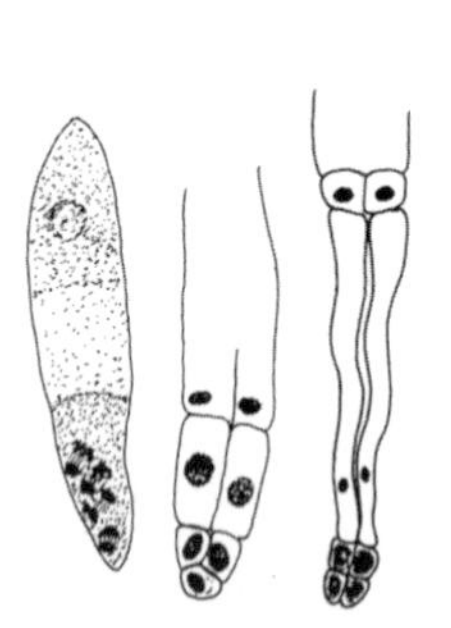

Figure 4.44. Embryogenesis in the cycad *Encephalartos*, at first coenocytic and then with tiered development.

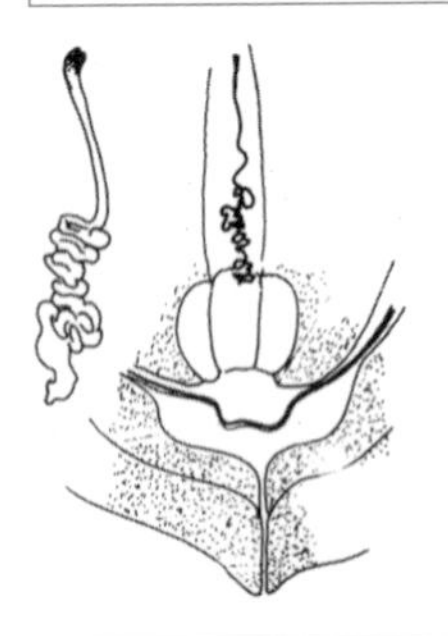

Figure 4.45. Embryogenesis in the conifer *Thuja*. The long suspensor pushes the embryo into the heart of the gametophyte.

Many diverse patterns of embryogenesis have been described. In *Ginkgo* and the cycads there is a period of free nuclear division, producing up to 256 nuclei, before cell walls are formed to create a suspensor and proembryo. In the conifers the free nuclear stage only produces four nuclei before cell walls are produced. Transverse division of these first four cells produce two or more layers of cells with four cells in each layer. These are a proembryo and a suspensor. The suspensor elongates to push the tiered proembryo into the tissue of the female gametophyte, which is then digested by the growing embryo. The four suspensor cells, and the embryo cells attached to them, may separate so that several individual embryos may be started (cleavage polyembryony). This is in addition to the potential existence of several different embryo developments, each the product of the fertilisation of a different archegonium. However, normally only one embryo reaches maturity in each seed.

Eventually the embryo differentiates an embryonic root, the radicle, at the suspensor end, and an embryonic shoot at the other end, called the plumule, which consists of a hypocotyl bearing two to several cotyledons. In flowering plants embryogenesis is said to be direct because there is no free-nuclear phase (Figure 4.43).

This is one of the several characteristics that allow angiosperms to complete the reproductive phase of their life cycle very rapidly. The zygote divides to produce a two celled proembryo. Cell divisions are regular, with cell walls formed at each stage. The terminal cell of the proembryo divides to form the embryo. The basal cell divides to produce the suspensor, which connects the embryo to the seed wall. In some angiosperms the basal cell also contributes to the formation of the embryo. This happens more often in species that produce large seeds.

4.4.4 Reproductive strategies

The dispersal of pollen and spores and even small seeds shares much in common. Irrespective of their different natures, the one thing that determines their potential for dispersal is their size. Small disseminules disperse easier and can be produced in larger numbers than large disseminules. However, large propagules, mainly seeds and fruit, have greater energy reserves and enable more secure establishment. Large seeds are associated with shade or drought where they allow rapid growth and establishment of a seedling.

Disseminules range in size across several orders of magnitude. On average, spores and pollen are smaller than seeds, and seeds smaller than fruits, but there is a large overlap in size across categories. For example, the dust-like seeds of orchids and parasitic flowering plants are closer in size and functionally more similar to the spores of mosses and pteridophytes than they are to the larger seeds produced by most flowering plants. For disseminules that are dispersed by wind their potential for dispersal is determined mainly by their size. Pollen is, on average, a little larger than spores because so much is dispersed by animals where buoyancy is not important. Seeds and fruits show a remarkable range of size; from the double coconut of *Lodoicea* that weighs up to 27 kg, or jak-fruit (*Artocarpus heterophyllus*) nearly 1m long and weighing up to 40 kg, to the seed of some orchids and parasitic plants more than 20 million times smaller (about 1 mg).

Diaspore, disseminule = organ of dispersal – spores, seeds, fruits.
Propagule = organ of multiplication and establishment – spores, seeds, fruits and also vegetatively produced organs.
Monocarpic = reproducing sexually once and then dying.
Polycarpic = reproducing sexually repeatedly.

Dispersal occurs in time as well as space. Different reproductive organs on the same plant mature at different times, enabling dispersal over an extended period. One example is provided by the filmy ferns where the sporangia within the indusium are carried outwards by an extending receptacle as they mature. In most plants the spores within each sporangium mature at the same time though they may not be released at once. The hornworts are exceptional in having a sporangium that continues to grow and releases spores over a long period. The peristome in many mosses can be regarded as an adaptation extending the period over which spores are released. The indehiscent sporangia and capsules of many lower plants also effectively extend the period of spore release because the capsule or sporangium wall takes a variable amount of time to break down.

r-strategists = plants adapted to reproduce rapidly and colonise new open and perhaps transitory habitats where competition is low, tending to be annual and monocarpic.
K-strategists = long lived, tending to be perennial and polycarpic, reserving some of their energy budget to maintain the vegetative growth.

Seed size is also correlated with regeneration strategy. K-strategists tend to have larger disseminules than r-strategists. There is a direct relationship between seed size and shade tolerance and the height to which seedlings can grow in reduced light conditions. Most shade-tolerant species have large seeds, often with food storage in cryptic cotyledons (*Persea* spp.). Fast-growing, light-demanding climax species of secondary tropical forest (*Artocarpus* spp., *Campnosperma* spp., *Endospermum* spp.) have seeds that are larger than most pioneer species, whereas pioneers frequently have tiny seeds that germinate rapidly and can only form tiny seedlings. With adequate photosynthetically active radiation (PAR), these have high relative growth rate (RGR) and soon catch up in size with bigger seedlings from larger trees. But, in the smallest gaps where PAR is insufficient these seedlings remain tiny and are suppressed by competition.

4.4.5 Germination and seedling growth

In the tropics there is no marked seasonality to compare with high latitudes, although there are usually distinct dry and wet seasons. Particularly in rain forests, germination of seeds is more closely linked to the light regime of the forest floor than to season. Here, the conditions for the germination of seeds are complex and may involve both stimulation and inhibition factors operating either simultaneously or sequentially. For example, dark dormancy requires certain conditions during the ripening of the seed or inhibition of germination may depend on the red/far-red light regime experienced by the imbibed seed, or in photodormant species germination occurs in the dark but is inhibited by light.

Photoblastic = seeds that require light for germination.
Hypogeal = describes germination in which the cotyledons remain below the soil.
Epigeal = describes germination in which the cotyledons emerge into the light.
Phanerocotylar = cotyledons emerge from seed coat.
Cryptocotylar = cotyledons remain within the seed coat.
Foliaceous = green, photosynthetic and leaf like.
Haustorial = absorptive cotyledons.
Reserve = storage cotyledons.

In forests light for germination may only be provided in forest gaps by tree fall, or by other disturbances such as ground-foraging animals. Photoblastic seeds will begin germinating after a brief exposure to white or especially red light. In sunlight the red:far-red ratio is about 1:1.15, but is reduced to lower levels in shade, and most light-requiring dormant seeds will not germinate under these conditions. For example, the forest canopy can inhibit germination in *Cecropia obtusifolia* and *Piper auritum* by its reduction of the red to far-red ratio of radiation. Both species need red light (660 nm) for germination, whereas far-red (730 nm) inhibits it. This accounts for the scarcity of seedlings on the forest floor and the flush that follows the appearance of a gap in the canopy. Germination by pioneers below closed canopies (presumably in response to the full light of ephemeral sun flecks) is followed soon by death at a tiny size.

Figure 4.46. Hypogeal seedling germination of *Pisum*: the cotyledons remain in the soil. Etiolated seedling grown in dark on right.

Various seedling types have been characterised based on mode of germination and cotyledon. Seeds germination is either hypogeal, with the cotyledons remaining in the seed and in the soil, or epigeal, in which case the cotyledons emerge from the seed above the ground. The radicle become the primary or tap root though in many plants, including most monocots, it is soon replaced by adventitious roots arising, especially, from the base of the stem. Phanerocotylar-epigeal-foliaceous seedlings are the most common (95% in temperate herbaceous floras and 33%–56% in woody tropical floras). In monocots cryptocotylar seeds are ubiquitous.

The percentage of cryptocotylar species increases along a moisture gradient across four forest types in Puerto Rico, from dry thorn to evergreen rain forest, but wetter mossy forest has a low percentage of cryptocotylar species. Here there is a strong association between dispersal mode and seedling type. Phanerocotylar-epigeal-foliaceous seeds are dispersed by the widest range of animals, but there are many exceptions. In Panama many of the lianas are wind-dispersed and have cyptocotylar-hypogeal-reserve seedlings.

Some seed plants are viviparous. The seed germinates precociously before it is released from the parent. A notable example is that of the mangrove *Rhizophora* where vivipary is an adaption for establishment in an intertidal environment. The hypocotyl and radicle protrude spear-like from the fruit. When the seedling is released it falls, and plants itself in the mud. More commonly, however, the seedling floats, and produces small roots near the base of the hypocotyl. When beached during low tide these get attached to the mud. Vivipary is also shown by other mangroves such as *Avicennia*, *Bruguiera* and *Aegiceras*. Some grasses, especially those of high altitude or high latitude habitats are pseudoviviparous. Florets are replaced by vegetative buds, from which plantlets are produced.

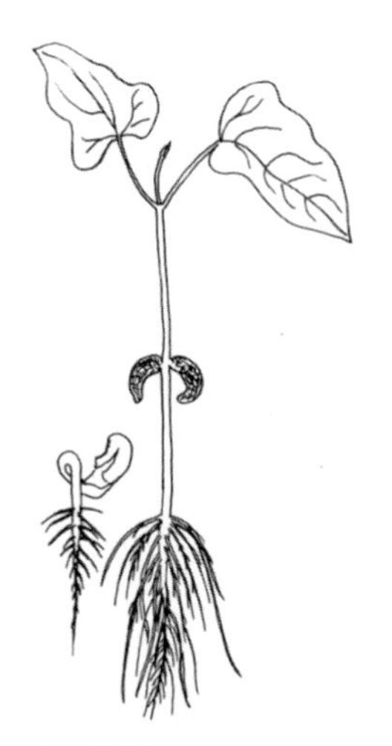

Figure 4.47. Epigeal seedling germination in *Vicia*: the cotyledons are raised above soil level.

4.4.6 Vegetative spread

Plants have a remarkable ability for vegetative reproduction because they grow by the replication of modules. The plant can fragment so that each part can grow into an independent plant. Leafy bryophytes with a delicate structure fragment easily, and in the moss *Dicranum flagellare* the shoot is modified to be especially fragile. Fragmentation of the plant as a regular reproductive strategy is also quite common in a diversity of aquatic plants.

Plants spread laterally by the production of rhizomes, stolons (runners), and invasive roots. The fern *Asplenium rhizophyllum* spreads by means of proliferating frond tips, which root and produce new plants where they touch the soil. Two types of clonal growth have been recognised. Phalanx growth is highly branched so that the plant advances relatively slowly as a set of closely packed modules. This is a highly competitive growth form. Guerilla growth is less branched with rapidly extending internodes. The plant seeks out more open areas where nutrients, water or light are more available, thus escaping competition.

The fern, *Pteridium aquilinum* (bracken), probably provides the best example of phalanx growth. It is one of the most successful species in the world. A single clone extending over an area of 474 × 292 m has been recorded and ages of 1400 years have been suggested for individual plants. Bracken invades by means of a deep rhizome, which may advance 1 m ahead of the fronds. The front guard fronds are supported nutritionally by the main wave of fronds behind. Grasses are shaded or buried in frond litter and, in addition, phenolic compounds leaching from the living plant and leaf litter slow the growth of competing grass roots.

This kind of chemical warfare between plants is called allelopathy. Other well-brown examples are the poisoning of the underflora beneath the American walnut, *Juglans nigra*. *Encelia farinosa*, a desert shrub from California, inhibits the growth of annuals within a circle of 1 m. There is a toxin, a benzene derivative, in its leaves that is released as fallen leaves decompose. In the Californian chapparal *Salvia leucophylla* and *Artemisia californica* have zones of bare soil around them and it has been suggested that volatile terpenes are responsible.

Ferns and horsetails provided the herbaceous vegetation before the evolution of herbaceous flowering plants. In the Jurassic, in some habitats *Equisetum* was dominant. Some horsetails such as *E. arvense* are still troublesome weeds. They have deep growing rhizomes that can grow over a metre below the soil surface and that fragment easily.

Grasses are now the overall champion as lateral spreaders. They are highly competitive sward formers. They produce a fibrous shallow adventitious root system and spread by rhizomes and stolons at the soil surface or rhizomes in the soil. These readily produce new shoots called tillers. Different species have different patterns of growth. The sand couch grass *Elymus farctus* and the sea lyme grass *Leymus arenarius*, which colonise embryonic dunes, have remarkable abilities to spread laterally but restricted in growing vertically if buried by sand, unlike the marram grass *Ammophila arenaria*, which can grow vertically and horizontally up to 1 m a year in growing dunes. The deeply buried parts of the plant die. The clover, *Trifolium repens*, has a guerilla kind of growth form. It produces stolons which can quickly ramble through the sward to find openings where it can root, and most importantly where it can get light. It is commonly overtopped by grasses.

Specialised vegetative propagation is particularly common in the bryophytes, many of which reproduce sexually only very occasionally. *Gemmae* are the asexual propagules of some liverworts and mosses. The thalloid liverworts *Lunularia* and *Marchantia* have gemma cups in which small disc-shaped gemmae are produced. Each gemma has a characteristic notch at the apical meristem. *Blasia* has a flask-shaped organ inside which gemmae are budded off. They are extruded from the flask when mucilage in the flask takes up water. Following release each gemma produces some rhizoids and grows into a new thallus. *Riccardia* produces two-celled gemmae within thallus cells. Each gemma is released when the containing cell wall breaks down. Some leafy bryophytes also produce gemmae. They arise on their leaves, at the apex or costa, or on the stems or rhizoids.

The moss *Aulocomnium androgynum* produces a stalk with a spherical mass of gemmae at its tip. Some pteridophytes also produce gemmae. They arise on the developing prothallus of *Psilotum*, *Tmesipteris* and epiphytic *Lycopodium*. Many ferns, some club mosses and flowering plants, produce bulbils that may arise in different positions, on the margin or lamina of the leaf as in *Kalanchoe* and many ferns, or on the axil of the leaves as in *Saxifraga granulata*. The bulbils may remain like small tubers or gemmae until after they are shed, or they may develop precociously into tiny plantlets with a swollen base and roots. In some *Allium* species bulblets are produced in an umbel on a scape, in the same position as florets. They either fall off or remain attached until the wilting of the frond or stem brings them into contact with the soil.

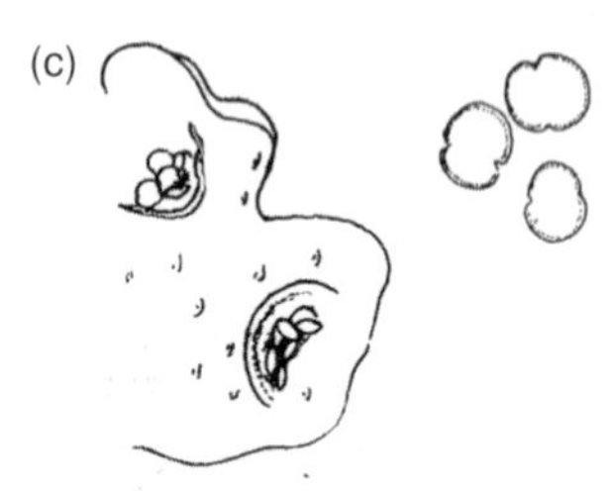

Figure 4.48. Vegetative propagules (gemmae) in bryophytes: (a) at the apex of shoots in leafy liverwort *Calopogeria*; (b) at tips of leaves in moss *Ulota*; (c) gemma cups in liverwort *Lunularia*; and (d) *Marchantia*.

4.5 Dispersal mechanisms

4.5.1 Wind dispersal

It is the evolution of air-dispersed spores of a particular kind, produced in tetrads, with a trilete scar and a spore coat containing sporopollenin, that has been used as one of the primary indicators of the origin of land plants. Wind dispersal is still important in many groups and is especially common in non-flowering plants. There are three phases of wind dispersal; release from the parent, the air-borne phase, and the settling or entrapment phase. The same physical rules influence the release and dispersal in the air and deposition of all disseminules whether they are spores, pollen, seeds or fruits.

Wind pollination is common in the plant communities that are relatively species poor and homogeneous like the forests or grasslands of temperate regions, but also in the savannas of the tropics, and in tropical forests where, either because of the uneven terrain or because some species are taller than others, some trees emerge above the main canopy. In temperate regions wind-pollinated trees flower in spring before the leaves are fully flushed.

Wind dispersal of pollen is probably not ancestral for flowering plants but several groups have reverted to it. This reversion occurred at an early stage in some lineages, for example in the relatively primitive tree *Trochodendron* from Japan. Although reversion to wind pollination has occurred independently in many different groups, wind-pollinated flowers show many convergent features:

- a dense, clustered inflorescence, especially in catkins (aments)
- small flowers, with a reduced perianth or apetalous
- lack of nectar
- pollen produced in large quantities in large anthers or many stamens
- anthers on long filaments or pollen liberated explosively
- expanded surface of the stigma or sometimes feathery
- sexes separated (dioecious or monoecious)
- pollen dry, rounded and smooth and generally smaller than that of insect pollinated species, 20–30 μm in diameter (cf. 10–300 μm in entomophilous plants)

Many familiar temperate tree species: oak, beech, birch, sweet chestnut and plane are wind pollinated. The origin of anemophily in most catkin-bearers is very ancient but some bear traces of their insect pollinated ancestors. There is also broad range of wind-pollinated monocots, the most important of which are the grasses.

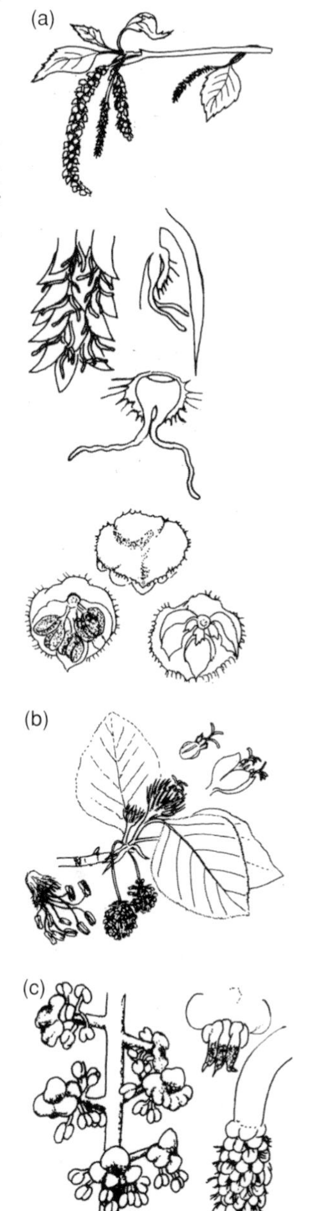

Figure 4.49. Catkins of wind-pollinated Fagales: (a) *Betula*, male and female florets detailed; (b) *Fagus*, male and female florets detailed; (c) *Alnus*, female floret detailed.

Spore and pollen release

The means by which the spores and pollen get into the air stream are very important aspects of wind dispersal. The majority of plants have a passive release. The position and arrangement of sporangia and anthers above or outside the boundary-layer effects of leaves, bracts or perianth segments allows the wind to shake spores and pollen from them. Many species have a male cone, strobilus or catkin or long dangly filaments. The vibration of these organs in the breeze may be very regular. Even under moderate airflow of 2–3 m/s grass inflorescences have been observed to shake harmonically.

Mechanical methods of spore release are powered by changes in water pressure as cells dry. Elaters that move hygroscopically within the spore mass are found in three groups: in liverworts, hornworts and *Equisetum*. As the elater dries and rehydrates, it twists and turns to fluff up the spore mass and allow slight air movements to catch the spores. Elaters are long and have a band or bands of thickening that makes them twist as the humidity changes. In some species of liverwort the elaters are attached to a small columnar elaterophore, either at the base of the capsule (*Pellia*) or at the apex (*Riccardia*). In *Cephalozia*, the elater is compressed and twists up as it dries until the column of water in it breaks. The elater then springs back, twisting to shake the spores off. In *Frullania* the elaters are attached to the top and bottom of the valves of the capsule. As the valves open they stretch the spring-like elaters so that they rip off at the base, flicking out spores. Of all groups of non-seed plants the mosses have the most diverse range of mechanisms for spore dispersal. There is considerable variation in the form of the peristome in mosses (see Chapter 5). The teeth may be multicellular and made of whole cells or constructed from the parts of adjacent cells that have broken down at maturity. The peristome may be twisted and the teeth may be forked or filamentous. The shrinkage of the moss capsule pushes the cap-like operculum off because it shrinks less than the rest of the capsule, and also pushes spores towards the opening where the peristome teeth lift the spores into the air stream by flexing and reflexing hygroscopically. In some species such as *Bryum* there is a double peristome. The outer ring interdigitates between the inner and collects spores on its toothed segmented tips, flicking them out. In the bog mosses, *Sphagnum* (class Sphagnopsida), the operculum explodes off, shooting out spores, because of the pressure that builds up inside as the capsule dries and shrinks.

In leptosporangiate ferns the sporangium has a special ring of cells, the annulus that helps to catapult the spores from the sporangium. The annulus cells shrink as they dehydrate. Eventually the sporangium breaks along the stomium, into two parts, one bearing the spores. Immediately, the liquid water inside the annulus cells vaporises under the tension, thereby releasing pressure so that the two parts spring back near to each other and the spores are catapulted into the air.

In the nettle *Urtica dioica* the stamens are curved inwards in bud and the filaments are bent back like springs. In dry weather the filaments snap back flinging pollen into the air. As the bud of *Broussonettia* opens the filaments spring out and pollen grains are thrown violently into the air. The annual mercury *Mercurialis annua* has pedicels that break, launching the flowers and their pollen into the air.

Explosive dispersal of seeds and fruits is called autochory. Notable examples are the dispersal of seeds by the jaculator, a modified funicle, in the Bignoniaceae, and the exploding capsules of *Viola* and *Impatiens*. *Ecbalium*, the squirting cucumber, is well named because it squirts seed out of an opening formed where the pedicel breaks off.

Figure 4.50. Action of *Frullania* elaters catapulting spores from the capsule. The capsule splits open in lobes; elaters attached to the capsule wall are stretched and then spring free, so catapulting spores out.

Bouyancy

The aerial motion of diaspores is related to the Reynolds number (*Re*) which is a calculated from the velocity of the object (v_p), the length of the object (l) and the kinematic coefficient of viscosity of the air (V); $Re = (v_p l)/V$. Pollen, spores and dust seeds are so tiny that they are well within the range in which the viscous forces of air are dominant. They have terminal velocities less than 100 mm/s. There is a change in the speed of airflow from zero at the surface of an object to ambient wind speed at a certain, quite small distance.

The morphology and density of spores and pollen grains may be affected by the humidity of the air. It seems that many grains become flattened as they lose water on leaving the sporangium or anther. Possibly they are adapted to collapse and reinflate on landing as they rehydrate prior to germination. Large and low-density spores or pollen grains will be more easily lifted from the sporangium or anther and will settle or collide more easily with a receptive surface but small, light spores and pollen grains will travel further before settling. These requirements are in conflict. These constraints explain the narrow size range in wind-dispersed spores or pollen grains.

Figure 4.51. Action of capsule in *Sphagnum* in spore release. The operculum explodes off, shooting out the spores when the capsule shrinks in maturity.

Wind-dispersed seeds and fruits are normally much larger than spores and pollen except for the dust-like seeds of orchids and parasitic flowering plants. The tiny seeds of parasitic plants such as *Orobanche*, which are produced in huge numbers, 2000 in each capsule, may be only 200 μm long. They have a testa with ridged reticulations. Orchid seeds range in weight from 0.3 to 14 μg. They are produced in huge numbers, up to several million per capsule.

Large wind-dispersed seeds and fruits generally have adaptations to slow their speed of fall; for example, the parachute-like pappus of dandelions, *Valeriana* and *Clematis*, or the unilateral winged fruits, the samaras that make the seed or fruit twirl or spin like a helicopter, as in many species of the families Aceraceae (maple, *Acer*) and Oleaceae (ash, *Fraxinus*). In the lime (*Tilia*) and hornbeam (*Carpinus*) there are bracts connected to the fruit, which act as the rotor. It is the legume itself and not the seed that is winged in the South American tipu tree (*Tipuana tipu*). The Pinaceae (pines, spruces etc.) have winged seeds with a papery wing derived from part of the ovuliferous cone scale. Flutterers tend to have a wing that forms a kind of skirt around the seed or fruit. Most remarkable perhaps are the symmetrical wings that allow the seed or fruit to glide like a plane. *Alsomitra macrocarpa*, an Asian climber, produces seed in great pendulant gourds. Each seed has a wingspan of more than 10 cm and glides through the air when it is released.

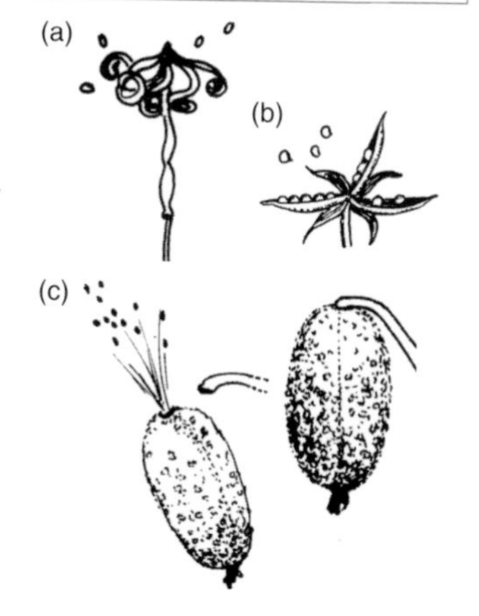

Figure 4.52. Autochory, explosive release of seeds: (a) *Impatiens*; (b) *Viola*; (c) explosive berry of *Ecbalium*, the squirting cucumber.

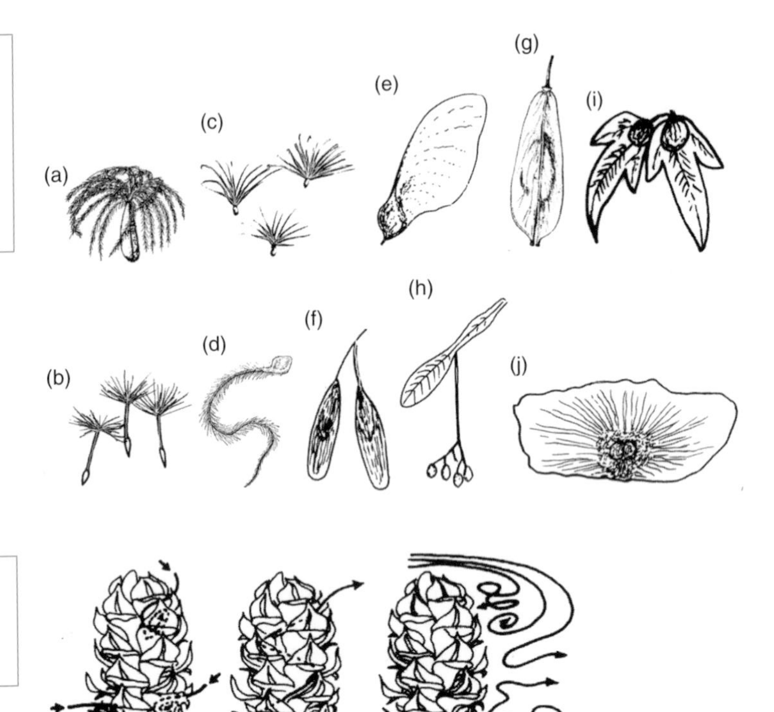

Figure 4.53. Anemochory, examples of winged and parachuted seeds and fruits: (a) *Valeriana*; (b) *Taraxacum*; (c) *Epilobium*; (d) *Clematis*; (e) *Acer*; (f) *Fraxinus*; (g) *Tipuana*; (h) *Tilia*; (i) *Carpinus*; (j) *Alsomitra*.

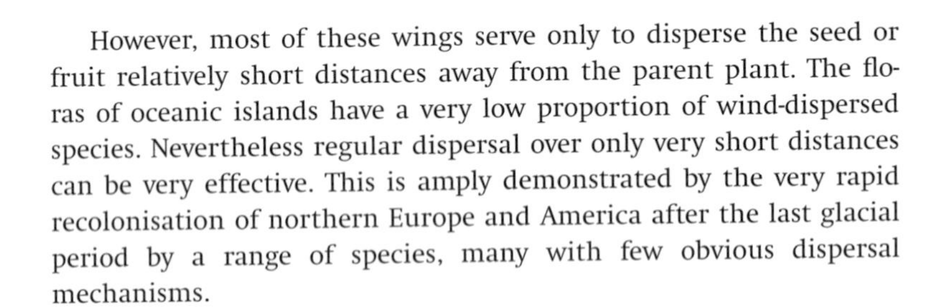

Figure 4.54. A female cone of *Pinus* showing entrapment of wind-borne pollen (after Niklas, 1985).

However, most of these wings serve only to disperse the seed or fruit relatively short distances away from the parent plant. The floras of oceanic islands have a very low proportion of wind-dispersed species. Nevertheless regular dispersal over only very short distances can be very effective. This is amply demonstrated by the very rapid recolonisation of northern Europe and America after the last glacial period by a range of species, many with few obvious dispersal mechanisms.

Deposition and entrapment

Landing of a wind-dispersed seed or fruit is not a significant problem but pollen has very tiny targets on which it must land in order to effect pollination. Wind-pollinated plants have specialised receptive surfaces for trapping pollen. Any surrounding structures or perianth are generally reduced. Stigmas are broad and folded or feathery. The feathery stigma of the grasses acts like a net, sieving a large volume of air as the inflorescence vibrates backwards and forwards in the wind. Some wind-pollinated flowers have a relatively simple stigma. Electrostatic attraction may play a part in pollen attraction to the stigma. However, vortices are created in the air stream in the area of the stigma or ovuliferous cone, because of the shape of surrounding bracts, so that pollen grains are caught in a vortex and settle (Figure 4.54). In most of the wind-pollinated gymnosperms, including cycads and conifers, pollination is aided by the production of a drop of fluid, the pollination drop, which captures the pollen grain.

4.5.2 Water dispersal

In several different lineages of plants there has been a return to dispersal by water. This form of dispersal was particularly significant for early heterosporous plants since the large megaspore is most efficiently dispersed by water. Even today heterosporous ferns are mainly aquatic. Several different kinds of heterosporous plants of the Carboniferous coal swamps presumably had water dispersal. *Lepidocarpon* is a fossil cone scale that acted first as a wing or parachute and then as a float or sail for the megaspore. The descendants of these Carboniferous trees, *Isoetes*, might also use water for dispersal, but the sporangia are often half buried in mud and it has been proposed that they may be dispersed by earthworms. There is, however, significant fragmentation of *Isoetes* swards by autumn and winter storms, so dispersal of megaspores may be effected in this way.

Many aquatic plants reproduce mainly by asexual fragmentation. When they have been introduced into new areas they have often spread rapidly asexually. For example a female clone of *Elodea canadensis* was introduced to the British Isles in 1836 and within a few decades was choking canals and other waterways. *Hydrilla*, from the Old World, introduced to the Americas, and *Lagariosiphon*, from South Africa, introduced to New Zealand, have a similar kind of history and have become nuisance weeds. The water-lilies *Nymphaea* and *Nuphar* have rhizomes with adventitious roots, thus aiding the vegetative spread of the plant and providing anchorage. Genera such as *Potamogeton*, *Vallisneria* and *Cryptocoryne* propagate by runners, or, in the case of *Sagittaria*, by side shoots. In a few genera such as *Aponogeton*, adventitious plantlets form on the inflorescence. In temperate regions most of these aquatics enter a period of dormancy in the winter. Food storage occurs in the rhizomes (*Nymphaea*), in tubers (*Sagittaria* and *Aponogeton*) or in turions (*Hydrocharis*, *Potamogeton*).

Water dispersal of seeds and fruits is termed hydrochory. Water-dispersed fruits and seeds are relatively large or even very large. The double coconut *Lodoicea maldavica* weighs up to 90 kg and contains a two-lobed seed 50 cm long. It floats but is killed by prolonged exposure to seawater. Coconuts of *Cocos nucifera* are one-seeded drupes. The fruit wall, the pericarp, has a skin-like exocarp, a fibrous mesocarp that provides buoyancy, and a hard endocarp with three pores. The seed testa, a thin brown layers adheres to the endocarp. The endosperm is abundant with a white outer zone, the 'meat', and a liquid centre, the coconut juice or 'milk'. The embryo is at the base. Another case of gigantism is *Entada gigas*, a woody climber that produces legumes 12 cm across and more than 1 m long. The fruit breaks into floating segments, each carrying a buoyant seed 6 cm × 5 cm, which drop into streams. It was the observation of the seeds on the beaches of the Azores that is supposed to have prompted Christopher Columbus to attempt to find their source the other side of the Atlantic. Water-dispersed seeds and fruits have buoyancy because of the presence of air chambers, either as part of the fruit as in *Xanthium*, or pseudo-fruit as in *Atriplex*, or in the testa as in *Menyanthes*. The box fruit

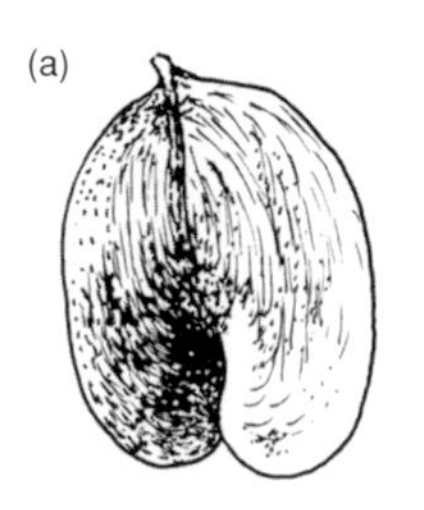

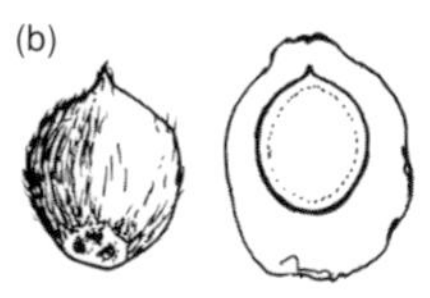

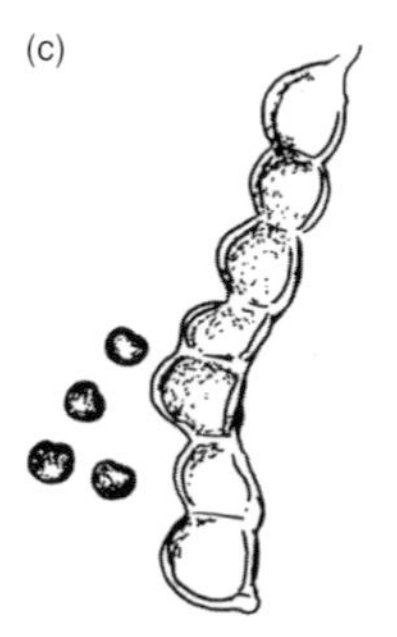

Figure 4.55. Hydrochory, dispersal of seeds and fruits by water: (a) *Lodoicea maldavica*, double coconut; (b) *Cocos nucifera*, ordinary coconut, showing inner seed surrounded by fibrous endocarp (left) and section through fruit (right); (c) *Entada gigas*, sea-bean fruit and seed.

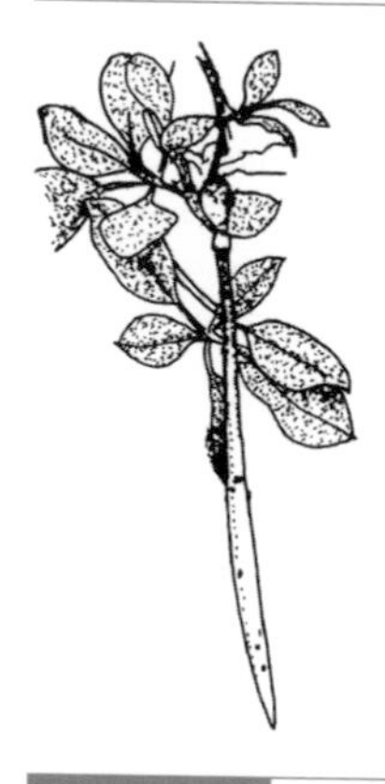

Figure 4.56. Vivipary in the mangrove *Rhizophora*: the embryo germinates precociously and the hypocotyls expand spear-like in order to plant the seeding.

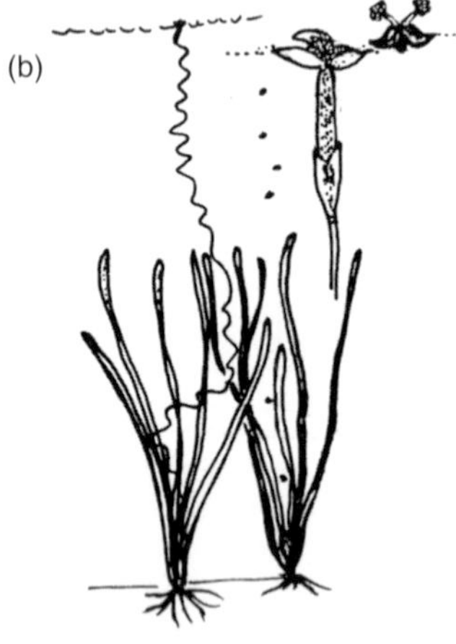

Figure 4.57. Pollination in the Hydrocharitaceae: (a) *Elodea* produces an elongated receptacle, and the male flowers shed pollen onto the water surface; (b) *Vallisneria* has female flowers on a long peduncle and male florets that break free and float to the surface aided by a gas-filled blister. Pollination occurs when the tiny (1 mm) male flowers directly contact the stigmas of the female flower.

(*Barringtonia asiatica*) of SE Asia can remain buoyant for at least two years and is even used for fishing floats.

It is estimated that only about 250 species are regularly dispersed to oceanic islands by water and it has been estimated that, of the 378 original plant colonisations of the Galapagos Isles, only 9% were by water (the rest were transported by the wind or birds). It is not just a matter of surviving dispersal in the sea. Establishment in the intertidal zone is precarious. As described above, *Rhizophora mangle*, aids this process by vivipary producing a long pendant seedling before it is released. A similar mechanism is present in *Aegialitis* (Plumbaginaceae). This was one of the first tropical drifters to reach Krakatau after the catastrophic volcanic eruption of August 1883. In the tropics, the coastal aroid, *Cryptocoryne ciliaris* has floating, partly-developed embryos, a precursor to true vivipary.

Some aquatics and aquatic margin plants are not strictly hydrochorous but ichthychorous, dispersed by fish, or avichorous, dispersed by waterfowl. The fruits of the aquatic grass *Glyceria* are consumed by carp and the olive-like fruits of the aquatic weed *Posidonia* are eaten by tuna in the Mediterranean. The perianth hairs on the fruit of *Typha* prevent wetting before liberation but later aid flotation on water. The fruits then dehisce and the seeds sink, but the pointed seeds have also been found adhering to the skin of fish (suggesting possible ichthyochory). *Nymphaea* seeds mature in the fruit under water but, when released, float to the surface in a mass. They are then attractive to waterfowl. The passage of seeds through the guts of waterfowl may act as a stimulant to germination. In *Nuphar*, the fruits mature above water, and it is the carpels containing the seeds that float on the surface. The seeds of *Nuphar* spp. are eaten by fish. Many fruits of aquatic plants are hard, with spiny projections, characteristics that may aid dormancy and dispersal, as well as anchorage in unstable substrates. However, the spines on the fruits of *Victoria amazonica* play a different role, that of protection from herbivores. Our knowledge of fossil aquatic plants such as *Ceratophyllum* and *Trapa* is often based on the persistence of such hard fruits.

Aquatic pollination

Pollination is a particular challenge to submerged aquatic plants. Some aquatic plants manage above-surface pollination by producing flowers on long pedicels so that they float on the water surface or are held in the air for wind (*Hippuris, Myriophyllum, Potamageton*) or insect pollination (*Nymphaea, Nuphar*). *Potamageton* is interesting because the spikes of small protandrous flowers, if held erect above the surface, may be wind-pollinated but when growing in deep water may be water-pollinated, and some species have lax spikes that trail in the water. Some species have flowers specifically adapted for aquatic pollination. In the Hydrocharitaceae for example, frogbit (*Hydrocharis*) has conspicuous insect-pollinated flowers whereas *Vallisneria* and *Elodea* are highly adapted for water pollination.

Water pollination occurs in 31 genera in 11 different families of flowering plants. There has been substantial convergent evolution between them. Most have tiny and highly reduced unisexual flowers, in some cases consisting of a single stamen or pistil. The breakdown of hermaphroditism is more prevalent among fully submerged aquatics than in any other group of flowering plants, and intermediate stages such as monoecy and dioecy (e.g. *Vallisneria* spp.), are present in many species. About half are dioecious. The flowers of *Ceratophyllum* are so small and reduced that it was once thought to be very primitive.

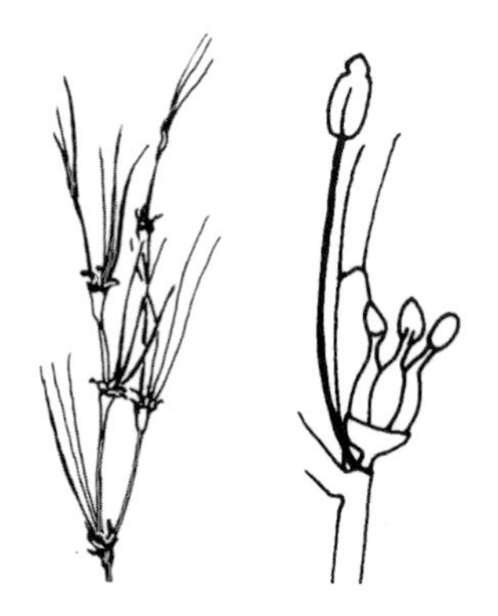

Figure 4.58. Inflorescence of aquatic weed *Zannichellia*: the male and female flowers are highly reduced.

Most water-pollinated species utilise the water surface for pollination, which has the advantage that, being just two-dimensional, it enhances the chances of the pollen contacting a stigma. The female flowers are produced at the surface on long thin stalks which often reach disproportionate lengths to reach the surface. In *Ottelia* spp. these elongated peduncles elevate both hermaphroditic and female flowers to the surface, which are then pollinated by insects. Thereafter, the peduncles spiral and pull the developing fruit under the surface. *Vallisneria* is dioecious, and, as in *Ottelia*, the peduncles may spiral to pull the developing fruit under water. *Callitriche* (Callitrichaceae) hedges its bets and has underwater surface or aerial pollination.

Figure 4.59. Inflorescence of *Zostera*: female flowers in detail showing simple forked stigma.

Floating male flowers occur in some genera of the frogbit family (Hydrocharitaceae). In *Lagariosiphon* and *Vallisneria* the erect male stamens act like miniature sails so that the male flowers are blown like little sailing ships towards the tethered female flowers. Near the female flower the suface of the water is depressed by surface tension, and the male flowers slide down the female, up-ending and depositing sticky pollen. *Enhalus* produces similar male 'boats' that become trapped within the perianth of the females as the tide falls. In *Hydrilla* the pollen is released explosively into the air to reach the stigmas of female flowers and is destroyed if it contacts the water. *Ruppia* (Ruppiaceae) has bisexual flowers and either underwater or surface pollination. *Ruppia* frees air bubbles from its submerged anthers, trapping the pollen on their surfaces. The flowing water waves the flowers across the stream surface to collect pollen. Anther bubbles are also released by water-pollinating *Potamogeton* species. The bizarre behaviour of the male flowers takes another turn in the hornworts (*Ceratophyllum* spp.). Ripe stamens detach themselves from the male flowers and float to the surface where the pollen is released. Pollen then slowly sinks to come into contact with the stigmas of female flowers. *Zannichellia* pollen also sinks onto the funnel-shaped stigmas of female flowers in this manner.

Pollen is normally positively buoyant and sometimes hydrophobic. At the surface pollen grains commonly adhere to each other in rafts. In three related families Cymodoraceae, Posidoniaceae and Zosteraceae pollen grains are elongated like noodles increasing their chance of catching on the stiff stigmas. As soon as they contact the stigma they curl around it. In the Hydrocharitaceae and Zannichelliaceae the pollen is strung together in mucilaginous chains or surrounded by a flat mucilaginous raft. *Thalassia* produces mucilagenous strands of pollen at the surface but these sink in the waves to reach the submerged stigmas.

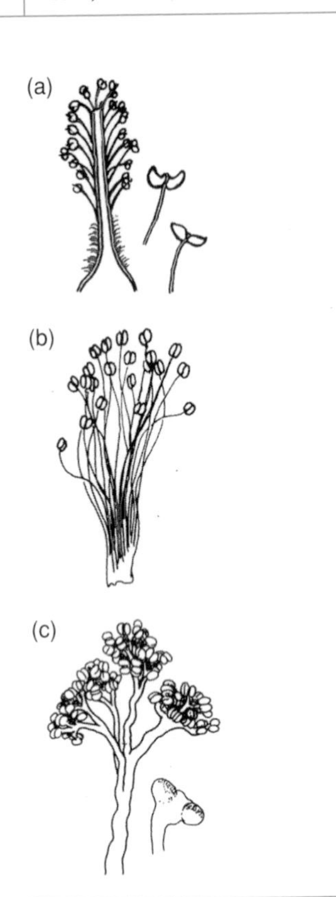

Figure 4.60. Multiple stamens of pollen flowers: (a) connate stamens forming a staminal tube in *Malva* (Malvaceae); (b) stamen bundle in *Hypericum* (Guttiferae); (c) branched stamen in *Ricinus* (Euphorbiceae).

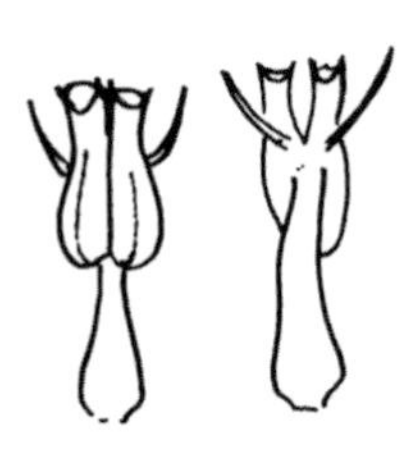

Figure 4.61. Poricidal horned stamens of *Vaccinium* (Ericaceae).

4.5.3 Animal dispersal

Animal dispersal often relies on the attraction of the potential vector to the plant. Most animals are attracted by the provision of food, which can be the spores, pollen, seeds, nectar, or even the whole plant. Dispersal of spores by animals is rare but takes place for example in the moss *Splachnum rubrum*, which has a capsule with a swollen, bright red, foul smelling apophysis that attracts flies. An alternative food to seeds may be provided by a fleshy aril or fruit, or an alternative source of nutrition, mainly nectar, or brood places, may be provided instead.

These primary attractants must be advertised by colour, shape and scent. The advertisements have the secondary advantage of encouraging fidelity of the pollinator or seed dispersal agent. Fidelity improves efficiency because there is less wastage. There has been extensive coevolution of plants and their animal vectors carrying pollen, seed or fruit.

Primary attractants

POLLEN AND SPORES

Pollen was probably the attractant in early flowers. Insect-dispersed spores and pollen grains may be sculptured and coated by a sticky material secreted by the tapetum and called the pollenkitt. Some species have pollen held together by viscin threads derived from the anther in various ways. This is sticky and helps it adhere to the bodies of insects. Many flowers produce an excess of pollen as a food for the pollinator. Pollen is a highly nutritious food and has been found to have 16%–30% protein, 1%–7% starch, 0%–15% free sugar, 3%–10% fat plus vitamins. The high nutritional value of the pollen provides the energy for pollen to germinate and grow but it is also a rich reward for effective pollinators.

Vectors attracted by the provision of pollen include hover flies, bees, beetles and bats and other small mammals. Pollen is the attractant in many 'primitive' magnolioid flowers but it is the only primary attractant even in some fairly advanced flowers, such as the rose, tulip and poppy. The pollen grains of pollen flowers are adapted in various ways. They are coloured by pigments called carotenoids and flavonoids that give the mainly yellow colour. The yellow colour is seen most effectively when it contrasts with a differently coloured corolla. The pollen of some species is deep red, purple or even blue.

Several different kinds of pollen flowers can be distinguished. Pollen flowers have many stamens and powdery pollen may be shed onto the perianth or collected directly out of the anthers as in *Argemone* and *Paeonia*. Some, such as *Magnolia* have numerous stamens and shed sticky pollen onto the perianth where it is eaten by beetles, whereas others have relatively few but showy stamens. Buzz-pollinated flowers in some species of Solanaceae, Ericaceae and Melastomataceae have anthers opening by terminal pores. Vibrations and rhythmic squeezing forces the pollen out. Pollen flowers often magnify the attraction of their stamens by producing accessory organs that mimic the stamens. Sterile stamens (staminodes) are quite common. The Melastomataceae have dimorphic stamens, the outer showy staminodes and the inner ones fertile. Some Lecythidaceae such as *Couroupita* have sterile stamens fused into a ligule arching over the pistil. Another feature often observed is a textured or bearded region of the corolla coloured like the stamens and advertising the availability of pollen.

Figure 4.62. Pollen flowers: (a) *Hypericum* (Guttiferae), (b) *Argemone* (Papaveraceae) (c) *Couroupita* (Lecythidaceae) with showy staminodes.

OVULES OR SEEDS

Another kind of primary attraction is the provision of ovules or seeds. This is very common where the visitor is being attracted to disperse the seed. Seeds are produced in much larger quantities than required and many are sacrificed to an animal visitor so that a few can be dispersed. It is a kind of tax on reproduction to ensure dispersal.

Only a few plants provide ovules as the primary attractant at the stage of pollination. These plants are said to have brood-place blossoms. In these species there is a rather constant relationship between pollinator and flower. The yucca has large flowers in which the stamens are kept well clear of the stigma. The female yucca moth *Tegeticula yuccasella* lands on a flower and hooks itself onto a stamen. It collects pollen with two prickly mouthparts and then flies to another flower. First, it establishes whether it has been pollinated, and if not, pushes some pollen into a tube formed by the three elongated stigmas. Then it lays an egg in one of the ovary locules. It repeats this process a few more times and then flies off to another flower. The moth larva develops within the ovary, eating some seeds, but a proportion of the seeds survive. Eventually the moth escapes by chewing a hole in the ovary wall.

The figs, *Ficus* (Moraceae), may represent the most extraordinary reversion to insect pollination in the angiosperms. Some primitive species in the family are wind pollinated. They have small flowers aggregated into a head. In the 500 or more species of fig, the head has invaginated to form a bottle-shaped syconium with the florets on the inside. Some specialised florets provide brood places for species-specific gall wasp larvae. Inside the syconium, there are male, female and sterile flowers. The sterile flowers have short styles enabling the female wasp to reach the ovary to lay its egg with its ovipositor. These flowers are on a longer peduncle so that to the gall wasp all stigmas appear to be at the same level. The wasp larvae develop to maturity inside the fig. They give rise to wingless males and winged females. The males never leave the fig but fertilise the females before they depart. On leaving the fig the female gall wasps collect pollen from the male flowers. They find another fig at the right stage and pollinate all the female flowers inside, before laying eggs in the short-styled ones.

Figure 4.63. The *Yucca* flower (Agavaceae) provides ovules/seeds to attract pollinators; it has a glandular stamen and grooved pistil.

Animal dispersal of seed and fruits is called zoochory. Zoochory is very common and predominates in some habitats such as moist and wet tropical forests, heathlands and grasslands. The evolution of the fruits of flowering plants may have provided an important new kind of high-energy food especially for small mammals and birds.

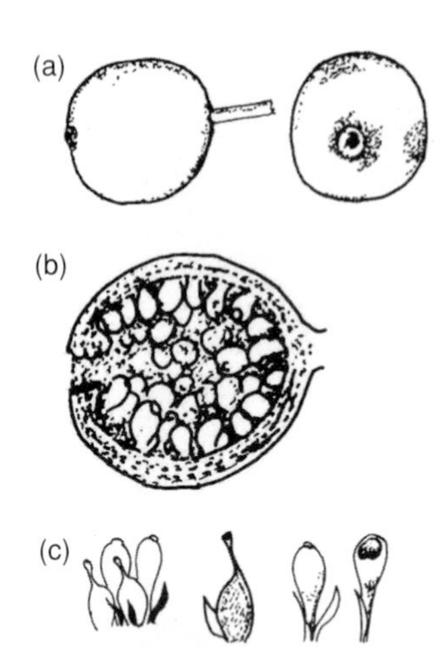

Figure 4.64. Pollination in figs: (a) exterior view of syconium showing the apical pore; (b) section showing florets inside syconium; (c) dimorphic female florets. Long-styled florets produce seed and short-styled florets provide brood places for fig wasps.

Figure 4.65. Ectozoochory, the hooked burr of burdock.

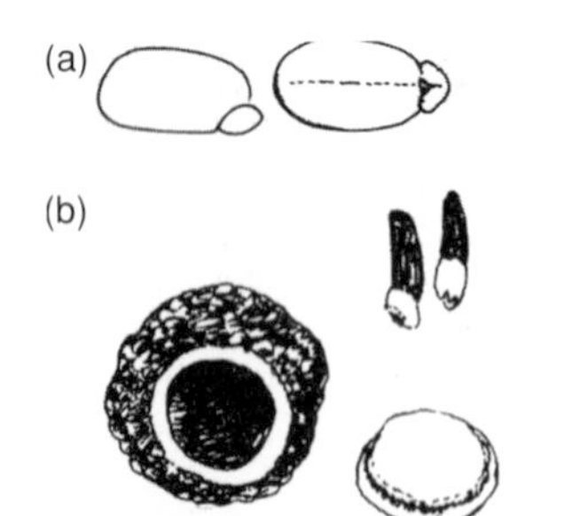

Figure 4.66. Arils in flowering plants: (a) *Ricinus* caruncle on the seed; (b) fruit of *Bertholletia* (Lecythidaceae) brazil nut with a woody capsule and hard seeds, each seed having a fleshy aril.

Dispersal by mammals and birds is either external, ectozoochory, or internal, endozoochory. Different fruits or seeds have characteristic retention rates on or in the animal. Zoochory is usually accompanied by mechanisms to protect the seed from destruction while in transport. For example multiple-seeded dry indehiscent fruits called loments or schizocarps can split into single-seeded segments (each segment is called a mericarp), so that the seed is also protected by the pericarp.

Ectozoochorous fruits and seeds are adapted by having hooks, burrs or being sticky though some are simple and dispersed, for example, in the mud clinging to the animals' feet.

A special example of zoochory is the dispersal of seed by ants, myrmechory. Myrmechory has been recorded in over 80 families of plants, and in some communities myrmechorous species account for 35% of all plants. It is especially common in dry heathlands. In some cases there is a close mutualistic relationship between the plant and the species of ant. Ant-dispersed seeds may possess a special oil-body, an elaiosome, or an aril, as a food supply for the ants. The elaiosome is clipped off in the nest. However, seeds lacking these structures may also be dispersed by ants. One limitation on seed dispersed by ants is that the seed has to be relatively small, but, ant-dispersed seed has the advantage of being abandoned/planted in or near the fine nutrient-rich tilth of an ant nest.

One disadvantage that endozoochorous seeds suffer is that they are concentrated in the animal's gut, and then there is very strong competition between seedlings growing out of faeces. However, endozoochorous seed can be quite large and may be retained longer in the animal's gut and therefore dispersed further. The large size gives the seedling a large energy source to enable it to establish itself rapidly. Endozoochorous agents include mainly mammals and birds but also fish, tortoises and earthworms. For example fish are especially important in the dispersal of river bank *Ficus* species. Usually the seed is surrounded by an attractive, sometimes succulent, nutritious fruit and has a digestion resistant testa or endocarp. In the tropical forest of Peru two thirds of the fruit species belong to one of two classes, adapted for dispersal by birds or mammals. Bird diaspores are small unprotected drupes, scentless, black, blue, green, purple or red, and rich in lipid or protein. Mammal (i.e. monkey) fruits are large arillate or compound fruits often protected by a husk. They are heavily scented and green, yellow, brown, orange or white. They are rich in protein, sugar or starch. Bat fruits are odourless or musky; large or small, white, whitish, green or yellow, and rich in lipid or starch.

Some species have fruits or seeds that are dispersed by a wide range of agents. It is not uncommon for a range of dispersal syndromes to have evolved in a single genus. In *Acacia* there is an interesting geographical pattern of dispersal syndromes. Myrmechory is found only in the Australian species, and the seed is arillate. Bird dispersal is common in the American species. They have seeds embedded in and contrasting in colour with a bright fruit pulp. In Africa the many species produce indehiscent or tardily dehiscent fruits, which are eaten by large herbivores.

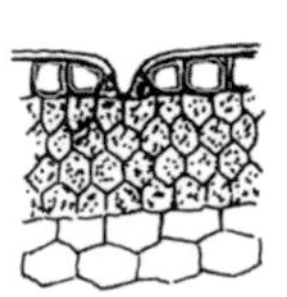

Figure 4.67. Nectaries have different forms, from breaks in the epidermis to modified stomata.

NECTAR

Many species pay an indirect tax on reproduction by providing nectar as an alternative food to pollen. In the early stages of the evolution of flowering plants the stigmatic exudate may have provided the attractant but in most nectariferous living flowers nectar is produced by specialised nectaries. At its simplest nectar is exuded onto the surface of the disk or petal but it may also be stored in various pouches or spurs. Nectar can be exuded from the epidermis through the cuticle, through thin cell walls and in places where the epidermal cell wall and cuticle is incomplete, or through special pores that are often situated at the end of the phloem strands.

Petal and tepal nectaries are commonly associated with a corolla spur or a pouch that may not be nectariferous itself but collect nectar produced by the disk or stamen. There is large variation in the length of any spur. Some orchids have extraordinary long spurs. Perhaps the longest are those of *Angraecum sesquipedale* from Madagascar which has spurs more than 40 cm long. It is visited by a hawkmoth with an extraordinary long proboscis. Stamen nectaries are provided as staminodes or as modified filament bases. The tubular honey leaves of *Helleborus* in the the buttercup family are fundamentally modified staminoids. In buttercup itself (*Ranunculus*) the staminoidal petals retain a cup-like nectary at the base. Nectariferous appendages of the stamens reach into the spur in *Viola*. Disk nectaries are very common in which the receptacle has a conspicuous ring between the stamens and the pistil that glistens with nectar. Septal nectaries are present in five different super-orders of monocots including the amaryllids, bromeliads and gingers. The nectariferous region is found in a groove or pocket between the septa of the locules of the ovary. The nectary connects to the outside by a slit or pore at the base of the style. They are absent from most lilies and orchids, which have a nectariferous perianth. Nectar may be exuded by specialised hairs (trichomes), which are often found in tufts. This type of nectary is relatively rare and is confined to a few families.

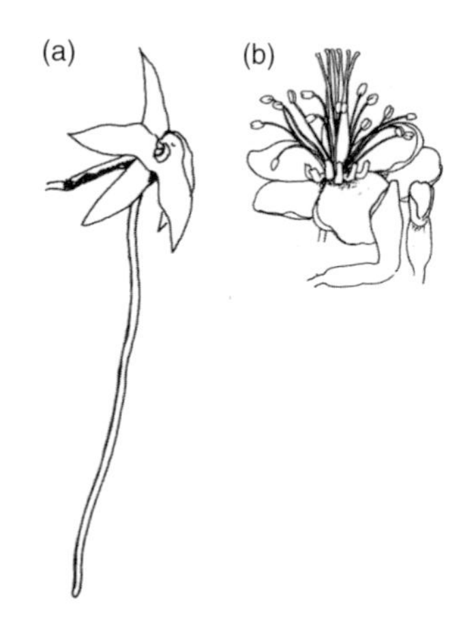

Figure 4.68. Flowers with spurs or honey leaves: (a) *Angraecum sesquipedale* with a spur up to 40 cm long; (b) *Helleborus* honey leaves are modified staminodes.

Nectaries outside the flower are commonly associated with maintaining a population of ants that protect the plant but in some specialised inflorescences like the cyathium of the spurges *Euphorbia*, the extra-floral nectaries do attract pollinators.

Sugar concentration varies between about a quarter and three-quarters by weight. There are three main types of menu on offer. Some plants like *Berberis* and *Helleborus* produce a nectar that has mainly sucrose. Others produced a balanced menu with sucrose, fructose and glucose in roughly equal proportion. *Abutilon* is like this. Some large and relatively advanced families like the cabbage, carrot and daisy families (Brassicaceae, Apiaceae and Asteraceae) produce nectar that has low sucrose but high fructose and glucose composition. Different species of rhododendrons produce each kind of menu.

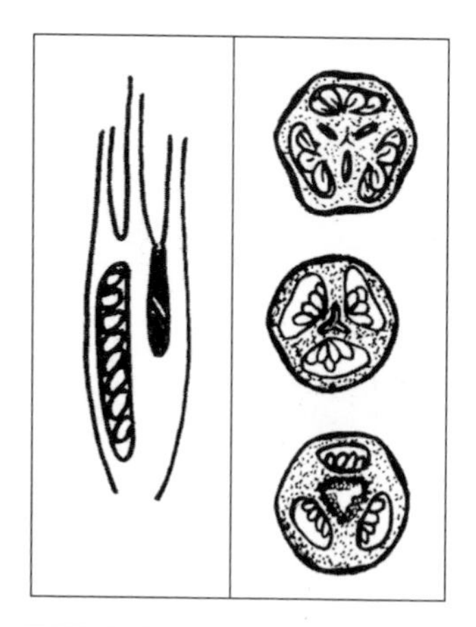

Figure 4.69. Septal nectaries in Asparagales, seen in transverse and longitudinal sections of the pistil.

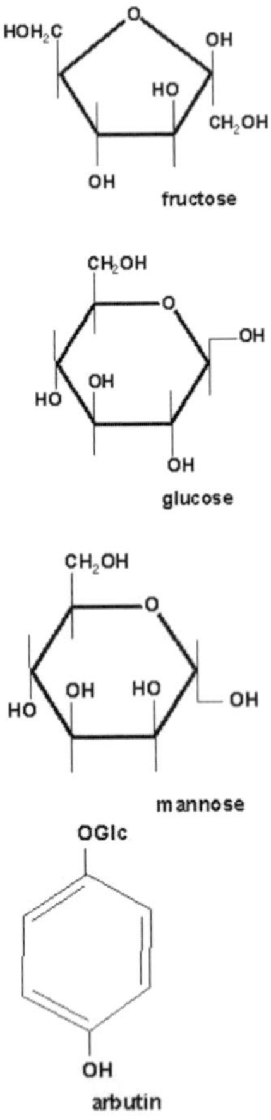

Different menus are favoured by different kinds of pollinator. Hummingbird and bee flower nectars have high sucrose contents. Sugar concentration and therefore viscosity of the nectar is low in flowers that are visited by hovering pollinators, which may only stay for a short visit. Concentrated viscous nectar could be a problem for a long-tongued moth. Most birds suck nectar, while bats quite commonly lap it, so that both may require relatively dilute nectar. Very concentrated, even crystalline, nectars are most suitable for pollinators like flies with relatively unspecialised mouthparts. Concentrated nectars provide more energy per unit volume. An important difference is between the total volume of nectar produced per flower by different species, which varies very greatly. Dilute nectars in hot dry weather are an important source of water. Flowers vary in the amount of nectar available as they age and during the course of a day. Commonly nectar is produced early in the morning and sometimes also in the evening. The nectar becomes more concentrated as it evaporates during the day. In mustard (*Sinapis*), butterflies and bees visit the dilute nectar at the beginning of the day whereas flies visit for the concentrated nectar later in the day.

Nectar also contains amino acids for protein building. Nectars with high concentrations of amino acids are especially common in butterfly-pollinated flowers and also flowers visited by carrion and dung flies. The amino acid concentration of flowers pollinated by dung flies and carrion beetles is up to twelve times more concentrated than other kinds of flowers. Tropical butterflies will also visit carrion, human sweat and putrescent fruit for the nitrogen it provides. Thirteen different amino acids have been recorded in the nectar of more than half the flowers that have been tested. Alanine is the most widespread.

A strange phenomenon is the presence of toxic compounds in the nectar of some species. Toxic compounds include alkaloids, phenolic substances and glycosides. In some cases it seems that these compounds are addictive to the pollinator. Hawkmoths get 'high' on the nectar of a species of *Datura* in the potato family. These alkaloids are the hallucinogenic compounds used by many indigenous peoples in their sacred rituals. The hawkmoths get dizzy and fly off very erratically after visiting the flowers. Honey produced from ragwort nectar (*Senecio jacobaea*) contains alkaloids and is bitter to taste. Arbutin from the strawberry tree (*Arbutus*) is toxic to bees, as is the sugar mannose in the nectar and pollen of the linden (*Tilia*). Nectar may not be quite as innocuous as it looks. Some species of *Rhododendron* (Ericaceae) have nectar that is toxic to humans. The function of this is unknown but it may play a role in reducing nectar-robbing and promoting greater pollinator specificity. In the Himalayas the exquisite Fire-tailed Myzornis (*Myzornis pyrrhoura*: Timaliidae), which has a bristled tongue, feeds on the nectar of *Rhododendron* spp. apparently without any ill effects. The sugar galactose in the stigmatal exudate of tulips is also toxic to bees. The function of this toxicity is unknown. Perhaps the natural pollinators are tolerant of the toxic compounds.

Figure 4.70. Constituents in nectar.

OIL, FOOD BODIES AND WAX

Small amounts of lipids (oils) are present in nectar but they are also produced by specialised glands called elaiophores. Oil production is relatively rare and is associated with pollination by solitary bees. The bees use the oil to feed their pupae. *Calceolaria* have dense mats of hairs that produce oil. Similar oil glands are found in some species of *Lysimachia* and some species of the Iris family.

Starchy food bodies are provided by some plants as a reward for ants that protect the flowers. Food bodies are also produced for pollinators like beetles, as in *Calycanthus*. Several rather primitive genera provide food bodies: *Belliolum*, *Zygogynum*, *Exospermum*, *Bubbia*, *Eupomatia* and some species in the custard-apple family (Annonaceae). The staminoids secrete sticky nutritious material or sometimes are eaten themselves.

Resin may be provided and collected from plants by bees for nest construction. Resin is normally collected from wounds in the plant but some species have specialised organs for its production. *Dalechampia* in the Euphorbiaceae has bright pink bracts and a brightly coloured inflorescence axis with small white flowers. Glands on the inflorescence produce resin. *Clusia* produces resin from its stamens and staminodes and elsewhere. The resin is collected by humans and used to caulk boats or as bird-lime. The orchid *Maxillaria* produces waxes as an attractant.

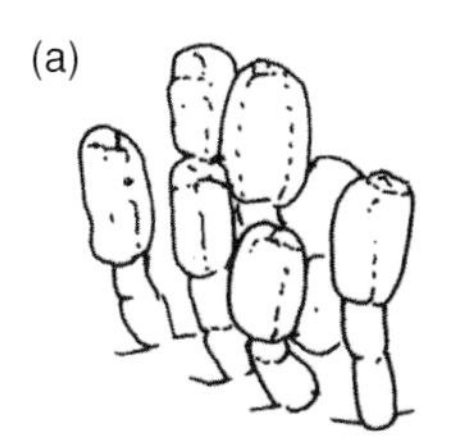

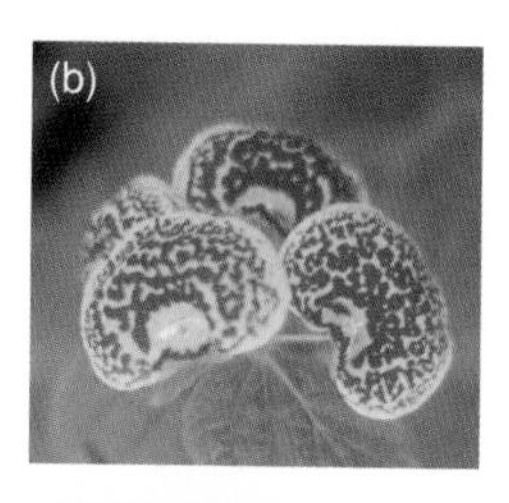

Figure 4.71. Oil-producing flowers: (a) stalked elaiophores of *Calceolaria*; (b) *Calceolaria* flowers.

FRUITS

Another kind of secondary attractant is the fleshy tissue present in fruits. Fleshy fruits of several different sorts have been categorised on the basis of the anatomy of the pericarp. Three anatomical regions are present in the pericarp: an outer exocarp, an inner endocarp and the mesocarp between. There are also specialised multiple pseudofruits in the flowering plants in which the fleshy tissue is at least in part composed of the receptacle, stem and/or bracts. Two contrasting examples are the pineapple and the fig.

Figure 4.72. Resin-producing flower of *Clusia*.

Fleshy fruit types	Structure, anatomy	Example
Berry	Fleshy mesocarp, thin endocarp	Tomato, banana
Hesperidium	Berry with segments and glandular mesocarp	Orange
Drupe	Fleshy mesocarp and stony endocarp	Plum
Drupecetum	Aggregate of small drupes	Blackberry
Pome	Inferior ovary with fleshy receptacle	Apple, pear

Secondary attractants – advertising

Advertising is an essential aspect of animal dispersal. The form, scent and colour of flowers, seeds and fruits are attractive. Fruits and arils are brightly coloured and scented to attract animals. The arils of *Taxus* are bright red. Seeding cones of cycads have gorgeous colours of purple and red and black. Bird-dispersed diaspores tend to be scentless whereas mammalian-dispersed fruits are richly scented often by esters.

Advertising for pollinators is largely responsible for the extraordinary diversity of flowers described below.

4.6 The diversity of flowers

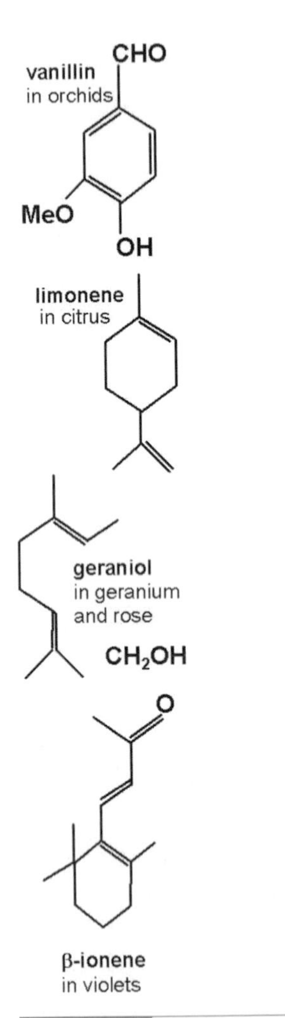

Figure 4.73. Components of flower scents.

There are 250 000 species of flowering plants. To a very large extent the diversity of flowers is expressed in the way they advertise themselves to animal pollinators. Vectors are initially attracted to flowers, and guided within them, by various kinds of advertising and sign posting. We can thank the secondary attractants of scent, colour, and shape for their wonderful beauty and diversity.

4.6.1 Flower scents

Scents are particularly important as long-distance adverts in insect and bat, but not bird, pollination. Scent can be highly specific. Each *Ophrys* orchid produces a scent that stimulates its own pollinator but not others: *Ophrys lutea* – *Andrena pubescens*, *Ophrys sphecoides* – *Andrena jocobi*, *Ophrys apifera* – *Eucera tuberculata*. Bees learn the scents of flowers much more rapidly than their colour and shape. In experimental tests they correctly choose a flower based on its scent 97%–100% of the time after a single exposure. Scents are ethereal oils, mainly mono- or sesquiterpenes or many other compounds. The scent is composed of several or many components which reinforce each other.

The whole plant may be scented, as in the Lamiaceae, which includes many herbs like the mints, *Mentha*, but such whole-plant scents are not necessarily attractive, and may be repellant to herbivores. Scents as adverts are usually produced from the surface of the petals in patterns of spots or streaks or in higher concentration towards the centre of the flower. In some flowers scent is produced in specialised organs called osmophores. These are 'scent-aerials' that include flat, whip, and paint-brush shapes. The orchid *Pleurothallis* has club-shaped osmophores on the lateral tepals.

Pollen may also be scented. The pollen of different species has different volatile constituents: 31 terpenoids were detected in 2 species of rose (*Rosa*). Peculiarly, some pollen is repulsive because of its scent and toxic constitution. Honeybees try to avoid the pollen of some species of cotton, *Gossypium*, while both bees and wasps try to avoid contact with *Kallstroemia grandiflora* pollen. Some flowers smell sweet, but aminoid odours, which are unpleasant to us, attract flies and other insects by mimicking the smell of dung or decaying meat. Scents are monoamines like methylamine, or diamines like putrescine and cadaverine whose names are testament to their pungency. The Asclepiadaceae have many species adapted to fly pollination: *Stapelia* species mimic the smell, colouration and texture of rotting meat. The world's largest flower *Rafflesia arnoldii*, which is up to a metre across, is pollinated by flies attracted by the smell of carrion and its blotchy appearance. The flies make their way to the underside of the disk where polysporangiate anthers smear a pollen mush on their backs.

Many aroids or arums produce scents of carrion, dung or urine, to attract the pollinators, which are mostly flies. The spadix has cells which are packed with mitochondria. Respiration can significantly raise the temperature, up to 16 °C above ambient in *Arum*. The raised temperature mimics that of the rotting animal products and also drives off the scents effectively.

Insects may be attracted by the warmth of a flower. In arctic and alpine regions flowers are heliophyllous so that the flower follows the sun, focusing its rays within the corolla. Mosquitoes bask at the focal point of the flower. Patterns of heat distribution within the flower may help direct the pollinator towards its centre after arrival at the flower.

Scent and heat production are often closely associated. The waterlily *Nuphar* raises the temperature of its flowers to 32 °C over a four-day period. Flies and beetles are attracted by an evaporating scent and then trapped in a kind of chamber roofed by the enclosing tepals. The insects are highly active on the warm dance-floor provided, and by accident first pollinate and then get covered by pollen before they are released. The flowers of species of *Annona*, the soursop and custard apple are also strongly warmth producing. Again, beetles congregate in the warm flowers at night-time to mate. Male flowers cease to be warm first, so that beetles move on to the female flowers to keep warm and thereby carry out pollination.

Figure 4.74. The orchid *Pleurothallis raymondii* with paired orbicular osmophores (scent wands).

Figure 4.75. Carrion flower of *Stapelia*. This flower produces a foetid smell of rotten carrion that is attractive to flies.

4.6.2 Flower colour and texture

Different colours are effective for different pollinators. Red attracts birds. Pale colours are effective for night pollinators such as moths and bats, and in deeply shaded forests. Insects see in a different spectrum of wavelengths to humans. Bees cannot see red but see yellow, blue and ultra-violet. Red poppies are ultra-violet to bees. The predominance of red in bird-flowers may be because birds have their greatest spectral sensitivity and finest hue discrimination towards the long-wavelength (red) end of the spectrum. Many bird-flowers are actually green, for example some species of *Centropogon* (Lobeliaceae). Hummingbirds have no inherited colour preference although this can be modified by conditioning. Individuals have different preferences for particular flower species in some areas and rarely visit other flowers. They pefer red but can also respond to near-violet light, which is invisible to humans.

Figure 4.76. *Nuphar* raises the temperature of its flowers to attract pollinators. An alcoholic scent is driven off to attract beetles.

Cyanidin, which is also found in gymnosperms, is probably the basic pigment (Figure 4.77). There are three basic anthocyanidins (*anthos* is Greek for flower); cyanidin (magenta), pelargonidin (orange-red) and delphinidin (purple). Different patterns of hydroxylation result in pigments like apigenidin (yellow) and petunidin (mauve). Other factors that modify colours are the pigment concentration,

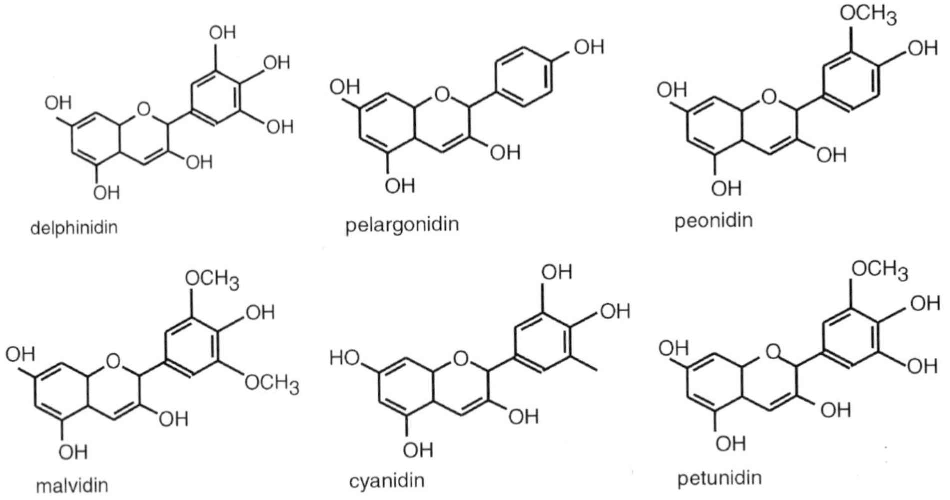

delphinidin pelargonidin peonidin

malvidin cyanidin petunidin

Figure 4.77. Anthocyanidin pigments are widespread in plants.

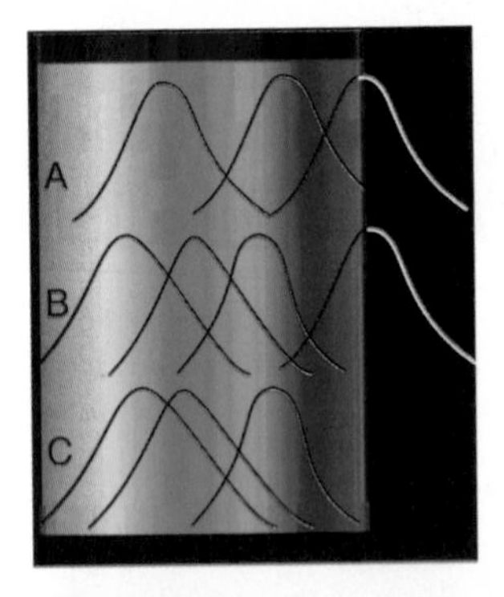

Figure 4.78. Visible spectrum of bees (A), birds (B) and humans (C). Both birds and bees are sensitive to the ultra-violet, but birds are more sensitive at longer wavelengths.

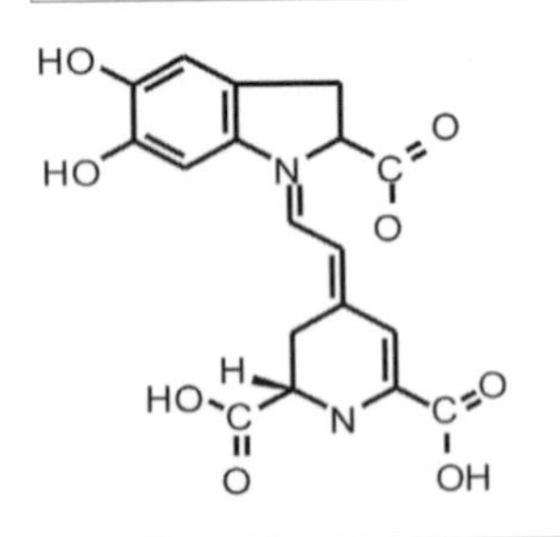

Figure 4.79. A betanidin pigment: betaxanthin. Betanidins are found in only a small group of Caryophyllid families.

the presence of flavone or flavonol co-pigments, a chelating metal, carotenoids, an aromatic acyl or sugar substitution, or a methylation. Differences in flower pigments exist at various levels in the taxonomic hierarchy. A well-known example is the presence of a unique set of pigments, the betalains, in 10 families of the Caryophyllales (Centrospermae), a characteristic that, among others, links such diverse families as the Cactaceae, Aizoaceae, Chenopodiaceae and Amaranthaceae.

Colour patterning helps direct pollinators to the centre of the flower; many are not visible to our eyes but show up in the UV spectrum. There are stripes that lead to the centre, like the lights leading to an airport runway. These 'honey guides' are particularly prominent in bee flowers. Alternatively, the centre of the flower where the nectar is provided may be marked out like a bull's-eye. Dark stamens may be silhouetted against a lighter-coloured corolla.

Petals have textural patterning, mattness or shininess or even directional brightness. A common pattern is the possession of conical-papillate epidermal cells, like blunt projecting cones which maximise the reception of low-angle light and evenly spread light reflected back from the mesophyll. Surface microsculpturing adds guide lines.

In some flowers other visual cues on the plant signal the status of the flower, for example in epiphytic *Columnea florida* (Gesneriaceae) the flowers are inconspicuous but just before flowering two translucent red spots or 'windows' appear on the leaves, and are conspicuous to hummingbirds approaching from below. In the West Indian *Marcgravia sintenisii* (Marcgraviaceae), which is visited by hummingbirds, honeycreepers (Coerebidae) and todies (Todidae), the nectaries are at first yellow but turn bright red when the flower opens.

Both insect and vertebrate eyes are attuned to detecting movement. One interesting kind of patterning is the chequer-board pattern seen in many fly-pollinated flowers such as *Fritillaria* and *Oncidium*. It is thought that this pattern creates an optical illusion of movement because of the insect's compound eye. Possibly the arrangement of florets in the umbel of the Apiaceae produces a similar effect. Movement in the breeze emphasizes the presence of a flower, and the location of a flower on a flexible peduncle may exaggerate the motion. The orchid *Bulbophyllum macrorhopalon* does it in a more straightforward but extraordinary way. It has strange tassel-like appendages that flicker in slight drafts of air.

Colour is sometimes used deceptively. For example *Lantana camara*, *Asclepias curassavica* (blood-flower) and *Epidendrum radicans* share a common colour scheme – a mixture of red and orange yellow. The flowers are very different in shape but from a distance the inflorescences create a strikingly similar impression. They are pollinated by the same kinds of monarch butterflies. *Epidendrum* actually exploits the other flower species because it is nectarless. There are several remarkable examples of orchids that mimic other flowers, which provide the pollinator with nectar or pollen, but fail to provide either for the fooled. For example in northern Thailand the orchids *Cymbidium insigne* and *Dendrobium infundibulum* mimic *Rhododendron lyi*.

Figure 4.80. Monosymmetric flower with guide line patterning: *Euphrasia*.

Large flowers are impressive but small flowers achieve the same visual power by being grouped together in an inflorescence. In some inflorescences the effect is of a single flower, a pseudanthium, with flowers in different parts of the inflorescence having different colours and shapes; the outer ones look like petals and the inner ones are less showy and are mainly reproductive.

Figure 4.81. *Verbena (Lantana)* showing colour variation in inflorescence; the inflorescence changes colour as it matures.

4.6.3 Flower symmetry

The silhouette of a flower or inflorescence is important for its attractiveness. Dissected, star-shaped flowers attract bees more effectively than those with smooth outlines. Usually the corolla determines the shape of a flower but sometimes the calyx is showy. A particularly important difference is between flowers that have radial symmetry (actinomorphic or polysymmetric flowers) and those with a single axis of symmetry (zygomorphic or monosymmetric flowers). In the irises the tepals are arranged in two whorls. The outer deflexed 'falls' are often bearded while the inner 'standards', which alternate with them are glabrous and erect. Petalloid stigmas hooded over the stamens and alternate with the standards. The flower is radially symmetrical but functionally divided into three zygomorphic parts. A bee that approaches and lands on a fall, pushes its way under the petalloid stigma, pollinates it, and then collects pollen from the anther as it passes. As it reverses out, a flap on the stigma is pushed over the receptive surface preventing self-pollination.

Figure 4.82. Chequer-board pattern in *Fritillaria* attracts flies, perhaps by simulating false movement.

Zygomorphic flowers, with an expanded lip or flag, are particularly associated with tall inflorescences where the flowers are held in a lateral position and approached in one direction only. The different sexual timing of flowers in an infloresecence enhances cross-pollination.

Two particular groups have highly zygomorphic flowers and specialised flowers, the gingers and their allies (the Zingiberales) and the orchids (Orchidales). In most orchids the lower median petal, has

Figure 4.83. *Aster* capitulum with a ring of showy strap-shaped florets surrounding the yellow tubular florets in the centre.

Figure 4.84. *Iris* is polysymmetric but behaves functionally like three monosymmetric flowers.

Figure 4.85. Monosymmetric flowers: the Indian-shot plant, *Canna*.

become adapted as a landing platform, a labellum, for the pollinator, giving the flower bilateral symmetry. In some cases the labellum has rough patches to aid landing. The style is fused to one of the stamens to form a column, the gynostemium and the median stigma lobe is a sterile flap, the rostellum, separating the anther from the fertile stigma lobes so that self-pollination is prevented. The pollen is aggregated into sticky masses called pollinia. The gingers, Zingiberaceae, have an analogous flower except that the style is free, and lies in a deep groove of the one surviving median stamen. The stigma protrudes beyond the pollen sacs. The other stamens are staminodes, some forming the labellum.

4.6.4 Flower architecture

There is a bewildering diversity of floral architecture and most flowers can be visited by a wide range of pollinators. In the endless evolutionary ballet between flowers and their pollinators it is wise to be able to attract different partners. Most flowers are polyphilic and attract a diverse range of pollinators, while most pollinators are polytropic and will visit a range of flower types. The diversity of flowers and pseudanthia is bewildering but they can be categorised into various architectural models either by their shape or by the kind of pollinators that visit them. Form is taken to the extreme in a few species of flowers, which are so specialised that only a single species of pollinator can effectively pollinate them, but these are rare.

Disk- and bowl-shaped flowers have relatively open access in many possible directions and often have unfused parts. They are pollinated by a wide range of pollinators including not only the relatively unspecialised beetles and flies, but also bees, syrphids, and even bats. Included in this category might be those species with flat umbels.

Bell- and funnel-shaped flowers have slightly restricted access and usually a fused perianth. They are pollinated by the more agile flies, bees and syrphids, which have to crawl inside the perianth. If the bell is large enough, bats may cling onto the rim. Most flowers are held erect or laterally but some are pendent, when the perianth protects the interior of the flower from getting wet.

Tubular flowers have a narrow tubular region of the perianth to which only long proboscis, tongue or beak can gain access. There is an obvious overlap between this kind of flower and the funnel- or bell-shaped flower but here the opening is more of a key-hole than a gullet. In most tubular flowers the perianth is fused and the tube is sometimes protected by hairs at the throat or within it. Some kinds of tubular flowers, (salverform) flowers, have a flat disk-shaped platform around the mouth of the tube. Others are monosymmetric and have a lip or a flag.

Spurred flowers have an alternative mechanism for concealing the nectar. The spur is part of the perianth projected back behind the receptacle. Butterflies, moths and birds such as hummingbirds are the most important pollinators.

Lipped flowers are usually monosymmetric. The lip provides a landing platform for the pollinator. They are usually tubular or funnel shaped. The perianth may be relatively open as in many other Labiates.

Resupinate flowers have the whole flower twisted so that a median upper tepal or petal can act as a lower landing platform. They are common in the orchids, and are also found in some Fabaceae.

Keeled flowers like the sweet pea are monosymmetric and have a large advertising flag above a boat-shaped keel that protects the stamens.

Revolver and roundabout flowers or inflorescences have multiple points of access to pollen or nectar arranged in a circle. They differ in the behaviour of the pollinator: in revolver flowers, such as various species of asclepiads, the insect rotates its body around the flower with its head at the hub of the wheel; in roundabout flowers, such as various species of passion flower, round the circumference of the flower the insect walks sideways.

Figure 4.86. Floral form: resupinate flower, *Clitoria*.

Brush blossoms have long and showy stamens. Some are pollen flowers only, but many also provide nectar. The flowers may be relatively large and solitary with many stamens but sometimes with the stamens fused together, as in *Hibiscus*. More commonly brush blossoms are actually inflorescences of many small flowers grouped together, each flower with exserted styles and stamens. Many Myrtaceae have brush-like flowers grouped to make the bottle-brush type inflorescence. Mimosas and acacias have small flowers aggregated into dense spikes or heads as in *Calliandra*.

Figure 4.87. Floral form: roundabout flower, *Passiflora*.

Trap blossoms are the physical manifestation of a particularly fascinating kind of relationship between flowers and their pollinators. A dance macabre has arisen in which the flowers temporarily trap their pollinators. When making its escape the insect is forced to pollinate the flower. There are several different kinds of trap blossoms.

Pit traps have evolved in several families. The orchid *Coryanthes* is a kind of trap orchid. It attracts bees with scent. The flower has a very complicated three-dimensional shape. Spurs at the base of the gynostemium drip water into a bowl-shaped lip. Bees are caught by the drops and fall into the watery trap from where they have to clamber out via the stigma and pollinia. The lady's slipper orchids (*Cypripedium*) also have a trap. Pollinators fall inside the inflated lower lip and can escape only by climbing out the back and along the column (gynostemium), depositing pollinia on the stigma before being reloaded by the flowers own pollinia. A translucent patch at the back of the flower acts like an exit sign.

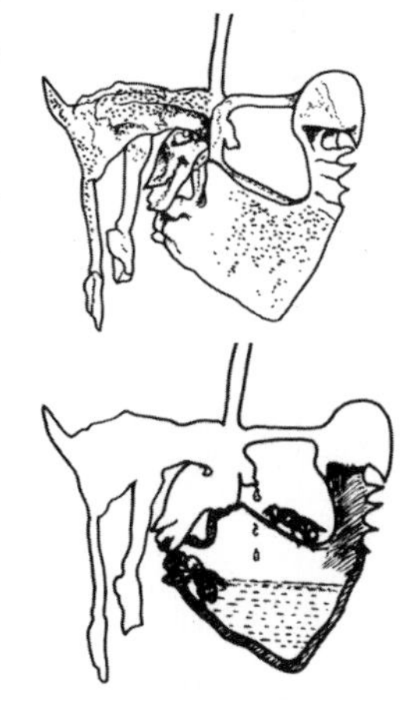

Figure 4.88. Trap blossom, water trap; *Coryanthes*. Pollinators slide into a pool of liquid in the lower lobe and have to climb out by squeezing past the fertile parts of the flower.

Aroid inflorescences are a special form of pit trap. The aroids have a spathe surrounding a spadix, which has around its base a curtain of many tiny unisexual flowers. In *Arum*, midges are attracted by the warm and odiferous part of the spadix above the curtain but slip into the chamber because of the oily, smooth surface the spathe. The only escape is by climbing up the spadix, but at first they are restricted to the lower part by a curtain of downward pointing sterile

Figure 4.89. Pit trap in *Arum maculatum*. A curtain of sterile flowers in the throat of the spathe prevents the escape of pollinators until the female flowers have been pollinated and the male flowers have shed their pollen.

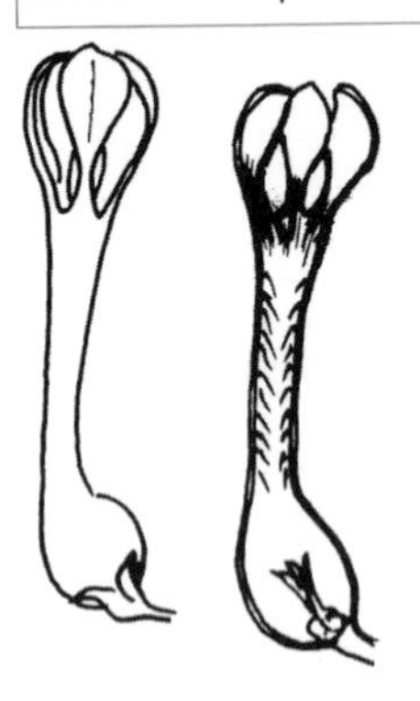

Figure 4.90. *Ceropegia* encourages pollinators to enter its chamber by patterns of light and shade, and discourages their escape with downward-painting hairs.

Figure 4.91. Lobster-pot trap; *Calycanthus*. The tightly overlapping tepals allow only entry into the flower until they reflex to open.

florets. In this region there are female florets that are receptive to the pollen the flies have carried from other plants. Later, the midges are allowed further up into a region where there are male florets. Here they are covered with pollen before the outer curtain of sterile florets withers to allow their escape. *Alocasia* and *Colocasia*, taro and cocoyam respectively, attract beetles and flies into the chamber and trap them there overnight when the neck constricts. *Amorphophallus titanum*, over 2 m tall, which is pollinated by large beetles, prevents their escape from the region of the spadix by an overlapping ledge around the rim of the spadix and by the smooth walls of the spathe. They only escape when the spathe disintegrates. The effectiveness of pit traps is often magnified by scented and optical trickery.

Optical traps have evolved in several different families often combined with rings of hairs that temporarily prevent exit from a pollinating chamber. They have a dark chamber whose entry is visible from the outside but is hidden in darkness from the inside. Instead, the pollinators are confused by translucent windows in the wall of the chamber. There has been remarkable convergence between *Aristolochia* (Aristolochiaceae) and *Ceropegia* (Asclepiadaceae) in this respect. In Dutchman's pipe, *Aristolochia*, the curved perianth tube attracts pollinators in but confuses them so that they fly or climb up towards the anthers and stigma. In *Ceropegia* the lantern-like upper part of the flower allows the pollinator in, and downward pointing hairs and a translucent lower chamber encourage it down to the anthers and stigmas.

Lobster-pot traps like that in *Calcyanthus* have many tightly overlapping tepals which form a cone like a lobster-pot trap. Tiny beetles are attracted in between the tepals by scent to eat the food bodies produced on the tips of the tepals and then find they cannot get out because of stiff downward pointing bristles. They are forced further and further in towards the receptive stigmas; 2–3 days later the staminodes bend to cover the stigmas, while the stamens release their pollen onto the beetles. Then the tepals then bend back to release the beetles.

A **liquid trap** is produced by the water-lily *Nymphaea*. The cup-shaped stigma exudes a pool of sugary liquid to form a liquid trap. Insects carrying pollen land on the upright immature stamens and slide off into the pool where some of them drown. The pool evaporates or is reabsorbed so that the bodies of the insects float down with their pollen onto the stigma. Later the stamens fold over the stigma and release their pollen.

Booby-trap flowers in the broom and gorse, *Cytisus* and *Ulex*, have a staminal tube and style forming a stiff spring depressed by the keel; the insect landing on the keel causes it to split, releasing the style and stamens explosively (Figure 4.92). One of the most bizarre pollination mechanisms is utilised by the tropical mistletoes (Loranthaceae). The petal lobes of the tubular corolla are joined at the tip. They become increasingly turgid as the flower matures until, at the slightest touch by a visiting bird (usually a flowerpecker: Dicaeidae), it bursts open

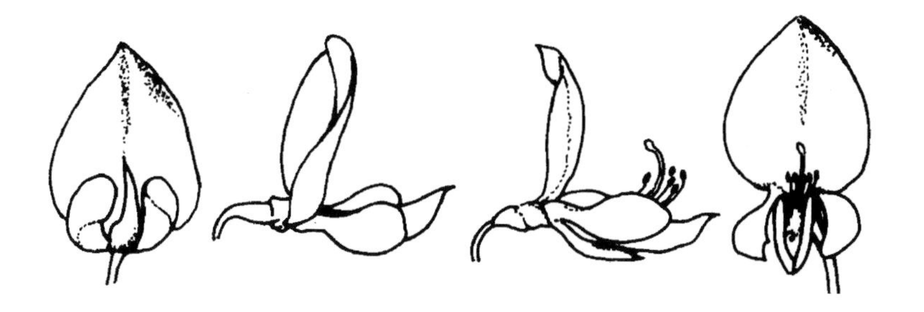

Figure 4.92. Explosive flower of *Cytisus*: the staminal tube and style form a stiff spring depressed by the keel. The insect landing on the keel causes it to split, releasing the style and stamens explosively.

violently, and the stamens, which are adnate to the corolla, inflex or recurve sharply, showering the bird's head with pollen. At the same time the style snaps to one side to prevent self-pollination.

Tender traps lure their victim with the rich scents of sex or food. Some species of orchid have encouraged faithful pollination by mimicking the female of their male pollinator so that the bee attempts to copulate with the flower. *Drakea*, the dragon orchid, is an Australian orchid that has a labellum mimicking a female wasp. As the male wasp lands on it the labellum swings forward banging the wasps head against the anther or stigma. Pseudo-copulation is a common pollination mechanism, well known in the European flora because of the different species of *Ophrys* (bee-orchid), each resembling a different pollinator. There is no food reward for the provision of nectar would encourage the bee to visit every flower in an inflorescence leading to self-pollination. A scent or wax is provided, which may act like an insect pheromone. Euglossine bees collect droplets of perfume from the surface of the flower and store it in their hollow hind legs.

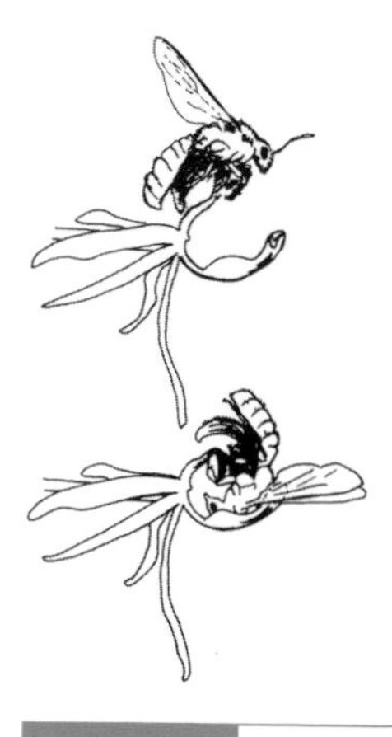

Figure 4.93. Trip flower; *Drakea* mimics a female wasp and flips the male against the anther and stigma.

Tender traps have the advantage that male bees searching for a female to copulate with are likely to range widely and are more discerning of shape and scent than they would be if looking for food, so that cross-pollination is encouraged. A different example of tender trap appeals not to the sexual appetite of the pollinator but to its hunger; for example in *Arisarum proboscideum* the spadix is spongy and white like the underside of a fungus, and gnats are attracted and congregate inside the spathe.

Figure 4.94. Tender trap; *Ophrys bombylifera* has a flower that mimics the form and scent of a female bumble bee.

4.6.5 Flower pollination syndromes

From within the vast diversity of 250 000 flower species a number of syndromes can be discerned that are associated with particular kinds of pollinators.

Melittophily or pollination by bees is the most important type of pollination. Bees collect both pollen and nectar. Most pollen flowers are pollinated by bees. Bee flowers tend to have ultra-violet, blue and yellow pigments. A single species of bee can potentially pollinate a broad range of flowers but on any one day an individual often shows constancy. Bees range in size from 2 mm to 4 cm, and vary in physical features such as the length of the proboscis and adaptations for collecting pollen. Pollen is collected by buzzing on or within the flower, vibrating pollen out of the anthers. Flowers with

Figure 4.95. Bee pollinating a lavender. Blue and mauve flowers are commonly pollinated by bees.

Figure 4.96. Buzz pollination in *Solanum* (Solanaceae). The anthers have terminal pores, and vibrations induced by the pollinator make the pollen squirt out.

long tubular anthers, which dehisce through an apical pore (poricidal) are particularly adapted to buzz pollination. Pollen is small, dry and produced in large quantity. It is trapped on the hairy body, attracted by electrostatic attraction, and in many cases brushed to specialised pollen-carrying areas. Lipped and keeled flowers like the snap-dragon *Antirrhinum* and *Lathyrus* are particularly adapted to pollination by large powerful bees, which can force their way into the flower seeking nectar.

The honey bee *Apis mellifera* is by far the most important individual species. There are about 20 000 species of bees. Colletids are the most primitive bees and have a broad, blunt or two lobed tongue. Hellectids or sweat bees have short tongues. They include the family Megachilidae which have a scopa, a special pollen collecting brush on the under surface of the abdomen. The Anthoporidae are long-tongued. There are three kinds: the Nomadinae, the large hairy Anthophorinae, and the carpenter bees Xylocopinae. Large carpenter bees of the genus *Xylocarpa* are very important pollinators of tropical trees. The Apidae have a pollen collecting area on the hind leg called a corbicula or pollen basket. The leg has a row of hairs which act as a comb, a projection like a rake and a smooth area where the pollen is packed. Pollen is dampened with nectar and saliva to pack it into place. There are four main kinds of Apidae: the Euglossini are important orchid bees in tropical regions of the Americas; the Bombini are hairy bumble bees; the Meliponinae, distinguished in being stingless, range from the smallest to some of the largest of all bees; and the Apinae include only five species, one of which is *Apis mellifera*. A broad contrast can be made between flowers pollinated by large bees and those pollinated by small bees.

The characteristics of bee flowers	Small bee flowers	Large bee flowers
size	small and regular	large and complex
colour and scent	white	bright and highly scented
reward	pollen, little or no nectar (short tube)	pollen nectar (long tube) and also scent
symmetry	mainly polysymmetric	monosymmetric
incompatibility	more commonly self-compatible but some dioecious	more commonly self-incompatible

Flowers present their pollen onto particular parts of an insect's body. Two distinct kinds are nototribic flowers, which put pollen on the upper surface or back of the pollinator and stenotribic flowers, which deposit pollen on the underside or belly of the pollinator. *Salvia* has hinged anthers, which have one theca sterile and one fertile. As a bee enters the corolla tube it pushes against the sterile theca thereby levering the pollen-bearing theca onto its back. Alternatively the architecture of a flower can be ineffective if the pollinator the robs flower by piercing the perianth.

About 60% of orchids are pollinated by bees although others are adapted for pollination by wasps, moths and butterflies (Lepidoptera), flies (Diptera) and even humming birds. Non-social bees, like bumble bees in the Northern Hemisphere, and other solitary bees are more important pollinators than social bees because they are more effective in pollinating widely dispersed populations. Few bee species visit only one or few species of orchid. In any one area the commonest and most widespread orchid is pollinated by the commonest and most widespread bee. However, the predominant orchid shares the bee with other orchids. Reproductive isolation is maintained because each species places its pollinaria on a different part of the bee's body.

Figure 4.97. Bees pollinating the lousewort *Pedicularis*. Bees enter the flower in different orientations.

Psychophily or pollination by Lepidoptera (butterflies and moths) is a very significant pollination syndrome. Most butterflies and moths have a long proboscis and collect nectar, but primitive micropterygid moths have chewing mouthparts and feed on pollen from open flowers like buttercups. Three major kinds are important pollinators, butterflies, noctuid moths and hawkmoths, each of which predominantly pollinates a different kind of flower.

Butterflies are active during the day. They pollinate many of the same kind of flowers visited by long-tongued bees but do not have the strength to open gullet flowers. They normally alight on the flowers they are visiting, and the flowers are held horizontally with a large lip or are small and held erect but are grouped together to provide a landing platform, as in *Buddleja*. The flowers are brightly coloured, often with contrasting colours, and reds, pinks or even orange are favoured.

Figure 4.98. Placement of pollinia of different species of orchid on different parts of a bee.

Noctuid moths have a relatively short proboscis. Flowers are light coloured, yellowish, greenish or purplish and scent is the main attractant. They normally alight on the flower. Some more open flowers like *Lilium martagnon* have grooves in the tepals which guide the moth's tongue to the nectary.

Hawkmoths, which are active at dusk or in the night, are strong fliers and hover in front of the flowers they pollinate. The flowers are slender and delicate, white or very pale and held horizontally or are pendent. The flower often has a deeply lobed or dissected silhouette to help guide the moth to its centre, but scent is the main attractant. Pollen may be deposited on the proboscis of the moth from flowers with anthers inside the floral tube, but it is commonly deposited on the moth's hairy body and wings by exserted versatile anthers. Commonly the nectar is well hidden at the end of a long tube or spur.

Figure 4.99. *Lonicera* (honeysuckle) is pollinated by moths and has very fragrant flowers.

Myophily is pollination by Diptera (flies). Flies visit relatively simple open flowers for nectar. Glistening nectaries are important attractions. Myophilous flowers have dull colours but attract pollinators by odour. Many work by deceit, mimicking the odour of carrion, urine or dung. The fungal-gnat flowers mimic the scent and surface pattern of mushrooms and toadstools. They include *Arisarum* and *Asarum*. The

Figure 4.100. Fly-pollinated flowers of *Deherainia smargdina* (Theophrastaceae) have a fetid smell and are green. The flowers are protandrous; the stamens lie close to the stigma at first but spring back before the stigma becomes receptive.

Figure 4.101. *Viburnum* has an umbel in which showy sterile flowers surround the fertile flowers.

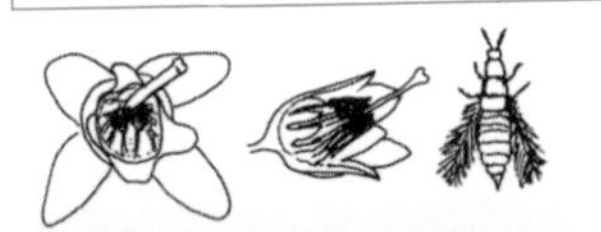

Figure 4.102. Thrip-pollinated *Calluna*. It is also pollinated by bees (producing a honey that flavours the liqueur 'Drambuie') and is also pollinated by wind.

flies attempt to lay their eggs on the flower surface. Syrphids (hover flies) have a long proboscis and mimic bees and wasps. They can gain nectar from tubular flowers and can also eat pollen when it is mixed with saliva. They are common visitors to the yellowish, whitish and pinkish umbels of ivy (*Hedera*), *Cornus*, *Viburnum*, and elder (*Sambucus*), and species of the Apiaceae (Umbelliferae). Bee flies mimic bumble bees and are common visitors to inflorescences with many tiny flowers like willow and thyme.

A range of other insects are pollinators. Heather (*Calluna vulgaris*) is visited by a wide range of large pollinators and is an important source of nectar for honey production but is also pollinated by a species of thrip, 1 mm long, which is so tiny it lives most of its life within the flower. Several can be found in a single flower. The females fly between flowers to mate effecting cross-pollination. They lay their eggs at the base of the corolla and the larvae mature there over winter but leave to pupate in the soil. *Calluna* can also be pollinated by wind.

Wasps will visit a wide range of relatively open flowers. A few (*Polistes*) have specialised mouthparts. Species which are wasp pollinated are often 'brown' and have a fruity scent. Gall wasps are important for fig pollination and some are fooled by *Ophrys* species.

Cantharophily or beetle pollination is relatively unspecialised. The beetles scramble around relatively large bowl-shaped flowers or inflorescences of small closely aggregated flowers and are as likely to eat and destroy floral parts as eat nectar or pollen. They are attracted by sweet scents. Flowers are pale or dull and floral adaptations include having many relatively broad stamens. Weevil-pollinated flowers are more specialised. They have a narrow entrance to a chamber within which the weevils may live. The flowers of *Victoria amazonica* behave in a fascinating, if bizarre, way. During the first night, at sunset, the flowers open. At this stage, the petals are shining white and attract scarab beetles which are trapped at around midnight when the petals close. One to two hours later, the flower begins to change colour to a delicate pink. By sunrise the following morning, it is closed completely. The colour gradually deepens to a rich purple and, by mid afternoon, the flower re-opens, although the staminode chamber containing the beetles remains closed. At sunset, on the second night, the flower reopens, the beetles fly off to repeat their nocturnal feasting and, in so doing, pollinate a fresh flower. During their exit they pick up pollen on their bodies now sticky from a night of gorging on the sugary exudates of the flower's carpels. Thereafter, the flowers submerge and the seeds ripen under water.

Bird pollination or ornithophily is especially important in the tropics. Hummingbirds, honeyeaters, sunbirds, lorikeets and others are all important pollinators. Flowers are often bright red or yellow but lack scent (Figure 4.103). Bird-flowers have a tremendous diversity of form, pattern and colour but, for descriptive purposes, they are often classified into five basic types: gullet, tubular, brush, capitate and spurred (Figure 4.103). Flowers are commonly simply tubular and more or less polysymmetric but some like *Erythrina* have a large flag. Nectar guides are not usually present but the perianth may have

Figure 4.103. Tubular and gullet bird-pollinated flowers: (a) *Bomarea* (Alstroemeriaceae) and (b) *Justicea* (Acanthaceae).

Figure 4.104. Laterally held long tubular flowers adapted for bird pollination are found in many genera of different families. The dogfish-form is seen in several genera of bird-pollinated flowers, shown here in the American *Columnea* and S. E. Asian *Aeschynanthus* (both Gesneriaceae).

grooves to guide the beak and tongue, and there has been extensive coevolution of beak length and corolla tube depth.

Flowers visited by hoverers, mostly hummingbirds, contrast with those visited by perchers. They are pendulous or are held horizontally. A common syndrome is for a group of tubular flowers each a source of nectar, or in *Aquilegia* provided by multiple spurs of a single flower. Inflorescences that encourage a bird to perch are frequently robust and long-lasting, particularly those species with coloured bracts such as the Bromeliaceae and *Freycinetia* (Pandanaceae). The projecting styles and stamens of many bird-pollinated flowers are hardened but elastic and a common flower shape has a projecting upper lip giving a dog-fish-like appearance.

Figure 4.105. *Banksia* brush blossom is visited by parakeets and also by honey possums.

Other bird-pollinated species have many smaller flowers packed sometimes together in a cone-like inflorescence. This pattern is especially common in Australia with *Banksia* and *Grevillea* in the Proteaceae, and *Eucalyptus* and *Callistemon* in the Myrtaceae. Most bird-pollinated flowers provide a copious volume of nectar as a reward. In *Aloe ferox* (Aloeaceae) almost one third of the corolla tube is filled with nectar. However, lorikeets have a brush-like tongue which enables them to collect pollen as well as nectar from brush blossoms.

Strelitzia is one rather unusual example of a bird-pollinated species; it produces flowers in sequence from within the stiff perch provided by the spathe, and pollination is carried out by the birds' feet.

Figure 4.106. *Strelitzea* is pollinated by the feet of visiting birds.

A strange example of pollination is that by mites which live within the flowers of various species visited by hummingbirds. The mites are transported between plants in the nostrils of the hummingbirds but act within the inflorescence carrying pollen between flowers causing self-pollination. There is a strong association between species of mite and species of floral host.

Bat pollination or chiropterophily is very important in the Tropics. Some important tropical families that have a high proportion of

Figure 4.107. *Cobaea* is a bat-pollinated species, although fly and bee pollination have also been observed.

Figure 4.108. *Strongylodon* (Fabaceae) has bat-pollinated bluish-green, almost luminous flowers.

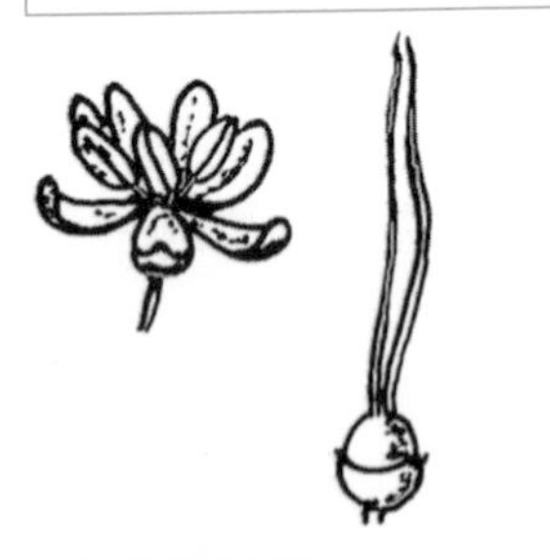

Figure 4.109. Dioecious flowers of *Humulus* (Cannabidaceae): male (left) and female (right).

bat-pollinated species are the Bignoniaceae, Bombacaeae, Cactaceae, Caesalpiniaceae, and Lobeliaceae. Bats are particularly valuable pollinators in the tropical forest because they will fly long distances between distant trees. Bat flowers tend to be white or dull greenish and brownish, they produce copious viscid nectar and lots of pollen, and often smell of rotten fruit. The flower form is either that of a brush blossom or it is bowl shaped. The flowers or inflorescence is rigid and strong to allow the bat to grip. Flowers are relatively large to allow the snout of the bat access. Alternatively some bats do not alight on the flower but hover in front. Some of the most familiar tropical trees and climbers are bat pollinated. They include, the kapok tree (*Ceiba*), the baobab (*Adansonia*), balsa (*Ochroma*) and the cup and, saucer vine (*Cobaea*). Bat flowers are commonly produced away from the foliage on long pendulous stems, like *Strongylodon*, or directly on the larger branches and trunk (cauliflory) so that they are readily accessible.

Various other mammals, such as marsupials, rodents and lemurs, are pollinators. There are stories that the huge aroids like *Amorphophalus* might be elephant or tapir pollinated by these animals pushing their trunks inside the spathe! *Ravenala*, or traveller's palm (Strelitziaceae), is bird pollinated but perhaps is also adapted for lemur pollination. *Dryandra* and *Banksia* in Australia and *Protea* in South Africa seem to have converged for pollination by mice and mouse-like marsupials. Australian *Tarsipes*, a shrew-like 'honey-mouse', is the most obviously adapted marsupial; it has reduced or absent teeth and a long brush-like tongue for retrieving nectar. *Rohdea*, from China and Japan, is a member of the lily family that smells of stale bread. It attracts snails and slugs, which feed on its fleshy perianth and is said to be pollinated by them.

4.6.6 Breeding systems

Most flowers are adapted to promote cross-pollination but not all flowers strictly apply it. Inbreeding species often have small flowers and save in the production of pollen. Pollen/ovule ratios for obligate inbreeding species are two orders of magnitude smaller than those for obligate outbreeding species. The ability to inbreed has evolved many times in flowering plants, especially in regions where there is a shortage of pollinators, or there is environmental uncertainty, or in weedy species where the rapid production of large quantities of seed is required. In those species that grow in small isolated populations, especially colonising species, mechanisms that promote out-crossing may be disadvantageous because they reduce the chances of any seed production.

Cleistogamous flowers never open. Pollen/ovule ratios for cleistogamous species are lower even than non-cleistogamous but obligate inbreeding species. The rate of cleistogamy may be affected by the environmental conditions. Several species such as ground ivy, *Glechoma hederacea*, produce cleistogamous flowers only at the end of the season. *Commelina forskalaei* produces subterranean cleistogamous flowers in the dry season. Several aquatic plants are regularly cleistogamous but at least one of these, *Subularia*, has been observed to produce open flowers on a dried up lake margin. In addition to vegetative propagation, *Lobelia dortmanna* is able to adopt several sexually reproductive strategies. Normal out-crossing aerial flowers are visited by numerous small insects. However, the inflorescences of those individuals that grow in deep water cannot break the surface, and therefore they develop cleistogamous flowers. In 1847, Sir Joseph Dalton Hooker reported on the strange mode of pollination found in the mudwort (*Limosella aquatica*) on Kerguelen Island. In winter, in two feet of water, beneath the ice, it was found to have fully formed flowers in which pollination had occurred. An air bubble was generated in the space between the over-arching petals, and, within this air pocket, the pollen was transferred to the stigma.

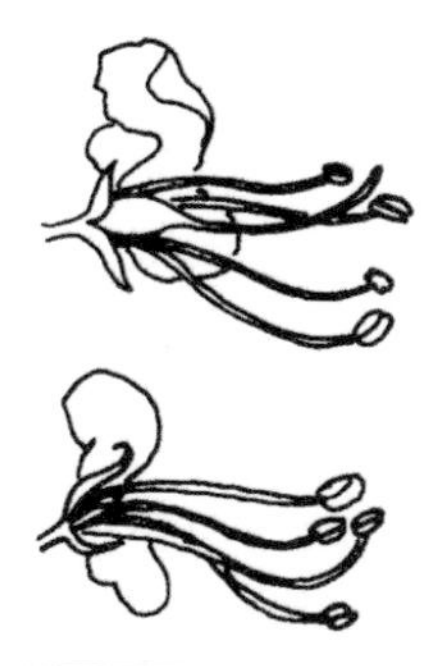

Figure 4.110. Andromonoecy in *Aesculus*: hermaphrodite and male flowers are produced on the same tree.

Reproduction in flowering plants is a two-stage process, pollination followed by fertilisation, while mechanisms to ensure cross-fertilisation are of two sorts: those that promote cross-pollination, usually by the physical or temporal separation of male and female parts of the flower, and those that prevent self-fertilisation by a chemical/physiological self-recognition after self-pollination called self-incompatibility.

Only about 4% of flowering plants are dioecious, with separate male and female individuals. There are also monoecious species, with separate male and female regions (diclinous). Interesting intermediate conditions like andromonoecy, male and bisexual regions on the same plant (as in *Aesculus hippocastanum*, horse chestnut) or gynodioecy, separate female and bisexual plants (as in some species of *Ficus*) are also found. Gynomonoecy is associated with the specialisation of different flowers within the capitulum in the daisy family, the Asteraceae: outer petalloid flowers are female (or sterile) and inner flowers are bisexual. There are regional and ecological differences in the rate of dioecy: only 2% of the British flora is dioecious and 12% of the New Zealand flora. In temperate regions wind-pollinated flowering plants tend to be dioecious or monoecious and diclinous.

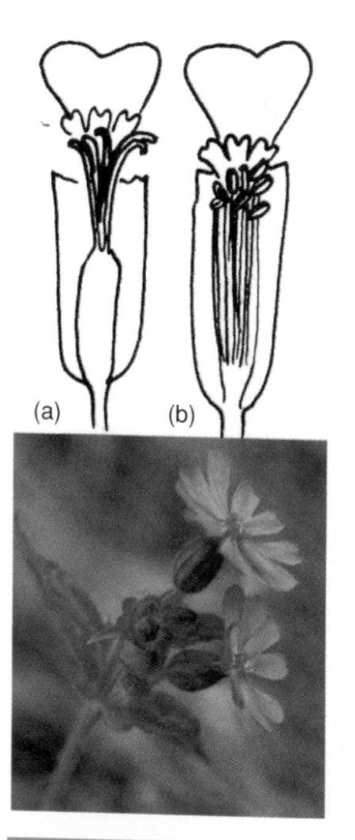

Figure 4.111. Dioecy in *Silene dioica*. The drawings show the female flower (a) and male flower (b).

Where dicliny is associated with insect pollination nectar is usually the attractant but *Decaspermium parviflorum* is a dioecious Indonesian species that provides sterile pollen in the female. It is important that the male and female flowers look identical to the insect so that it does not discriminate between them. Sterile stamens are found in female *Silene dioica*, which also compensates for its lack of pollen by the production of more nectar. In the Cucurbitaceae, in which there are closely related monoecious and dioecious species, the stigma is lobed and somewhat reminiscent of the three stamens of the male flower. Dioecious species in the tropics are more often insect pollinated than those in temperate regions. They tend to have large animal-dispersed seeds and it is possible that dioecy in them is the result of competition between the sexes within plants. In dioecious species male plants sometimes produce more flowers and grow

Figure 4.112. Monoecious but diclinous *Cucurbita*: male (above) and female (below) flowers mimic each other.

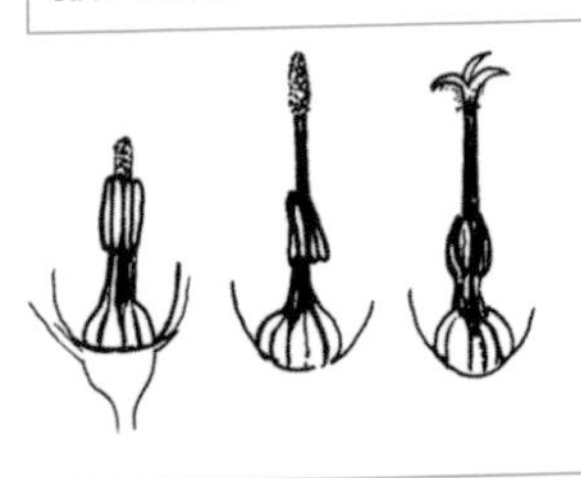

Figure 4.113. Pollen presentation mechanisms in *Campanula*. Pollen is shed on to a hairy portion of the style (presenter region) while still in bud. The style elongates as the flower opens and the stamens wither. When an insect visits the flower, the hairs invaginate, like the fingers of a glove, to release the pollen. Only later do the stigmatic lobes open and become receptive to cross-pollination.

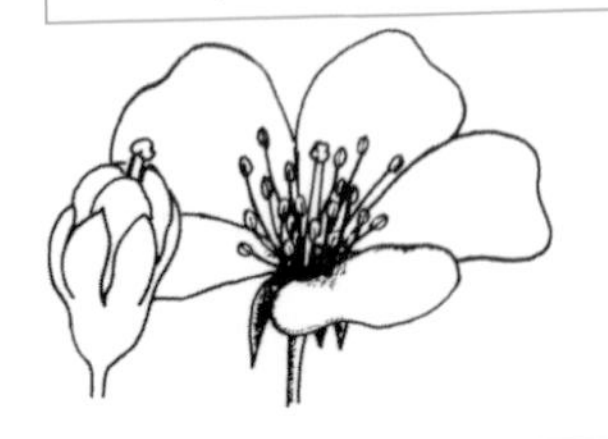

Figure 4.114. Protogyny in *Prunus*: the stigma protrudes from the bud before the flower opens.

vegetatively more vigorously, perhaps reflecting the extra energy cost of producing pistils and seed.

The important family Euphorbiaceae is entirely diclinous. They illustrate several interesting aspects of sexual niche separation, with each sex growing in slightly different environments, in for example *Mercurialis perennis*. In diclinous *Ricinus* the male flowers have large numbers of branched stamens while the females have below them numerous conical nectaries. In the genus *Euphorbia* effective bisexuality has evolved. The inflorescence, mimicking a flower, called a cyathium, has one female flower at the centre of a cup, which bears on its inside a number of male flowers, so reduced that they each look like a single stamen. Around the margin of the cyathium there are large nectariferous glands like a perianth.

Even though over 70% of all angiosperms have only bisexual flowers, the sexes nevertheless may be functionally separated by different timing of development, by protandry or protogyny. Pollen presentation mechanisms commonly include protandry. In *Campanula*, pollen is shed onto a hairy style while in bud. The hairs first trap the pollen but later they invaginate like the fingers of a glove to release it. The style elongates as the flower opens and the stamens wither. The lobed stigma then opens to permit cross-pollination. *Prunus* is protogynous, the mature stigma is poked out of the bud to allow cross-pollination. The anthers shed their pollen later when the flower opens properly. In this case, as in many other examples of protogyny and protandry, this is a mechanism to ensure efficient pollen transfer not to prevent self-pollination, since a self-incompatibility system is also in operation.

Self-incompatibility (SI) is present in at least half of all flowering plant species that have been tested. Self-incompatibility is a chemical/genetic self-recognition between the pollen and stigma so that a plant's pollen will not germinate on its own stigma or, if it germinates, the pollen tube will not successfully grow down the style. In some cases the incompatibility reaction is late-acting, even after fertilisation. The closure of the carpel and the evolution of a differentiated style and stigma has enabled the evolution of self-incompatibility by providing tissues where the incompatibility reaction can act. The process of pollination is complex incliuding pollen entrapment/adhesion to the stigma, pollen hydration and germination, growth of the pollen tube and penetration of the stigma cuticle, and navigation of the pollen tube down the style to the ovule. At each stage a complex system of signals integrates the process. Self-incompatibility has evolved by subverting these signals.

A large number of species, including the families Asteraceae and Brassicaceae, have a system in which the sporophytic anther tapetum produces a polymorphic soluble signal called **SCR** (S-locus Cysteine-Rich protein), which is carried within the pollen exine. The self-incompatibility locus codes two other tightly linked genes **SLG** (S-Locus Glycoprotein), which encodes part of a receptor present in the cell wall of the stigma, and **SRK** (S-Receptor Kinase), which encodes a transmembrane protein embedded in the plasma membrane of the stigma cell, the other part of the receptor. Interaction with the SCR protein produces a cascade of physiological reactions that prevents the self-pollen from germinating successfully. Self-recognition, in an alternative and equally widespread mechanism, is mediated by the gametophytic tissue of the pollen tube. Incompatible pollen tube growth is blocked by the activity of S-locus-encoded ribonucleases (RNase), which are synthesised within the style. The RNase molecules contain a hypervariable region which confers S specificity (S1, S2, S3, etc.). The RNase enters the pollen tube and then destroys specific RNAs only in 'self' tubes. Because of extensive polymorphism of the self-incompatibility locus, plants cannot self-fertilise but can fertilise most other plants in a population.

Several kinds of self-incompatibility are particularly interesting because they are associated with morphological differences between flowers. In these heteromorphic systems there are either two or three flowers morphs with pollinations only possible between different morphs. In *Primula* there are two morphs. The pin morph has a long style, anthers located low in the corolla tube, and produces small pollen. The thrum morph has a short style, anthers at the top of the corolla tube and produces large pollen. The incompatibility reaction either prevents the pollen tubes of self-pollen penetrating the stigma or inhibits their growth a short way down the style. The heterostylous condition helps to promote efficient cross-pollination between the morphs but does not mediate the incompatibility. The reciprocal positioning of the anthers and stigmas helps to prevent the stigmas becoming coated with pollen from their own flower. The pollen in pin flowers is less effectively dispersed from the hidden anthers, which is compensated for by a greater production of smaller grains. In *Lythrum salicaria*, which is also heterostylous, there are three morphs.

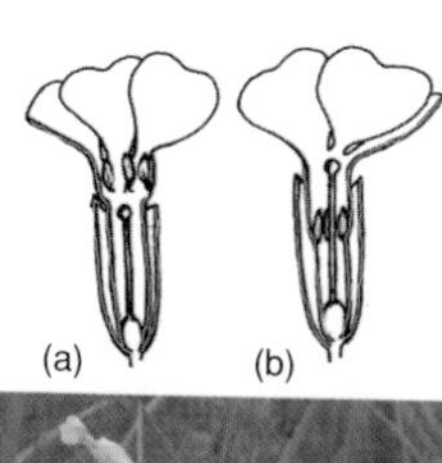

Figure 4.115. Heteromorphic self-incompatibility: distylous *Primula*. The drawings show a thrum flower (a) and a pin flower (b).

In the family Plumbaginaceae the different morphs may actually play a part in mediating the incompatibility reaction. There are two morphs with different stigma morphologies and different patterns of reticulation on the pollen. 'A' pollen, which is produced in flowers with 'Cob' stigmas, germinates only on the 'Papillate' stigmas of the alternative morph. The 'Papillate' morph produces 'B' pollen, which will germinate only on 'Cob' stigmas. The 'A' pollen has a pattern of reticulations that allows it fit closely onto the 'Papillate' stigma. 'B' pollen is relatively smooth like the surface of the 'Cob' stigma. Since close connection of pollen and stigmatic cells is necessary for rehydration of the pollen grain prior to germination this may be an important part of the incompatibility reaction. In some species self-compatibility has evolved by a cross-over within the self-incompatibility gene so that it is monomorphic with the self-compatible combination of 'A' pollen and 'Papillate' stigma. A similar breakdown of incompatibility has been observed in many other groups, which may also exhibit trimorphy/tristyly.

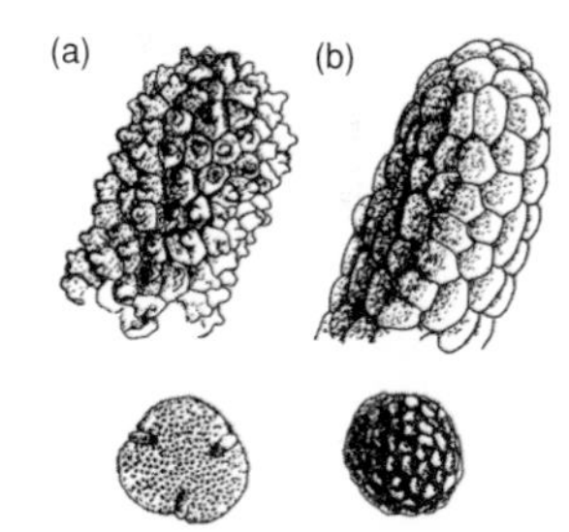

Figure 4.116. Dimorphic stigma and pollen in *Limonium*: (a) Papillate and (b) Cob morps.

In contrast, perhaps the most diverse of all flowering plant families, the Orchidaceae, lacks self-incompatibility. In the orchids it is the vast diversity of flower shapes, colours and scents that encourages effective cross-pollination. An important factor in their speciation has been pollination with pollinia. This very effectively multiplies and stabilises new variants. Orchid pollination biology, with its ability to seemingly make endless forms, mirrors one of the most effective methods of old-fashioned plant breeding to create new cultivars; crossing two genetically different variants and then creating a diverse range of distinct cultivars by selection and inbreeding. The evolutionary dance is made manifest and in this family there do seem to be endless forms of flowers!

Further reading for Chapter 4

Barth, F. G. *Insects and Flowers: The Biology of a Partnership* (Princeton, NJ: Princeton University Press, 1991).

Endress, P. K. *The Diversity and Evolutionary Biology of Tropical Flowers* (Cambridge: Cambridge University Press, 1994).

Faegri, K. and van der Pijl, L. *The Principles of Pollination Ecology*, 3rd edition (Oxford: Pergamon, 1979).

Foster, A. S. and Gifford, E. M. *Comparative Morphology of Vascular Plants*, 3rd edition (San Francisco: W. H. Freeman, 1990).

Niklas, K. J. The aerodynamics of wind pollination. *Bot. Rev.*, **51**, 328–386 (1985).

Proctor, M., Yeo, P. and Lack, A. *The Natural History of Pollination* (London: Harper Collins Publishers, 1996).

Richards, A. J. *Plant Breeding Systems* (London: Allen and Unwin, 1986).

Chapter 5

Ordering the paths of diversity

They are all bound, each to each by powers that are virtues; the path of each is traced and each one finds its own path.

André Gide, 1897

5.1 The phylogeny of plants

One of the best ways to understand variation is by comparison among related groups. Perhaps the greatest early success in this approach was that of Hoffmeister in the nineteenth century when he realised that the evolution of ovules and seeds could be best understood by understanding the variations of heterosporous and endosporic non-seed plants. The availability of an independently-derived phylogeny, from DNA sequence data, has vastly increased the power of this comparative approach.

Looking at phylogeny it is clear that particular forms have evolved repeatedly. Time and again similar morphologies and anatomies have evolved separately in distinct lineages. These examples provide a key to understanding the evolution of plants not just in terms of adaptation, say in understanding a convergent feature as one that has evolved to fit a similar function, but perhaps more importantly in understanding the shared environmental and developmental processes that have constrained or permitted certain evolutionary pathways.

5.1.1 The paths of diversity

Imagine the map of diversity as if it were a city plan. There are the city blocks, at different longitudes and latitudes of morphology, anatomy, physiology and chemistry. These are the archetypes. They are connected by the paths that represent the developmental pathways between them. The new phylogeny has allowed us to place an arrow on the paths; some are two-way, but many are mainly or only one-way. Some paths are busy avenues of evolution that have been explored time and again. Other streets are quieter and have been explored much more rarely.

But the metaphor of a city plan of plant form is too static. E. J. H. Corner in *The Life of Plants* writes about charting the lost channels of the delta of biology. Borrowing that image, we can visualise plant form as channelled between the twin banks of development and adaptation. The channels converge and diverge. There are eddies and back currents. The flow is slow in some channels, in others fast. Different lineages end up travelling down the same channel. There are two great rivers of vegetative form, either towards greater woodiness and growing as a tree, or towards being a herb. There are many tributaries towards specialisations such as being an aquatic, a climber or a succulent. These channels and tributaries have been explored again and again in different orders, families, and even genera of plants. It is this pattern of exploration that represents the archetypes of plant evolution, not any particular ideal form.

Occasionally the phylogenetic flow breaks through the twin banks of adaptation and development and floods the plain. When this happens there is, for a while, no limit on where the flow can go until new channels emerge or the flood retreats to particular isolated pools. What we shall see is that some lineages are circumscribed, confined to particular pools of adaptation and development, but others have a history of breaking their banks, and exhibit a great diversity of form. The challenge is to understand the historical reasons for this difference and to characterise the developmental and adaptive constraints on change, but the first step is to delineate the lineages themselves.

5.1.2 The archetypes of plants

The archetypes are forms and patterns that result from major channels of development. Examples include the following:

- woodiness and herbaceousness
- succulence
- liane (woody climber) habit
- arborescence in herbs
- reduction in herbs
- evolution of evolvability
- tubular flower
- floral symmetries
- centrifugal (basipetal) and centripetal (acropetal) development
- zygomorphy
- central placement of pollen – presenters and indusia (Malvaceae Campanulaceae, Rubiaceae, and Goodeniaceae, etc.)

There are many others in the following account.

5.1.3 The taxonomic hierarchy

There are perhaps 350 000 living species of plants, a vast array that seems impossible to comprehend. The family tree of plants has a dense and tangled crown, with the final twigs representing the living species, and dead-ends in the centre of the crown the extinct lineages. Trying to make sense of this tangle is no easy task. It is a task that has occupied plant taxonomists for several centuries, and, although great strides have been made recently, there is still a long way to go.

Things can be named and from which concepts can be formed. If the names that can be named are correctly chosen, they somehow come close to existence – even if only as 'guests of reality', not as reality's master. They can serve in some way to create order, to pass on tradition and thus preserve the continuity of human activity.

Lao Tzu, *The Phenomenal World*

Humans have a natural ability to discern similarity and measure relationship. We can see kinds of things and give them names. Even so-called 'primitive' cultures of indigenous peoples have developed a sophisticated taxonomy of plants, especially of the ones they use. For example the Maoris recognised 53 different varieties of New Zealand flax *Phormium tenax* that provided them with different qualities of fibre. However, over the past few centuries western botany has attempted something a little different, to circumscribe groups that are natural, that represent the similarities and relationships between plants in all their features, not just the parts of plants that humans use.

It was one of the triumphs of seventeenth and eighteenth century botany to develop a hierarchical system of classification with taxa at different ranks. Taxa at higher ranks group those at lower ranks and include species of more distant relationship. Species are just one rank in the taxonomic hierarchy, but a very special one because they represent the final twigs in the evolutionary shrub. It was the works of Linnaeus that established that the taxonomic name of a species was treated in a special way as a two-word phrase, a binomial, that includes their genus name (written with an initial capital letter) with a specific adjective, both of them written in Latin form and either printed in italics or underlined. There are taxonomic ranks below species (subspecies, variety and form) but these represent different aspects of the variation of a single species (see Table 5.1).

Taxon = the general term for a taxonomic group of any rank (plural = taxa). Each taxon is given a special name. The naming of taxa is governed by a set of rules called nomenclatural rules.

The taxonomic hierarchy can be represented graphically as a tree. Part of such a tree flattened to show it on the page and only including the main branches is shown in Figure 5.1. This, of course, mirrors the way we represent family trees, and after the theory of evolution was proposed, taxonomists used the taxonomic hierarchy to represent different branches of a phylogenetic or evolutionary tree. The largest diameter oldest branches were recognised at high taxonomic rank and the smaller diameter branches low taxonomic rank.

There are two important points here. Because we can see only the surface of the crown of the evolutionary tree it is sometimes unclear to which branch each twig connects, and worse, hidden deep in the crown, it is often far from clear to what major branch the smaller branches connect. There is only very fragmentary fossil evidence from when the tree was smaller and the branches that are now hidden

Notice how the names of groups above species each have their own particular ending so that the taxonomic rank can be recognised. Note too how superior ranks have the stem name of one of the taxa they include at a lower rank. However, this is not compulsory and can lead to confusion because different authors may name the same group at different ranks. Alternative names tend to be used at higher ranks, but this too can get confusing as different names are used for exactly the same group. In the text below some of the most commonly used alternatives are listed. Above all, we should avoid mistaking the menu for the meal. Classification has its limitations and pitfalls.

Table 5.1 The hierarchy of taxonomic ranks for plants with the example of a species of wheat and maize showing to which group (taxon) each belongs at different ranks

Kingdom	Viridiplantae	
Subkingdom	Plantae (or Embryobionta)	
Division (Phylum)	Tracheophyta	
Subdivision	Spermatophytina	
Class	Magnoliopsida (or Angiospermopsida)	
Subclass	Monocotyledonidae (or Liliidae)	
Order	Poales	
Family	Poaceae	
Subfamily	Pooideae	Panicoideae
Tribe	Triticeae	Andropogoneae
Subtribe	Triticinae	Tripsacinae
Genus	*Triticum*	*Zea*
Subgenus	*T.* subgenus *Triticum*	
Section	*T.* section *Triticum*	*Zea* section *mays*
Species	*T. aestivum*	*Z. mays*
Subspecies	*T. aestivum* subsp. *aestivum*	*Z. mays* subsp. *mays*
Variety	*T. aestivum* var. *aestivum*	

deep in the tree were near the surface and visible. This difficulty is being overcome utilising DNA sequence data variation and computer based methods of analysis, but it has led to differences of opinion and different classifications, among taxonomists. The tree represented in Figure 5.1 is only one version of what is currently thought to be the arrangement of the major phylogenetic branches of plants. It is possible, indeed likely, that some of the branches will move position as more plants are sampled.

The second point is that it is impossible to give a name to every branch of the evolutionary tree. There simply are not enough taxonomic ranks in the taxonomic hierarchy. So taxonomists have to decide relatively subjectively which branches to honour with a name and at what taxonomic rank. This is the source of many differences in the classifications you will observe in different books.

With these points in mind you are now ready to embark on a journey. It is actually a journey that we should start in the present day on the smallest twigs, all 350 000 of them, one for each living species, and like an army of ants, travel down into the evolutionary tree back in time, but for simplicity let us imagine it the other way starting on the trunk and climbing in turn up each branch.

5.1.4 Kingdom Plantae or subkingdom Embryobionta: plants

The name Embryobionta refers to the multicellular embryo produced at the earliest stage of the growth of the diploid sporophyte in plants. Another feature that distinguishes plants from almost all algae is the presence of complex reproductive organs, the archegonium, antheridium and sporangium. Other shared features of plants include the following: haplodiplobiontic life cycle and multicellular sporophytes; sporopollenin in the spore wall (also detected in some algae); cuticle (waxy outer layer of the epidermis).

In this book we restrict the term plants to photosynthetic organisms that have a complex multicellular body differentiated into complex tissues *and organs adapted for the land*. So here we exclude all the algae, but in other texts some, or all, of the algae are often included as kinds of plants.

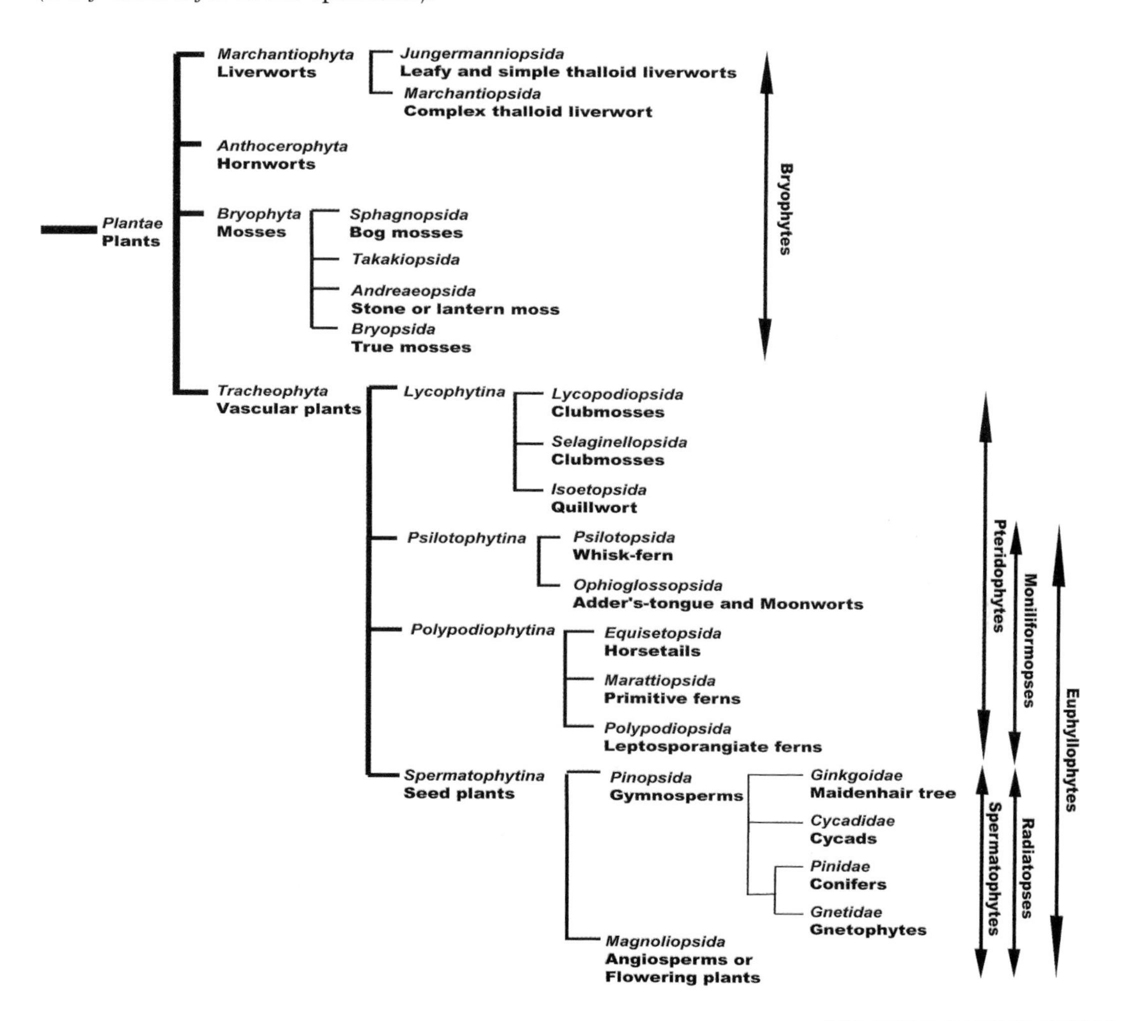

Figure 5.1. A phylogenetic arrangement of living plants.

5.1.5 The main groups of plants

These are the four major phylogenetic lineages. In order of numbers of species, smallest first, they are the hornworts (Anthocerophyta), the liverworts (Hepaticophyta), the mosses (Bryophyta) and the vascular plants (Tracheophyta). It is not clear yet which branch arose first. Relationships inferred from DNA sequence data provide weak support for a relationship between mosses and liverworts, as if they arose from the same branch separate from hornworts.

Figure 5.2. Leafy liverwort: *Diplophyllum.*

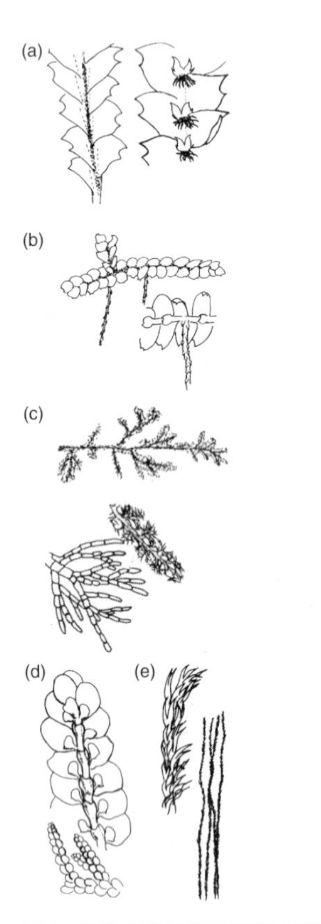

Figure 5.3. Examples of leafy liverworts: (a) *Lophocolea*, showing amphigastria and rhizoids; (b) *Bazzania*, with wick-like branches; (c) *Tricholea*, with finely divided leaves; (d) *Frullania*, with helmet-shaped lobed underleaf; (e) *Herberta*, an upright form.

Liverworts differ from all other plants in lacking stomata. Indeed all other plants have been put in a group called the 'stomatophytes' in recognition of this. The stomatophytes may also be linked by another feature that was subsequently lost from most of them, a columella or central sterile zone, in the sporangium, but still present in the hornworts and mosses. Another intriguing shared character is the ability of stomatophytes to distinguish between two isomers of methionine. Methionine is a sulphur containing amino acid and has an important role in some proteins because it forms stabilising di-sulphide bridges between different parts of a polypeptide chain. It is also a precursor to other important compounds like ethylene. It can occur in two mirror image forms, as D- and L-methionine. In the liverworts and algae the metabolism of both D- and L-isomers is identical but in stomatophytes the D-isomer is treated differently. Mosses, but not liverworts or hornworts, share with the vascular plants a range of other features such as the emission of isoprene (2-methyl-1, 3-butadiene) and other gaseous hydrocarbons like monoterpenes. We are most familiar with the monoterpenes α- and β-pinene that cause the pine scent in coniferous forests. Mosses also have similar heat shock proteins to the vascular plants.

The vascular plants are by far the most diverse of these lineages and seem to have dominated the landscape since the earliest times. In contrast the fossil history of hornworts, liverworts and mosses is unclear. In many respects they represent a more primitive grade of organisation than the vascular plants. Indeed they are loosely called together 'the bryophytes'. They are small and relatively simple plants that share a superficial similarity, ether because they share ancestral (plesiomorphous) features or because they have become adapted to a similar niche. Most of them have a limited ability to control their water content (poikilohydric). The vascular plants are characterised by a well-developed ability to control their water content (homoiohydric) with the possession of a specialised xylem tissue for transporting water. This kind of tissue is not entirely absent from the bryophytes: some species of mosses and liverworts have hydroids that are analogous to the tracheids of vascular plants but moss hydroids have smooth, not unevenly thickened, walls, thinner than that in tracheids. Perforated water conducting cells are present in the moss *Takakia* and are also found in the thallus of some thalloid liverworts. In fact the water-conducting cells of the earliest known plants are of various sorts, some more like hydroids and others like tracheids. Bryophytes lack roots but have rhizoids, elongated cells that penetrate the soil. These are a general feature of plants although they have been lost from some more specialised plants. They differ from roots in their simplicity; they are mainly unicellular or, if multicellular, they are usually uniseriate.

Bryophytes also share a kind of life cycle where the most obvious plant of the life cycle is the gametophyte. The sporophyte is dependent on the gametophyte and consists only of a capsule (sporangium) on an unbranched stalk (seta). In contrast the vascular plants have a sporophyte that is independent of the gametophyte, with sporangia produced on a branched stem. For this reason non-bryophytes have been named as polysporangiophytes by some workers.

5.2 The non-flowering plants

5.2.1 Division Marchantiophyta – liverworts

Class Jungermanniopsida

SUBCLASS JUNGERMANNIIDAE (LEAFY LIVERWORTS)

The leafy liverworts number about a third of all liverworts and are very diverse, especially in the tropics. They normally have three rows of leaves attached to a thin stem. Commonly two of the rows of leaves are held laterally to produce a flattened frond-like structure. The third smaller row of leaves is situated below the frond and is variously modified for water uptake and transport. Water is conducted externally over the plant, by capillarity, and absorbed or lost directly through the whole of the plant surface. The plant body is fine and delicate. The leaves are usually only one cell thick, allowing the easy transmission of water from the exterior to all parts of the plant. Most leafy liverworts belong to the order Jungermanniales.

The sex organs are either lateral in axils of leaves or apical with a kind of perichaetium, called the perianth, surrounding them. Various kinds of perianths are found in different species. A few genera such as *Calyopogeia* have a subterranean pouch, called the marsupium that protects the archegonia.

There is one genus of leafy liverworts, *Haplomitrium* (12 species) that is possibly endohydric, This moss-like liverwort has all leaves its about the same size and even has radial symmetry. It is sometimes placed in its own order, the **Calobryales**. It has a branched subterranean rhizomatous system. Rhizoids are absent but there is a fungal associate that grows outside the rhizome as well as within the cortex, where there are abundant mucilage cells. These cells may encourage the fungus to grow or they may directly help in the uptake of water.

Alternative names: Marchantiopsida, Hepatophyta, Hepaticae ~8000 species

Distinguishing features: small simple gametophyte flattened thalloid or leafy growth form, bearing unbranched, smooth, unicellular, rhizoids; oil-bodies in which terpenoid lipids are accumulated; reproducing by spores and gemmae; sporophyte simple capsule and seta; the spores are mixed with sterile elaters that aid their dispersal, both gametophyte and sporopyte lack stomata.

Life-form/ecology: three main kinds – complex thalloid, simple thalloid and leafy liverworts; found in damp habitats on soil and rocks or growing epiphytically.

SUBCLASS METZGERIIDAE (SIMPLE THALLOID LIVERWORTS)

Pellia is a common thalloid liverwort. It has a relatively uniform thallus that lacks air spaces although it may have regions of elongated water conducting cells. The thallus is thin at the margin and is one cell thick, but has a kind of thickened midrib, several layers thick. There are numerous rhizoids arising underneath, especially from the midrib region. The rhizoids are usually unicellular and smooth, although the tip of the rhizoid may be swollen or branched. *Metzgeria* lacks rhizoids. All cells of the thallus may have chloroplasts.

There are also colonies of *Nostoc*, the blue-green bacterium, present in some. Most species have a close association with fungi; the tips of the rhizoids are often occupied by fungal hyphae. *Cryptothallus*, which is subterranean, except when sporing, lacks chloroplasts and is completely mycotrophic.

(a)

(b)

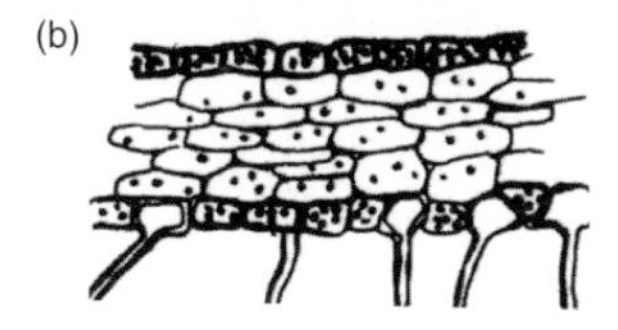

Figure 5.4. *Pellia*. (a) Plant; (b) a section through the *Pellia* thallus.

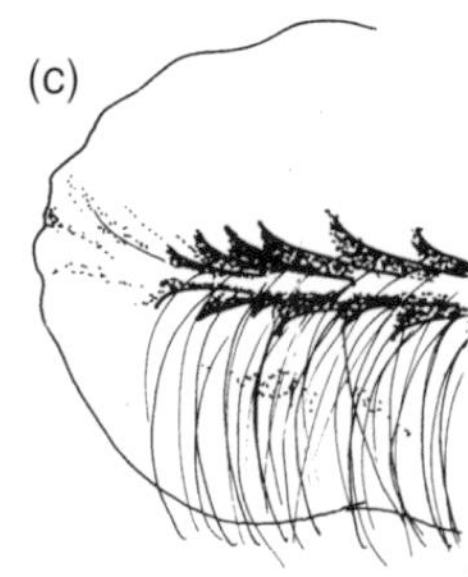

Figure 5.5. *Marchantia* thallus with (a) antheridio- and (b) archegoniophores, and (c) under-surface showing midrib, scales and rhizoids.

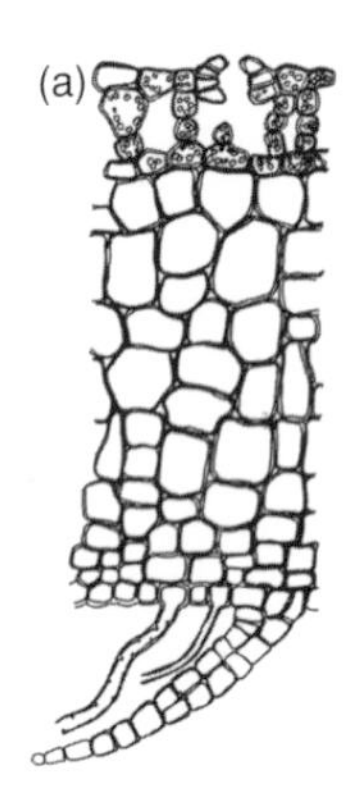

Figure 5.6. *Marchantia.* (a) Section through *Marchantia* thallus showing two different types of rhizoids, ventral scales, water storage section, chambered upper surface with chlorophyllous cells and pores; (b) side and surface view of areole and pore.

The thallus grows from an apex that branches irregularly occasionally. Lobed leafy forms within the subclass illustrate the relationship to leafy liverworts of the order Jungermanniidae. *Monoclea* from New Zealand and South America has a large thallus up to 20 cm long and 5 cm wide rather like *Pellia*, although the uppermost cells have many chloroplasts while, lower layers have few but many starch grains. Brown oil bodies are present.

Class Marchantiopsida

The Marchantiopsida are thalloid liverworts with a differentiated and chambered thallus.

SUBCLASS MARCHANTIIDAE (COMPLEX THALLOID LIVERWORTS)

Marchantia and *Preissia* are examples.

Of all liverworts they are the best adapted to high light levels, are most tolerant of drought and tolerant of high nitrogen and phosphorous levels. The upper part of the thallus is photosynthetic and has air chambers which connect to the outside by complex pores which are analogous to stomata but which do not normally open and close. In each chamber there are columns of photosynthetic cells. The lower part of the thallus is a storage tissue and may contain a few oil bodies. In *Preissia* the pore is surrounded by several cell layers and is barrel shaped. The lowest cell layer projects into the pore and gapes open when the thallus is turgid. When the thallus loses water the pore shrinks and this cell layer seals off the chamber. The surface of the plant is cutinised. The thallus branches pseudo-dichotomously. Rhizoids and rows of scales arise from the ventral surface of the thallus. The rhizoids of the Marchantiidae are of various sorts, which may be present together in the same plant. They are either smooth, like those in *Pellia*, or have internal peg-like projections of the cell wall that serve to increase their surface area for water uptake. The smooth ones penetrate the soil. The tuberculate ones run together like a wick, which is held in place by the ventral scales.

SUBCLASS SPHAEROCARPIDAE (BOTTLE LIVERWORTS)

These have peculiar upright thalli of various sorts with very little internal differentiation. *Sphaerocarpos* has a multi-lobed thallus, each lobe forming a rounded ball with an opening at the top. *S. texanus* (Texas balloonwort) is a dioecious winter ephemeral. *Geothallus tuberosus* grows on damp soil in grass and has a swollen base. In *Riella* there is a spirally wound thallus attached on one side to a thickened stem. *Riella* is a submerged aquatic plant of fresh or brackish water in transient pools or streams in semi-arid regions.

Figure 5.7. *Sphaerocarpos* a bottle-liverwort (Marchantiopsida, Sphaerocarpidae).

5.2.2 Hornworts – division Anthocerophyta

Small simple thalloid gametophyte plants, several cell-layers thick and shaped as a thin ribbon-, heart- or disk-shaped structure, and an attached columnar sporophyte. The sporophyte has stomata but *Notothylas*, *Dendroceros* and *Megaceros* lack fully-developed stomata. In some species the chloroplasts have pyrenoids, a feature which they share with the green algae.

Alternative names: Anthocerotopsida, Anthocerotae ~300 species.

Distinguishing features: a single large chloroplast per cell; thallus with mucilage chambers that may become occupied by the cyanobacteria *Nostoc*.

Life-form/ecology: thalloid, growing in damp relatively open areas such as stream-sides or arable fields or on bark. They have a strong mycorrhizal association (vesicular-arbuscular type) and symbiotic associations with the nitrogen-fixing blue-green bacterium *Nostoc* are also present.

Fossil record: The evolutionary origin of hornworts is obscure. The oldest certain fossils are spores from the late Cretaceous that are similar to those of the living genus *Phaeoceros*. Some features of hornworts, like the presence of a central sterile column (columella) in the sporangium, link the hornworts to the mosses and to the fossil *Horneophyton* from the Devonian.

Sister groups: according to some workers they are sister to all other plants.

Order Notothylales

The sporangium of *Notothylas* is short and capsule-like, sometimes remaining surrounded by an involucre formed by the gametophyte.

Order Anthocerotales

These have an elongated horn-like sporangium. Spores are produced from a basal meristem, mixed with sterile pseudo-elaters. They include the genera *Anthoceros*, *Phaeoceros*, *Dendroceros*, *Megaceros* and others. *Dendroceros* is an epiphyll (growing on the leaves of other plants).

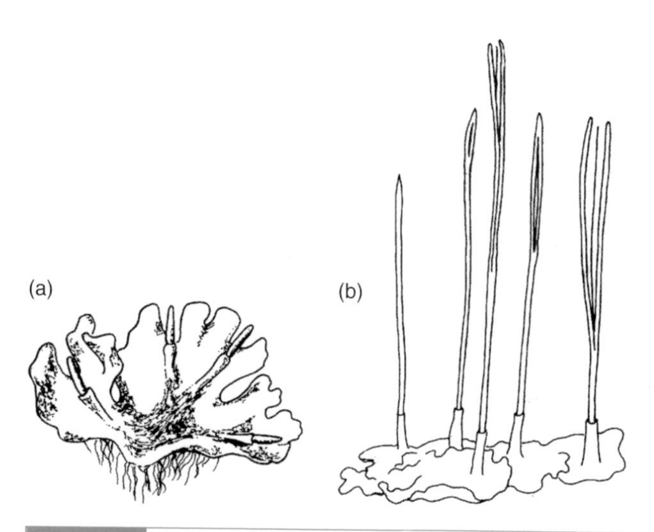

Figure 5.8. Hornworts: (a) *Notothylas*; (b) *Anthoceros*.

Alternative names: Bryopsida, Musci ~10 000 species

Distinguishing features: spore producing plants; sporophyte is dependent on gametophyte but photosynthetic with stomata; capsules with diverse adaptations for dispersal of spores. Leaves are small (microphylls) and lack the kind of complex internal structure of intercellular air spaces connecting to the exterior via stomata of other plants, though this kind of internalised air-space including stomata is present in the apophysis of the sporangium (capsule) of some.

5.2.3 Mosses – division Bryophyta

The small leafy gametophyte exhibits a broad range of adaptations for the retention and transport of water including, in some species, specialised internal water conducting tissues. The bog mosses, *Sphagnum*, dominate large areas in cool wet climates. The largest moss *Dawsonia superba* can grow to a height of 70 cm or more. Mosses grow terrestrially or as epiphytes. In cloud forests the living biomass of mosses and liverworts may exceed that of all other plants. Mosses are common in damp shady places but can also survive long periods of desiccation in hot deserts and tundras. Linking together all bryophyte gametophytes is the fact that they are all to a greater or lesser extent poikilohydric. They have only a very limited ability to control their uptake and loss of water. Another feature that links them is the absence of stomata, which are only present in some of their sporophytes. They have several life-forms that intergrade with each other.

Particular forms characterise particular habitats. Epiphytes tend to be mat, weft or pendant forms. Turf forms tend to be found in open habitats. Some species are plastic in form. One major distinction between upright and adpressed mosses correlates with the position of the reproductive structures: acrocarpous mosses are upright with, usually single terminal sporophytes; pleurocarpous mosses are mostly prostrate with several 'lateral' (i.e. terminal on short side branches) sporophytes. Pleurocarpous mosses are more commonly the mat and weft formers. Some, the endohydric mosses, do have a vascular system, which is directly comparable to that of the tracheophytes. Other bryophytes are called ectohydric because water is not conducted internally. Intermediate forms between ecto- and endohydric, the mixohydric mosses are also defined by some bryologists.

Class Sphagnopsida (peat or bog mosses)

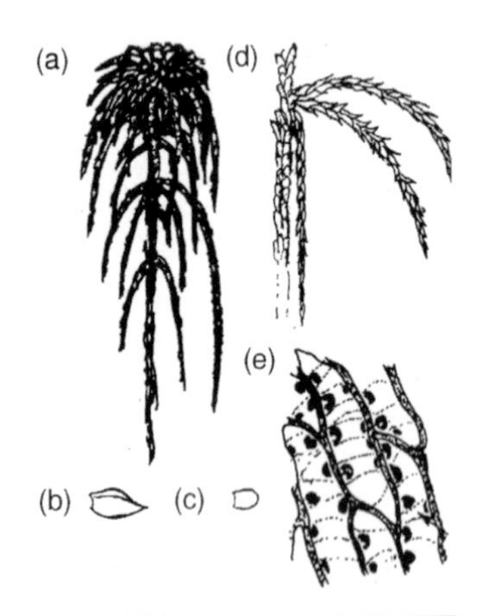

Figure 5.9. Bog moss *Sphagnum* (a) upright stem with branches; (b) stem leaf; (c) branch leaf; (d) branches; (e) arrangement of hyaline cells with pores alternating with sinuous photosynthetic cells.

On the basis of molecular data the sister group to all other mosses is the Sphagnopsida. Paraphyses are absent, perichaetia are present and antheridia are found individually at the base of each leaf in the upper third of a male branch. They have a simple rounded capsule, which explodes to release spores. They are highly branched with many overlapping leaves in which photosynthetic cells alternate regularly with empty hyaline cells (colourless and transparent) for water storage. The lower part of the plant is dead but is still functional as a wick. The hyaline cells have a pore to allow any air bubble that forms to escape, enabling them to fill with water. There are also large hyaline retort cells on the stem, each with a single pore. The water holding capacity of a *Sphagnum* plant is up to 20 times its dry weight. Branches are produced in bundles and there are two kinds. One is held horizontally and is photosynthetic. The other hangs down as a wick. On the stem and pendulous branches the leaves are small and clasping, providing capillary pathways. Different species of *Sphagnum* differ in the relative development of the leaf hyaline cells and in the number and size of pores they possess, correlating with how aquatic they are.

Sphagnum is the main component, the familiar bright green or reddish clumps, and peat former of blanket bogs that are found at high latitudes in both hemispheres. Blanket bogs usually have a water table at or slightly above the surface, and well-developed bogs often have a central lake. *Sphagnum* rarely occurs in areas where the pH is greater than 6.0. Water percolating through living *Sphagnum* becomes more acidic because, in common with other bryophytes, *Sphagnum* exchanges protons (H^+ ions) for nutrient cations. The cation exchange capacity is correlated to the uronic acid content of the species.

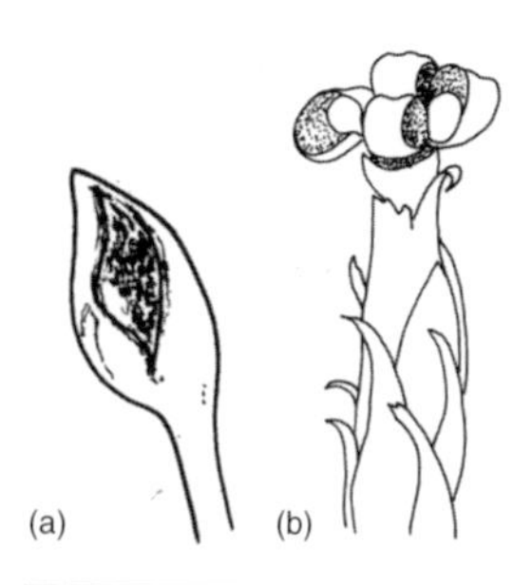

Figure 5.10. Capsules of (a) *Takakia* and (b) *Andreaea*.

Class Takakiopsida (*Takakia*)

Takakia was at one time thought to be a kind of leafy liverwort. It has cylindrical leaf like appendages and a capsule that opens by a single longitudinally split. Branching, root-like rhizomes penetrate the soil. *T. ceratophylla* has an interesting disjunct distribution and is found in Sikkim, Nepal, Tibet, Yunnan and the Aleutian Islands.

Figure 5.11. *Polytrichum* peristome.

Class Andreaeopsida ('granite', 'lantern' or 'stone' mosses)

The 'stone mosses', Andreaeopsida, have a capsule lacking a seta and splits longitudinally to form a chinese-lantern shape. Perigonia are on short lateral branches, and perichaetia at the shoot apex. There are only two genera, *Andreaea* and *Andreaeobryum*, found in arctic/alpine conditions on rock faces.

Class Bryopsida ('true' mosses)

The true mosses have a peristome that is adapted in different ways to control the release of spores or to actively disperse them. They are the most diverse kind of moss, and are especially common in shady and humid conditions.

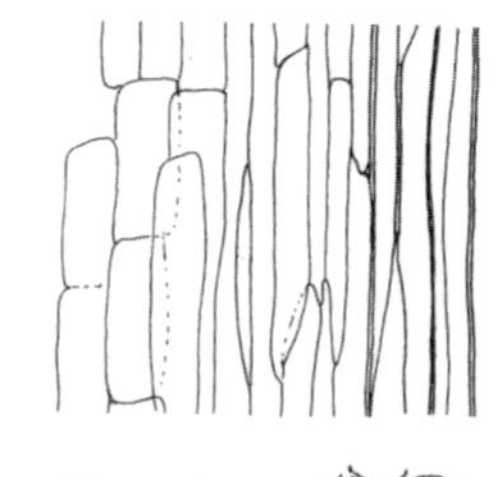

Figure 5.12. *Polytrichum* stem l.s. and t.s. showing leptoids and hydroids.

SUBCLASS POLYTRICHIDAE (HAIR-CAP MOSSES)

They have the peristome as a pepper-pot-type structure, a membranous epiphragm joins the teeth together and dehiscence is through gaps between the teeth. They include several of the most robust types of mosses like *Polytrichum* and *Dawsonia*. They have an extensive network of capillary rhizoids arising from the epidermis, which feed water into the cortex of the stem. The rhizoids are covered with small papillae, which help the uptake of water. In the centre of the stem is a conducting strand or stele composed of elongated hydroids and leptoids in water and solute conducting tissues (Figure 5.12).

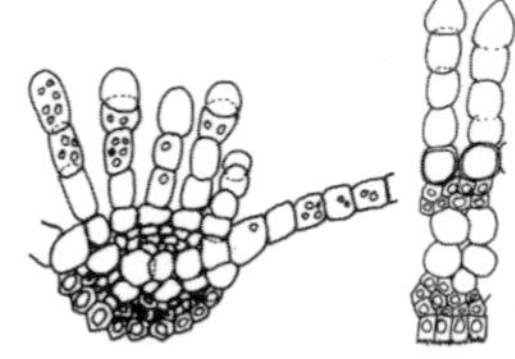

Figure 5.13. Portions of leaf section showing complex lamellate leaf structure in (a) *Atrichum* and (b) *Polytrichum*.

Figure 5.14. *Dawsonia superba* can grow to a height of 70 cm.

Figure 5.15. Capsule of *Tetraphis* showing four-toothed peristome.

The presence of a conducting system evidently permits some mosses to achieve a large size. One of the advantages of a conducting system is that it allows photosynthates to be translocated easily to the growing apex of the plant. It also allows the body of the plant to be thicker since diffusion alone from the surface is not the only source of water and nutrients. However, as well as absorbing water through their rhizoids, endohydric mosses also absorb water through the surface of the leaves and are therefore more properly called mixohydric. They have extensive decurrent leaf bases providing an external capillary path for water. In the leaf a system of lamellae, cutinised in the upper region, may function to maintain a humid atmosphere analogous to that in the mesophyll of other plants.

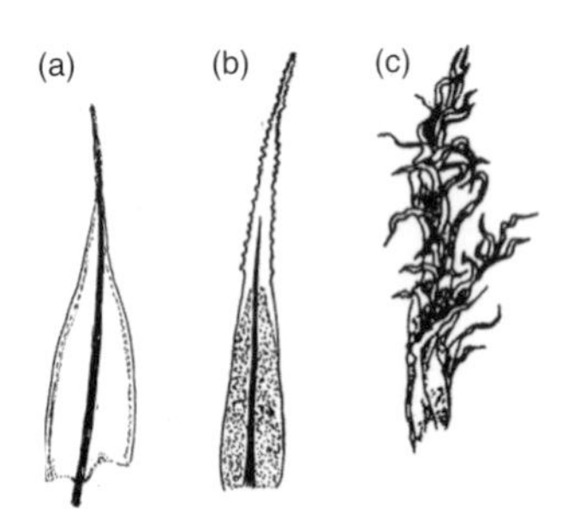

Figure 5.16. Poikilohydric moss: (a) *Bryum* with a leaf trace and excurrent arista; (b) *Rhacomitrium*; with extensive bristle-like apex; (c) dried *Tortula* plant.

SUBCLASS TETRAPHIDAE

They have a simple peristome with four teeth and are endohydric. There are only two genera: *Tetraphis*, with two species, grows in the conifer and mixed forests of the Northern Hemisphere on rotting wood, and *Tetrodontium*, with one species, on siliceous rocks in the Northern Hemisphere and New Zealand. They have a long-lived, branched, green protonema and a tiny leafy stem up to 1 cm tall, mainly just a splash cup for the archegonia and antheridia.

SUBCLASS BRYIDAE (ARTHRODONTOUS MOSSES)

They have the most complex peristomes. The jointed teeth (arthrodontous) are formed from fragments of cells, cell walls and move with changes in humidity to release the spores.

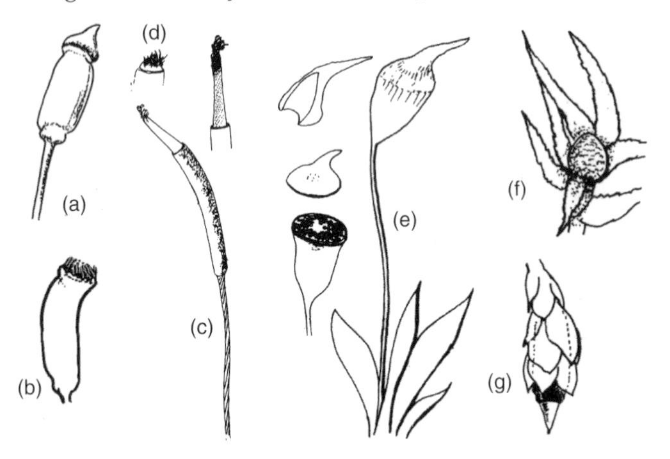

Figure 5.17. Arthrodontous mosses (Bryophyta, Bryopsida, Bryidae) showing different kinds of capsule and peristome: (a) *Hypnum* capsule with operculum in place; (b) *Dicranella* with 18-toothed peristome; (c) and (d) *Tortula* with a twisted peristome; (e) *Pottia* showing calyptra and operculum; (f) *Ephemerum* releases its spores by disintegration; (g) *Fontinalis* with short seta.

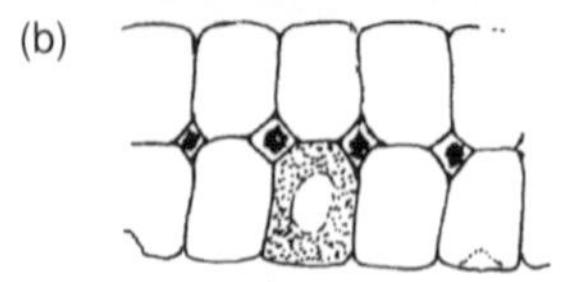

Figure 5.18. *Leucobryum*: (a) cushion growth form, the pale colour is a result of the extensive network of hyaline cells; (b) transverse section of leaf showing two layers of hyaline cells with smaller photosynthetic cells between them.

They are diverse leafy plants ranging in form from upright tufted or cushion formers to straggling or creeping kinds. Leaves range from multistratose with a thickened midrib to unistratose. Some are endohydric but most are ectohydric or mixohydric. Various adaptations can be observed to help the transmission of water. These include the folding of leaves (*Fissidens*), and the overlapping of leaves and leaf bases. Various structures act as wicks, including hairs, divided leaves (paraphyllia), tufts of rhizoids and tufts of branches. In *Aulocomnium* there is a dense felt coating the stem surface. Many ectohydric mosses have conical leaf cells, which project from the lamina increasing the surface for wetting. *Tortula* has papillae that are cutinised and which shed water into the spaces between them where the water is absorbed. Many ectohydric mosses are amazingly tolerant of desiccation. *Tortula muralis* can survive for 10 months without water and then revive within a few hours. Tolerance of desiccation is helped if drying is slow; growing in tight clumps and cushions helps slow drying. The presence of long, hyaline leaf tip hairs (aristae), as in *Rhacomitrium*, also helps. The aristae form points on which dew can form, and, when dry, they spread out and reflect the sun. Hyaline cells are prevalent especially in *Leucobryum*, which has the photosynthetic cells surrounded by hyaline cells in an analogous way to *Sphagnum*.

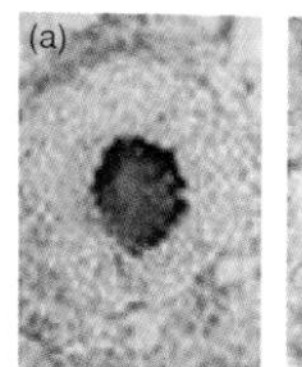

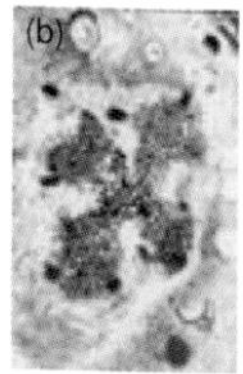

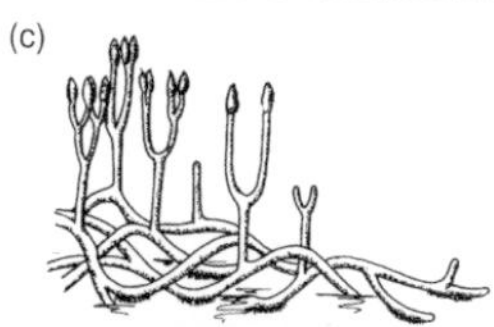

Figure 5.19. Rhyniophytes: (a) *Rhynia* vascular strand; (b) *Asteroxylon* vascular strand; (c) *Aglaophyton*; (d) Reconstructions of *Zosterophyllum* and *Asteroxylon* (background).

5.2.4 Vascular land plants – division Tracheophyta

Tracheophyta possess specialised water-conducting cells called tracheids. There are two main branches, the extinct rhyniophytes and the eutracheophytes, which includes all living lineages of Tracheophyta, as well as some extinct lineages (Table 5.2). Also called polysporangiophytes, they have a sporophyte generation that grows independently of the gametophyte. Much of the evolution of plants took place in the sporophyte; it became larger and more and more complex.

The evolution of a rooting system, an axis that can penetrate the soil, was an important advance for land plants. Rooting structures of many fossil plants are poorly known but these earliest land plants produced a shoot that grew from only one end, a unipolar system. Dichotomous branching of the stem produced rhizomes. In early stages these were no more than stems that grew horizontally or were positively geotropic; in other words, rhizomes. This kind of rhizome can be seen in the Devonian fossils *Asteroxylon* and *Aglaophyton*. The latter had a dichotomously-branching horizontal axis that rose and fell to give it knee-like joints, as if it were a tiny species of mangrove. Another, *Horneophyton*, had swollen corm-like structures at the base of

Table 5.2 The main lineages of Eutracheophytes

Lycophytina	Euphyllophytina
G-type tracheids with a thick decay-resistant inner layer.	P-type tracheids with bordered pits and strands of secondary wall crossing the pores and surrounding the pits. The secondary wall is laid down discontinuously separated by areas of primary wall.
Dichotomously branched.	Monopodial or pseudomonopodial, with helically arranged branches.
Kidney-shaped (reniform) and flattened sporangia, singly on short stalks in the axils of leaves, with cellular thickening along the line of dehiscence.	Sporangia in pairs in terminal bunches dehiscing through a single slit on one side.

the stem, with many rhizoids on the lower surface. These early rooting structures are different from true roots because they lack root hairs and a root cap. In rhizomorphic lycopods the first branching of the embryonic shoot produces one upright branch and one positively geotropic branch called the rhizomorph bearing modified leaves as 'rootlets'.

Rhyniophytes have a number of features that distinguish them from the eutracheophytes: a kind of adventitious branching in which the vascular strand of the branch is not connected to the main strand; an abscission layer at the base of the sporangium; S-type tracheids with a thin, inner, decay-resistant layer in the cell wall; and a spongy outer layer.

Alternative names: Lycopodiaceae ~400 species.

Distinguishing features: small spore-producing plants with sporophyte dominant; homosporous; gametophyte small mycotrophic and tuberous.

Life-form/ecology: clubmosses are widespread from the tropics to the arctic. They grow terrestrially in wet ground, grassland and as hanging tropical epiphytes (*Phlegmarius*).

Sister groups: Selaginellopsida and Isoetopsida.

Fossil record: the clubmosses are in some respects our most primitive living plants. Different forms look very similar to fossils from the Devonian. The simplest living kinds are species of *Huperzia*, like the extinct *Asteroxylon* from the Devonian They have fertile leaves, sporophylls, that are identical to vegetative leaves but have sporangia in their axils.

5.2.5 Subdivision Lycophytina

There are three main groups: various genera of clubmosses (Lycopsida), *Selaginella* (Sellaginellopsida) and *Isoetes* (Isoetopsida). The latter two groups are sometimes placed together in a group called the Ligulatae because they have a peg-like extension of their leaves called a ligule that is absent from the clubmosses. They are also heterosporous and have similar endosporic gametophytes.

Class Lycopsida – clubmosses

In all clubmosses the gametophyte is small and relatively insignificant. In some it is lobed and green, in others it is subterranean, tuberous or carrot-shaped and mycotrophic (= living in a close association with fungi). They have tracheids concentrated usually in a central column of xylem surrounded by phloem (protostele) which may sometimes be lobed (actinostele) or pleated (plectostele). The sporophyte plant bears multiple sporangia on a dichotomously branched leafy axis. In *Palhinhaea* it looks like a small branched tree. The stems are either held vertically upright or hanging, or there is a horizontal stem at the soil surface (stolon) or in the soil (rhizome) from which vertical stems arise. Dichotomously branching roots arise at the stele and, in forms with upright stems, travel through the stem cortex before emerging some distance from where they originated.

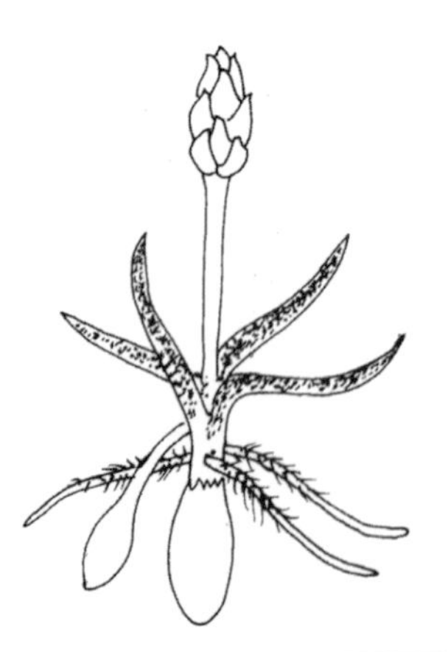

Figure 5.21. *Phylloglossum.*

Figure 5.20. Lycopods: (a) *Huperzia;* (b) *Phlegmaria;* (c) *Lycopodium.*

There are about 15 genera that are sometimes placed in two families: Huperziaceae (*Huperzia, Phylloglossum*), and Lycopodiaceae (*Lycopodium, Phlegmariurus, Diphasiastrum, Palhinhaea, Lycopodiella, Pseudolycopodiella, Phylloglossum*) that differ in the degree of differentiation of the fertile shoot. *Lycopodium* and *Diphasiastrum* have modified fertile branches (strobili or cones) in which the sporophylls are smaller and more closely overlapping. This condition is seen at its extreme in the Australian *Phylloglossum* where the vegetative part of the stem is shortened to a tuft of elongated leaves from which the fertile strobilus arises.

Class Selaginellopsida (*Selaginella*)

There is only one genus *Selaginella*. Some species look superficially similar to some Lycopsida, so they are sometimes called clubmosses, but they are heterosporous with both mega- and microsporangia and have ligulate leaves. Leaves are normally in four ranks with two larger than the others (anisophyllous). Some produce leaves spirally, in which case they are identical in size (isophyllous). A peculiar feature of *Selaginella* that it shares with *Isoetes* is the presence of a scale-like growth, or ligule, at the base of each leaf. The stems may produce positively geotropic aerial structures called rhizophores from which dichotomously branching aerial roots are produced. When the latter reach the soil they branch more profusely and produce root caps. *S. lepidophylla* is a resurrection plant, curling into a ball when dry, and able to survive for months in this state; the molecule trehalose acts as a

Alternative names: Selaginellaceae (selaginellas) ~700 species.

Distinguishing features: heterosporous leafy scrambling plant. Each megasporangium produces only four megaspores. The stem is branched and creeping, producing a frond-like appearance.

Life-form/ecology: widespread and particularly diverse in tropical regions, growing on wet soils and epiphytically, especially in shady conditions. Many have a remarkable ability to withstand drought.

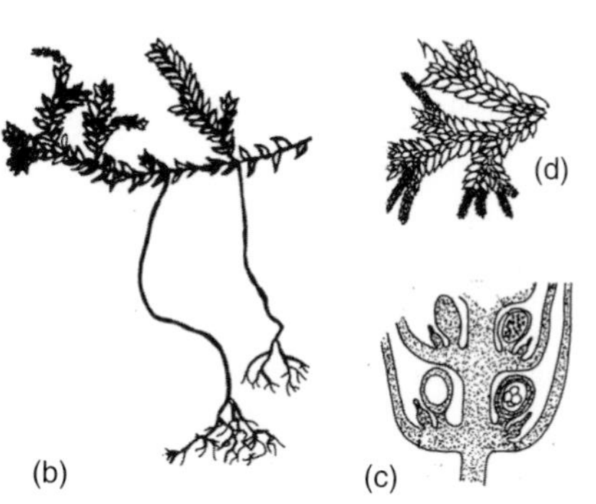

Figure 5.22. *Selaginella*: (a) frond with strobili at tips of shoots; (b) lateral view showing rhizophores; (c) *Strobilus* showing mega- and microsporangia and ligules; (d) portion of plant showing anisophylly.

Selaginella subgenera	subgenus *Selaginella*	subgenus *Tetragonostachys*	subgenus *Stachygynandrum*
Leaves	isophyllous	+/− isophyllous	strongly anisophyllous forming flattened 'frond'
Rhizophores	absent	present	present
Vessel elements	absent	present	absent
Strobili	cylindric, sporophylls only slightly different from vegetative leaves	quadrangular, sporophylls different from vegetative leaves, adpressed	quadrangular, sporophylls different from vegetative leaves, spreading
Number of species	2	About 50	600–700

drought protectant helping to maintain the integrity of the plant's physiological system. About 80% of the dry mass of *S. densa*, an inhabitant of short-grass prairie, consists of a tangled mass of roots in the top few centimetres of the soil, allowing it to rapidly soak up water when it is available. *S. wildenowii* is a climber; its striking iridescent colour is a shade adaptation.

Class Isoetopsida, *Isoetes* – Quillworts

Alternative names: Isoetaceae ~150 species.

Distinguishing features: small rosette and tussock formers with dichotomously branching stems (normally as a rhizome or corm, forming pedestals or stilts in *Stylites*), and unbranched roots.

Life-form/ecology: *Isoetes* is an aquatic or semi-aquatic plant, or grows in areas subject to seasonal inundation.

There are two genera of quillworts. They are most commonly small rosette plants with long-tapering lanceolate leaves, arising from the apex of a branched woody corm. Most plants are small but one species (*I. engelmannii*) has leaves up to ~50 cm long. Leaves are microphylls, narrow with a single vein and four air chambers running longitudinally, and with a short projection called a ligule at the base. The corm has two or three lobes. The outer layer of the corm is shed regularly as the corm grows. The cambium produces vascular tissue on the inside and a new cortex on the outside. Dichotomously branching roots arise from the grooves between the lobes of the corm. Each root has an air cavity surrounding the central vascular strand and separating it from the outer cylindrical cortex except for ridges of tissue. *Isoetes* is heterosporous. Megasporangia and microsporangia are located at the base of some of the leaves and contain many spores.

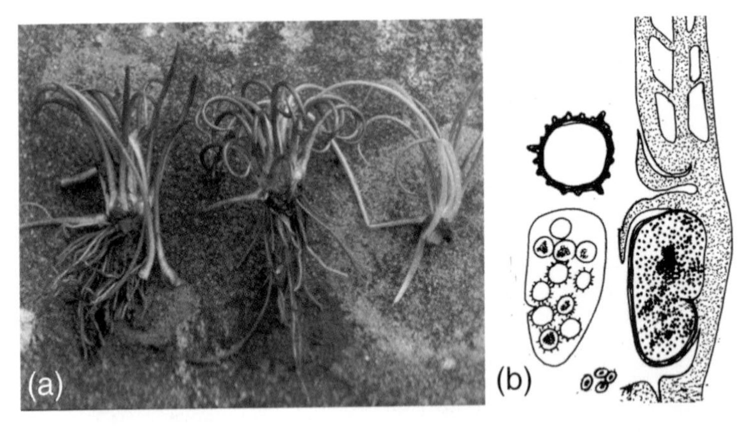

Figure 5.23. *Isoetes*: (a) plants showing rosette form and swollen corm at base of leaves; (b) section through the base of a leaf showing micro and megasporangia and spores.

Figure 5.24. *Lepidodendron* and *Sigillaria* Carboniferous trees.

Small stature with highly compressed internodes, long cylindrical pointed leaves, and relatively massive root systems has evolved convergently in some other plants of cold oligotrophic lakes and has been termed 'isoetid' (see Chapter 6). Some *Isoetes* plants lack stomata but they have the ability to take up CO_2 from the substrate and exhibit CAM photosynthesis as a response to daytime carbon deficit in oligotrophic lakes.

There are about 150 species, many recognised on the basis of differences in their spores. A separate genus called *Stylites*, discovered in 1940 in the Andes in Peru is sometimes included in *Isoetes*. It has an elongated pedestal-like corm.

5.2.6 Euphyllophytes

The earliest known euphyllophyte is the fossil *Psilophyton*, a small plant about 50 cm tall, that had profusely branching side branches. *Pertica* was similar but taller and more regularly branched.

There are three main subdivisions of living euphyllophytes: the whiskferns and adder's-tongue ferns (Psilotophytina), the ferns and horsetails (Polypodiophytina) and the seed plants (Spermatophytina). The first two have been placed in a group called the Moniliformopses because they have a vascular tissue that develops like a necklace, with the earliest xylem (protoxylem) confined to lobes of the necklace. The Moniliformopses include extinct early fern-like plants, the Cladoxylidae, Stauropteridae and Zygoteridae. Recently a new arrangement of living Moniliformopses has been suggested by molecular data, separating off the Adder's-tongue ferns from other ferns but including the horsetails as sister to the remaining ferns. The Spermatophytina are placed in a group, the Radiatopses, so-named because the vascular tissue develops in a radiating pattern outwards from the centre.

An important group of Radiatopses were the progymnosperms, so called because they share several features with, and seem to prefigure, the earliest seed plants. For example, *Archaeopteris* had webbed side-branches forming leaves or pinnae, the ability to undergo thickening growth, enabling it to grow as a tree, and heterospory. However, unlike the tree lycophytes, they have a vascular cambium that produces new tissues on both sides (bifacial).

Figure 5.25. *Archaeopteris* with detail of foliage.

5.2.7 Subdivision Psilotophytina

Class Psilotopsida – whiskferns

There are about 3–8 species of whiskferns.

Distinguishing features: they are extraordinary rootless plants bearing rhizoids only on rhizomes. Branching is by equal forking of the stems and rhizomes. *Psilotum* has scale-like leaves. In *Tmesipteris* the leaves are broad and flat and attached to the stems in a peculiar sideways manner giving the whole plant a frond-like appearance. The sporangia are fused together (synangia) and produced on short lateral branches. They are homosporous and produce a tuberous gametophyte not unlike a portion of the sporophyte's rhizome

Life-form/ecology: they grow as epiphytes or on rocks in the humid tropics and subtropics. *Tmesipteris* clothes the trunks of *Nothofagus* and tree ferns in New Zealand and in similar habitats including cycads in temperate Australia.

Figure 5.26. Psilotopsida: (a) *Psilotum*; (b) synangia (groups of sporangia in the axils of microphylls); (c) *Tmesipteris*.

Alternative names: moon fern, moonwort, grape fern, ~75 species.

Distinguishing features: eusporangiate, homosporous ferns with sporangia in a fertile spike like an upright bunch of grapes in *Botrychium*, in groups on branches in *Helminthostachys*, and in two fused rows in *Ophioglossum*.

Class Ophioglossopsida – Adder's-tongue ferns

Three genera of small plants, growing in grasslands, rocky soils and in open areas of tropical forests. A few, such as the pantropical *Ophioglossum palmatum* are epiphytic. Leaves arise from a short erect stem. The frond in *Ophioglossum* ('Moonwort') is simple and entire, highly dissected pinnately in *Botrychium* ('Grape fern') and palmate leaves arranged in two ranks along a rhizome in *Helminthostachys*. Unlike other ferns they do not produce their frond by unfolding a tightly coiled primordium that looks like a bishop's crozier (circinate vernation). The spike is the fertile frond derived from two basal pinnae, one sterile and one fertile.

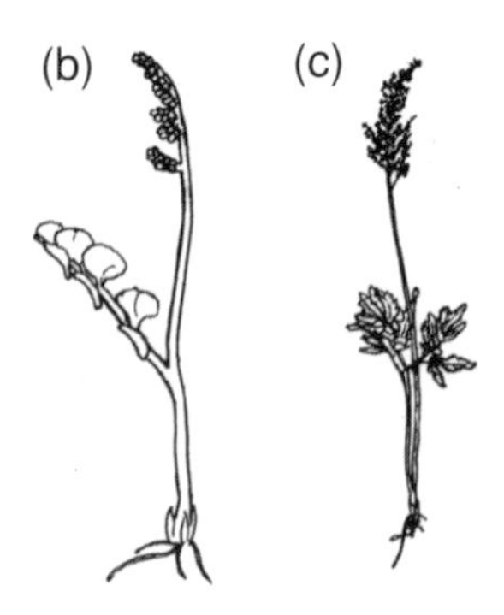

Figure 5.27. Ophioglossopsida: (a) *Ophioglossum*; (b) *Botrychium*; (c) *Helminthostachys*.

5.2.8 Subdivision Polypodiophytina

Class Marattiopsida – Eusporangiate or primitive ferns

Alternative names: Marattiaceae, Angiopteridaceae, Christenseniaceae, Danaeaceae, Kaulfussiaceae ~200 species.

Distinguising features: ferns with eusporangia. The sporangia are fused in synangia in *Marattia* and *Danaea*.

Life-form/ecology: tropical and warm temperate regions, mainly found in the southern continents.

Fossil record: the Marattiopsida have a fossil history dating back to at least the Middle Carboniferous when some species in the family Psaroniaceae had erect trunks several metres tall. Other fossil species are very similar to *Angiopteris*.

Sister groups: Polypodiopsida.

These are ferns with either massive short truncated stems, rarely longer than 60 cm, or creeping rhizomes (*Christensenia* and some species of *Danaea*) (Figure 5.28). The fronds unfurl from a curled crozier. They are simple or once pinnate in many *Danaea* but twice pinnate in *Marattia* and *Angiopteris* and palmate in *Christensenia*. Leaves can be huge, reaching up to 5 m in *Angiopteris*.

Figure 5.28. *Marattia* showing its huge compound leaves.

Class Equisetopsida, *Equisetum* – Horsetails

Perennial homosporous herbs with aerial stems arising from rhizomes bearing coarse fibrous roots (Figure 5.29). The stems have a very characteristic jointed appearance with leaves fused to form a toothed collar at each joint. Some have unbranched aerial stems but others produce whorls of branches at each node. The stem is ridged and strongly hardened with impregnated silica. The stem anatomy is peculiar with vallecular and carinal canals, and a large air space in the pith region as well as the normal vascular tissue.

Alternative names: Sphenophytina, Equistopsida, Sphenopsida, Equisetatae, Equisetaceae; 15 species.

Distinguishing features: nodal structure and whorled leaves and branches.

Life-form/ecology: they are inhabitants of moist habitats.

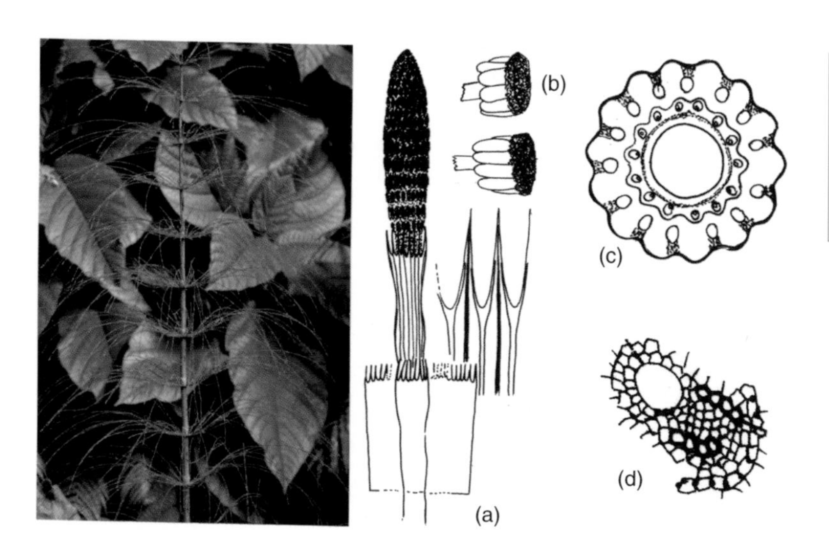

Figure 5.29. Diagrams showing (a) fertile stem with leaf sheath and strobilus; (b) peltate sporangiophores; (c) stem; (d) detail of vascular strand showing carinal canal.

Horsetails are homosporous. Sporangia are grouped on peltate sporangiophores in strobili, which are either produced on special non-photosynthetic stems or at the end of a green aerial stem. The outer wall of each spore peels back to form a pseudo-elater that helps fluff-up the spore mass. The gametophyte prothallus is somewhat heterothallic.

Two main kinds have been recognised as different subgenera. In subgenus *Equisetum* vegetative stems are normally branched and they die back to the ground producing new aerial stems each year. Stomata are scattered or in bands and are flush with the surface. In subgenus *Hippochaete* the aerial stems are normally unbranched, although branch primordia are produced, and they normally over-winter. Stomata are in single lines on each side of the furrows, sunken below the surface. Sporangia are rounded in subgenus *Equisetum*, pointed in subgenus *Hippochaete*. Stem dimorphism is only found in subgenus *Equisetum*. It is fascinating that both subgenera have $2n = 216$, although the chromosomes are larger in subgenus *Hippochaete*.

Figure 5.30. *Calamites*: a tree that grew in the Carboniferous swamps is an extinct relative of *Equisetum*.

Alternative names: Filicophytina, Filicopsida ~9000 species in a number of orders/families.

Distinguishing features: they share the possession of a stalked, thin-walled sporangium (leptosporangium).

Life-form/ecology: very diverse, including trees, epiphytes, climbers, rooted and free-floating aquatics as well as rhizomatous kinds like the cosmopolitan bracken (*Pteridium aquilinum*). Most are homosporous but aquatic ferns are heterosporous.

Class Polypodiopsida – Leptosporangiate or modern ferns

More than 95% of leptosporangiate ferns belong in a single lineage called the Polypodiaceous ferns. Other groups of ferns are basal to the polypodiaceous ferns and differ from them in some fundamental way.

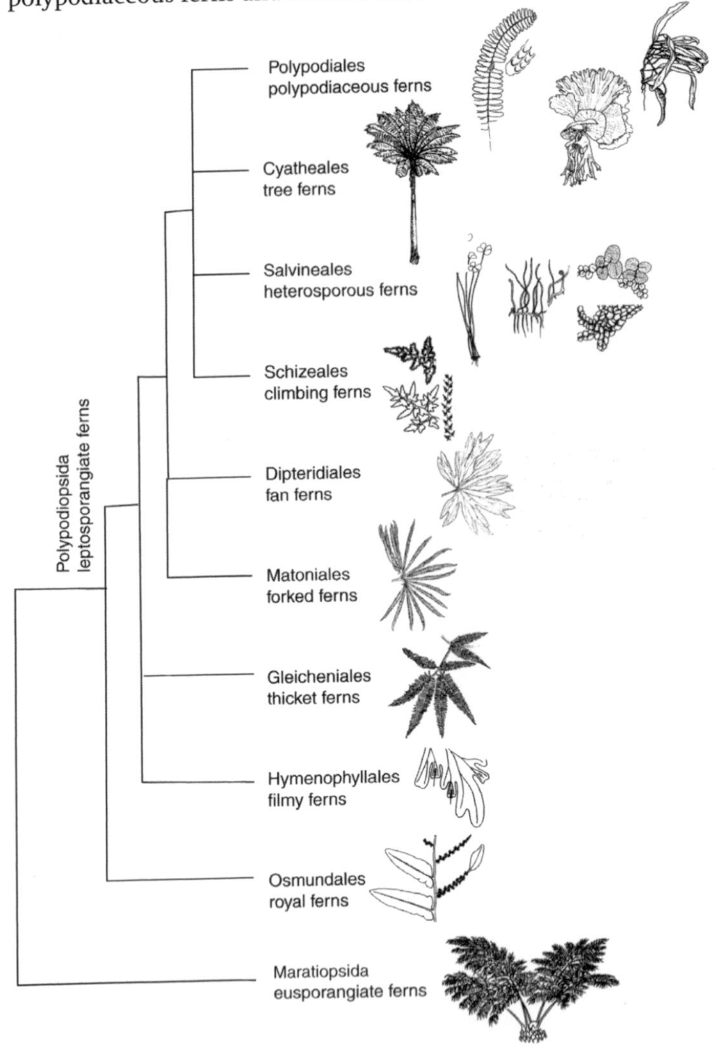

Figure 5.31. The modern ferns (Polypodiopsida) and their relationship with primitive ferns (Marattiopsida).

Figure 5.32. *Osmunda regalis.*

OSMUNDALES, THE ROYAL FERNS

Osmunda, *Leptopteris* and *Todaea* have beautiful plume-like fronds and some eusporangiate features: the sporangia are short stalked, have a weakly developed annulus and produce many spores. Sporangia are found either in sori under leaves, but lacking an indusium or on specialised fertile fronds or tips of fronds.

HYMENOPHYLLALES, THE FILMY FERNS

The filmy ferns are usually epiphytic or grow on wet rocks. They have fronds usually only 1 cell thick and a weakly developed vascular

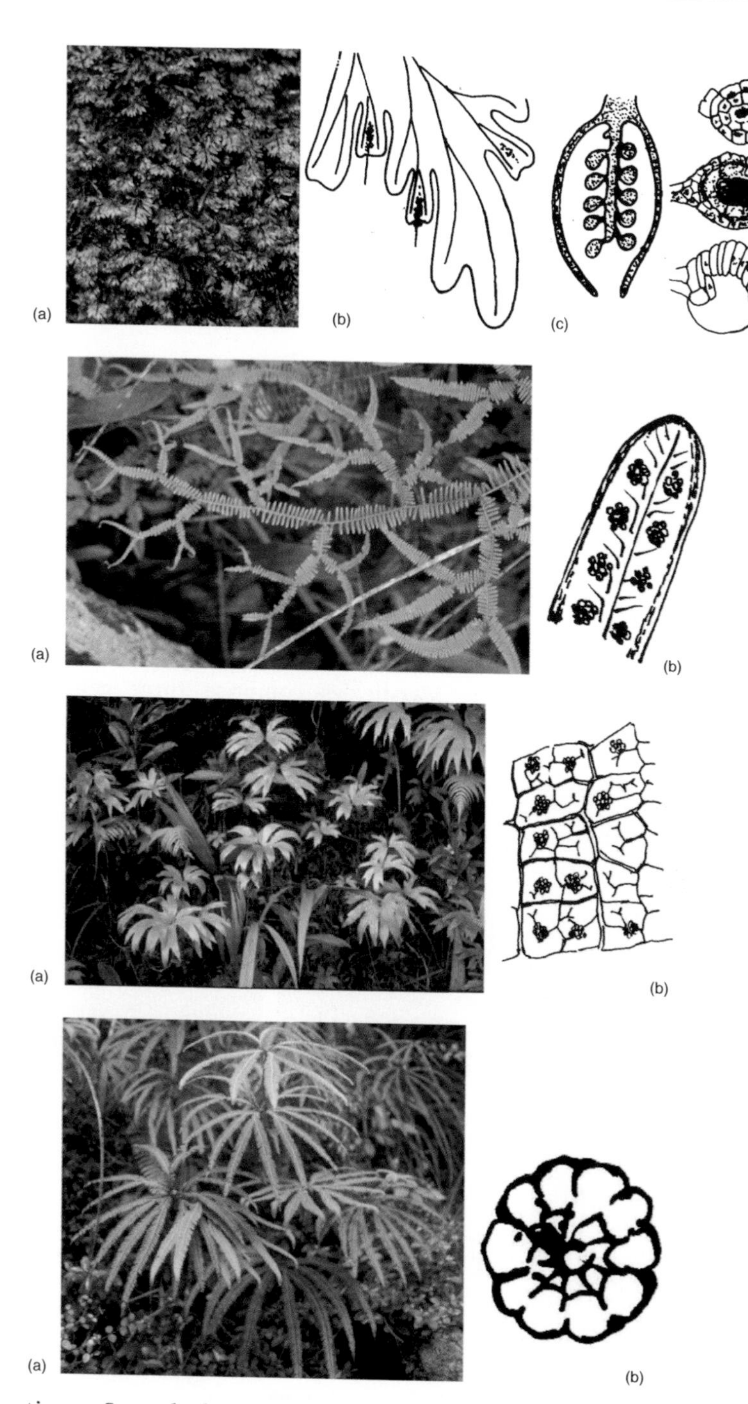

Figure 5.33. Hymenophyllales (filmy ferns): (a) translucent fronds; (b) cup-like sori with extending receptacle bearing sporangia; (c) sorus with sporangia of different ages.

Figure 5.34. Gleicheneales: (a) mature fronds showing pseudo-dichotomous branching; (b) reproductive frond (underside) showing naked sori and a few large sporangia.

Figure 5.35. Dipteridiales: (a) mature fan-like fronds; (b) reproductive frond (underside) showing naked sori with numerous small sporangia.

Figure 5.36. Matoniales: (a) mature fan-like pinnatifid fronds; (b) reproductive frond (underside) showing peltate indusium.

tissue. Some lack roots. Sporangia are produced on a receptacle that elongates as they mature, carrying them out of the indusium.

GLEICHENEALES, THE THICKET FERNS

Gleichenia and *Dicranopteris* are mainly tropical and have forked fronds (Figure 5.34), up to 10 m long arising from long creeping rhizomes. Sori commonly have 2–4 large sporangia and lack indusia.

Figure 5.37. *Lygodium* (Schizaeales) (a) climbing frond; (b) fertile pinnule.

DIPTERIDIALES, THE FAN FERNS

The fan ferns have leaves with two fan-shaped halves (Figure 5.35). There are only two genera in the order with dimorphic leaves *Dipteris* and *Cheiropleuria*, which are sometimes placed in a separate family.

MATONIALES, THE FORKED FERNS

Matonia and *Phanerosorus* are relict tropical plants from South East Asia (Figure 5.36). The fronds are forked and pinnatifid and an umbrella-like (peltate) indusium protects the relatively few sori in each indusium.

SCHIZAEALES, THE CLIMBING AND FRANKINCENSE FERNS

These have sporangia borne singly, and an annulus that is a group of thick-walled cells (Figure 5.37). *Lygodium*, the climbing fern, has an indeterminate frond that grows continuously from its tip producing pinnae to either side.

MARSILEALES AND SALVINEALES, THE HETEROSPOROUS AQUATIC FERNS

These are either rooted (Marsileales) or free-floating (Salvineales) (Figures 5.38 and 5.39). The Marsileales comprise three genera, *Marsilea*, *Regnellidium* and *Pilularia*. All of these plants live along pond

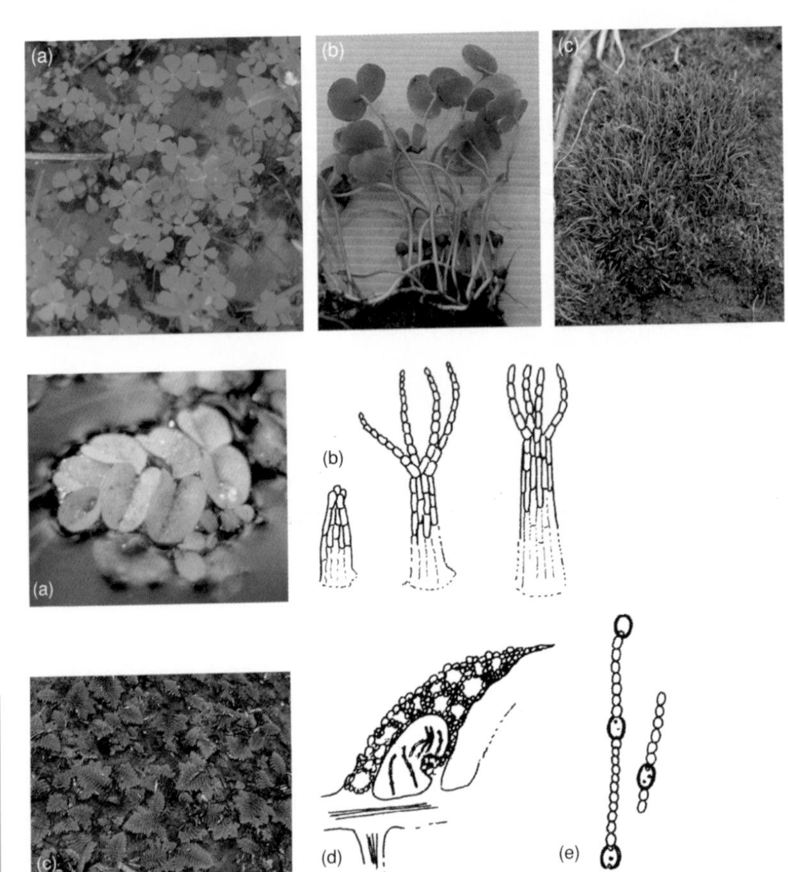

Figure 5.38. Rooted aquatic ferns: (a) *Marsilea*; (b) *Regnnellidium*; (c) *Pilularia*.

Figure 5.39. Free floating aquatic ferns. (a) *Salvinea* floating frond; (b) water-repellent hairs; (c) *Azolla* floating fronds; (d) t.s. leaf showing air-spaces and chamber containing *Anabeana*; (e) *Anabaena*.

or lake margins and produce a creeping rhizome that bears the fronds and roots at nodes. *Marsilea* has a four-lobed pinnule and resembles a four-leaf clover, while in *Regnellidium* the frond is two-lobed. *Pilularia* lacks a lamina to its frond and has a narrow cylindrical leaf. The Marsileales have bean-shaped sporocarps that are homologous to a reduced frond. When moistened the sporocarp produces a branched gelatinous structure bearing sori. The two genera of Salvineales are *Azolla* and *Salvinea*. *Azolla* has leaves tightly overlapping and fitting into each other so that they trap a buoyant film of air. *Salvinea* has leaves in threes on a branched rhizome: one is finely dissected, looking like a branched root, and acts as a stabiliser. The remaining two have an upper surface covered with water repellant hairs. They produce modified sori (sporocarps), producing either microspores or megaspores that float because of the presence of a frothy massula. *Azolla* harbours the blue-green nitrogen-fixing alga *Anabaena* in a symbiotic relationship, and thus has been economically important, especially to rice production.

CYATHEALES, THE TREE-FERNS

The main genera of tree ferns are *Cyathea* and *Dicksonia*. They grow up to 20 m tall and have thrice-pinnate leaves forming a huge umbrella (Figure 5.40). They differ from polypodiaceous ferns in having a complete annulus in the sporangium. There are about 600 species of *Cyathea*, inhabitants of montane forests and colonists of steep slopes. It has sporangia underneath the leaves in the forks of veins and has scales as well as hairs. *Dicksonia* (25 species) has sporangia at the margins of pinnules on vein-tips, protected by a two-valved indusium, and has only hairs, not scales. Some species have a thick mantle of adventitious roots surrounding the trunk.

Figure 5.40. Tree ferns (Cyatheales) growing in beech forest in New Zealand. The mantle of roots clothing the stem is visible in the fern on the left.

POLYPODIALES, THE POLYPODIACEOUS FERNS

These are very diverse (Figure 5.41). Most have sori covered by a flap of tissue called the indusium. Those like *Dennstaedtia* have marginal sori. In others the sorus plus the indusium has migrated back on to the under-surface of the lamina. The shape of the sorus and indusium is often rounded but others, including *Asplenium*, have linear sori. Some like *Pteridium* have lost the indusium.

Most have creeping rhizomes. There are terrestrial forms, either like *Dennstaedtia*, a thicket former, or like the cosmopolitan bracken, *Pteridium aquilinum*, the most successful of all ferns, which can dominate large expanses of the landscape. Bracken is a highly polymorphic species with many named subspecies and varieties. It is fire-adapted; its dead leaves and litter permit fire to take hold but its deep rhizomes allow it to survive where other plants are destroyed. Others are inhabitants of rocks such as many species of *Asplenium*, some of which have dimorphic leaves, but in this genus and many others there has been an adaptive radiation of epiphytes. *Asplenium nidis* is the bird's nest fern that encircles the trunks and branches and collects dead leaves to make its own humus garden into which its roots grow upwards. These gardens are often occupied by ants. Other polypodiaceous ferns such as *Lecanopteris* are ant-plants that provide a home to ants in a hollow swollen rhizome. *Platycerium* has fronds either forming a shield against the host tree trunk or fertile and hanging down. Diversification of polypodiaceous ferns as epiphytes appears to have occurred quite late in geological history along with the diversification of flowering plant trees.

Figure 5.41. Diversity of frond form in polypodiaceous ferns: (a) *Asplenium trichomanes*: (b) *Platycerium bifurcatum*: (c) *Asplenium ruta-muraria* (d) *Phyllites scolopendrium* (e) *Pyrrosia* sp. (f) *Hypolepis ambigua*.

5.2.9 Subdivision Spermatophytina – seed plants

Of all the lineages of plants that had its origin in the Devonian, the seed plants are by far the most successful today. They include the four living groups of gymnosperms (*Ginkgo*, cycads, conifers and gnetophytes) and the flowering plants, as well as several extinct lineages. One has been called *Archaeosperma*. *Lyginopteris* from the Lower Carboniferous was a small slender tree-fern-like plant with ovules attached to the fronds. *Moresnetia* was a diverse and important element in the Upper Devonian and Carboniferous vegetation as a smallish tree or scrambling member of the underflora. The Medullosales was one group with rather large seeds, up to several centimetres long. They grew like tree-ferns or as liana-like plants. Several different kinds are distinguished by their foliage: *Neuropteris*, *Alethopteris* and *Sphenopteris*. The Callistophytales and Peltaspermales were important seed-ferns distinguished by the form of their ovule-bearing structures. Early seed plants were the seed-ferns or Pteridosperms, so-called because of their frond-like foliage. The Glossopterids were a diverse group of seed-ferns that became abundant in the southern hemisphere in the Permian.

A seed develops from a fertilised ovule. The seed plants have a bipolar body. There is not only a primary shoot, but also a primary root with its own apical meristem. They have true roots with a root cap and which branch by producing laterals endogenously from inside another root. In the early embryo of seed plants there is a clear differentiation between the embryo root or radicle, which develops into the primary root, and the embryo shoot, the plumule which develops into the primary shoot. It is from 'seed-ferns' (pteridosperms) that the two living seed plant lineages, the gymnosperms (Pinopsida/Gymnospermopsida) and the flowering plants (angiosperms, Magnoliopsida/Angiospermopsida) originated.

5.2.10 Class Pinopsida – the gymnosperms

There was a tremendous diversity of primitive gymnosperms from the Carboniferous to the Cretaceous and it is not exactly clear from which groups the living gymnosperms (the Cycadidae, the Ginkgoidae, the Coniferidae, and Gnetidae) are derived. Data from DNA sequence variation indicate that they share a common ancestor and that they have a distinct origin from the only other surviving seed plants, the flowering plants.

Cordaites was an early type of gymnosperm closely related and possibly even ancestral to all living gymnosperms. The Cordaitales originated in the Devonian, and were most diverse in the Upper Carboniferous–Permian. They were very impressive plants with heights of up to 30 m and were clearly one of the more significant components of the Late Palaeozoic flora. *Cordaites* was possibly a mangrove or peat-swamp genus since it had stilt roots. It had broad, long strap-like leaves. Its ovuliferous regions are compound with terminal,

Figure 5.42. Reconstruction of the Carboniferous seed-fern *Caytonia*.

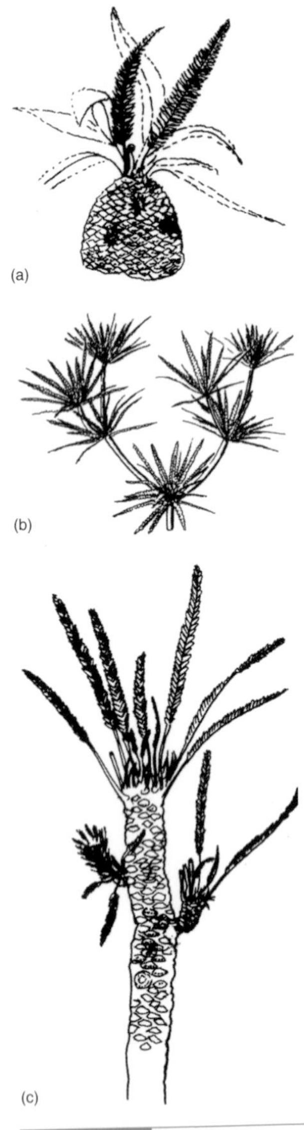

Figure 5.43. Three examples of Bennettites, an extinct lineage of seed plants: (a) Cycadioidea; (b) Wielandiella; (c) Williamsonia.

erect or recurved ovules. The pollen-bearing organs were erect pollen sacs, terminally attached. Pollen grains had air bladders, so they probably had a pollination drop. The seeds were flattened (platyspermic) and extended as a wing.

The Bennettitales evolved in the Triassic in parallel with the cycads, to which they were rather similar vegetatively. The Benettitales are recognisable by very distinct epidermal cells and stomata. Although not regarded as pteridosperms, they have been regarded as important precursors, if not ancestors, of the flowering plants. For many millions of years they were more diverse and more abundant than the cycads but during the Cretaceous they died out. Some (Cycadeoidaceae) were stout short trees like the living cycads but others had slender branches with tufts of leaves (Williamsoniaceae). Many show adataptions for dry habitats or fire: a thick spongy bark, thick leathery or deciduous leaves, sunken stomata, hairiness. A possibly related group were the shrubby Pentoxylales from Gondwana.

The ovules/seeds are attached in a variety of ways to the sporophyte in different lineages of seed plants. The gymnosperms are so named because they are supposed to bear their ovules/seeds unprotected (gymnos = naked in Ancient Greek). This is true for some of them, so, for example, in *Ginkgo* and the conifer *Taxus*, the ovules are exposed on short stalks or peduncles, in pairs in *Ginkgo* and singly in *Taxus*. In *Phyllocladus* they are borne in on lateral shoots called phylloclades surrounded by scales, and in *Cephalotaxus* on short lateral fertile shoots, again surrounded by bracts but still rather exposed. Cycads have cones, which in *Cycas* consist of loosely arranged megasporophylls with a number of ovules at the base, all with a dense covering of hairs. Another cycad, *Ceratozamia* has just two ovules on each megasporophyll. The close relationship between vegetative and ovule/seed bearing leaves is shown by them both being pinnate. Conifers got their name because they normally have megasporophylls arranged in a tightly packed cone. The cones protect the developing ovules but gape open in the right season to allow the pollen to reach the ovules. In the conifer family Pinaceae the ovules are situated on the abaxial side and near the base of a tough woody ovuliferous scale. Each megasporophyll has an accompanying bract scale. Some fossil gymnosperms like the Caytoniales have a cupule, a spherical structure containing ovules. Cupules were arranged opposite each other, like pinnae on a sporophyll. The cones of modern conifers can be traced back through the so-called transition-conifers, the Walchiaceae or the Majoniaceae to the Cordaites.

In the living gnetophytes, *Ephedra* has a rather conifer-like female cone although with usually only two ovules. In *Gnetum* the ovules are exposed on a kind of female catkin and in *Welwitschia* they are surrounded by a 'perianth' of fused bracts with each 'floret' hidden within the cone.

5.2.11 Subclass Ginkgoidae – Ginkgo

These are small dioecious trees growing up to 30 m (Figure 5.44). Their leaves, which arise on short side branches, are fan shaped with dichotomous venation. The sporangia arise in axils of scale leaves on short shoots; the male microsporangiate strobilus resembles a catkin. Each sporangiophore is peltate and has two sporangia. The ovules are borne in pairs on a stalked sporangiophore. The flat (platyspermic) ovules have a three-layered integument and a swelling around the base, the 'collar'. The female gametophye contains chlorophyll and produces two to three archegonia. The pollen produces a prothalial tube that releases two motile sperm. After fertilisation the seed develops a stony inner layer and a fleshy outer layer that stinks of rancid butter.

Alternative names: Maidenhair tree; one species – *Ginkgo biloba*.

Distinguishing features: fan-shaped leaves.

Life-form/ecology: it is now restricted naturally to a small part of China.

Fossil record/sister groups: several other extinct genera are from the Permian including *Trichopitys* and *Polyspermophyllum*. Fossils from the Jurassic have been assigned to *Ginkgo*.

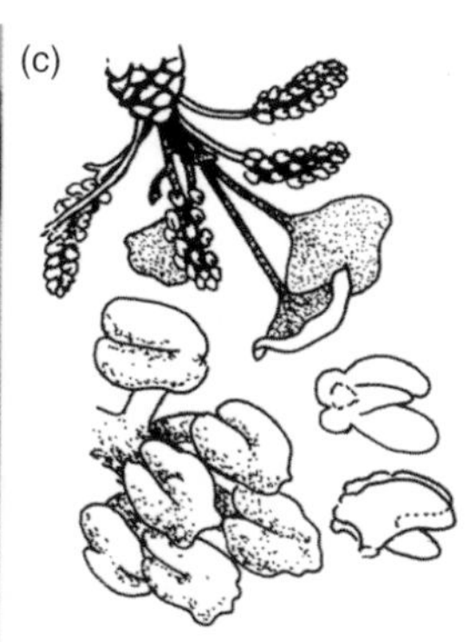

Figure 5.44. *Ginkgo*: (a) tree showing characteristic form; (b) foliage showing long and short shoots; (c) male catkins; (d) foliage of extinct Ginkgoidae.

5.2.12 Subclass Cycadidae – Cycads

Early cycads were commonly taller, more slender, and more branched than living cycads, which are short, thick-stemmed and +/− unbranched (pachycaul) (Figure 5.45). They produce massive cones and are mostly insect pollinated by weevils. They have coralloid roots that have symbiotic cyanobacteria.

Cycadaceae	Stangeriaceae	Zamiaceae
1 genus (*Cycas*) 17 species Leaves with a midrib but without subsidiary veins	1 genus (*Stangeria*) 1 species (*S. eriopsis*) Pinnae with a midrib and secondary veins	8 genera 125 species Pinnae without midribs but with dichotomously-branched longitudinal veins

Alternative names: none ~140 living species in three families.

Distinguishing features: dioecious trees and shrubs with massive, rarely branching starchy trunks (pachycaul) and large compound leaves produced in a terminal rosette.

Life-form/ecology: grow in the tropics on well-drained sites in tropical forests and savannas.

Fossil record/sister groups (extinct and living): a very ancient lineage, the earliest fossils are from the lower Permian of China 280 million years ago. Fossils over 225 Ma old in the late Triassic are known from Antarctica.

Figure 5.45. Cycads with cones: (a) *Encephalartos*; (b) *Stangeria*; with megasporophylls (c) *Cycas*.

5.2.13 Subclass Pinidae – Conifers

Alternative names: Coniferidae, Coniferae, Coniferales, Coniferopsida, Pinopsida ~600 species.

Distinguishing features: branching, mainly evergreen trees, sometimes with resinous wood and a long-shoot/short-shoot system.

Life-form/ecology: growing throughout the world but dominant in cold-temperate latitudes and at high altitudes.

Both leaves and reproductive structures vary considerably in the Pinidae (Figure 5.46). They generally have scale-like or needle-like leaves but a few like *Agathis* have broad leaves; or frond-like, as they are in the branch systems of *Phyllocladus*, or tightly overlapping scales as they are in the cypresses (Figures 5.47–50). Paradoxically, the cone-bearers 'the conifers' have a great diversity of reproductive structures including the fleshy arils of the yew (*Taxus*), the 'berries' of juniper (*Juniperus*) as well as diverse male structures. The wood (soft-wood) has tracheids only and no vessel elements or fibres.

Triassic conifers clearly display compound ovuliferous cones of modern conifers. An earlier example of a modern-type conifer was *Utrechia* from the Permian. It had a branching pattern and leaves like the living Norfolk Island pine *Araucaria heterophylla*. It was in the Triassic that there was a major radiation of modern conifers so that most modern families were in existence by the beginning of the Jurassic. The Auracariales and Podocarpales were the earliest to appear in the fossil record and include forms very similar to living genera. *Dacrycarpus* (Podocarpales) is known from fossils in the Upper Triassic. Araucariales was widespread in both hemispheres, and *Araucaria* itself was present in the Jurassic, but the family began to decline in the Tertiary.

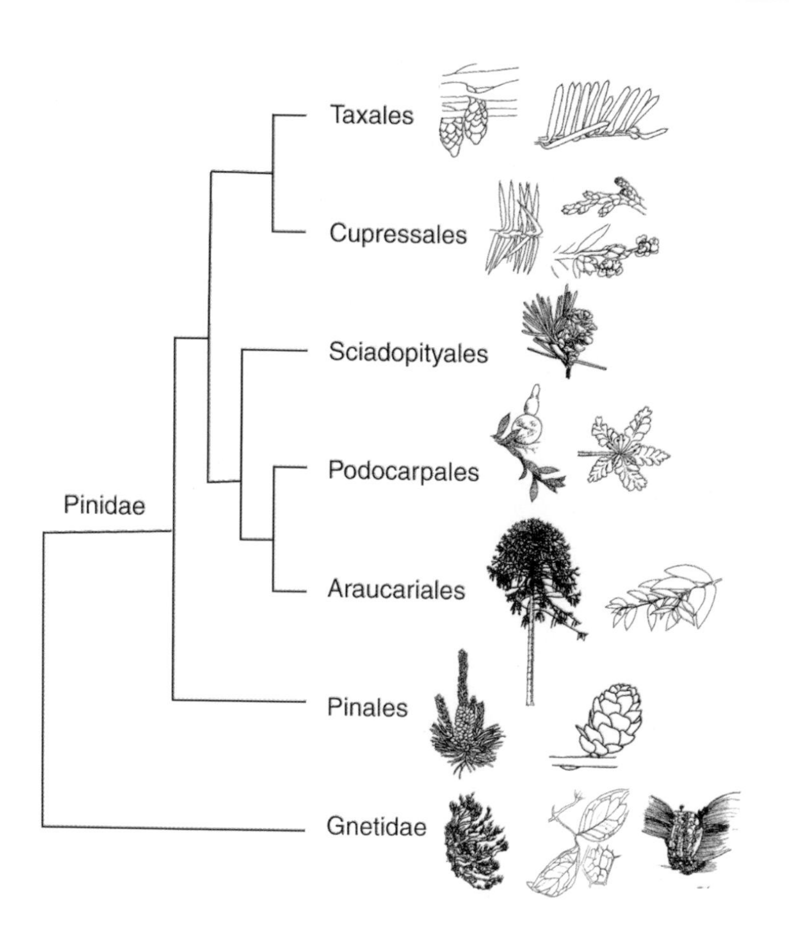

Figure 5.46. Phylogeny of living conifers.

PINALES (PINES AND FIRS)

Pine (*Pinus*), and its allies spruce (*Picea*), cedar (*Cedrus*), larch (*Larix, Pseudolarix*), hemlock (*Tsuga*), *Cathaya, Keteleeria,* fir (*Abies, Pseudotsuga*) with their needle-like leaves and long cones are the most abundant living conifers, especially in the Northern Hemisphere where they dominate the boreal forests. Most are evergreen but some, like larch, are deciduous. They have pollen with air bladders and woody cones that release winged seeds.

Figure 5.47. Pinales: (a) *Abies koreana*; (b) *Pinus sylvestris*.

Figure 5.48. Conifer foliage: (a) *Araucaria*; (b) *Agathis*; (c) *Podocarpus*; (d) *Phyllocladus*.

ARAUCARIALES (MONKEY PUZZLES)

These are the most primitive living conifers, the most similar to the Voltziales. As well as *Araucaria*, the monkey puzzles and bunya-bunya, kinki, hoop and Norfolk Island pines, there are the massive trunked *Agathis* species like Kauri (*Agathis australis*) from Australasia. *Wollemia*, discovered in 1994 near Sydney, is probably the most primitive member of the order.

PODOCARPALES (PODOCARPS)

The podocarps are a sister lineage to the Araucariales and together with them constitute the majority of the conifers of the Southern Hemisphere. The female cones generally bear only one or two ovules surrounded by bracts. The name *Podocarpus* ('foot-fruit') comes from the fleshy stalk bearing the ovule that becomes fleshy (arillate) in seed. The male cones are like catkins. Characteristically, podocarps have a very simple and primitive pattern of branching but this does not prevent them growing as very tall trees emerging above the canopy of *Nothofagus* forest in New Zealand, for example. Today they are not normally dominants but on the rolling fertile lowlands podocarps such as rimu and totara once formed extensive forests in New Zealand before they were cleared by the Maori and European colonists. Podocarps are also an important component of the shrubby vegetation found at higher altitudes and in areas of disturbance. The related *Phyllocladus*, with flattened frond like branches, is normally placed in its own family the Phyllocladaceae.

SCIADOPITYALES (UMBRELLA-PINES)

The only species *Sciadopitys verticillata* (Japanese Umbrella Pine) is a very distinct conifer with whorls of long deeply grooved leaves. Now confined to Japan, its fossils are found in Tertiary rocks of Europe; see Figure 5.49.

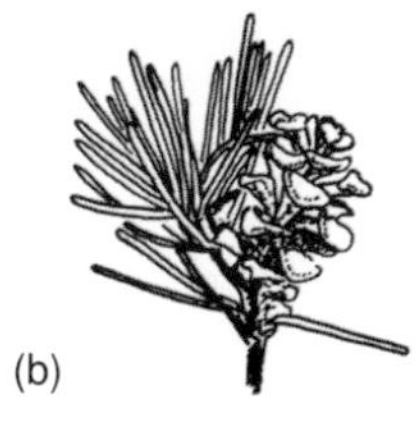

Figure 5.49. Sciadopitys: (a) foliage; (b) cone.

CUPRESSALES (REDWOODS, CYPRESSES AND JUNIPERS)

The Cupressales have scale-like or needle-like leaves although some species with scale-like leaves have needle-like juvenile leaves. Pollen lacks air bladders. Two families, the Cupressaceae and Taxodiaceae, are normally recognised.

The female cone scales either become woody releasing the seed or, as in *Juniper*, become fleshy to produce an indehiscent 'berry'. The Taxodiaceae include *Sequoia sempervirens* and *Sequoiadendron giganteum*, which are, respectively, the tallest (110 m) and most massive (trunk diameter to 11 m and 90 m tall) individual living organisms. Both were once widely distributed in the Northern Hemisphere.

Figure 5.50. Cupressales: (a) *Sequoia*; (b) *Cupressus*; (c) *Callitris*.

Cupressaceae	Taxodiaceae
Leaves opposite or whorled	Leaves borne spirally
Female cone scales in opposite pairs or threes	Female cone scales spirally arranged

TAXALES, YEWS AND CALIFORNIA NUTMEG

The Taxales are sister to the Cupressales (Figure 5.51). The Taxaceae have solitary ovules at the ends of branches. In *Taxus* the seed becomes surrounded by a fleshy aril. In *Torreya* the edible aril is fused to the seed and the whole structure looks like a small green plum. The Cephalotaxaceae have similar arillate seeds to the Taxaceae, looking like olives but produced in cones, although commonly only one seed will fully develop per cone. They are sister to the remaining Taxales.

Figure 5.51. Taxales: *Taxus* foliage and arillate seed.

Alternative names: Gnetopsida, Gnetaceae.

Distinguishing features: they have compound cones that resemble in some cases flowers.

Life-form/ecology: they are trees, shrubs or lianes of tropical forests (*Gnetum*), shrubs of arid regions (*Ephedra*) or giant leafy ground plants of the desert (*Welwitschia*).

Fossil record/sister groups: there is no unambiguous fossil record of the group older than the Cretaceous.

5.2.14 Gnetidae, the gnetophytes

There are three very distinct genera: *Ephedra, Gnetum* and *Welwitschia* (Figure 5.52).

Gnetaceae	Ephedraceae	Welwitschiaceae
1 genus *Gnetum* 28 species	1 genus *Ephedra* 40 species	1 genus *Welwitschia* 1 species, *W. mirabilis*

Gnetum has petiolate leaves with reticulate venation, and wood tissue with vessel elements that resemble those in flowering plants. However, flowering plants' vessel elements are thought to have evolved from tracheids with scalariform thickening whereas those of the Gnetales are derived from pitted tracheids. In some species the secondary phloem (unlike most gymnosperms) contains companion cells and sieve tubes, but these also develop in a different way from those in flowering plants. These features have encouraged some botanists to suggest that they may be the ancestors or sister group to flowering plants, but recent molecular data indicate that they are gymnosperms, sister to the conifers.

Figure 5.52. Gnetophytes: (a) *Gnetum*; (b) *Ephedra*; (c) *Welwitschia*.

The reproductive structures of the gnetophytes have many peculiar features. The process of female gametophyte development in the Gnetophytes is different in *Ephedra, Welwitschia* and *Gnetum*. It is most similar to other gymnosperms in *Ephedra* and the number of nuclei varies between 256 and 1000 in different species. In *Gnetum* the female gametophyte remains free-nuclear with about 7200 nuclei and no archegonia are formed. The micropyle (part of the inner integument) projects as a long tube. A free nucleus at the micropylar end functions as an egg. In *Welwitschia* multinuclear cells are formed that, at the time of fertilisation, produce tubes, which grow up through the nucellus. A nucleus in one of these tubes functions as an egg. Many of these tubes meet the downward-growing pollen tubes. At point of contact, both tubes break down and several female nuclei enter the pollen tube where fertilisation takes place. No other plant is known to achieve fertilisation in the pollen tube. In *Ephedra, Welwitschia* and *Gnetum* the generative nucleus in the male gametophyte gives rise directly to the sperm.

5.3 Class Magnoliopsida – flowering plants

There are more than 250 000 species of flowering plant. Alternative names for them include the angiosperms, Angiospermopsida.

5.3.1 Distinguishing features

DNA sequence data indicate that the flowering plants are a monophyletic group with a separate origin from all gymnosperms. Characteristics that most flowering plants share but which are rare or absent in other groups and which also support the view that they are monophyletic (synapomorphies) include the following:

- differentiated xylem tissue including fibres, parenchyma and usually vessel elements
- phloem sieve-tube elements and companion cells formed from a common mother cell
- reaction wood produced in branches as a response to tension is made up of gelatinous fibres in an adaxial part of the xylem (in contrast to the abaxial rounded tracheids produced as a response to compression in conifers)
- leaves broad and flat with a distinct petiole
- leaves with pinnate secondary veins and fine veins reticulate and veinlets ending blindly
- a high degree of plasticity in vegetative growth
- bisexual reproductive axis (-flowers) with male organs situated below the female
- insect pollination
- pollen wall chambered (tectate)
- male gametophyte with three nuclei only
- carpel or pistil enclosing the ovules and fruit surrounding the seed, although in primitive flowering plants the carpel is sealed by a secretion only
- female gametophyte normally with eight nuclei only
- double-fertilisation and formation of the triploid endosperm
- 'direct' development of the embryo from the zygote, i.e. with no intervening free-nuclear proembryo phase

Not all of these features are found in all flowering plants and some are found in a few other plant groups.

Figure 5.53. Floral diversity: (a) Houttuynia; (b) *Aristolochia*; (c) *Allium*.

5.3.2 Fossil record and origin

The origin of flowering plants, Darwin's 'Abominable Mystery', is conjectural. There is scarcely any hard evidence of their origin before the Cretaceous (135 million years ago) but molecular data indicate a much more ancient origin for the lineage that eventually gave rise to flowering plants. There are intriguing fossils of flower-like structures like those of *Archaefructus*, an aquatic plant of 124 million years ago, but the first undoubted flowering-plant fossils are pollen from equatorial latitudes of the Late Early Cretaceous, 125 million years

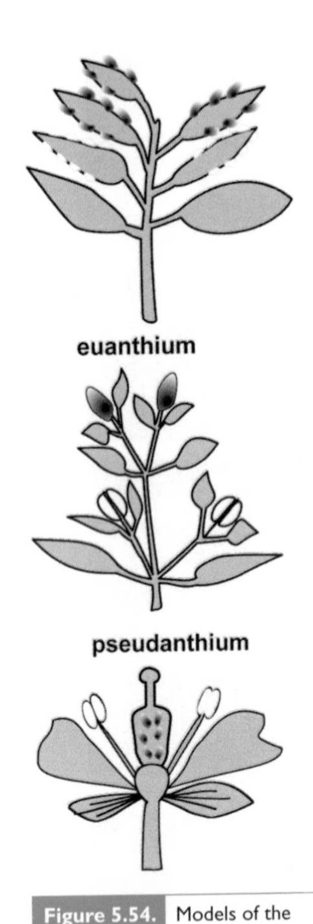

Figure 5.54. Models of the origin of the angiosperm flower.

ago. The flowering-plants probably originated in the tropics; extant primitive families are tropical or subtropical and early fossils show no adaptations to temperate conditions.

There have been a number of theories about the origin of flowers, homologising in different ways floral structures with the reproductive structures of other plants. One theory, the anthophyte theory, related flowers to the bisexual flower-like axes present in groups such as the extinct Bennettitales and living Gnetidae. Two contrasting theories are the euanthium theory, which derives a flower from a uniaxial cone bearing both micro- and megasporophylls, and the pseudanthium theory, which derives the hermaphrodite flower from a complex inflorescence of unisexual male and female flowers. However, these different theories place undue emphasis on a fundamental distinction between organ types. It has also been suggested that a flower-like structure could have arisen by a change of sex of some of the microsporophylls of a male cone, a process called gamoheterotropy. This process is not hard to imagine in plants that are hermaphrodite and where the determination of sex is a developmental phenomenon that is only rarely associated with sex chromosomes. A similar transfer of sex has been proposed in the more recent evolution of the maize cob. This theory has received recent support from a study of the genes of reproductive development. The gene determining the sex of floral organs in flowering plants is a homologue of a gene active in the male axis of conifers but not the female axis.

The first known leaves of flowering plants appear in fossils dated at 125 Ma and the first fossil inflorescence has been dated at 120 Ma. fossil has a mosaic of characteristics that are found in a range of basal groups of flowering plants. This has a female inflorescence, with a bract and two bracteoles at the base of female flowers that lack a perianth (achlamydeous flowers). The flowers are effectively just tiny pistils, less than 1 mm in diameter.

5.3.3 The evolutionary radiation of flowering plants

There are many threads to the evolutionary diversification of flowering plants:

- specialisation for pollination, including a reversal to wind pollination as well as more and more bizarre adaptations to attract specialist animal pollinators
- specialisation in fruit and seed dispersal, such as fleshy or otherwise elaborated fruits, and the evolution of seed dormancy
- adaptations for seedling establishment and growth in competition with other plants
- specialisation for growing in different habitats from aquatic to arid terrestrial habitats, surviving heat, cold and low light levels, or growing as epiphytes and parasites
- adaptations conferring resistance to or prevention of herbivory

Insect pollination is closely associated with the origin and subsequent diversification of flowers. However, it is important to remember that insect pollination is associated with several other groups of seed plants, both living and extinct: Bennettitales, Gnetales, Cheirolepidaceae (extinct conifers), Cycadales and Medullosales (seed ferns). Insects grew in diversity with the origin of seed plants in the Late Devonian and this increase of diversity with the origin of flowers is just part of a continuous trend. Indeed there is some evidence to indicate a temporary decline in insect diversity as flowering plants became more abundant in the Cretaceous. Nevertheless, flowers diversified in parallel with particular groups of pollinator specialists, the bees (Apoidea/Apidae), the pollen wasps (Vespidae: Masarinae), brachyceran flies (Acroceridae, Apioceridae, Bombyliidae, Empididae, Nemestrinidae, Stratiomyidae and Syrphidae) and the moths and butterflies (Lepidoptera). The evolution of a bisexual reproductive axis was a crucial event.

Several trends in floral evolution can be discerned. Primitive flowers either lack a perianth or have one in which there is a single whorl of tepals. In the perianth there has been a trend from having a perianth in which distinct whorls are not clearly differentiated to clear specialisation of a distinct calyx and corolla. The calyx protects against drought, temperature shock and predatory insects and the corolla attracts and controls pollinators. Both calyx and corolla may have been derived from tepals, but it is likely that in some groups, such as the buttercups, the petals originated from stamens, to which they are anatomically similar. In the peony, *Paeonia*, there is a gradual transition from leaves, through modified leaves on the flowering stem called bracts, into the perianth with parts at first sepal-like and then petal-like. Generally there has been a trend for the greater integration of the floral parts with greater precision in number and placement as flowers have become specialised to particular patterns of pollination.

An important aspect of the diversification of flowering plants and their evolutionary success has been their vegetative flexibility. They have evolved into a bewildering range of forms through the activity of sub-apical and intercalary meristems. Flowering plants also have a greater capacity for elongation of cells, including root hairs and trichomes. One of the most important vegetative specialisations has been their possession of vessels in their wood, permitting more efficient water transport and hence greater photosynthetic rates, fast growth rates and earlier maturation. There have been significant adaptations in seasonal habitats; for example, although a few non-flowering plants are deciduous this has been a particular flowering-plant trait that has adapted them to high latitudes or seasonally-dry environments.

One aspect to the burgeoning biotic diversity of flowering plants from the Cretaceous onwards was their great expansion of chemical diversity. New ranges of what have been called secondary compounds provided attractants for pollinators or seed dispersers, and repellants and toxic compounds to inhibit herbivores.

Figure 5.55. Specialisation in the perianth in the Asparagales: (a) *Belamcanda*; (b) *Kniphofia*.

5.3.4 The phylogeny of flowering plants

Ideas about the phylogenetic history of the flowering plants have been revolutionised in recent years by the cladistic analysis of DNA sequence data (Figure 5.56). What is presented here is as up to date as at time of press. Classifications are achieved by consensus and gain currency by use.

Figure 5.56. Phylogeny of the flowering plants from the Angiosperm Phylogeny Group (APG). In some respects the basal clades of flowering plants are only linked together by what features they lack. These were called by some the ANITA group from the initial letters of their family names. The last families belong in the same order (the Schisandrales), and the first two perhaps ought to be recognised as orders on their own (the Amborellales and Nympheales).

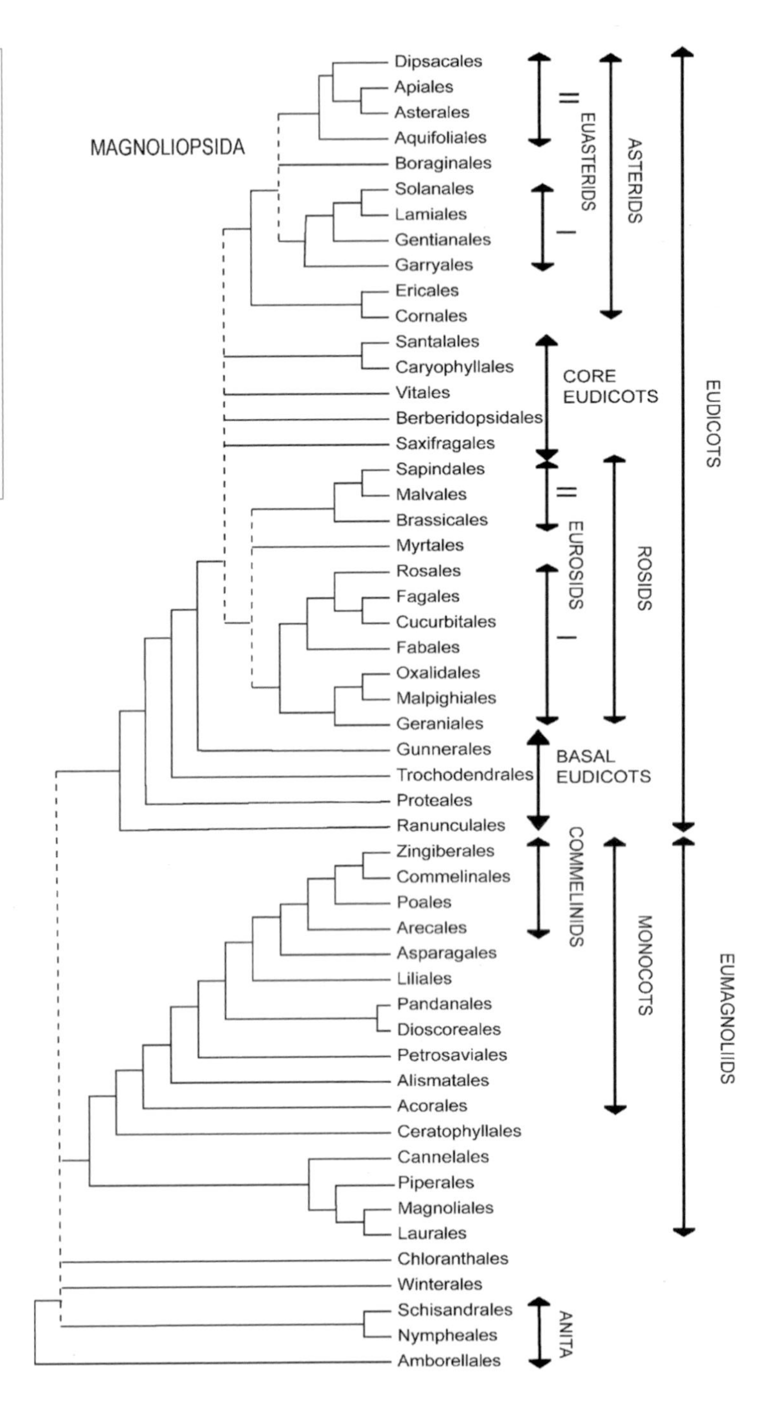

5.3.5 Basal flowering plants and Eumagnoliids

At the base of the flowering-plant phylogenetic tree there is a diverse group of families a 'grade rather than a clade' with a high proportion of primitive or unspecialised character states, fascinating because they represent the relicts of an early stage of flowering plant evolution.

Table 5.4 The distribution of features of flowering plants

Features in high frequency	
In basal groups (primitive/plesiomorphic/unspecialised)	In derived groups (advanced/apomorphic/specialised)
Small shrubs, lianes or rhizomatous perennial herbs, or aquatics and semi-aquatics	Various but including annuals and tall trees and herbs
Vesselless	Vessels present
Parts in whorls or spirals of variable numbers	Parts in whorls of three, four or five
Actinomorphic	Actinomorphic or Zygomorphic
Parts free	Parts connate or adnate
Stamens broad with poor differentiation between filament and anther	Stamens with well-differentiated filament
Apocarpous	Syncarpous
Unsealed stigma and poorly differentiated style	Sealed stigma with well-differentiated style
Follicles	Fruits (various)

AMBORELLALES

DNA sequence data places a plant called *Amborella trichopoda* from New Caledonia in a basal position as a sister to all other flowering plants. This is not to say that it is the ancestor of all other flowering plants. Rather that it is the closest living relative of the ancestor and like all other living flowering plants it exhibits a mixture of primitive and derived features. It is a shrubby evergreen plant with simple leaves that may be lobed. It has tracheids but no vessel elements. It is dioecious with flowers grouped in axillary cymose inflorescences. It has a perianth consisting of five to eight undifferentiated segments that are weakly joined at the base and are arranged in a spiral. The male flower has numerous (10–25) laminar stamens, the outer fused to the base of the perianth segments. Pollen is aperturate to non-aperturate and sulcate with a granulate outer wall and is possibly not tectate, a feature of possibly great significance. The female flower has five to six free carpels in a whorl. Carpels are open at the tip and have one ovule. Seeds are endospermic and the embryo has two cotyledons.

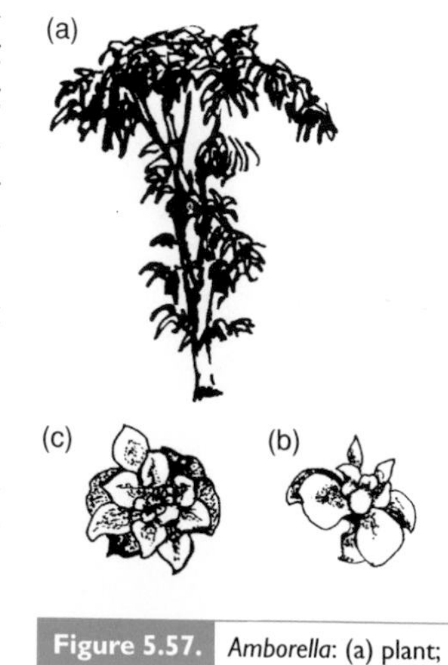

Figure 5.57. *Amborella*: (a) plant; (b) male flower; (c) female flower.

Figure 5.58. Nymphaeales: (a) *Nymphaea*; (b) *Cabomba*.

NYMPHAEALES (WATER-LILIES)

The next most-basal flowering plants are either six genera and about 40 species of water-lilies, with large flattened floating leaves and large bowl-shaped flowers in the Nymphaeaceae, or two genera (*Brasenia* and *Cabomba*) of waterweeds with floating stems and relatively small simple and unspecialised flowers in the Cabombaceae (Figure 5.58). The Nymphaeales have a mixture of unspecialised, probably primitive features, and specialised features such as abundant aerenchyma tissue, adapting them to their aquatic habitat. Vessel elements have been recorded in some species but these are not like those of other flowering-plants. The pollen has a tectum of sorts but its inner layer, the endexine, is compact and lacks the columellate appearance of all other flowering-plants. Some features are shared with the monocots: the primary root is soon aborted and the stem has scattered closed bundles. The showy petals have originated as sterile stamens (staminodes). The families differ in the degree of fusion of the carpels; laterally fused in Nymphaeaceae and free in Cabombaceae. Flowers are normally hermaphrodite with only a weak distinction between sepals and petals. Sepals and petals are arranged in a spiral. The large-flowered water-lilies *Nymphaea*, *Victoria* and *Nuphar* are specialised for beetle pollination. It is remarkable that similar looking aquatic plants have evolved convergently in the distantly related *Nelumbo* (Proteales) and *Nymphoides* (Asterales).

SCHISANDRALES (INCLUDING ILLICIALES AND AUSTROBAILEYALES)

The Schisandrales include four families of small trees and scrambling shrubs or lianes. AUSTROBAILEYACEAE: These are lianes. *Austrobaileya scandens*, one of only two species in this family, grows in NE Australia, and has flowers that smell of rotting fish. The flowers have 12 perianth segments, 6–11 laminar stamens, with sterile stamens (staminodes) inside surrounding the 6–9 free carpels. TRIMENIACEAE: There are only two genera, *Trimenia* and *Piptocalyx*, with a total of five species, of small trees and scrambling shrubs found from SE Asia to Australia. They are monoecious with small wind-pollinated flowers. The female flower has a single carpel and the male flower 6–25 stamens in pairs. ILLICIACEAE: There is only one genus, *Illicium*, with 42 species of trees and shrubs found in SE Asia and USA, Mexico and the Caribbean. Flowers may have 12 or more perianth segments, and 7–15 carpels. *Illicium* has peppery tasting leaves, and produces a star-shaped unripe fruit called star-anise that is used as a spice. SCHISANDRACEAE: There are only two genera, *Schisandra* and *Kadsura*, and a total of 47 species of lianes and twining shrubs, found in East Asia and eastern North America. The Illiciaceae and Schisandraceae share some chemical features, a primary vascular cylinder and tricolpate pollen. Unlike the previous two orders some members have clearly formed vessel elements although of a primitive sort, with sclariform (ladder-like) perforations.

Figure 5.59. *Schisandra*.

EUMAGNOLIIDS

The eumagnoliids include several dicot orders as well as the monocots (see below). They include many species that exhibit some primitive features they share with basal monocotyledons: whorls of floral parts in threes (trimerous), monosulcate/uniaperturate pollen, apocarpous flowers.

Like the ANITA grade of flowering plants many of these eumagnoliids are aromatic and include, for example, *Lindera* – allspice, *Piper* – pepper, *Cinnamomum* – cinnamon, *Aniba* – bois-de-rose oil, and *Sassafras* with its scented insecticidal oil. Another useful plant in this group is *Persea*, the avocado.

Figure 5.60. *Chloranthus.*

CHLORANTHALES

The Chloranthaceae (Figure 5.60) is the only family. It has about 75 species, most in the genus *Hedyosmum*. They are shrubs, lacking vessel elements. Their wood is soft and their swollen internodes sometimes collapse on drying. Flowers are very small and unisexual, with a single stamen or carpel and with or without a single whorl of three perianth segments.

Figure 5.61. *Laurus.*

PIPERALES (PEPPERS AND BIRTHWORTS)

It is hard to believe that *Aristolochia*, with its extraordinary tubular trap blossom belongs in the same order as *Piper* with its spikes of tiny flowers that lack a perianth altogether, along with *Hydnora*, which has flowers that arise from buds in the roots. The Aristolochiaceae are woody vines, the Piperaceae are herbs and the Hydnoraceae are root parasites that lack chlorophyll.

LAURALES (LAURELS)

There are seven families, most of which are trees, shrubs or lianes but *Cassytha*, in the Lauraceae, is a genus of twining, almost leafless plant parasites like the dodders (*Cuscuta*). They have flowers that vary from small to large (Figure 5.61).

Figure 5.62. *Magnolia.*

MAGNOLIALES (MAGNOLIAS)

Generally small trees or shrubs and lianes, most having relatively large showy flowers (Figure 5.62). The large flowers have a poorly differentiated perianth and quite often a variable number of segments. In addition some have partially sealed carpels and diverse but commonly broad stamens with a weakly distinguished filament and valvate anthers.

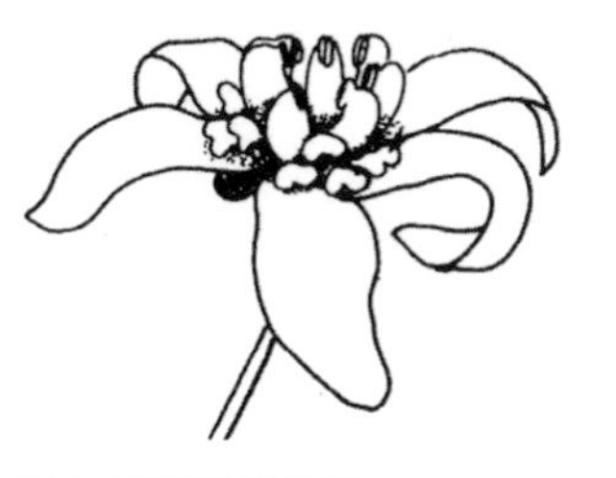

Figure 5.63. *Drimys.*

CANELLALES

There are only two families of evergreen shrubs and trees, the Canellaceae and Winteraceae, in the order. *Drimys*, in the Winteraceae, has a peppery taste. The flower has variable numbers of perianth segments, flat stamens with a poorly differentiated filament and only weakly sealed carpels. The Canellaceae have the stamens fused in a ring.

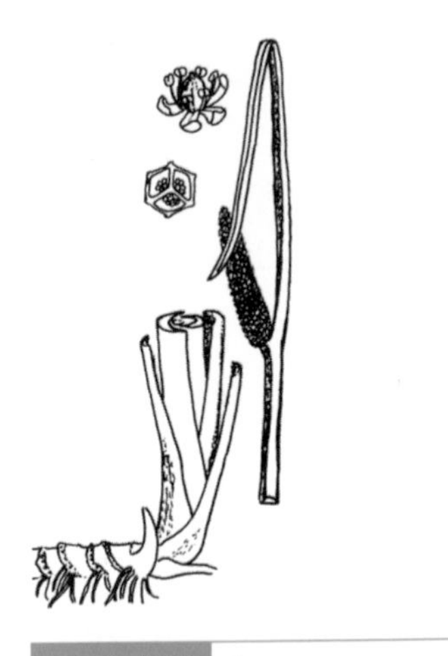

Figure 5.64. *Acorus.*

5.3.6 Monocots

The monocots represent by far the most evolutionarily successful lineage of the eumagnoliids and represent about 25% of all flowering plants, about 50 000 species. They are very diverse ranging from tall palm trees to tiny aquatic plants like *Lemna* (duckweed). Numerically they might be considered less important than the eudicots, but they include the grasses that provide the great majority of food for humans, either directly (wheat, rice, millet, etc.) or by feeding domestic grazing animals. They are clearly a monophyletic group and most have:

- single cotyledon
- sympodial growth (the palms Arecales are monopodial)
- linear, parallel-veined leaves in which the leaf base surrounds the stem
- primary root soon aborts and a wholly adventitious root system develops
- closed vascular bundles and lack of interfascicular cambium
- flower parts in threes
- pollen development (microsporogenesis) successive
- monosulcate pollen

The basal monocots are aquatic or semi-aquatic plants. As the seedling germinates the single cotyledon elongates to push the primary embryonic root, the radicle, out of the seed and down into the wet mud. It is because of their lack of a vascular cambium that monocots are well able to undertake elongating growth but poor at thickening (secondary) growth. Elongating growth adapts them for the aquatic, climbing and epiphytic niches where they predominate, and also permits them to re-grow rapidly after grazing. The parallel venation of their leaves is a consequence of the extension of the leaves from a basal meristem. Indeed it is likely that parallel-veined monocot leaves are homologous to the petiole of leaves in other flowering plants. Monocot trees and shrubs undergo various different and amorphous kinds of stem thickening, which is sometimes described as anomalous, and they are usually either unbranched or only weakly branched.

Figure 5.65. Alismatales: (a) *Butomus;* (b) *Lysichiton.*

ACORUS (SWEET-FLAG OR CALAMUS)

At the base of the monocot lineage are the two species of *Acorus*, a rooted aquatic. It has a tiny bisexual flowers crowded in a spadix. The carpels are primitive, intermediate between the ascidiate of the ANITA group and folded ones of other flowering plants. It has been utilised for centuries as a rush for floor covering because of its spicy scented properties.

ALISMATALES (WATER WEEDS AND AROIDS)

There are several distinct aquatic families in this order linked by many adaptations to the aquatic habitat (floating stems, aerenchyma) as well as tricolpate pollen, and several have stamens produced in pairs, but even among these there are distinct rush-like forms (Butomaceae) as well as free-floating and submerged ones (especially the Alismataceae, Potamogetonaceae and Hydocharitaceae). Unlike most monocots, several of these aquatic families have a leaf that is clearly petiolate, a condition that is also present in the main terrestrial family of the order, the Araceae. Here also are found the few flowering plants to have entered the marine habitat, in the Zosteraceae and Posidoniaceae. The Alismataceae and Limnocharitaceae produce latex. The Juncaginaceae have a fascinating reversal of floral whorls with an inner perianth whorl inside the outer stamens. It is likely that the Araceae have an aquatic origin but they are mainly terrestrial now. The few aquatic Araceae are very specialised, free-floating ones in the subfamily Lemnoideae, including the smallest flowering plant *Wolffia*, as well as *Lemna* and *Pistia*. Most Araceae are vines and epiphytes and form an important component of tropical and subtropical forests. Like *Acorus* the Araceae have a spadix associated with a large spathe.

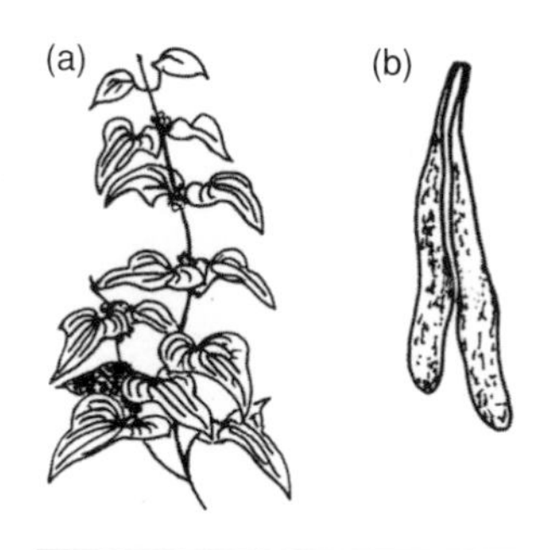

Figure 5.66. *Dioscorea*: (a) twining stem; (b) root tubers.

ASPARAGOIDS DIOSCOREALES (YAMS)

The Burmanniaceae are mycotrophic, effectively saprophytic by utilising fungi to garner nutrients and energy. Some even lack chlorophyll and have only scale-like leaves. The Dioscoreaceae are yams, from the African word nyami, twiners with thick rhizomes or tuber-like swellings, sometimes many kilograms in size, and net-veined leaves. They are exceptional in some having secondary thickening, although it occurs in the tubers! There are only four genera but about 900 species.

Figure 5.67. *Pandanus*.

PANDANALES (SCREW PINES)

The Pandanaceae, increase their girth as each node is produced so that the 'trunk' is balanced on a conical base, but supported by profuse, thick, strut-like roots. Because of the spiral way they produce their leaves they seem to screw their way up towards the forest canopy. The young fibrous leaves of *Carludovica* are divided into strips and bleached with lemon juice to be made into Panama hats. *Pentastemona* in the Stemonaceae stands out as a monocot with three whorls of five parts. They also have petiolate leaves. The Pandanales, like several other monocot orders, have their own family (the Triuridaceae) of echlorophyllous mycotrophs.

Figure 5.68. *Lilium.*

LILIALES (LILIES)

In contrast to the superficially similar Asparagales, the Liliales tend to have nectaries at the base of the floral parts, spots on the petals and stamens with anthers opening to the outside (Figure 5.68). Other features like the cellular structure of the seed coat also link them. Like the amaryllids they include many geophytes, producing bulbs (Liliaceae, Melanthiaceae) or corms or rhizomes (Colchicaceae). Some are shrubby or vines (Philesiaceae, Smilacaceae). They also include chlorophyll-lacking mycotrophs in the Corsiaceae. There are four main lineages: (1) Smilacaceae with Liliaceae; (2) Alstroemeriaceae with Luzuriagaceae and Colchicaceae; (3) Campynemataceae; and (4) Melanthiaceae.

Figure 5.69. Veltheimia bracteata (Eastern Cape Province, South Africa).

ASPARAGALES (AMARYLLIS, IRISES AND ORCHIDS)

The Aparagales is the largest order of monocots and includes many beautiful flowers (Figure 5.69). The Asparagales have a seed coat in which cellular structure has become obliterated and there is a black crust of phytomelan. Most Asparagales are rhizomatous herbs but a few such as the Ruscaceae are shrubs (*Ruscus*) or even trees (*Dracaena*). Many of the most beautiful flowers are in a lineage of bulb-formers that includes the Amaryllidaceae, Agapanthaceae and Alliaceae. Many of these produce flowers in an umbel with a spathe at its base. The Iridaceae (irises and crocus) are in a distinct lineage and are distinguished by their divided style. Another interesting feature is that they do not produce root-hairs and rely entirely on mycorrhizae for garnering nutrients from the soil. The irises have a flower that provides three distinct entry points for pollinators, and a petaloid stigma overarching each of the three stamens. Another interesting lineage is one including tussock formers and trees in the Xanthorhoeaceae (grass-trees) and Asphodellaceae the aloes (*Aloe*) and red-hot pokers (*Kniphofia*).

Figure 5.70. (a) *Watsonia*, (b) *Tulipa*, (c) *Xanthorhoea*, (d) *Moraea*.

The orchids are the largest family of the order and comprise the most diverse and remarkable of all flowering plant families with between 800 and 1000 genera and up to 20 000 species, rivalled in numbers only by the Asteraceae. Orchid flowers have a complex architecture and are exceedingly diverse in the structure and arrangement of the various parts of the column, rostellum and pollinaria and in the shape, colour, and scent of the perianth.

Figure 5.71. Floral diagrams of the three main patterns of orchid architecture. The Apostasioideae are a small SE Asian subfamily with three stamens. The Cyprepediodeae are the slipper orchids. The monandroid orchids are by far the most numerous and include several families of both terrestrial orchids (mainly Orchidoideae and Spiranthoideae), and epiphytic orchids (mainly Epidendroideae) and even lianes (Vanilloideae).

Figure 5.72. *Oncidium* sp. are epiphytic orchids commonly cultivated for their beautiful flowers.

Figure 5.73. Three kinds of terrestrial orchids: (a) *Anacamptis laxiflora*; (b) *Neottia nidus-avis* lacks chlorophyll and is mycotrophic; (c) a slipper orchid, *Phragmipedium* sp.

Figure 5.74. Palm inflorescences.

Commelinids

ARECALES (PALMS)

The Arecaceae palms are the most important group of monocot trees but, like other monocot trees, they are unbranched or only weakly branched and lack a vascular cambium (Figure 5.30). Rather, secondary growth occurs by the expansion, and occasional dichotomous splitting, of a large apical meristem. The leaves of palms are large and complex, often palmate or pinnately lobed, and highly folded. Their flowers are relatively simple, following a standard monocot pattern with parts in threes, but often they are grouped together in massive, profusely branched inflorescences. About six subfamilies have been distinguished based on the form of the leaf (fan or feather) and its folding (rib down – induplicate, or rib up – reduplicate).

POALES (GRASSES, SEDGES AND RUSHES)

The Poales dominate most ecosystems where the growth of trees is limited. The Typhaceae, Juncaceae and Cyperaceae are common dominants of semi-aquatic and water-logged conditions. In contrast the Poaceae dominate moist or dry habitats in many climates, everywhere tree growth is restricted for any reason. They tolerate fire and grazing, even mowing, regrowing rapidly after they have been damaged. Grasses have a diverse of photosynthetic mechanism; C4 photosynthesis is common and has evolved many times over, permitting them to grow rapidly in the tropics.

Figure 5.75. Bromeliad inflorescence, showing the disticous arrangement of bracts and flowers common in the Poales.

The Xyridaceae have showy flowers. The Bromeliaceae have showiness, but it is mostly provided by colourful bracts. The bromeliads are very important epiphytes in the Americas as tank-plants or air-plants. Their flowers are relatively unspecialised, but commonly each flower is produced at the base of a brightly coloured bract. Most are epiphytic and show many adaptations for their epiphytic life-style such as water storage tissue or water-absorbing peltate scales. The air-plants *Tillandsia*, including Spanish moss, are perhaps the most remarkable in their ability to absorb moisture from the air. Most of the ground bromeliads have adaptations for arid environments.

Most other families in the order have reduced and relatively inconspicuous, often unisexual, flowers adapted for wind pollination by having an inflorescence held on a long leafless scape above the leafy rosettes. This trend has occurred in several of the constituent lineages. For example, sister to the Xyridaceae, are the mostly wind pollinated Eriocaulaceae, with similar dense heads of flowers. A few species in the Eriocaulaceae are insect pollinated with nectariferous petals. Some species are monoecious with female only marginal flowers in the head and central ones male. Other species are dioecious. In the Typhaceae the male part of the spadix is above the female. The Juncaceae have a typical monocot perfect flower, but it has a green or brown and papery perianth. They are wind pollinated, and a few are dioecious, but some have become secondarily insect-pollinated although they lack nectaries. In their sister family the sedges, Cyperaceae, the perianth is reduced to scales or bristles or is absent. Some genera, such as *Scirpus*, have hermaphrodite flowers. Others, including the largest genus *Carex*, have unisexual flowers with male and female flowers in different parts of the inflorescence. The male flowers consist only of three stamens with a bract called the glume on the abaxial side. A glume is also present in the female flower. In addition, two inner glumes have become fused to form a bottle-shaped perigynium or utricle surrounding the pistil. The style protrudes through the opening in the utricle.

Figure 5.76. Xyridaceae.

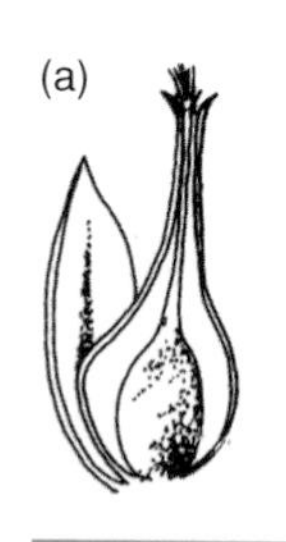

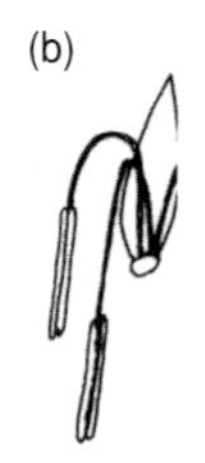

Figure 5.77. Cyperaceae florets: (a) female; (b) male.

In the Poaceae, the grasses, the flowers are similarly reduced and adapted for wind pollination (Figure 5.78). They are grouped alternately side-by-side in spikelets. The primitive floral condition for grasses is retained by the bamboos (Bambusoideae). They have simple, often trimerous spikelets and they may have three lodicules and three stigmas. Progressive reduction has given rise to the standard grass floret pattern. *Ampelodesmus*, a primitive member of the advanced subfamily Pooidae also has three lodicules. At the base of the spikelet are two glumes that protect the spikelet in development. Each floret in the spikelet is enclosed by two other bracts; the lower is the lemma and the upper is the palea. Within the floret there are usually three stamens and a pistil with two feathery stigmas. At the base of the pistil there are two tiny lodicules, remnants of the perianth, which swell to push open the floret or shrink to allow it to shut. Cross-pollination is ensured not by separation of the sexes but by the different time of pollen release and stigma receptivity and, most importantly, by a unique form of self-incompatibility. The fruit is an achene with the seed fused to the fruit wall (a caryopsis). It is usually shed enclosed within the lemma and palea, which may be modified to aid dispersal. Threshing releases the grain from this chaff. The Poaceae is one of the most successful of all flowering plant families because of its ability to spread laterally by rhizomes and stolons or the production of herbivory, drought- and fire-resistant tussocks.

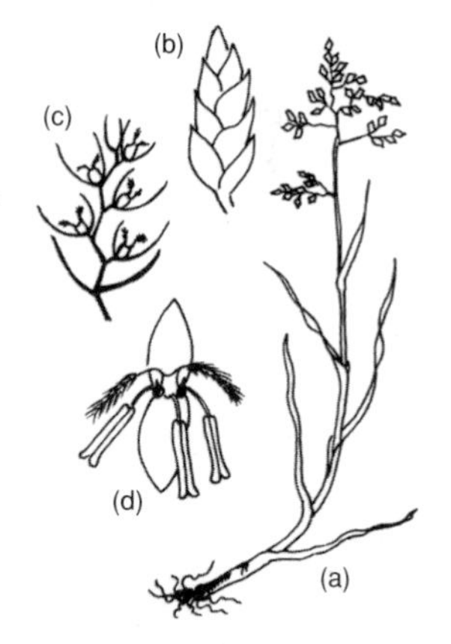

Figure 5.78. Poaceae: grass inflorescence: (a) plant with paniculate inflorescence, (b) spikelet, (c) distichous arrangement of florets in spikelet, (d) single floret with three dangling stamens and two feathery stigmas.

Figure 5.79. Commelinaceae: *Tradescantia gigantia*, an upright form (Enchanted Rock, Texas).

COMMELINALES (SPIDERWORTS)

The Commelinales are mainly herbs like the Haemodoraceae and Commelinaceae (Figure 5.79) but include the aquatic Pontederiaceae. Frequently the flowers are zygomorphic (monosymmetric). Heteromorphic flowers are found in the Commelinaceae (enantiostyly) and Pontederiaceae (heterostyly).

ZINGIBERALES (GINGERS)

The Zingiberales may be regarded as plants that do not produce an aerial stem except when flowering (Figure 5.80). They have showy, often strongly zygomorphic flowers especially adapted to bird pollination. The form of the flower with a large lip (labellum) is only exceeded in complexity in the monocots by the orchids. Sepals and petals are fused (connate) and sterile stamens (staminodes) are petaloid. The different families illustrate a great variety of specialisations for pollination in the tropics. The bananas, Musaceae, are the largest of all herbs, though they look like trees. The gingers, Zingiberaceae, include more than a thousand species of tropical herbs. In some ways the flower is analogous to that of the orchids with its pronounced zygomorphy and adnation of a single stamen to the perianth with other stamens converted into tepals and the anther supporting a slender style. However, the filaments and style are long and the anthers and stigma are exposed so that the pollinator is not as effectively 'controlled' as in the orchids. The Costaceae have five staminoids connate as a labellum and the stamen is broadly petaloid. Pollinators include hummingbirds and large bees. As in the gingers, the anther supports the slender style. The arrowroots (Marantaceae) have an asymmetrical flower with a single median stamen, which is half petaloid and all others are staminoidal and petaloid. The style is under tension and triggered to scoop pollen from the bee pollinator. *Canna* (Cannaceae), or Indian shot, has two whorls of three tepals, and up to five petaloid staminoids, which are showier than the perianth. In addition the fertile stamen and style are petaloid.

Figure 5.80. Zingiberaceae: (a) *Zingiber*; (b) *Heliconia*.

Basal Eudicots

Basal eudicots such as the Proteales and Ranunculales have a pattern of leaf venation in which the lateral veins terminate at the margin in a small tooth (craspedodromus). They have a well-developed perianth but this is poorly differentiated into a calyx and corolla, and has a variable number of tepals spirally arranged or in whorls of three. Stamens and carpels are numerous and varying in number. The carpels are free to connate and have a sessile stigma.

Figure 5.81. Fumarioideae (*Corydalis*): (a) plant; (b) detail of a single flower.

RANUNCULALES (BUTTERCUPS, POPPIES AND BARBERRIES)

Evolutionary trends in the Ranunculales include changes in the symmetry and the increasing complexity of the flower. Floral parts, especially the numerous stamens and carpels are commonly spirally arranged and the fruit is usually a follicle or an achene. The flower is apocarpous with superior pistils. The poppies, Papaveraceae (~660 species) are derived from the buttercups from which they differ by having a syncarpous gynoecium and only two to three sepals. There are two subfamilies, the actinomorphic Papaveroideae, which produce latex, and the strongly zygomorphic Fumarioideae, which have a clear sap. Another large family in the order, the Berberidaceae (~570 species) is distinguished from the Ranunculaceae by having stamens opposite the petals and a single pistil which becomes a berry. They have their flower parts in whorls rather than a spiral with two whorls of stamens and one to two whorls of nectaries.

Figure 5.82. *Protea*: the long slender flowers are grouped in heads surrounded by showy bracts. Each flower has a single pistil surrounded by four petals (three fused and one free) with anthers adnate to the petals.

PROTEALES (PROTEAS, BANKSIA AND GREVILLEAS)

The Proteales are well represented in the Southern Hemisphere. The family Proteaceae in particular shows a distribution which records the old Gondwanan supercontinent. The Proteaceae are one important family (Figure 5.82) where brush blossoms have evolved. Some relationships discovered by the analysis of DNA sequence data are distinctly odd. For example, in the Proteales, *Nelumbo*, the sacred lotus, and *Platanus*, the plane tree, are sister groups. If this sister relationship is true it provides an astonishing reminder of how little we know about the evolutionary history that connects living plant groups. What kind of shared ancestor did these two lineages have and what were the circumstances that led to one becoming aquatic and the other a tree? The existence of such differences in sister lineages demonstrate the potential evolutionary fluidity of morphological characters, and emphasises the fact that living plants are only the tips of a highly branched phylogenetic bush. Within the bush many branches end blindly and do not reach the surface so that intermediate linking kinds between living plant groups do not now exist.

Figure 5.83. *Trochodendron.*

Figure 5.84. *Gunnera.*

Figure 5.85. Polygonaceae: *Reynoutria.*

Figure 5.86. Amaranthaceae: *Ptilotus.*

TROCHODENDRALES

This order has only two species of evergreen trees from east and South East Asia each in its own family, *Tetracentron sinense* and *Trochodendron aralioides* (Figure 5.83).

GUNNERALES

There are only two genera in the order. *Gunnera* has the familiar, massive, palmate and deeply ribbed leaves. Usually grown beside water it is also a colonist of land-slips. One advantage it has is the fixed nitrogen it gets from the symbiotic blue-green bacteria (*Nostoc*) that live in its exposed roots and rhizomes. It produces large strobiloid inflorescences of tiny flowers, either bisexual or unisexual. *Myrothamnus*, from tropical Africa and Madagscar, is a resurrection plant, appearing to dry out but able to revive and start growing again when water becomes available.

Core Eudicots

Core eudicots have predominantly flowers with parts in fives (pentamerous) with a clear distinction between calyx and corolla. There are two main lineages, the Rosids and Asterids, three large basal lineages, the Caryophyllales, the Santalales and the Saxifragales, and a number of others of uncertain relationship like the Berberidopsidales and Vitales. These basal orders are crassinucellate (see below).

CARYOPHYLLALES (CATCHFLIES, STONECROPS AND CACTI)

The Caryophyllales is a large, and an interesting group of about 4% of all flowering plants that exhibits a unique set of characters (see Figure 5.86). For example it seems to lack mycorrhizae. It comprises mostly herbs, but others are lianes or twiners and shrubs. They have a peculiar pattern of secondary growth with the production of diffuse or successive cambia, which is commonly associated with succulence and CAM photosynthesis. Many families that have a high frequency of succulence have species that are either xerophytic like the cacti (Cactaceae), stonecrops (Aizoaceae) or halophytic like the sea-lavenders and thrifts (Plumbaginaceae).

Most have a campylotropous ovule in which the inner integument protrudes, and a peripheral embryo surrounding a nutritive central perisperm tissue and so they were previously called the Centrospermae. Another shared character is a peculiar type of sieve-tube plastid, and their chemistry is distinct. The core caryophyllid families are Amaranthaceae, Aizoaceae, Cactaceae, Caryophyllaceae, Didiereaceae, Molluginaceae, Nyctaginaceae, and Phytolaccaceae. Most of these have a shikimic acid biosynthetic pathway as a starting point for the synthesis of nitrogen-containing benzylisoquinoline alkaloids and the betalain pigments. The latter are utilised instead of the anthocyanins used in other flowering plants.

Non-core caryophyllids are the Plumbaginaceae, Polygonaceae, Tamaricaceae, and Frankeniaceae in one clade and four families of insectivorous plants in a sister clade. The former include many halophytic plants, which are also found in core caryophyllids such as the Amaranthaceae and Chenopodiaceae (*Salicornia*). Many have epidermal glands but in different families the glands are adapted to produce either mucilage (in Polygonaceae), excrete salt as in the halophytes, or digestive enzymes in the insectivorous Droseraceae and Nepenthaceae. In the latter, the gland-type is shared but the subsidiary insect-trapping apparatus is quite diverse ranging from pitchers (*Nepenthes*), sticky traps (*Drosera*, *Drosophyllum*) to spring traps (*Dionaea*, *Aldovandra*).

Several families in the order have a peculiar flower development that starts in a polymerous way but becomes organised into a pseudodiplostemonous way. Stamens appear to be produced in pairs. In the Caryophyllaceae and Plumbaginaceae stamens are antipetalous and arise with the petal as a unit.

Some families in the Caryophyllalaes (Caryophyllaceae, Chenopodiaceae) generally lack mycorrhizae perhaps because they tend to occupy nutrient rich fresh soils. In contrast, the insectivores in the Nepenthaceae, Droseraceae and Drosophyllaceae can inhabit nutrient-poor soils.

Figure 5.87. Aizoaceae: *Mesembryanthemum*.

Figure 5.88. Grossulariaceea: *Ribes*.

SANTALALES (SANDLEWOODS)

All five families of the Santalales (Santalaceae ~500 species, Olacaceae ~200 species, Opiliaceae 28 species, Misodendraceae 8 species, and the Loranthaceae ~940 species) include tropical parasitic species. The Loranthaceae are the most specialised parasites. The shrubby, liane or twining habit is common throughout the order.

SAXIFRAGALES (SAXIFRAGES, CURRANTS AND STONECROPS)

The Saxifragales is a very diverse order and includes: trees (Hamamelidaceae ~100 species of witch hazel and sweet gum, Cercidiphyllaceae – katsura, Altingiaceae), shrubs (Grossulariaceae ~325 species of currants and gooseberies) and showy ornamentals like the Paeoniaceae (~34 species of peony), but the two largest families are herbs, mainly rosette-forming Saxifragaceae (~475 species of *Astilbe* and saxifrage) or leafy succulents, the Crassulaceae (~1280 species of *Sempervivum*, *Echeveria*, *Sedum* and *Kalanchoe*). Many of these rosette formers in the Crassulaceae, Saxifragaceae and some other small families, are linked by a similar look to their flowers and a number of characters such as a persistent scarious calyx and cellular endosperm.

Figure 5.89. Paeoniaceae: *Paeonia*.

VITALES (VINES)

This order is of supreme importance to us as the source of wine from the grape vine *Vitis*. Many other genera are twining vines with or without tendrils (*Rhoiocissus*, *Cissus*). *Cyphostemma* is a caudiciform and *Leea* a shrub and small tree.

BERBERIDOPSIDALES

This tiny order of two families, one with only one species *Aextoxicon* from Chile and the other with only two genera *Berberidopsis* and *Streptothamnus* from Chile and eastern Australia.

Figure 5.90. Berberidopsidales: *Berberidopsis*.

Rosids and Asterids

The remaining eudicots form two great lineages, the crassinucellate (thick nucellus) Rosids and the tenuinucellate Asterids, each split into two sister lineages (Eurosid 1 and 2, Asterid 1 and 2). The tenuinucellate condition is a derived feature of the ovule where the tissue layer, the nucellus, surrounding the developing megasporangium/embryosac is thin. There are various kinds of crassinucellate condition. Each main and sub-lineage is very variable but exhibits particular evolutionary tendencies. For example, all the nitrogen-fixing families are found in the Eurosid I group: Casuarinaceae, Myricaceae (Fagales), Eleagnaceae, Rhamnaceae, Ulmaceae (Rosales), Fabaceae (Fabales) and Coriariaceae (Cucurbitales). There are well defined families like the Fabaceae, almost stereotypical with its characteristic fruit, the legume, and characteristic forms of flowers and inflorescence, or large and diverse families, like the Rosaceae, that have proved evolutionarily flexible.

5.3.7 Rosids

Rosid orders or even families are very diverse in their floral structure. Most of the ecologically (and economically) important trees from forests and savannas around the world are Rosids. They include the tallest flowering plants such as the eucalypts, the mahoganies in the tropics, the savanna acacias, and oaks, maples and beech from temperate forests.

Percentage of families	Rosids	Asterids
stipules	66% with	16% with
stamens	58% with two whorls	79% single whorl
corolla	97% free	75% fused
integument(s)	94% two	87% one
nucellus	92% crassinucellate	91% tenuinucellate
endosperm	96% nuclear	75% cellular
iridoids	5% present	61% present

Figure 5.91. Geraniales: *Geranium.*

Basal Rosids

GERANIALES (GERANIUMS)

The Geraniales include one large family, the Geraniaceae, and several very small ones. One interesting feature they share is the presence of glands on the margin of the leaf. They have pentamerous obdiplostemonous flowers with a persistent calyx. The two largest genera *Pelargonium* and *Geranium* produce similar beaked fruits but differ in the former having monosymmetric flowers and a nectariferous pedicel (Figure 5.91). The Geraniaceae are commonly herbs with jointed stems but the other families in the order include shrubs and trees.

CROSSOSOMATALES

These are a small order of shrubs and small trees adapted to dry habitats.

MYRTALES (EUCALYPTS AND MYRTLES)

The Myrtales have an uncertain relationship to either main Eurosid clade. Many have flowers with a large number of stamens (Figure 5.92). However, not all show this pattern. In the Melastomataceae (4750 species) the stamens are dimorphic with showy colourful outer stamens and short pollen-producing inner ones. The Onagraceae (650 species) have only four or two stamens and a long hypanthium tube. These families are significant as herbs or shrubs but the Myrtales includes the Myrtaceae and Combretaceae, which are highly significant as trees in the tropics especially in semi-arid areas of Australia (*Eucalyptus* ~450 species) and Africa (*Combretum* ~250 species, and *Terminalia* ~150 species).

Figure 5.92. Myrtales: *Eucalyptus*.

Eurosid I

ZYGOPHYLLALES (CREOSOTE-BUSH AND LIGNUM-VITAE)

There are several species important in arid and saline soils like *Larrea*, *Balanites* (thorny) and *Zygophyllum*. *Larrea divaricata*, the Creosote bush of the deserts of USA and Mexico is strongly allelopathic.

CELASTRALES (SPINDLE-TREE AND EBONY)

A member of the Celastraceae familiar to us is the widely planted garden plant *Euonymus*, the spindle-tree, with its characteristically angled fruit. *Maytenus* (ebony) is an important tree in warmer areas throughout the world. The flowers generally have a broad disk with the ovary submerged in it.

Figure 5.93. (a) *Viola*; (b) *Linum*.

MALPIGHIALES (SPURGES, VIOLETS, WILLOWS AND PASSION-FLOWERS)

The diversity of the order is illustrated by a comparison of the families Violaceae (pansies), Passifloraceae (passion flowers), Linaceae (flax), Salicaceae (willows) and Clusiaceae (*Hypericum*), each with a very different kind of flower. Perhaps its most interesting family is the latex-producing Euphorbiaceae that includes such important genera as *Ricinus*, *Euphorbia*, *Manihot* and *Hevea* (rubber) (see Figure 5.94).

OXALIDALES (BERMUDA BUTTERCUP AND WOOD SORREL)

The order includes trees, shrubs, lianes and herbs (*Oxalis*) and also the insectivore *Cephalotus* (Cephalotaceae).

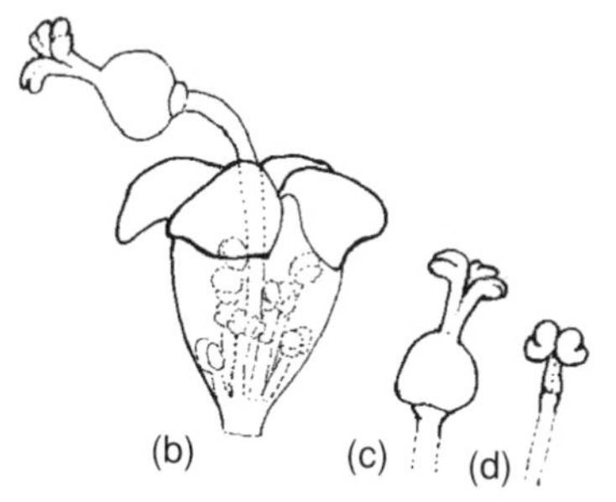

Figure 5.94. Euphorbiaceae: the inflorescence (a cyathium) mimics a flower: (a) succulent euphorb with a crown of leaves and cyathia; (b) a cyathium showing the arrangement of a central female floret surrounded by male florets and an involucre of bracts; (c) a single female floret; (d) a single male floret.

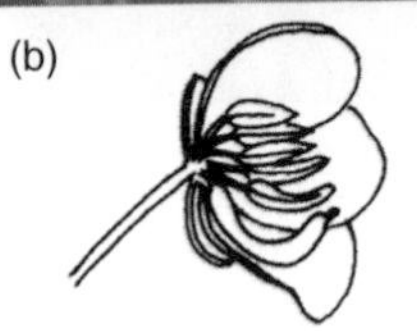

Figure 5.95. Cesalpinioideae: (a) *Delonix*; (b) *Cassia*, half-flower; and (c) side and front view.

FABALES (LEGUMES)

The Fabales include the Fabaceae, sometimes recognised as a single family, with three subfamilies, or as three families (Figure 5.95). They all share the characteristic of a kind of fruit, the legume that gives them their alternative name, the Leguminosae. The Caesalpinioideae (~2000 species) have large showy, more or less regular flowers. They are basal in the family and the other two families show different patterns of specialisation.

The Mimosoideae (~3100 species) have small regular flowers in dense spikes or heads, brush blossoms with numerous exserted stamens. Two evolutionary trends are observed in the Mimosoideae. One trend is for an increase of the number of stamens although each has a tiny anther. The filaments are long and the stigma is small and cup-shaped. Pollen is released as a polyad. In *Acacia* only one polyad can fit on each stigma and the number of seeds produced in the legume is directly related to the number of pollen grains in the polyad. Another evolutionary trend shows a reduction of the number of stamens but specialisation of the flowers in the head to form a kind of pseudanthemum. For example, in *Parkia* the lower florets are showy and scent-producing but sterile, the intermediate ones sterile but nectar producing and the upper fertile.

The Caesalpinioideae and Mimosoideae are mainly trees and shrubs but most of the third subfamily, the Faboideae (~11 300 species) are herbs. They have the zygomorphic flower that gives them their alternative name (Papilionoideae). The flag blossom is pollinated by large bees which land on the keel. The nectary is at the base of a staminal tube. In forcing its proboscis into the staminal tube the heavy insect pushes the keel petals (alae) down and the stamens and stigma rub against its ventral surface. A similar type of papilionate flower has evolved in parallel in the Polygalaceae, also in the Fabales. The Faboideae includes many agriculturally important species, peas and beans of all sorts. Clovers are especially important in pasture.

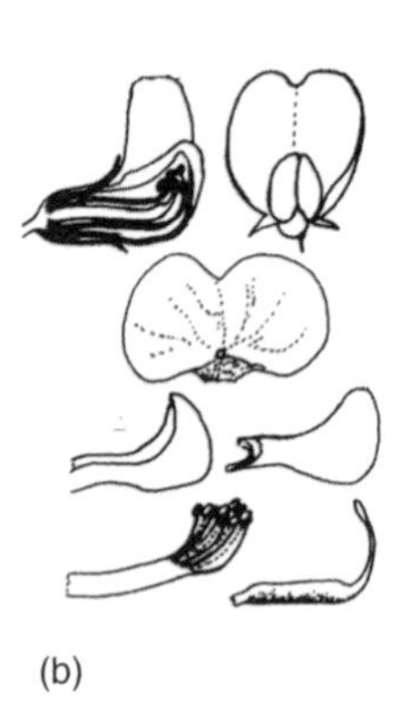

Figure 5.96. Faboideae: (a) *Trifolium* (keeled flowers in a head); (b) *Vicia* (half-flower and dissected flower).

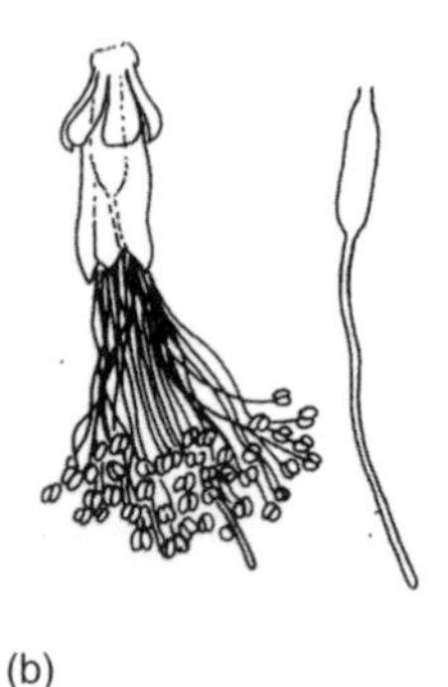

Figure 5.97. Mimosoideae; *Acacia*: (a) florets in spherical heads; (b) single floret and a pistil.

The stereotypical nature of the Fabales/Fabaceae is emphasised because about one third of all species in the order/family belong to one of a few very large genera: *Acacia* (1200 species), *Mimosa* (400 species) (both Mimosoideae), *Cassia* (540 species) (Caesalpinioideae), *Astragalus* (2000 species), *Crotalaria* (600 species), *Infigofera* (700 species) (all Faboideae/Popilionoideae). All these large genera are important in open habitats, especially in the arid and semi-arid tropics.

ROSALES (ROSES AND ALLIES)

The Rosales include several interesting families such as the figs and mulberries (Moraceae ~1200 species), the elms (Ulmaceae ~140 species), the nettles (Urticaceae ~1050 species) and buckthorns (Rhamnaceae ~880 species) but Rosaceae (~3000 species) is the largest family. It is central to the evolution of many other families in the Rosales but relatively difficult to circumscribe. For example the ovary may be hypogynous, perigynous or epigynous and perhaps the only widespread feature is the possession of a hypanthium, a floral cup that is primitively small and saucer- or cup-shaped, or has evolved to become large and ultimately connate with the carpels. Traditionally the family has been divided into four subfamiles differing in the form of the fruit (the Spiroideae – an aggregate of follicles, the Rosoideae – an aggregate of achenes, the Prunoideae – a drupe, the Maloideae – a pome). One of the smallest families in the order, with only three species, but certainly not the least significant is the Cannabaceae because it contains both hops and cannabis.

Figure 5.98. Rosaceae: *Rosa.*

CUCURBITALES (GOURDS AND BEGONIAS)

The Cucurbitales include three families interesting for different reasons. The Coriariaceae have only one widely distributed genus, *Coriaria* (Mexico to Chile; Mediterranean to Himalayas/Japan; New Guinea to New Zealand and W Pacific) but it is interesting because it has a nitrogen-fixing association with *Frankia* in root nodules. The Begoniaceae have only two genera, *Symbegonia* with 12 species from New Guinea and *Begonia* with over a thousand. *Symbegonia* differs from *Begonia* in having a corolla fused in a tube. *Begonia* are succulent herbs and shrubs. Water conservation is aided by Crassulacean Acid Metabolism (CAM) and they have stomata in clusters. The Cucurbitaceae are climbers with tendrils and are important commercially as gourds, squash, melons, etc. These families share a tendency to fleshy or juicy tissues.

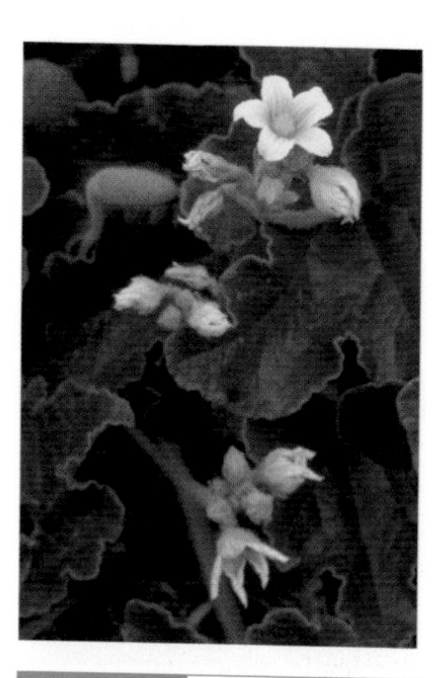

Figure 5.99. Cucurbitaceae: *Cucurbita.*

FAGALES (OAKS AND BEECHES)

The Fagales include many familiar wind-pollinated trees of temperate regions. The related families of trees, Fagaceae (beech, hornbeam and oak), Betulaceae (birch and alder), Casuarinaceae (she-oaks) and Juglandaceae (walnuts) exhibit a wide range of adaptations for wind-pollination. Different genera show different degrees of reduction of the flower and its aggregation into unisexual catkins. Wind-pollinated species are concentrated in the temperate regions but some genera like the oaks and hornbeams have insect-pollinated species in the tropics and here have stiff erect catkins. Walnut (*Juglans*) and wingnut (*Pterocarya*) have a tiny but well-formed perianth. In Oak (*Quercus robur*) male florets have a six-lobed perianth and seven to eight stamens but female florets have a minute perianth but a scaly cupule. Hazel (*Corylus*) has male florets consisting of two bracteoles and four stamens only with a bract at the base and female flowers that are surrounded

Figure 5.100. Fagaceae: *Alnus.*

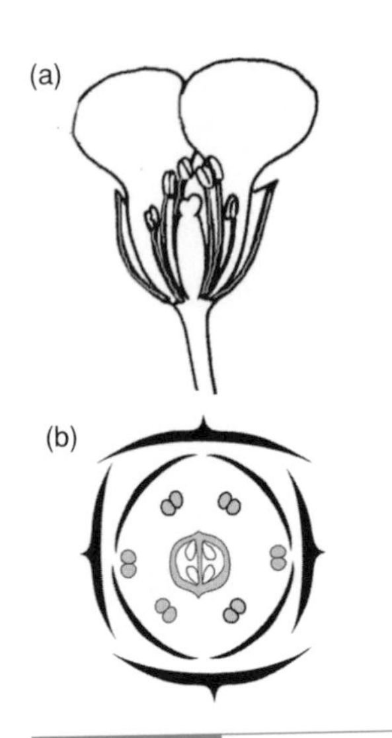

Figure 5.101. Brassicaceae: (a) half flower; (b) floral diagram.

by scales and have a minute perianth. The strange she-oak (*Casuarina*) of South-East Asia and Australia was once considered to be very primitive because of its very simple inflorescence. The flower consists only of a bract with two scale-like bracteoles with, in the male, a single stamen and, in the female, a single pistil. Male flowers are aggregated into catkins and the hard bracteoles of the female flowers form part of a woody 'cone'.

Eurosid II

The Eurosid II clade includes these important orders: Brassicales, Malvales, and Sapindales.

BRASSICALES (CRUCIFERS)

The Brassicales include sister lineages so distinct and without intermediates that one wouldn't guess their close relationship. The families Brassicaceae (crucifers), Resedaceae (mignonette), Limnanthaceae (poached-egg flower), Batidaceae (saltwort), Koeberliniaceae (allthorn), Setchellanthaceae, Moringaceae (Bennut), Caricaceae (papaya) and Tropaeolaceae (nasturtium) are very distinct in their floral morphology. For example, the Brassicaceae is also called the Cruciferae because of its cross-shaped flowers of four petals and usually six stamens. In contrast the Resedaceae usually has six fringed petals and the Tropaeolaceae has five and also has a long hairy claw. One floral feature that is present in several families of the order is a nectariferous portion of the axis below the stamens (androgynophore) or pistil (gynophore). One of the most significant features these families share is the possession of mustard oils (glucosinolates). This seemingly obscure chemical character provides protection against herbivory and fungal attack. Another interesting feature is the usual lack of mycorrhizae in the Brassicaceae, perhaps because they tend to occupy relatively nutrient rich early successional situations.

Figure 5.102. Malvaceae: *Hibiscus*.

MALVALES (HIBISCUS AND MALLOWS)

The Malvales are linked by a chemical characteristic of obscure significance, the presence of mucilage cells, or canals and cavities. The Malvales include the important family of tropical trees, the Dipterocarpaceae. Many exhibit a common rosid trait of showy polypetalous flowers with many stamens. The sequence in which the stamens mature is centrifugal, a pattern of development that was thought to be significant enough to warrant separating them from the rosids (centripetal development) in a group called the Dillenidae, but this pattern of development is difficult to see in the Malvaceae where the stamens are united to form a tube around the style.

Figure 5.103. Anarcardiaceae: *Pistacia*.

SAPINDALES (MAHOGANIES)

The Sapindales also include several very important tropical and subtropical families of trees and shrubs such as the Meliaceae (~575 species), Sapindaceae (~1350 species), Anacardiaceae (~850 species) and Burseraceae (~540 species). They frequently have pinnate or tri-foliolate leaves. Many of these families are highly resinous. The resin is toxic and protects them to a degree from leaf-browsing animals and also protects the wood from wood-boring insects. *Azadirachta* in the Meliaceae is a source of insecticide. The Sapindaceae often have saponins present. We are familiar with the Sapindales as the source of fruits and seeds (litchi, longan or rambutan – Sapindaceae; mango, cashew, pistacia – Anacardiaceae) and as timber trees (mahogany *Khaya*, *Swietenia* – Meliaceae).

Figure 5.104. Cornaceae: *Cornus*.

5.3.8 Asterids

The Asterids are tenuinucellate. Most also have a pentamerous sympetalous corolla, and most have an equal number of epipetalous stamens, alternating with the five corolla lobes. This set of attributes has long been recognized as those of a group called 'Sympetalae' (for a tubular corolla of connate (fused) petals). The Asterids contain the most advanced members of the Eudicots, and the most recently evolved. They have diversified especially in having specialised pollination mechanisms. Floral architecture and behaviour show many individual adaptations to particular kinds of pollinator.

Basal Asterids

CORNALES (DOGWOODS)

The Cornales exhibit a tendency, seen more fully developed elsewhere in the euasterids, towards the possession of a pseudanthium, a compound inflorescence of small flowers grouped together in a flat head and made showy in different ways. In *Hydrangea*, for example, flowers are in a cymose inflorescence with marginal ones sterile and showy, and fertile central ones. An alternative pattern is seen in *Cornus* (Cornaceae), which has large, showy, outer bracts like petals around the inflorescence. Although some Cornales have a synsepalous calyx most have free petals. There are three large families in the order, the Cornaceae, Hydrangeaceae and Loasaceae, ranging from trees and shrubs to robust herbs. The Loasaceae have barbed stinging hairs.

Figure 5.105. Theaceae: *Camellia*.

ERICALES (HEATHERS)

The Ericales are a diverse order and include, as well as the heathers (Ericaceae), other very distinct families such as the balsams (Balsaminaceae – fleshy herbs), the Marcgraviaceae (lianes), the Polemoniaceae (mainly herbs, especially of arid areas, but some shrubs and lianes), the camellias (Theaceae – shrubs and trees with thick leaves), and primulas (Primulaceae – herbs). The brazil-nut family Lecythidaceae, and the Sapotaceae, another important tropical family, are sister families in the order. The latter produces latex and gums, and includes species such as the chewing-gum plant *Manilkara* and gutta-percha plant *Palaquium*. One lineage of Ericales includes the insectivorous pitcher-plant family Sarraceniaceae (*Sarracenia*, *Darlingtonia*, *Heliamphora*) and *Roridula* with sticky resin secreting hairs (but not insectivorous) and sensitive stamens, as well as the Actinidiaceae (the kiwi-fruit or Chinese gooseberry).

Figure 5.106. Primulaceae: *Dodecatheon*.

Figure 5.107. Ericaceae: *Rhododendron.*

The Clethraceae and Cyrillaceae, two small families sister to the Ericaceae, are all trees or shrubs with tough, spirally-arranged leaves and pendulous flowers.

There are about 4000 species of Ericaceae. They are strongly mycorrhizal, with different subgroups having different forms of mycorrhizae: broadly identified as arbutoid, ericoid and monotropoid types. *Erica* and *Rhododendron* are by far the largest genera, each with about 800 species. The flowers of Ericaceae range from small and relatively inconspicuous *Calluna* type to large and highly colourful *Rhododendron* blossoms, and from radial to bilateral symmetry, representing a great diversity of pollination mechanisms involving wind, thrips, bees and other insects, birds and other pollinators.

Euasterid I

The Euasterid I group shows a trend towards large zygomorphic flowers, held laterally, exemplified by snapdragon, salvias and deadnettles.

Figure 5.108. Garryaceae: *Garrya.*

GARRYALES (SILK-TASSEL)

Aucuba is the commonly grown yellow spotted evergreen 'laurel'. *Garrya* (Silk-tassel Bush) is another commonly cultivated shrub with showy catkins. It produces highly toxic alkaloids. One rather peculiar feature of this order is the presence of petroselenic acid as a major fatty acid in seeds.

GENTIANALES (GENTIANS AND BEDSTRAWS)

The order Gentianales has the most generalised flowers in the Euasterids I with five epipetalous stamens, but they also exhibit secondary pollen presentation mechanisms, which in the asclepiads, have resulted in one of the most peculiar floral morphologies and pollination mechanisms of any flowering plants. The Gentianaceae (~1200 species) are regular, sympetalous and actinomorphic with five normal epipetalous stamens. The Rubiaceae (~11 000 species) are similar, although frequently they have floral parts in fours and much smaller flowers in cymes. Many shed their pollen onto a club-shaped stigma while in bud. When the bud opens the pollen is presented to the pollinator and only later does the stigma mature. The Gentianaceae are mainly herbs and are especially common in temperate conditions, while the Rubiaceae are important as tropical trees, shrubs and lianes.

Figure 5.109. Gentianaceae: *Gentiana.*

The Apocynaceae (~5000 species) ranges from genera such as *Vinca* (completely united carpels, and anthers, which are distinct and fully fertile) to those like *Nerium* (carpels separated up from the base and united only by their style and stigma, and anthers in which only the top part produces pollen and grouped closely together in depressions around the top of the expanded style). The asclepiads were formerly recognised as a separate family. In *Asclepias* the anthers are adnate to the style to form a structure called the gynostegium and the pollen in each theca is a compact mass, a pollinium. Pollinia from adjacent thecae are united by an acellular yoke called the translator and the whole structure is released as a pollinarium. Evolution of the pollinium has been accompanied by a merging of the pollen sacs of

Figure 5.110. Apocyancaeae: *Nerium.*

each theca so that each anther is bisporangiate. The translator clips the pollinarium to the pollinator and it is later pulled off by being caught in a groove in the stigma of another flower.

SOLANALES (POTATO AND MORNING GLORY)

The Solanales have two large families, the Solanaceae (~2600 species) (potato, tomato, tobacco, petunia) and the Convolvulaceae (morning glory, bindweed). Most species are polysymmetric with five equal corolla lobes but some genera of the Solanaceae are zygomorphic. *Schizanthus* has even lost one stamen and, of the remaining four, only two are fertile. The small family Nolanaceae has five carpels, the primitive condition, but other families in the Solanales have fewer carpels; either two- or pseudo-four-locular. In some of the Convolvulaceae (*Dichondra*) there are two carpels and the ovary is deeply two- or four-lobed with a gynobasic style, which resembles the pattern of the Boraginaceae and Lamiaceae.

Figure 5.111. Solanaceae: *Brugmansia*.

LAMIALES (DEAD-NETTLES AND GESNERS)

The Lamiales are mainly characterised by their monosymmetric tubular flowers. From an ancestral pattern of five more or less equal corolla lobes and five stamens there has been a shift to either four (with the loss of the posterior stamen) or two stamens (with in addition a loss of one lateral pair) and a strongly lipped flower. Patterns of floral evolution are complex though. For example the Oleaceae (olive, ash, privet, jasmine), close to the base of the order, are polysymmetric in their corolla but usually have four corolla lobes and only two stamens. *Calceolaria* (Calceolariaceae), also in a basal family, is one of the most strongly monosymmetric with two stamens and a corolla having a pouch-like lower lip.

The order includes plants of many different life-forms. The Gesneriaceae are mainly herbs or shrubs with many epiphytes and lianes and have many evolutionary novelties (see Chapter 3). Dispersal of the seed in this epiphytic niche has been accompanied by the evolution of unilocular ovaries with many seeds and parietal placentation. An alternative adaptation is exhibited by the Acanthaceae (~4350 species). They have the funiculus (the stalk that attaches the seed to the fruit) modified into a hook-shaped jaculator, which flings the seed out. Another family with many lianes but also some large tropical trees are the Bignoniaceae. They have large flowers and fruits variously adapted for pollination and seed dispersal especially by birds and bats.

Figure 5.112. Acanthaceae: *Justicia*.

The Scrophulariaceae was a large diverse family of many showy monosymmetric flowers that has been split and circumscribed in new ways as a result of molecular information (see Chapter 8). Less controversially and perhaps preserving features more representative of an earlier stage in the evolution of the order are the Verbenaceae (~1900 species) (verbena, teak). The two-carpellate pistil has each carpel divided by an extra wall. Each locule has a single ovule. The corolla is only slightly zygomorphic but bilateral symmetry is emphasised by the loss of one stamen in most species. The Lamiaceae (~5600 species), the labiates, so-named because of their two-lipped flower,

Figure 5.113. Lamiaceae: *Thymus*.

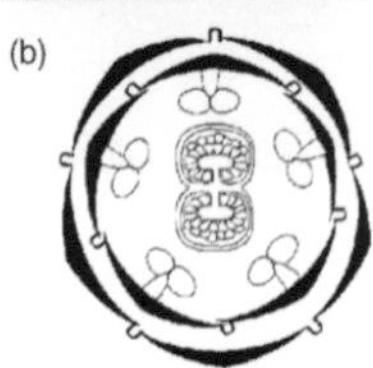

Figure 5.114. Scrophulariaceae: (a) *Digitalis*; (b) floral diagram.

have flowers that are usually very strongly zygomorphic and have either four or two (*Salvia*) stamens. The four-loculate ovary is divided into four segments from the top. The style reaches down to the base of each locule (gynobasic). Each segment is dispersed as a separate nutlet when the seed is mature. In the Lamiaceae the inflorescence is a verticillaster with cymose verticals on a main raceme.

The Orobanchaceae, contains most of the parasitic, while the Lentibulariaceae and Byblidaceae are insectivorous and have sticky and digestive glands.

BORAGINALES (BORAGES)

Parallel evolutionary trends can be seen in the related family, the Boraginaceae. Some borages have an entire ovary like the Verbenaceae whereas others are like the Lamiaceae with a gynobasic style and four nutlets. Most have polysymmetric flowers arranged in cymes. *Echium* is exceptional with its zygomorphic flower.

Euasterid II

The Euasterid II lineage shows varying degrees of aggregation of flowers into a head in which different flowers may become specialised for showiness (pseudanthium). They tend to have small epigynous flowers grouped together in a flat head, exemplified by the umbel of the Apiales (umbellifers) and the capitulum of the Asterales (daisies). Some are woody shrubs or climbers but the majority are herbs.

They have a remarkable chemical diversity especially in compounds that act as herbivore deterrants.

Figure 5.115. Aquifoliaceae: *Ilex*.

AQUIFOLIALES (HOLLIES)

The Aquifoliaceae is by far the biggest family in the Aquifoliales because of the genus *Ilex* (holly) with 400 species of small evergreen trees or shrubs. Flowers are normally small and unisexual. Most species are dioecious. *Helwingia* (three species) and *Phyllonoma* (four species) have epiphyllous inflorescences.

DIPSACALES (TEASELS AND HONEY-SUCKLES)

The Dipsacales exhibit various stages in the evolution of a pseudanthium. The valerians (Valerianaceae ~400 species) have cymose inflorescences of small florets. The honey-suckles and elders (Caprifoliaceae ~365 species) have cymose umbels. The teasels and scabiouses (Dipsacaceae ~250 species), have the most highlly developed capitulate heads with an involucre of bracts surrounding the head that in some species have the outer florets modified to be more showy (*Scabiosa*, *Knautia*). In the Dipsacales the head is not bracteate and is basically cymose.

Figure 5.116. Dipsacaceae: *Scabiosa*.

APIALES (UMBELS AND IVY)

There are two families in the order (Apiaceae ~3100 species and Araliaceae ~800 species). The Apiaceae is a family that has many species cultivated for food (carrot) or more frequently for spice (coriander, cumin, etc.). The compound umbel is so characteristic of the Apiaceae that the family was one of the first flowering plant families to be clearly recognised as the Umbelliferae (Figure 5.117). However, the compound umbel is confined to subfamily Apoideae. In some genera like *Hydrocotyle*, in subfamily Hydrocotyloideae, the compound umbel is reduced to a single floret; and in *Eryngium*, subfamily Saniculoideae, the inflorescence is rounded rather than flat. The umbel is normally visited by a range of different insects, which move freely over the platform of the umbel. In each floret, the ovary has a disc at the base of the paired styles. The styles are swollen, together forming a nectariferous stylopodium. Flowers in different parts of the umbel may be specialised. Those in lateral umbels are sometimes female and sterile with abortive ovaries and shorter stylopodia. The ovary has two carpels and matures into a dry fruit, a schizocarp, which splits into two mericarps joined by a carpophore. Some species have fruits with thick corky walls so that they float, have wings for wind dispersal, or have hooks for animal dispersal. The Apiaceae have a particularly effective combination of anti-herbivore chemical repellants such as polycetylenes and sesquiterpene lactones. The Araliaceae share many similarities with the Apiaceae but are mainly tropical and rarely form regular compound umbels. They also produce a fleshy fruit.

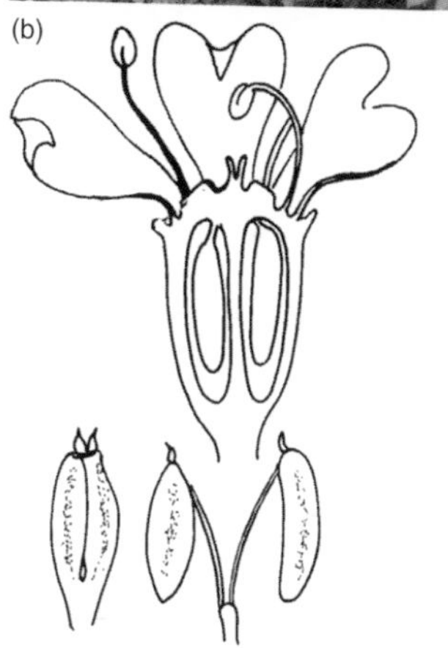

Figure 5.117. Apiaceae: (a) *Heracleum*; (b) floret in section showing stylidium and schizocarp.

ASTERALES (DAISIES, LOBELIAS AND BELLFLOWERS)

Most species of Asterales are in one family, the Asteraceae, which is widely recognised as one of the most advanced families of flowering plants (Figure 5.121). Its origin is relatively recent but it has 1100 genera and 20 000 species. All species have a capitulum, a head with many florets on a common and usually flattened receptacle, and surrounded by an involucre of bracts. The involucre is particularly obvious in 'everlasting' flowers like *Helichrysum*, where it is showy. Florets are tubular and epigynous with a single ovule. They are protandrous and mature centripetally (i.e. the capitulum is racemose). The florets are of different sorts, either polysymmetric or monosymmetric. The latter have either three corolla lobes that are connate and greatly expanded to form a strap, or all five corolla lobes extending in one direction. Florets may be pistillate, hermaphrodite or functionally staminate. Heads are made of a single kind of floret or a combination of kinds, commonly with central disk of hermaphrodite, polysymmetric florets and a margin of showy, monosymmetric ray florets, with the latter pistillate or sterile, as in the daisy and sunflower. One group, which includes lettuce and dandelion, have all florets with all five corolla lobes forming the strap.

Florets are protandrous. Anthers form a tube. Pollen is shed into the anther tube and then the immature style elongates and pushes

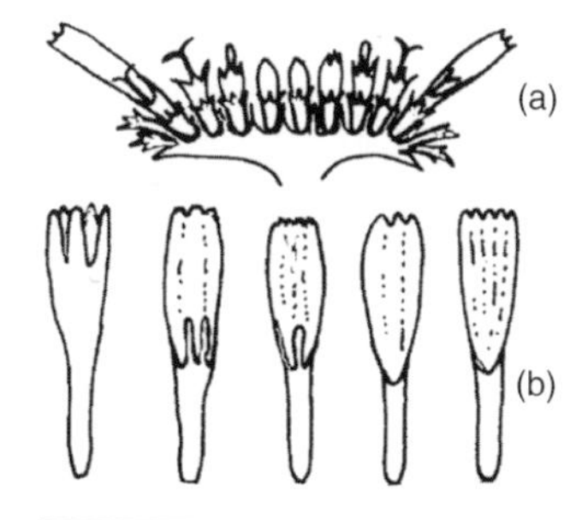

Figure 5.118. Structure of a capitulum (a) capitulum t.s. showing central and marginal florets; (b) florets of diverse sorts.

Figure 5.119. Tubular central floret showing inferior ovary and the calyx converted into a bristly pappus.

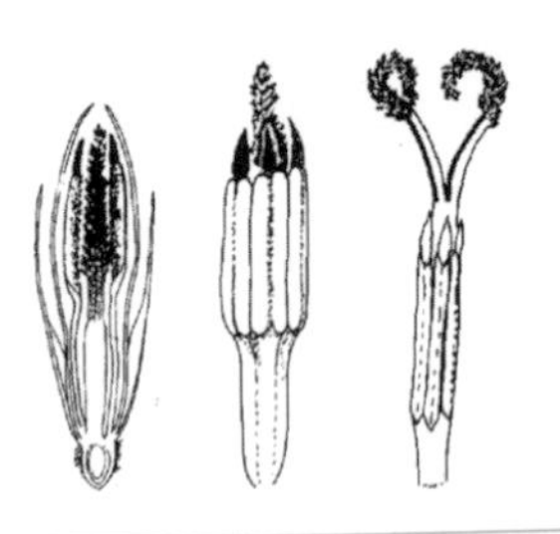

Figure 5.120. Pollen presentation mechanism. Florets are protandrous and the piston like style pushes pollen on to the surface before the stigma lobes open to become receptive.

the pollen onto the surface of the capitulum. Later the stigmatic lobes open. The fruit of the Asteraceae is usually crowned by a pappus derived from the calyx. The pappus in Asteraceae is very variable, either absent or cup-like, or with scales, bristles, simple or feathery hairs, which are barbed, or glandular. The fruit, called a cypsela, is a kind of achene of an inferior ovary. The dispersal adaptations of the fruit contribute to the success of many species as weeds.

The features described above have evolved in many groups outside the Asteraceae. Many of the structures of a capitulate infloresecence found in the Asteraceae are paralleled in other families in the Asterales such as the Goodeniaceae (~300 species), which has an indusium, the Calyceraceae (~55 species), Lobeliaceae (~1200 species) and Campanulaceae (~600 species). *Phyteuma* and *Jasione*, in the Campanulaceae, both have capitulate inflorescences surrounded by an involucre of bracts. *Jasione* has a kind of primitive 'pseudo-indusium' formed by swollen stigmatic lobes. The one species of the monotypic Brunoniaceae, which is remarkably similar in appearance to *Jasione* and is sometimes put in the Goodeniaceae, shows a further parallel in having an involucrate head, though the head is cymose and the florets are hypogynous. The piston-like mode of pollen presentation in the Asteraceae also has parallels with that in the Campanulaceae (*Physoplexis*) and Lobeliaceae.

The repeated evolution of these features argues strongly that they are adaptive. One advantage of having a capitulum, for example, is the protection given to the ovule and seed. Functionally it provides a large showy target for pollinators and yet each ovule is packaged separately, as a defence against predators and for dispersal. There is a lot of diversity in the size and number of florets that capitula contain. There are the familiar huge capitula of the sunflowers, which have been selected by plant breeders. At the other end of the spectrum many species have capitula containing very few florets.

Figure 5.121. Diverse Asteraceae: (a) *Bidens* with marginal strap florets mimicking a 5-petalled flower; (b) *Centaurea* with marginal expanded disk florets; (c) *Galactites* with a showy involucre of bracts; (d) *Cynara* with a head of disk florets and a spiny involcre; (e) *Solidago* spikes of small ligulate heads; (f) *Echinops* with a head of capitula each with a single floret.

Frequently species with small capitula have the capitula grouped in some way. In *Solidago* the capitula are arranged in spikes. In *Achillea* they are grouped in a corymb. In *Echinops* each capitulum only has a single floret and they are grouped in a globular head with its own involucre.

However, because these features are also present outside the Asteraceae they do not explain the particular evolutionary success of the Asteraceae. The Asteraceae have an unusual multiallelic, homomorphic, sporophytic, self-incompatibility that has maintained high levels of genetic diversity among individuals and populations. However, they are also flexible in their breeding system and many weedy species are secondarily self-compatible and self-pollinating. One example is the ubiquitous groundsel, *Senecio vulgaris*, which, no longer needing to attract pollinators, lacks the ray florets of its relatives. Polyploidy, sometimes following hybridisation, seems to play a significant role in the evolution of the weedy species, especially since it can destroy the self-incompatibility. Another aspect of their evolution is the chemical diversity associated with the deterrence of herbivores, especially in alkaloids. One of the largest genera of all plants, the ragworts (*Senecio*), with over 1500 species, has its own peculiar type of alkaloid.

Figure 5.122. Campanulaceae (a) *Campanula*; (b) *Jasione*.

Further reading for Chapter 5

Beck, C. B. *Origin and Evolution of Gymnosperms* (New York: Columbia University Press, 1988).

Bold, H. C., Alexopoulos, C. J. and Delevoryas, T. *Morphology of Plants and Fungi*, 5th edition (New York: Harper and Row, 1987).

Heywood, V. J. (ed.) *Flowering Plants of the World*. (Oxford: Oxford University Press, 1978).

Mabberley, D. J. *The Plant Book* (Cambridge: Cambridge University Press, 1987).

Schofield, W. B. *Introduction to Bryology* (New York: Macmillan, 1985).

Shaw, A. J. and Goffinet, B. *Bryophyte Biology* (Cambridge: Cambridge University Press, 2000).

Stewart, W. N. *Paleobotany and the Evolution of Plants* (Cambridge: Cambridge University Press, 1983).

Takhtajan, A. *Evolutionary Trends in Flowering Plants* (New York: Columbia University Press, 1991).

Thomas, B. A. and Spicer, R. A. (1987) *The Evolution and Palaeobiology of Land Plants* (London: Croom Helm, 1987).

Willis, K. J. and McElwain, J. C. *The Evolution of Plants* (Oxford: Oxford University Press, 2002).

Chapter 6

The lives of plants

There is not a 'fragment' in all nature, for every relative fragment of one thing is a full harmonious unit in itself.

John Muir, 1867 (*A Thousand-Mile Walk to the Gulf*, 1916)

When we try to pick out anything by itself, we find it hitched to everything else in the Universe.

John Muir, 1869 (*My First Summer in the Sierra*, 1911)

6.1 Plant diversity around the world

A complete treatment of vegetation around the world would be impossible in a whole book let alone a single chapter. Instead, we concentrate on plants that inhabit different environmental extremes. We bring into focus the biotic relations of plants and, in addition, we consider some aspects of plant evolution in relation to the Earth's history and climate change, by looking at plants of islands.

The greatest omission this chapter is an account of the forests of the world. Every botanist should visit a tropical rainforest at least once. No vegetation formation on Earth can compare to tropical rainforest in its staggering wealth of life forms, its diversity of species. It is the 'Ultima Thule' of the botanical world, after which everything else falls into perspective.

. . . I measured my insignificance against the quiet majesty of the trees. All botanists should be humble. From trampling weeds and cutting lawns they should go where they are lost in the immense structure of the forest. It is built in surpassing beauty without any of the necessities of human endeavour; no muscle or machine, no sense-organ or instrument, no thought or blue-print has hoisted it up.

E. J. H. Corner, 1963

6.1.1 The kinds of plants

There are several ways in which the life-forms of plants may be categorised. One of the most useful ways has been the life-form classification of Raunkiaer, which divides vascular plants into different categories on the basis of where the buds are situated in the harsh season and how they are protected (Table 6.1).

Raunkiaer (1934) carried out extensive surveys of the geographical and ecological distribution of life-forms. The different proportion of

Table 6.1 The life-forms of plants based on Raunkiaer's system

Life-forms
Phanerophytes (trees and shrubs >25 cm tall)
Evergreen trees without a bud covering
Evergreen trees with a bud covering
Deciduous trees with a bud covering
Shrubs 25 cm–2 m tall
Chamaephytes (woody or semi-woody perennials, with resting buds < 25cm above ground)
Shrubs or semi-shrubs, which die back to the resting buds
Passively decumbent shrubs
Actively creeping or stoloniferous (procumbent) shrubs
Cushion plants
Hemicryptophytes (die back in harsh season, with resting buds at the soil surface)
Protohemicryptophyte
Partial rosette plant
Rosette plant
Cryptophytes (with buds below ground or in water)
Geophytes
Rhizome geophyte
Bulb geophyte
Stem tuber geophyte
Root tuber geophyte
Helophyte (marsh plants with resting buds in water saturated soil)
Hydrophyte (buds in water)
Therophytes (annuals, which survive as seed)

Classification of woodiness

1. Holoxylales
   – the whole plant is lignified
2. Semixylales
   – plants with the lower branches lignified and the upper herbaceous
3. Axylales
   – herbaceous plants

Leaf size

subleptophyll	<0.1 cm^2
leptophyll	0.1–0.25 cm^2
nanophyll	0.25–2 cm^2
nano-microphyll	2–12 cm^2
microphyll	12–20 cm^2
micro-mesophyll	20–56 cm^2
mesophyll	56–180 cm^2
macrophyll	180–1640 cm^2
megaphyll	>1640 cm^2

Leaf consistency

Malacophyll (soft)
Sclerophyll (hard)
Resinous/succulent
Water/succulent

different life-forms is an indication of the spectrum of life strategies that have been selected in different areas.

In parallel with Raunkiaer's classification, one of the simplest, and most useful system of categories for plants divides them into woody plants and herbaceous plants. Du Rietz devised a slightly more precise system of categories based on the extent of lignification. Categorising plants by the seasonality of growth can be made more precise or adapted to suit particular geographical areas. In temperate deciduous woodlands most plants are leaf shedders but, for example, in Negev desert communities, there are no leaf shedders, and all phanerophytes are branch shedders.

The problem with these systems is that they were developed in temperate countries of Europe and North America, and in places colonised by Europeans, and do not work well elsewhere. Even the relatively sophisticated system of Raunkiaer fails to cope adequately with the diversity of tropical life-forms. For better or worse, the plants of temperate countries were used as a yardstick. These are mainly mesophytes, plants 'moderate' in there requirements, adapted to environments that are neither extremely wet nor extremely dry, yet mesophytes are just as finely adapted as any other plants. Throughout, we describe plants adapted in different ways, including xerophytes

Classification of plants based on the nature of the shed organ(s)

Whole plant shedders which survive as seeds (annuals).
Shoot shedders, seasonally renewing the whole shoot.
Branch shedders, seasonally shedding the upper parts of their branches.
Leaf shedders.

(plants adapted to dry habitats), halophytes (plants adapted to salty conditions) and epiphytes (plants adapted to grow on other plants). These categories are not exclusive: sea-grasses are hydrophytes and halophytes, while epiphytes may be xerophytes too. The Podostemaceae, a peculiar family of aquatics, are more easily characterised by a system designed for the bryophytes.

Different plant groups with different evolutionary histories have tackled the problems of surviving in diverse ways but these ways are not unlimited and there are many convergences of form and physiology among unrelated groups. We also find plants of radically different appearances living side by side, an aspect that delights the eye and makes them so fascinating.

6.2 Aquatic and wetland plants

At least 53 families of plants have aquatic or wetland representatives but very few of the so-called lower plants have re-invaded aquatic habitats. Except for a few aquatic mosses, liverworts, ferns, quillworts and horsetails, the majority of aquatic plants are seed plants, and almost all of them are flowering plants. Among the gymnosperms, only the swamp-cypresses (Taxodiaceae) are closely associated with wetland habitats. However, the ancestry of aquatic flowering plants is ancient. Recent molecular studies have shown that the Ceratophyllales and Nympheales evolved very early in the history of flowering plants. Fossil material of nymphealean seeds, resembling the extant water-shield *Brasenia*, from Portugal and eastern North America, has been dated as Early Cretaceous. Many aquatics are monocots including *Acorus* and Alismatales whose lineages originated close to the origin of the monocots themselves.

Several different forms of wetland and aquatic plants may be distinguished. Maritime plants adapted to salt-water (halophytes), for example sea-grasses, salt-marsh and mangrove, are treated separately below. Halophytes are also found in some areas of temporary inundation where high rates of evaporation give rise to salt-pans and salt-deserts. Fresh-water submerged, emergent and free-floating aquatics have distinct adaptations, whereas herbaceous fringing plants can be distinguished from woody rheophytes. The features that determine aquatic associations are many and varied, but ultimately they are the physical parameters such as the geology and soil, the climate and weathering processes, and tidal influences at river-mouths. These factors result in water quality that varies significantly in terms of pH, salinity, nutrient status, and dissolved oxygen. Almost double the amount of oxygen can be found dissolved in water at 0 °C compared with 30 °C. Periodic flooding, silting or drying-up also affects plant-life, so the climate of the region is a strong determinant of those plants that are permanent occupants of wetlands and rivers and those that are temporary.

Swamp is most accurately applied to wetland vegetation of the tropics and marsh to temperate conditions. A zonation of different plant associations from the landward side to the water's edge can usually be seen at most fresh-water lakes (cf. hydrosere) (see Figure 6.1). Depending on the microenvironment of a particular body of water, there may be several different plant communities or associations according to the degree and frequency of soil inundation, distance from shore and depth of water, degree of water turbulence, and the presence or absence of other plants. For example, the reed-grass, *Phalaris arundinacea,* often gives shelter to small colonies of water-starworts, *Callitriche* spp. There may be a transition to more terrestrial habitats such as bogs and mires (Figure 6.2).

Figure 6.1. Cyperaceae at the aquatic margin.

Fresh-water plants, particularly the strictly aquatic species, may be distinguished by various adaptations that enable them to function normally in water, but the differences between terrestrial and aquatic plants are often slight. 'Hydrophytes' are best described as herbaceous 'plants growing in water, in soil covered with water, or in soil that is usually saturated'. Broadly speaking, rheophytes are plants which typically grow on riverbanks, or between channels of river beds, and can tolerate temporary inundation during periods of flooding. They can also include species that occur away from rivers, at the margins of lakes, ponds, marshes and bogs.

Rheophytes occur in diverse families, so, not surprisingly, they display extremes of morphology from the tiny thalloid Podostemaceae to large trees. They include temperate woody genera such as willows (*Salix* spp.) and alders (*Alnus* spp.), and the cottonwoods (*Populus* spp.). Other well-known examples are the desert catalpas (*Chilopsis* spp.) of western North America; and the oleanders of the Mediterranean (*Nerium* spp.). Many species that are not aquatic or wetland plants in a strict sense, can tolerate inundation and grow in seasonally dry rivers. They are often classified as facultative rheophytes, since they are not strictly confined to such habitats (e.g. *Tamarix* spp.).

Figure 6.2. The fen succession: there is a transition from open-water plants through sedge and reed (*Carex* and *Phragmites* respectively) fringing communities to wet woodland of alder (*Alnus*) and oak (*Quercus*).

Numerous river systems in the tropics such as the Amazon and the Rio Negro in South America, and the Fly and Sepik Rivers in New Guinea, are seasonally inundated and have large areas of swampy margins. In such areas the number and diversity of wetland genera is staggering. The most familiar of the tropical wetland genera include *Melaleuca*, *Excoecaria*, *Metroxylon*, *Pandanus*, and *Sesbania*, etc. Many genera such as *Pandanus* possess stilt roots that increase stability on a muddy substrate, or aerial roots, as in *Metroxylon*, that increase aeration. In the southern United States and northern Mexico, the swamp or bald cypresses of the genus *Taxodium* are rather unique among conifers in living in areas that are seasonally or permanently swampy, or along wet river margins. The roots have knee-like projections that emerge above the soil and presumably perform the same aerating functions as the pneumatophores of mangrove plants (see Figure 6.16).

The riverweeds (Podostemaceae) are obligate rheophytes, and resemble aquatic mosses, liverworts, and even marine algae. As the most unusual family of aquatic herbs, it deserves special mention.

Figure 6.3. The Everglades in Florida: the activity of alligators clears vegetation from large holes in the underlying limestone, maintaining an aquatic oasis when the rest of the glades dry out in the dry season.

Species in this largely pantropical family display great morphological diversity (46 genera and 260 species), even though most occupy a similar niche on limestone rocks in swift-flowing rivers. Some genera such as *Podostemum* are pantropical, although one species extends north to Canada, but others are much more localised, occurring in just a few rivers, for example *Torrenticola* on volcanic rocks of rivers in Papua New Guinea and north Queensland.

6.2.1 General adaptations of aquatic plants

The range of conditions that fresh-water plants endure is narrower than those experienced by terrestrial plants. This has constrained their evolution, with the result that there is a tremendous amount of morphological convergence between unrelated families that have to cope with very similar conditions. Stiff supporting tissues and woody tissues are rare. Many of the differences between aquatic families can be found in the diversity of their reproductive systems. Here, there are many bizarre adaptations indeed, some of which are described in Chapter 4.

The aquatic environment may not necessarily be as severe as a terrestrial one. Diurnal and seasonal temperature ranges are considerably less. Some, such as the water-soldier, *Stratiotes aloides*, and hornworts, *Ceratophyllum* spp., avoid freezing conditions by sinking to the bottom, while others overwinter as turions.

Nutrient and energy requirements are often at a premium. The plant-life of marshes and fens is often similar although, in marshes, the substratum may be inorganic or muddy, whereas in fens it is usually organic (peat), but relatively nutrient rich with the water pH neutral or alkaline. However, bogs overlie deep acid peat and are nutrient poor (oligotrophic), resulting in water with a low pH, and vegetation that is uniquely different. Nutrient conditions differ, depending whether the community is rain-fed (ombrogenous) or mainly ground-fed (topogenous). Flowing fresh-waters run through a diversity of rock types with their own characteristic soils enriching water systems with different mineral nutrients that helps to determine the range of aquatic species present. Lowland ponds, lakes and rivers with lush vegetation usually have a water input from surrounding areas rich in nutrients. Freely floating plants that must obtain all their nutrients from the surrounding water are frequently absent from oligotrophic lakes. The nitrification of waterways by the use of nitrate fertilisers on surrounding farmland clearly highlights these relationships. In recent years acid-rain pollution has been detrimental to many northern lakes and rivers by altering the delicate chemical balance of the ecosystem. In addition, the kind of vegetation that lines the banks of streams, and subsequently the degree of shade cast, is important in determining which plants grow in a particular stretch of water.

Carbon dioxide and oxygen in water are less freely available than in air, and terrestrial plants usually experience difficulty in obtaining adequate gaseous exchange for respiration and photosynthesis when they are immersed in water. In contrast, aquatic plants are able to absorb water and exchange gases over all surfaces of submerged tissues. The permanently submerged species usually have the most modified forms of gaseous exchange. In the majority of genera, the aquatic leaves lack stomata altogether whereas in partly submerged ones stomata are present only on the upper surface. The marine species of Potamogetonaceae lack stomata, but have openings at leaf apices.

Some bottom dwellers of cold oligotrophic lakes may have special methods of carbon assimilation. They are small in stature with highly compressed internodes, long cylindrical pointed leaves, and relatively massive root systems. This suite of morphological characters has been termed 'isoetid'. Carbon acquisition from sediment is recorded for temperate species of *Isoetes* and other isoetids such as *Lobelia dortmanna* and *Littorella uniflora*. Crassulacean acid metabolism (CAM), which is usually characteristic of xerophytes and many epiphytes, has evolved in aquatic *Isoetes* as a response to daytime carbon deficit. Apparently this plastic response is absent in terrestrial species or lost in individuals stranded in drier habitats.

The other group of submerged aquatics that frequent cold oligotrophic lakes are also highly reduced in stature, but are quite different in morphology from the isoetids. They generally conform to a *Ceratophyllum* type of body plan, for example the water-milfoil (*Myriophyllum alterniflorum*), but include highly reduced examples of rushes such as *Juncus bulbosus* and pondweeds such as *Potamogeton filiformis*. Also ubiquitous in these cold-water lakes are the charalean algae, *Nitella* and *Chara*.

The aquatic fern genus *Azolla* (Azollaceae) has a symbiotic relationship with the bacterium (*Anabaena azollae*) that is capable of assimilating atmospheric nitrogen. This alga lives in pits at the base of the lobes. They have an increased surface area in relation to volume because of the need to assimilate carbon dioxide directly from the water. Oxygen, as a by-product of photosynthesis, is retained and distributed by an elaborate aerating network (aerenchyma). In the aroid *Pistia* (Araceae), 71% of the volume of the leaves is occupied by air. Some of the free-floating aquatic aroids have root systems that hang in the water (e.g. *Lemna*) but others (e.g. *Wolffia*) are rootless. In an aquatic environment there are lower light levels but chloroplasts are concentrated in the epidermis. Despite the ability to assimilate over the entire surface, aquatic plants have very active transpiration, a necessary component of the transport of materials around the plant. Since submerged plants cannot lose the water of transpiration through evaporation they do so by guttation. This takes place via specialised 'water-pores' or hydathodes at nerve endings, and on the undersides of floating leaves, and is regulated in a layer of tissue known as the epithem.

There is a small group of mainly monocot families that comprises some of the most highly reduced of all the submerged species (e.g. the horned pondweeds, Zannichelliaceae), and the tasselweeds, Ruppiaceae). They are all very slender, almost filiform. The naiads

Figure 6.4. Submerged aquatics: (a) *Lilaeopsis*; (b) *Egeria*; (c) *Hygrophila*.

Figure 6.5. *Victoria amazonica* and *Pistia stratiotes* (water lettuce).

(Najadaceae), are slightly more diverse morphologically and may grow in dense clusters. One characteristic that they all share is their ability to live in brackish water, although species of tasselweeds of the genus *Ruppia* can also tolerate salt-water.

There are very few flowering plants that are completely adapted to life in the marine environment (about 12 genera and less than 50 species). All are monocotyledons and are collectively known as sea-grasses. They probably invaded the sea via brackish-water ancestors. Seagrass communities often comprise immense monospecific beds, forming bright green meadows in the lower intertidal zone when exposed at low tide. The most familiar species of temperate regions are the eel grasses or grass-wracks of the genus *Zostera* (Zosteraceae) that grow in flat sandy areas and quiet bays. Sea-grasses are all halophytes and are able to tolerate high salinities by their ability to be at equilibrium osmotically with the surrounding seawater.

The largest aquatic plants are non-woody, tropical species which combine floating leaves with bottom anchorage, for example the giant water-lily (*Victoria amazonica*) (Figure 6.5). All attached aquatics that root in mud obtain nutrients from it, and the size of the plant usually reflects the richness of the substrate as well as temperature. In colder, temperate climates, the largest rooted aquatics are the water-lilies (*Nymphaea* spp.), whereas the smallest such as the awlwort, *Subularia aquatica*, grow in cold oligotrophic lakes.

Numerous flotation devices have evolved independently across many lineages of aquatic plants and often seem to be derived from the aerenchyma system. The water hyacinth, *Eichornia crassipes* (Pontederiaceae), and *Pistia*, have petioles that are inflated as floats. In *Ludwigia* (Onagraceae) the vertical shoots have aerenchyma in the basal regions, and in the roots. In addition, there are spongy pneumatophores that arise from the rhizome and these probably increase aeration. The floating sensitive-plant, *Neptunia olearacea* (Leguminosae), is rather odd-looking with white floating tissue at the internodes that resembles polystyrene (styrofoam). In a closely related aquatic genus, *Aeschynomene*, the flotation tissue develops from the secondary xylem. This was used in the manufacture of pith helmets. *Salvinia*, an aquatic fern, has modified hairs in its aerial fronds that trap air and aid buoyancy. One of the fronds acts as a stabiliser and morphologically resembles a root. Smallest of all flowering plants, the tiny floating plants such as the duckweeds, *Lemna* spp. and *Wolffia* spp., rely entirely on dissolved nutrients to increase bouyancy. The sinking and rising of the water-soldier, *Stratiotes aloides*, is noteworthy. Mostly, it lives in calcium rich waters and, during the summer months, the leaves gradually build up an encrustation of calcium carbonate. This increases the specific gravity of the plant until it eventually sinks to the bottom in the autumn. With the growth of fresh leaves in the spring, the specific gravity of the plant lessens, and the plant rises to the surface.

Figure 6.6. *Hottonia palustris* (water violet) has finely dissected submerged leaves but produces an aerial inflorescence on a long peduncle.

The actual shape of aquatic leaves appears to be directly influenced by the physical properties of water, such as temperature, turbidity and light transmission. The amount of light passing through water drops dramatically, even after a few metres. Most aquatic leaves are either long and ribbon-shaped, or are finely dissected, offering little resistance to the flow of water. In floating leaves of genera such as *Nymphaea* and *Potamogeton*, the cuticle is waxy, preventing waterlogging and aiding runoff.

An abundant supply of water is the norm for many aquatic plants but it may not be continually available. Many ponds and rivers are drastically reduced in size during periods of drought or dry up completely. Cuticular development of the epidermis, and lignification of the xylem of submerged leaves is highly reduced or absent. Like many coastal seaweeds, fresh-water aquatic plants often have a covering of mucilage and are slimy to the touch. This may retard excessive entry of fresh-water into the plant tissues by osmosis, and subsequent loss of assimilates, but it also may serve to prevent injury in flowing water, or slow the process of dessication during periods of low water.

The rooted species have aerenchyma in the rhizomes and roots, and normally have parts of the main axis submerged. Leaves and flowers are either held above the water or floating on it. Such leaves frequently have retrograde development of the palisade and parenchyma layers and are relatively thin. However, they have a large surface area, and tend towards a peltate shape that maximises light interception. In the arrowheads, *Sagittaria* (Alismataceae), the aerial leaves are arrow-shaped, whereas in most others they are oval to rounded, with the petiole either angular to the plane of the lamina, or inserted centrally. Most of them rely on water for some degree of mechanical support, although the strongly ribbed leaves of *Victoria amazonica* can support weights of up to 50 kg. The leaves of this genus also possess folded margins up to 15 cm high. Water that may collect on the lamina drains through minute pores. Long flexible petioles allow leaves to float on the surface of still or slow-flowing water.

Leaves and petioles have well-developed aerating tissue or aerenchyma. Many species in this group also possess submerged leaves that are often finely divided and linear, for example the water-shields, *Cabomba* spp. (Cabombaceae). In such cases, the cuticle is absent or weakly developed and there are no stomata, thus increasing the

Figure 6.7. *Phragmites australis* rhizome system.

Figure 6.8. Pond weed (*Potamageton*) showing floating leaves on long petioles.

surface area for gaseous exchange. Aquatic plants of fluctuating water levels frequently display plasticity of leaf form (see Chapter 3) or heterophylly, for example, the water marigold (*Bidens beckii*) (Asteraceae), the water-crowfoots (*Ranunculus* spp.) (Ranunculaceae) and the water-dropworts (*Oenanthe* spp.) (Apiaceae). Tropical genera that display heterophylly include the water-shields (*Cabomba* spp.) and the water fern genus *Ceratopteris* (Parkeriaceae). Sometimes the physiology of a plant is disrupted by fluctuating water levels. *Nymphaea alba* will not produce floating leaves or flowers if the water is more than 2.5 m deep.

The sea-grasses may experience strong and fluctuating currents or wave-action and may be exposed at the lowest tides. All sea-grasses possess a well-developed anchoring system of creeping rhizomes, and long fibrous roots that allow them to endure tidal movements, and unstable substrates (see analogous rhizome system in *Phragmites*, Figure 6.7). They have the ability to compete successfully with other organisms of the marine environment, such as marine algae. They have a convergent and simplified morphology of mostly tough, strap-like or ribbon-like leaves that may occasionally be membranous. These are arranged alternately in two ranks, and have sheathing bases. In *Posidonia* (Posidoniaceae), *Cymodocea* (Cymodoceaceae) and in *Enhalus* (Hydrocharitaceae), fibres of the sheath survive and form a protective covering for younger leaves. Exceptions to this general morphology are species of *Halophila* (Hydrocharitaceae) that have broad, petiolate leaves, and *Cymodocea isoetifolia* with awl-shaped succulent leaves (Arber, 1920). The internal structure of the leaves is more diverse, even to the species level, suggesting a fine-tuning of their environmental requirements.

6.3 Halophytes

6.3.1 General features of halophytes

Most plants are non-halophytes ('halophobes') and cannot tolerate sodium chloride salt. Generally, they have ineffective regulatory mechanisms and are unable to adapt osmotically to the presence of salt in their environment. However, there is a gradient of tolerance, and some plants are more able to cope with salt than others (Figure 6.9). Tolerance also varies with age. Seedlings are often less tolerant, while many salt-tolerant species actually require dilution of the soil by rain before germination can occur (e.g. *Pancratium maritimum*, *Cakile maritima*, *Triglochin maritima*, etc.). The roots of dune, coastal-cliff, and foreshore plants are not immersed in salt-water, nor grow in salt saturated soils, and their shoots do not accumulate salt. Although these plants are able to tolerate moderate levels of salt (less than 0.5%), they do better on non-saline soils. They are sometimes called pseudo-halophytes or glyco-halophytes. Their growth is usually reduced when salt concentrations reach about 1/32 of seawater. In contrast, halophytes are plants that are able to grow on mildly to strongly saline soils (halobiomes). Halophytes which tolerate or endure high levels of

Figure 6.9. A saltmarsh showing the zonation of plants; the grass *Spartina* in the lower marsh is being inundated by a high tide.

salt are known as euhalophytes. In euhalophytes, growth is reduced only when concentrations reach from 1/8 to 1/2.

Growth of euhalophytes is stimulated to a certain extent by salt, and they usually grow best where salt levels are high. They can also grow on non-salty soils, but cannot compete well in such environments. Although most halophytes are maritime plants, they include species that occur thousands of miles from the nearest open sea. Characteristically, these inland species usually grow around salt lakes, salt pans and salinas, in arid places such as the Dead Sea, the Aral Sea, the Great Salt Lake of Utah, and over much of inland Australia, but they also occur in highly evaporative climates such as hot and cold deserts. The sea-pink or thrift (*Armeria maritima*) (Plumbaginaceae), an abundant coastal plant of western Europe, also grows in mountainous regions far from a salty environment.

Figure 6.10. Saltmarsh dominated by the grass *Spartina* with *Atriplex* in the foreground.

Halophytes and salt-tolerant plants belong to many different families and are very diverse (see Figure 6.10). The only common factor among them is their ability to tolerate either salt deposition, salty soils or survive immersion in the sea (Figure 6.11). Surprisingly, many are succulents, and most have physiological adaptations to cope with a high salt content in their environment. Those halophytes that absorb salt from soil water and then secrete it actually cause salinity to increase in the soil, thus profoundly influencing their environment. When the so-called 'alkali' halophytes (see below) of arid inland areas die, the sodium ion is returned to the soil in the form of soda (Na_2CO_3), contributing to the sodification of lakes.

Figure 6.11. Salt-flat with salt-tolerant grasses and shrubby *Suaeda*.

Seawater usually has an osmotic potential that is negative enough to cause water to diffuse out of plant tissues. Since plants need to absorb water, their osmotic potential must be more negative. The osmotic potential of cells of a halophobic species is on the order of −0.4 to −2.0 MPa, whereas those growing in saline soils may have osmotic potentials as low as −2.0 to −3.0 MPa without wilting or, in euhalophytes, it may be as low as -4.0 to -8.0 MPa. One way to

(a)

(b)

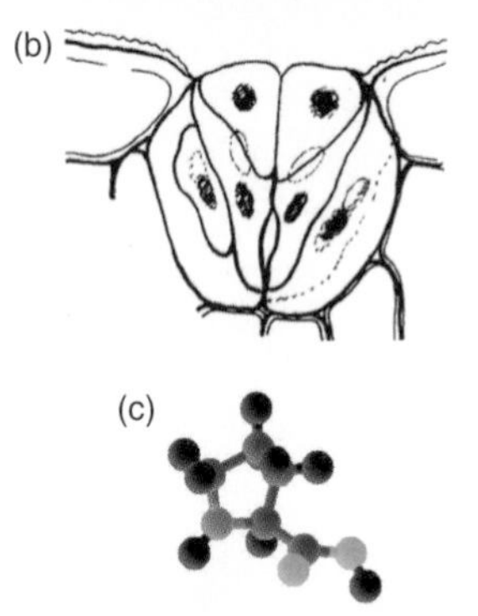

(c)

Figure 6.12. Salt tolerance: (a) thrift (*Armeria maritima*) in flower; (b) transverse section of gland, a complex of 16 cells (from *Armeria*); (c) proline molecular structure.

Figure 6.13. *Arthrocnemon*, a perennial relative of the glasswort *Salicornia*.

achieve a more negative osmotic potential would be for the cells to absorb salt to a higher concentration than seawater. This does not happen because increasing salt levels would lead to the denaturation of enzymes. For active metabolism of the cell, it is essential that the cytoplasm remains free from excess salt.

When salt stress is due simply to deposition on foliage, as takes place in foreshore habitats, cliffs and forward dune systems, most plants attempt to resist the uptake of excessive amounts by means of passive removal, or by various exclusion, and secretion processes. As the degree of salt stress increases, such regulation mechanisms are inadequate, and other mechanisms such as succulence come into prominence. The aquatic sea-grasses, and mangroves have the greatest number of such adaptations. Dune and foreshore plants usually have some resistance to salt spray, and possess some form of short term avoidance of salt injury, or are annuals. Many have greater xerophytic characteristics such as tough cuticles and sunken stomata. In contrast, halophytic species of salt-marshes have few xerophytic adaptations, have thin cuticles, and stomata that are often more frequent and unsunken. By transpiring freely, these plants make use of a large throughput of water to alleviate exposure to high salt. Some, such as *Salicornia* spp. (Chenopodiaceae), can also reduce the demands of the shoots by absorbing rainwater directly through the cuticles.

6.3.2 Salt exclusion mechanisms

Salt is excluded from the roots of many salt-tolerant species, including *Atriplex* spp., but the salt exclusion mechanisms vary. Salt may be excluded simply by the thick, tough epidermis, but tolerance to moderate exposure usually relies on an endodermal barrier in the roots. In salt-tolerant species of *Puccinellia*, the development of a double endodermis retards the passage of salt into the xylem. Halophytes such as *Salicornia* can separate fresh water from seawater by a simple, non-metabolic ultrafiltration process combined with ion transport. In some mangrove species salt is also excluded by ultra-filtration of the endodermal layer of the root, driven by high xylem tension. This results in a xylem sap that is almost pure water with an osmotic potential of nearly zero.

In many species, particularly those of permanently saline environments, some form of active exclusion is usually present. Salt accumulation in the shoots and leaves is prevented by an active transportation of sodium ions from the roots that requires the expediture of energy. Xylem parenchyma cells can function as 'pumps', which actively transport sodium ions against the concentration gradient out of the xylem via the symplast. Chloride ions are retained in the older parts of the roots. Conversely, potassium, which is in low concentration in the environment, is actively transported in. Carrier subunits of the root membranes selectively bind to the ions and convey them across the membrane, releasing them as free ions. The endodermis is cutinised and suberised, preventing re-entry of sodium.

6.3.3 Osmoregulators and osmoconformers

Salt-tolerant plants can be classified according to their tissue osmotic potentials and divided physiologically into osmoregulators and osmoconformers. Generally foreshore and dune plants are osmoregulators. The degree of osmotic adjustment (osmoregulation) is a function of the degree of stress caused by salt in the surrounding environment. In truly saline habitats such as salt-marsh where roots grow in permanently salt water, plants are able to exist only by coming quickly into equilibrium with the salt concentrations in the soil. Such plants are osmoconformers. Mangroves are special cases and exclude most of the salt in seawater. The amount of salt that accumulates in their tissues depends primarily on the efficiency of the salt-exclusion mechanism. Salt accumulation may subsequently be disposed of, either by the secretion of special glands, or by passive removal. Those species (mainly Rhizophoraceae) that do not secrete accumulated salt have a xylem sap that is still about 10 times more concentrated (about $<1/100$ seawater) than that of non-mangrove species, but they have no specialised mechanism for secreting salt. Salt may be lost by cuticular transpiration, or by shedding of parts, but very little is known about this in non-secreting mangroves. *Lumnitzera* accumulates salt in a large-celled hypodermis, and similar processes may occur in other non-secretors such as *Sonneratia* and *Rhizophora*.

In marked contrast to the metabolically active salt glands, salt filtration is a physical process, and probably occurs predominantly at the endodermis, since it is the effective absorbing surface. Positive pressure in the xylem is not possible with plants rooted in seawater. Mangrove species do not guttate, and a negative hydrostatic gradient in the xylem is achieved by transpiration alone. Stomata are not particularly specialised, being scarcely sunken or not at all. There seems to be a close correlation between the high negative pressures of the xylem, and the great density of mangrove wood. Vessels in mangrove woods are very small, and greater in number in proportion to the frequency of inundation. This is an adaptation to counter the effects of embolism that is liable to occur in vessels with large diameters. The Aizoaceae is an exception, while *Limonium vulgare* and *Plantago maritima* behave as intermediate types, and are also intermediate in their ecology.

In many osmoregulators, salt accumulates in the tissues and is not secreted. It is usually concentrated in the vacuole (up to 10 times more concentrated) rather than in the cytosol. Up to 95% of cell volume is taken up by the vacuole. The compartmentation of sodium chloride into the vacuole requires an equivalent lowering of cytoplasmic water potential. Salt is absorbed by osmosis, and the negative osmotic potential continues to increase throughout the growing season as salt becomes more concentrated. In pseudo-halophytes salt merely accumulates in the vacuoles of the cortical cells and in the xylem parenchyma of the roots.

However, compartmentation of salt is of limited duration, and is only a temporary refuge from salinity. Many of the plants that adopt this strategy are annuals, while others are perennials and annually shed leaves or die back in winter to a rhizome. At the end of the growing season so much salt has accumulated in their tissues that passive removal of salt by leaf shedding is a constant feature of *Juncus maritimus*, *J. gerardii*, and other foreshore and salt-marsh plants such as *Limonium* spp. Stem shedding also occurs in perennial species of *Salicornia*. Senescence through salt accumulation in perennials may account for the high number on annuals on foreshores. On British coasts, the most common strandline and foreshore species are all annuals, for example *Salsola kali*, *Cakile maritima* and *Matricaria maritima*.

Other, more tolerant halophytes (both osmoregulators and osmoconformers), accumulate ions that are translocated to other parts of the plant such as shoots and leaves. From there, it may be secreted either by salt glands, through the cuticle, in guttation fluid, retransported through the phloem to the roots (*Salicornia* spp.), or else concentrated in special leaf hairs (*Atriplex* spp.). Salt-secreting osmoregulators have a higher salt tolerance than those which don't secrete. Actively secreting families include the Chenopodiaceae, Tamaricaceae, Frankeniaceae, Plumbaginaceae, Poaceae, Primulaceae, and Convolvulaceae, etc., and mangrove families such as Rhizophoraceae, Avicenniaceae and Acanthaceae, etc. Often such halophytes feel wet or sticky to the touch. Salt glands which actively secrete salt are found in the leaves, stems, or root cells of many genera such as *Limonium*, *Frankenia*, *Glaux*, *Spartina*, and *Tamarix*. Salt gland cells lack vacuoles and are rich in mitochondria. Usually, they are not accumulating organs, but the saltbush, *Atriplex*, has two-celled hair-like glands with a large bladder-like vacuole that ruptures to release accumulated salt. The sunken leaf glands in *Spartina anglica* are also hair-like.

Many halophytic plants, including mangroves, adjust their osmotic potential by synthesising compounds in their cytoplasm that can then exist at higher salt concentrations without denaturing the enzymes essential for metabolic processes and maintain higher negative water potential in their xylem elements so that water is not lost to the substrate via the roots by reverse osmosis. These compatible solutes or 'compatible osmotica' such as proline (amino acid) and betaines (methylated ammonium compounds) build up in the cells and thus maintain a more negative osmotic potential, as a substitute for salt. Plants that produce compatible solutes must not only prevent the external salt from entering the cell but also prevent the compatible solutes from leaking out.

The exact mechanism of compatible solute synthesis is not completely known, but it is linked to the degree of cell turgor and abscisic acid (ABA). This has a general role as an endogenous growth regulator and is involved in adaptation to water deficit stress. Many methylated onium compounds are known to increase in response to salt stress, for example betaines (*Atriplex*, *Suaeda*); related sulphonium compounds (*Spartina*); proline (*Aster*, *Mesembryanthemum*, *Salicornia*, *Triglochin*);

sorbitol (*Plantago*); and the amides, asparginine, glutamine, serine and glycine (*Puccinellia*). Carbohydrates also play a role in osmoregulation, especially where salt is leached from the leaves by the action of rainwater and osmotic potential of the leaves is raised. In winter the leaves of *Halimione portulacoides* and *Limonium vulgare* both maintain a negative osmotic potential by synthesising more glucose and fructose than in summer. This phenomenon is also recorded in *Cochlearia anglica*, where an additional carbohydrate is raffinose.

In succulent halophytes the tissues of leaves or shoots swell by absorbing water and, consequently, salt concentrations do not increase much. Succulence is due to increase in volume of the cells in the spongy mesophyll, which also has fewer inter-cellular spaces and chloroplasts. This thickening increases volume but reduces surface area:volume ratio. Ecological distinction reveals two different types of succulents, halo-succulents and true succulents. Unlike true succulents, halo-succulents have a cell-sap of lower than −5.0 MPa. Halo-succulence is mainly in dicots and is greatest in species without salt excretion glands. It develops in response to salt concentration and is species specific. Some genera such as *Suaeda* are leaf-succulents, while a few such as *Salicornia* are leafless shoot-succulents.

The stimulus to take up salt is caused by chloride ions that cause swelling of the proteins, hence the succulence. The higher the chloride content of the cell sap the greater the succulence. These halo-succulents are sometimes called chloride halophytes. In xerophytic species true succulence develops from germination and provides a ready water reservoir for the leaf in times of drought. The Aizoaceae show a transition from true succulence to halo-succulence, combined with a dense covering of leaf bladder cells (extra-epidermal solute storage). The walls of these cells are highly permeable and provide a water and salt reservoir that is readily exchangeable with the subcuticular cells of the leaf. This is an additional water and salt buffer with which to protect the photosynthetic tissue, and is an example of a halophytic adaptation in a succulent species pre-adapted to arid environments. In *Mesembryanthemum* spp., salt is taken up from the soil into the cell until equilibrium is formed. Sodium and chloride ions are transported into the vacuole and the cytoplasm remains hydrated due to the synthesis of additional osmotically active, compatible solutes. It is this additional hydration that causes the succulence of the organs.

Certain halophytes that store larger quantities of sulphates as well as chlorides in the cell sap are not succulent, or only slightly succulent, for example *Tamarix* spp. Sulphate ions have the effect of decreasing water content of proteins. In contrast to the chloride halophytes, these plants are sometimes called sulphate halophytes. Both types can exist simultaneously on the same soil. A further distinction must also be made. In the so-called 'alkali' halophytes, the sodium ions in the cell are at a higher equivalent concentration than the sulphate and chloride ions together. The sodium ions must therefore be balanced by anions of organic acids such as oxalic acid.

Figure 6.14. Mangrove palm *Nypa fruticans* (a) Fringing community in SE Asia: aeration is increased by the massive spongy petioles of the leaves; (b) germinating seedling.

Figure 6.15. Pneumatophores are frequent and diverse. Aerial roots hanging from branches increase lenticel number. In *Avicennia* and *Sonneratia* they are finger-like or cone-like, simple and upright, and have chlorophyll in the subsurface layers. *Avicennia* spp. are pioneers of more sheltered areas, have great ecological amplitude, and have facultative pneumatophores. *Sonneratia* is found on more or less exposed rocky or gravelly shores but also in brackish water on deep mud. In *Bruguiera* the pneumatophores are knee-like projections. This genus prefers stiff clays of the inner mangrove, but it may be a pioneer along tidal rivers.

6.3.4 Mangroves

Mangroves conjure up images of hot, impenetrable, mosquito-infested swamps where walking is difficult and every step in the sticky mud produces a stench of hydrogen sulphide and the way is barred by stilt roots, but this is misleading. Many mangrove communities are relatively open forests and may occur on rocky coasts. Mangroves are mostly tropical trees that grow on exposed or submerged soils with a high salt content in the inter-tidal zone, and from where the water table is 50 cm or more beneath the soil surface to where the soil is covered by 150 cm or more of water. In adjacent communities mangroves can penetrate extensively along river banks, while, in sheltered estuaries and lagoons, they may form forests up to several kilometres wide. They are one of a number of usually discrete coastal communities, although many constituent species also occur in non-halophytic communities behind the mangroves. They comprise mainly rhizomatous perennials, and all produce aerial reproductive organs, but they have little or no capacity for vegetative regeneration. They belong to a wide range of families, but there are really only four cosmopolitan mangrove families, for example the Rhizophoraceae, Sonneratiaceae, Avicenniaceae and Combretaceae. There are some marginal genera such as the palm *Nypa* (Arecaceae), which occasionally forms large stands of mangrove sub-types due to competitive exclusion of other species by rhizomatous habit (Figure 6.14). Mangroves often blend into fresh-water swamps on the landward side, or to other shoreline communities such as strandline vegetation, or marine sea-grass meadows.

It has been hypothesised that mangroves are unlikely to have existed before the evolution of the seed habit, because the establishment of free-living independent gametophytes may not have been possible in seawater. There is a fossil leptosporangiate fern, *Weichsella*, which is reputed to have formed extensive back-mangal communities in the Early Cretaceous. Today, species of the fern genus *Acrostichum* will tolerate some salt (Figure 6.17), but are generally to be found in the back communities. It often grows on the mounds of the burrowing lobster, *Thalassina anabaena* (Thalassinidae).

Figure 6.16. *Rhizophora* spp., with their looping aerial roots, are usually the most abundant components of mangrove and, together with *Sonneratia* spp., are usually the pioneer species. The stilt roots combine as both buttresses and pneumatophores for aeration.

Figure 6.17. *Acrostichum* spp. (mangrove fern).

The mangrove *Avicennia* synthesises compatible solutes such as betaine, and the ureides, allantoin and allantoic acid, and these increase with salt exposure. Salt accumulators such as *Lumnitzera* also probably synthesise compatible osmotica in order to store salt safely in the hypodermis of its succulent leaves, but very little appears to be known about this. In secreting genera such as *Aegiceras*, *Aegialitis* and *Avicennia*, the concentration of salt in the xylem sap is relatively high, but still only about one tenth that of seawater. Salt is only partially excluded at the roots of these genera, but is excreted by highly metabolically active glands. Multicellular salt glands occur in *Acanthus*, *Aegiceras*, *Aegialitis* and *Avicennia*. Bud secretions occur in the Rhizophoraceae, and in other genera such as *Osbornia, Aegiceras* and *Aegialitis*, the function of which is unknown. The production of mucilaginous secretions coating delicate, developing meristems probably has some protective function, either to combat the effects of dessication, salt, or both.

Mangrove distribution is clearly associated with sea surface temperatures, especially the 24 °C (75 °F) isotherm. There are outlying communities in southern Florida, South Africa, Victoria (Australia), southern Japan, and in New Zealand as a consequence of warm ocean currents, and the presence of fossil *Nypa* in the Eocene of western Europe provides evidence of the warmth of the Tethys Sea that once extended from the Caribbean to SE Asia.

The unstable environment caused by wave action and thixotropic (semi-fluid) soils has led to the evolution of a high root: canopy ratio and trunks with buttresses or stilt roots. In *Bruguiera* there are buttresses to the base of the trunk. Mangrove trees can obtain heights of up to 40 m, but are characterised by low diversity. Because conditions are severe, intense selection pressure has produced a remarkable convergence of form. Leaves are relatively uniform in shape and size, and are frequently leathery in texture, glossy, and with a cutinised epidermis. There is very little structure to mangrove forest after development, because there is no further succession. There is an absence of climbers because their slender stems have relatively wide vessels, and their xylem is highly vulnerable to cavitation due to extreme water tension.

Mud is fine-grained and nutrient levels are high, with abundant calcareous and humic materials (peat accumulates from underground portions of the root system), but low in oxygen or anaerobic. Because of the low oxygen content of the mud, gaseous exchange and nutrient absorption in mangrove plants is mainly in roots near to the surface. Most mangroves have a large percentage of aerenchyma in their stems and roots, which are also covered with lenticels. Aerial roots increase aeration capacity. In addition to these more obvious structures, mangrove species have extensive subterranean root anchoring systems that involve shallow-spreading cable roots, from which descending branches act as anchorage and for absorption, while ascending branches are for aeration.

6.3.5 Salt-marshes and mudflats

Salt-marshes develop on sediments deposited in sheltered estuaries, inlets and bays, and in the lee of islands and offshore reefs. They are subject to periodic inundation and are usually intersected by numerous creeks. They are mainly coastal but they also can occur inland where environmental parameters and species composition are radically different. In South and Western Australia, salt-marshes commonly grade into salt-deserts in more arid areas, while relict communities of salt-marsh plants occur up to 500 km from the coast. Salt-marshes are usually dynamic and successional. Soil salinity is by no means uniform throughout the salt-marsh, and in coastal salt-marshes, the most saline soils occur at about the mean high water mark. Salinity also varies with the season, increasing in depth in wet months, and rising during dry periods. Salt-marshes are dominated by salt-tolerant herbs and grasses and there is often a zonal pattern running parallel to the shore corresponding to gradients in environmental conditions.

There is an isolated colony of *Acrostichum aureum* next to hot springs in southeast Zimbabwe at an elevation of 550 m and 400 km from the nearest mangrove colony in Mozambique, which suggests that this area was formerly coastal.

In many places, salt-marshes have two or sometimes three distinct zones: an upper one dominated by *Salicornia* spp. and other succulents, and a lower one dominated by cordgrass (*Spartina* spp.). In the regions of the upper marsh there may be a transition to glycohalophytes of drier soils, or to hydrophytes where salt-marsh passes imperceptibly into fresh-water reed swamp. Here, plant growth may be limited by salinity and waterlogging. Submergence is often less than six hours in the higher marsh and exposure to dessication may last weeks or months, only to be alleviated by extreme high tides. Salt-marsh species at the seaward side may be inundated twice daily. The length of submersion is a critical factor. Tidal flood of the lower marsh may last more than six hours, whereas exposure to dessication is limited to the fifteen day period between spring tides.

Figure 6.18. Low-nutrient heathland communities: (a) fynbos, South Africa; (b) heathland in England.

In the upper zones of salt-marshes, species diversity may be quite high. In relatively stable salt-marsh communities many species are co-dominant. Competition for water and nutrients is at a maximum during flowering, and therefore flowering times are staggered.

On mud-flats the strange stem-succulent glassworts *Salicornia* spp. are vigorous invaders that almost exclusively dominate and help stabilise muddy soils. Glassworts include annual and perennial species, but many salt-marsh plants, including *Spartina*, are perennials with creeping rhizomes. They grow so prolifically that they eventually cause the salt-marsh to dry up and thus 'reclaim the sea', and so are vitally important in coastal plant succession. They have been used for this purpose in Holland since 1924, and subsequently in many other parts of the world. One of the best known colonisers of salt-marsh is *Spartina* × *townsendii*, a hybrid between *S. maritima* and *S. alterniflora*. Above the zone of glassworts on European salt-marshes may be found a rich community of grasses and other halophytic herbs, dominated by *Puccinellia* species that spread vegetatively by creeping stolons up to 50 cm long. Further inland, this vegetation blends into a community often dominated by *Juncus gerardii* and *J. maritimus*, and ultimately into non-halophytic vegetation.

6.4 Plants of low-nutrient conditions

In nutrient-deficient ecosystems heterotrophy represents an extreme adaptation represented by hemi-parasitic, parasitic, mycotrophic and carnivorous plants. Plants growing on low-nutrient soils show a wide range of adaptations (Figure 6.18). Most species have a root system that develops plastically, the density of laterals varying at different depths in the soil. In the very low-nutrient conditions of tropical forest soils, roots grow upwards into the trunks of dead but still standing neighbours to scavenge nutrients. Root hairs increase the epidermal surface of the roots many times over; 5.8 times in *Leucadendron laureolum* from the fynbos of South Africa. Root hair production is suppressed in high-nutrient conditions. Plants of low-nutrient soils generally conserve nutrients by remaining evergreen and commonly have small hard leaves. Such sclerophyllous plants are widespread and especially common in heathlands and regions with a Mediterranean-type climate. In nutrient-deficient conditions of heathlands some plants have specially adapted rootlets. Proteoid rootlets of the Proteaceae are dense clusters of rootlets with a dense covering of long (6 mm) root hairs. In *Banksia* and *Dryandra* they form a dense mat at the soil surface, trapping nutrients as they enter the soil and prolonging nutrient release. *Personia* lacks the clusters of rootlets but has the dense cover of root hairs. Restioid rootlets of the family Restionaceae are similarly clustered and have long root hairs. Cyperoid (dauciform or carrot-like) rootlets in the family Cyperaceae are densely covered with long root hairs when growing in low-nutrient situations. Several other types have been recognised. The common feature of all these rootlet types is their ability to synthesise polyphosphate from orthophosphate, which is released in the decomposition of litter.

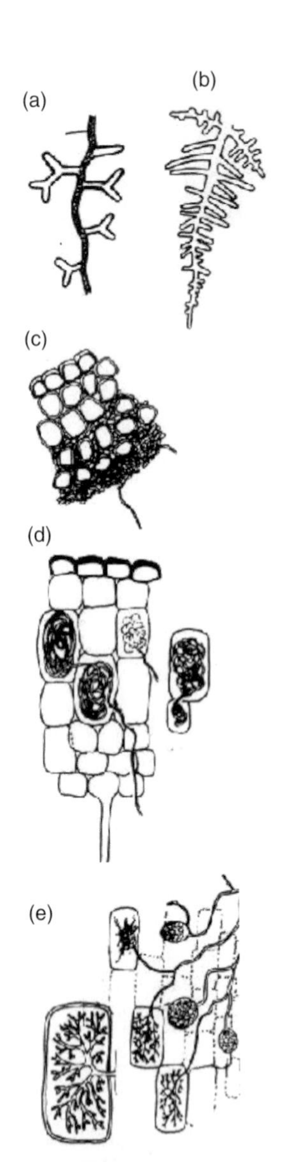

Figure 6.19. Mycorrhizae: (a) sheathing in *Pinus*; (b) sheathing in *Pseudotsuga*; (c) detail of a sheathing mycorrhizae; (d) vesicular–arbuscular; (e) orchid type.

6.4.1 Mycorrhizal associations

Most higher plants have a mycorrhizal association with soil fungi. The exceptions, or those that are only weakly mycorrhizal, are particularly interesting. They include some important but largely ruderal families such as the Cyperaceae, Polygonaceae, Brassicaceae, and Poaceae, especially cultivars grown in high-nutrient conditions. A mycorrhiza is a root infected with a symbiotic fungus. Most mycorrhizal fungi are obligate symbionts. The most important advantage for plants with mycorrhizal associations is that the fungus forages for, and stores, rare and localised nutrients such as phosphate. Ectomycorrhizae are important in obtaining soil organic nitrogen especially in stressed and infertile soils. The uptake of water is also aided. Vesicular–arbuscular (VA) mycorrhizae enhance recovery from wilting, render the root more resistant to pathogens, and stabilise unstable soils. In return the green plant provides the fungus with soluble carbohydrates as an energy source.

Ectomycorrhizae or sheathing mycorrhizae are found in the majority of trees of northern temperate areas especially in the order Fagales, which includes the oaks and birches, and in the conifers. In the Southern Hemisphere they are present in *Nothofagus* and the Casuarinaceae (*Casuarina* and *Allocasuarina*) and Myrtaceae. Ectomycorrhizal fungi mainly comprise a broad range of basidiomycetes and ascomycetes. There are many different patterns and colours of mycorrhiza depending on which plant and fungal species are involved, and a range of different symbionts may be present on a single root system. In *Eucalyptus* there is a high degree of host specificity in the symbiont.

The mycorrhizae are variously black and club shaped, pinnately branched, tuberculate, Y-shaped (in *Pinus*), or coralloid. The fungi are capable of producing auxin and cytokinin and so they may modify the morphology of the host root. The fungal sheath isolates the young root from the soil, and root-hair development is suppressed. The fungal mycelium penetrates between the epidermal cells and ramifies in the inter-cellular spaces of the cortex. The network of hyphae in the cortex does not penetrate past the epidermis and is called the Hartig Net.

Endomycorrhizae involve a more intimate relationship between fungus and host. Vesicular–arbuscular mycorrhizae (VA mycorrhizae) are very widespread in many families including gymnosperms, ferns and bryophytes, both in the tropics and temperate regions. The fungal symbiont belongs to a single zygomycete family, the Endogonaceae. Usually there is no fungal sheath. The fungal hypha flattens slightly where it touches the root surface and then penetrates into or between the epidermal cells. In the cortex the hyphae ramify between the cells and vesicles are produced between the cortical cells, pushing them apart to make space. Other hyphae penetrate the cell walls, producing branched arbuscules surrounded by the plasmalemma or tonoplast. The arbuscule, the site of nutrient exchange, is a transient structure which eventually lyses.

The order Ericales (heathers, rhododendrons, blueberries, etc.) often grow on acid soils that are very low in phosphate. They have peculiar and varied mycorrhizae involving septate fungi. The fungal symbiont, which provides the host with amnio acids, usually has a wide tolerance of different ericaceous hosts in species-rich heathlands. Ericaceous mycorrhizae have been divided into two main types, 'ericoid' and 'arbutoid'. Ericoid mycorrhizae are endomycorrhizal, and are more common. *Calluna* produces thin 'hair-roots' that lack an epidermis, and the single cortical cell-layer becomes heavily infected with intra-cellular hyphal coils. More cortical cell-layers are present in other genera. In arbutoid mycorrhizae, the fungus is symbiotic with trees such as *Arbutus*, and is ectomycorrizal. A sheath and extra-cellular haustoria are developed, but the Hartig Net is restricted to the outer layers of the cortex and some cortical cells are also filled with hyphal coils.

Many orchids have an obligate mycorrhiza. The tiny seeds cannot germinate successfully in the absence of the fungus. A hypha enters the seed through the suspensor region and penetrates the germinating embryo. As the embryo grows, new infections are made. The fungi are species of *Rhizoctonia*. There are two layers to the root: an outer fungal host layer within which there is a digestion layer where fungal hyphae penetrate and grow within cells. The hyphae form intra-cellular coils or irregular structures called pelotons that are digested by a process resembling phagocytosis, but repeated. In orchids that die back to a stem-tuber, a new mycorrhiza is established each season.

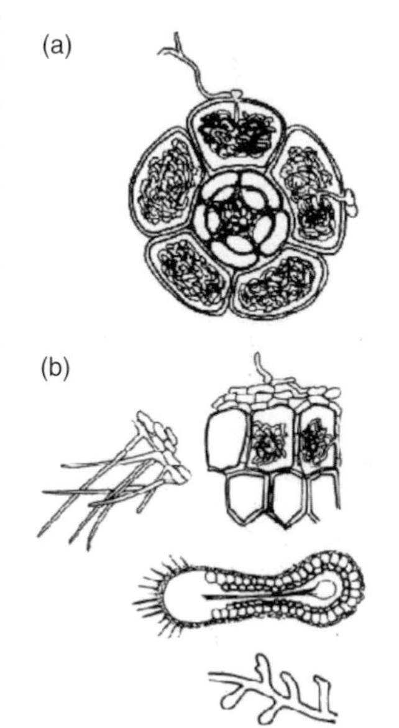

Figure 6.20. Ericoid mycorrhizae: (a) *Erica*; (b) *Arbutus*.

6.4.2 Nitrogen-fixing symbionts

The rhizosphere is the area of the soil around each root that is modified by the presence of the root. A particular microflora may be encouraged by exudates such as mucigel from the root apex or

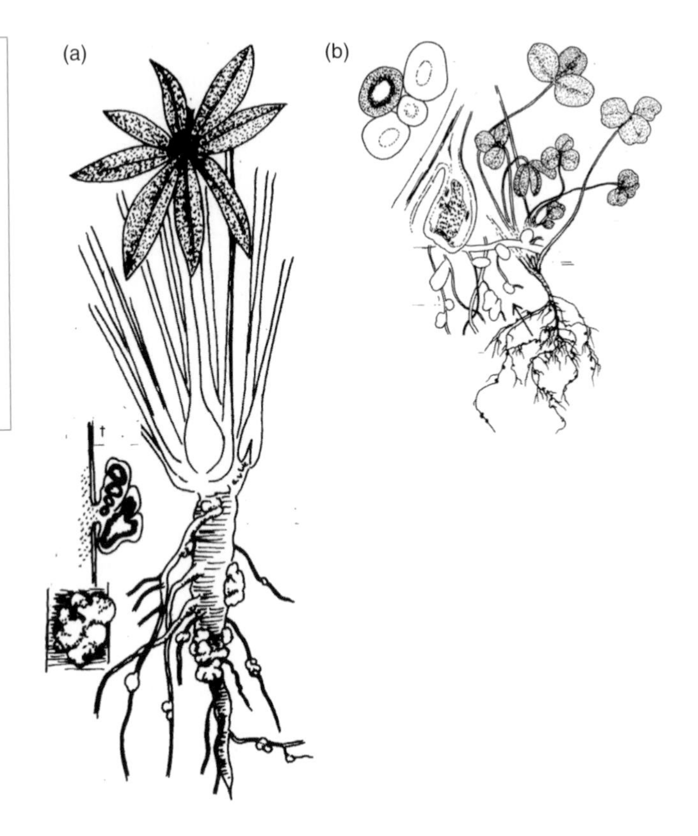

Figure 6.21. Nitrogen-fixing legumes with *Rhizobium*-containing root nodules. The nodule is a complex structure which protects the nitrogenase enzymes from atmospheric oxygen. The centre of the nodule is pink with leghaemoglobin which has a high affinity for free oxygen and maintains the low concentrations of oxygen necessary for nitrogen fixation: (a) *Lupinus* type; (b) *Trifolium* type.

secretions from other root cells. Simple leakage of nutrients from root cells may encourage microbes, and a special relationship may be present between the plant and the nitrogen-fixing root-surface microbes such as *Azotobacter*. The plant provides a carbon source for the microbe which releases ammonia into the rhizosphere.

Of huge significance but present in a rather restricted range of the Rosids is a symbiotic relationship with nitrogen-fixing microbes. Nitrogen-fixing nodules are found in the rootlets of Casuarinaceae, Fabaceae, and Zamiaceae. In the legumes and some other plants the presence of nodules is a manifestation of this relationship. In all three sub-families of legumes, the associate is the bacterium, *Rhizobium* (see Figure 6.21). One non-legume, the genus *Parasponia*, in the elm family Ulmaceae, also has *Rhizobium* in its nodules. The association may be very important in determining the success of the legumes in the semi-arid tropics, where *Acacia* is often the dominant tree. All Faboideae and Mimosoideae are nodulate, while 30% of the Caesalpinioideae are nodulate. In this family nodulation is uncommon in rainforest species. Nodulated plants may also have a mycorrhizal associate.

Root nodules are usually located in the upper 10 cm of the soil and may be renewed each year, especially in areas of seasonal drought. The plant produces a substance that attracts *Rhizobium* and, in turn, the *Rhizobium* makes the root hairs curl before infecting them. Infection threads proliferate into the root tissue. Flavones produced by

Figure 6.22. *Alnus* root nodules. The root nodules are clustered. The nitrogen fixing associate may be the actinomycete, *Frankia*.

the plant switch on the nodulating genes of the *Rhizobium* and, in turn, the cortical cells of the plant proliferate to produce the nodule. In the centre of the nodule there are swollen cells that contain strangely shaped bacterial cells called bacteroids. These cells are rich in nitrogen-fixing enzymes.

Non-rhizobial nodules are found in a range of non-legumes, mostly woody species, including *Casuarina* and *Myrica*. This has been called the *Alnus*-type of nodule from the best-known example (Figure 6.22).

Root nodulation has evidently evolved and has been lost many times. It is not even constant within a single species. The nitrogen fixing blue-green algae *Anabaena* and *Nostoc* are common associates of land plants. *Nostoc* is found in the massive coralloid nodules of the surface roots of cycads. *Gunnera*, the huge-leaved flowering plant that is grown beside water in ornamental gardens, has nodules containing *Nostoc* at the base of the leaves (Figure 6.23). Cyanobacteria are also associated with the aerial roots of orchids. In the tiny water fern, *Azolla*, there is a chamber at the base of the leaf which contains filaments of *Anabaena*. *Nostoc* is commonly found among the rhizoids and scales underneath thalloid liverworts. Free-living cyanobacteria are important fixers of nitrogen in aquatic environments such as rice paddy fields. Nodules on the roots of plants may indicate a mycorrhizal association rather than a bacterial or blue-green algal association, as in the Podocarpaceae.

Figure 6.23. *Gunnera*: (a) base of a plant showing an inflorescence; (b) leaf nodules.

6.4.3 Mycotrophic (saprophytes) and mycorrhizal plants

The majority of land plants are photo-autotrophs, i.e. they can manufacture their own energy requirements through photosynthesis, while those that do not are usually parasitic, and obtain water and nutrients from their hosts. However, some plants of the forest floor are devoid of chlorophyll and do not photosynthesise, but are not strictly parasitic, at least not on a plant host. They achieve this by means of a heterotrophic intermediary such as a fungus, usually those that form root mycorrhizae (Basidiomycetes and Zygomycetes). Hence the name 'mycotroph' has been given to these plants.

Mycotrophic plants were formerly known as saprophytes, but this term is more accurately applicable to the fungi that obtain their energy requirements by the breakdown of the complex organic compounds of dead organisms. In a sense mycotrophs are parasites of the fungus. Mycotrophic plants evolved independently of parasitic plants. Those flowering plants of forests that do not occur in gaps and along streambanks tend to be reduced in stature and have reduced chlorophyll in keeping with their shady, low-energy habitat, although the aroids are a conspicuous exception. Because light is critically low, many are lacking in chlorophyll and are highly dependent on mycorrhizal relationships while some have become completely mycotrophic. Because the potential clearly exists in all mycorrhizal plants for the assimilation of some organic compounds, there is strictly no hard and fast line between a mycotroph and a non-mycotroph, but some depletion of photosynthesis usually is the determining factor. Many are only hemi-mycotrophs and have some chlorophyll, and the evolution of holo-mycotrophic groups can be traced from them, for example the holo-mycotrophic habit in the Monotropaceae is probably derived from the hemi-mycotrophic Pyrolaceae, both of which are related to the Ericaceae, a family which also has unique mycorrhizal relationships.

Similarly the evolution of holo-mycotrophic orchids can be traced from hemi-mycotrophic groups. Most orchids have a mycotrophic stage in their seedling development, and the evolution of mycotrophy has occurred in numerous lineages of the family making the Orchidaceae the most mycotrophic of plant families. In temperate Eurasia the most familiar mycotrophic orchids are the bird's nest orchid (*Neottia* spp.), and the coral-roots (*Epigogium aphyllum* and *Corallorhiza trifida*). Holo-mycotrophs usually are found in shady situations whereas hemi-mycotrophs occur in relatively brightly lit areas. Most species are perennials although a few annuals are known.

Mycotrophic and parasitic plants bear a strong resemblance to fungi and therefore, in a sense, may be said to have a fungal life-form. Originally such plants were confused with fungi or were thought to be aberrations of the host plant. In appearance most mycotrophs are generally rather small and lack chlorophyll, although hemi-mycotrophs may be 'normal-looking', but have reduced roots and root hairs. One orchid genus, *Galeola* is an achlorophyllous climber, reaching to 50 m. In holo-mycotrophs, structure in general is reduced more or less to reproductive organs. Although most mycotrophic and parasitic forest plants have reduced flowers, there are exceptions. In contrast to parasitic plants, mycotrophic plants do not form haustoria but they have highly distinctive mycorrhizae, frequently in combination with an unusual root morphology. Such roots are termed 'coralloid'. Often there is strong host specificity, for example in orchids such as *Neottia* and *Corallorhiza* and dicots such as *Monotropa*.

Originally they were confused with fungi or were thought to be aberrations of the host plant. For example, Traffinnick wrote about *Rafflesia* and *Balanophora*:

We have no choice but to cast them together as oddities, into their own category, much as in an asylum we bring together the mentally ill, whose mania are extremely varied, but of whom no one is really what he pretends or imagines to be.

L. Trattinnick, 1828

Where the source of carbon is ultimately from a photosynthetic host plant (in addition to decomposing organic matter) the term 'epiparasite' may be applied to mycotrophs. For example, *Monotropa sylvatica* obtains fixed carbon from the forest tree *Fagus sylvatica* via the latter's mycorrhizal system. Perhaps *Monotropa* has a mutualistic or symbiotic relationship with the fungus, because it is known that fungal development is stimulated by *Monotropa*.

6.4.4 Parasitic plants

In appearance most parasites are generally rather small, lack chlorophyll, and have reduced roots and root hairs. The holoparasites are generally reduced, more or less to reproductive organs, whereas there are hemiparasites that may be 'normal-looking' and facultatively parasitic, with simple haustoria or with dimorphic roots. Such dimorphic roots may be haustorial or non-haustorial. The haustorium is a chimaeric structure, combining tissues of both host and parasite, and which connects with the host vascular system. Generally, all parasites produce many tiny seeds.

Figure 6.24. *Cytinus* (Rafflesiaceae): showing reduction of a plant to an inflorescence.

Holoparasites do not require light and they obtain all their carbohydrates, water and mineral nutrients from the host. Hemi-parasites are photosynthetic, although their photosynthetic rates may be low and respiration rates high so that they may rely on the host for significant quantities of fixed carbon. There is a traditional distinction between obligate and facultative parasites. Facultative parasites may be grown independently in experimental conditions. They have an effective, though poor root system. The more obligate parasites have weaker root systems even lacking root hairs. Parasites may also be divided into stem and root parasites.

Parasitism has evolved many times over. It is present in 17 families of flowering plant and over 3000 species. Parasites include many herbaceous species but also some large trees. The principal adaptation is the organ of penetration and attachment, the haustorium. This is a complex structure formed from the intimate union of host and parasite tissue. Solute transfer is aided by the parasite having a higher transpiration rate than the host, creating a suction pressure between parasite and host. The haustorium taps into the xylem or the xylem and phloem of the host. Haustoria are very variable in morphology between species. Even within the single genus *Striga* the haustorium may be simple as in *S. hermontheca* or massive and swollen as in *S. gesnerioides*. The host–parasite relationship has been studied in detail in *Striga*, which is a parasite of many crops including sorghum and maize. *Striga* seed is stimulated to germinate in the presence of exudates from the host roots. Haustoria form in response to chemical stimuli, possibly bark or root exudates (defensive compounds) such as the sesquiterpene strigol and benzoquinones. Sometimes lateral roots establish secondary haustoria. Root parasites often show a measure of specificity between the host and the parasite. *Rafflesia*, is parasitic only on tropical climbers of the family Vitaceae. Even within individual *Striga* species, there is specificity between cultivars of the crop host and biotypes of the parasite.

Figure 6.25. *Orobanche*, the broomrape, a parasitic plant lacking cholorophyll.

In many groups similar evolutionary trends in parasitism can be observed. They involve reduction of the parasite, the loss of its photosynthetic ability, a reduction in leaf size, increased self pollination and greater production of smaller seeds. In the Scrophulariaceae and related families, or even within single genera, there are parasites that

Figure 6.26. The dodder *Cuscuta*: (a) in flower on a host branch; (b) section through haustorium showing how the host vascular tissue is penetrated.

look scarcely any different from non-parasitic plants, while others are obviously highly parasitic. *Striga hermontheca* is a leafy outbreeding plant with showy flowers, *S. asiatica*, is an inbreeder but also leafy and *S. gesnerioides* has small scale-like leaves with reduced chlorophyll. The related genus *Orobanche* is devoid of chlorophyll. *Lindenbergia* is not parasitic and there are several hemi-parasites such as *Rhinanthus*, *Bartsia*, *Euphrasia*, *Pedicularis*, and *Melampyrum*.

In species of the Balanophoraceae, reduction has gone further. *Cytinus* spp. produce a short stem directly from the root of the host. It consists of a few bracts and a large inflorescence only. The record-holder must be the bizarre *Rafflesia arnoldii* from Sumatra, which has the largest flower of any plant (often up to 80 cm diameter), the undoubted extravagance of a parasitic life. The unisexual flowers are grouped on a swollen receptacle, which is all that is ever seen above ground. Plants are dioecious or diclinous, the female flowers consisting of only a pistil with a peltate stigma. The rest of the plant is part of the haustorium within the root of the liane host. In the peculiar and related Hydnoraceae of Africa and Madagascar, only the apex of the flower appears above the soil.

Figure 6.27. The hemi-parasite *Rhinanthus*.

This kind of reduction has led to the remarkable example of convergence between the dodders *Cuscuta* and *Cassytha*, which are taxonomically isolated. *Cuscuta* is often placed in its own family, the Cuscutaceae. It is closely related to the Convolvulaceae, which includes many twining and climbing herbs, but there are no linking forms between *Cuscuta* and other twiners in the Convolvulaceae. *Cassytha*, too, is particularly isolated in the Lauraceae. The remarkable similarity between these two genera is only superficial, a result of the reduction of the plant to a yellow twining stem with tiny scale-like leaves. Numerous haustoria are produced from the stem (Figure 6.26). The stems may be produced so profusely that the host is covered. Both *Cassytha* and *Cuscuta* are holoparasites, since the stem normally lacks chlorophyll. However, the seed germinates in the soil and there is a short green twining phase before parasitic contact is made. Like the tropical mistletoes and many tropical epiphytes they produce fleshy berries, which are attractive to birds. *Cassytha* is a perennial, but if the host enters a dormant phase the stems become green, or it may follow the host into a period of dormancy (Figure 6.28). *Cuscuta* is mainly an annual, and it mostly parasitises herbs. *Cuscuta nitida* can perennate within the host's tissue, sending out new shoots in Spring. There is little host specificity in either genus.

Figure 6.28. *Cassytha* covering a host tree (*Melaleuca*: Myrtaceae), Papua New Guinea.

There are two main kinds of stem parasite: the large bushy aerial hemi-parasites of trees, especially the mistletoes in the order Santalales (families Loranthaceae, Viscaceae and Eremolepidaceae); and the dodders *Cassytha* and *Cuscuta*. Stem parasites have some characteristics of root parasites, in the development and form of the haustoria, which links them to the host and in the reduced photosynthetic capacity of some of them. Like root parasites, their host provides them with water and nutrients. They have numerous stomata so that a high transpiration pull draws water and nutrients from the host. However, it is possible that many stem parasites have evolved from and share many features with ordinary epiphytes. Some epiphytes, such as *Aeschynanthus hildebrandtii* (Gesneriaceae), have roots that penetrate the bark of the host tree and are suspected of being at least partially parasitic.

Figure 6.29. Tropical mistletoe Loranthaceae.

The distinction between stem and root parasite would seem fairly arbitrary, except that different taxonomic groups have specialised in either habit. This may indicate different evolutionary origins for each kind of parasitism. The possibly greater specificity and wider taxonomic distribution of root parasites is interesting. Root parasites may have evolved either from direct root fusion or by sharing a mycorrhizal association, both of which require a close physiological compatibility, whereas stem parasites may have evolved from epiphytes that generally do not have species-specific relationships.

The Santalaceae, which are related to the mistletoes, are particularly interesting because they are root parasites, except for one genus of stem parasite, *Dendrotrophe*. The Santalaceae are hemi-parasites with no host preferences. *Thesium* has wandering roots that traverse the soil at a depth of a few centimetres, establishing multiple contacts with hosts, rather like the epicortical roots of the mistletoe *Plicosepalus*. Perhaps the Santalaceae illustrate one example of the origin of stem parasitism from root parasitism or vice versa.

6.4.5 Mistletoes . . . hemi-parasites of the canopy

Mistletoes are sometimes confused with epiphytes, but they are hemi-parasites, i.e. they are photosynthetic plants that obtain their water and nutrients from the host tree. In times of drought there is no buffering effect due to stomatal closure and transpiraton shut-down since, at such times, they transpire even more to obtain essential supplies from, and increasing the stress of, the host, This is the opposite of normal epiphytic behaviour.

In the mistletoes, the hypocotyl elongates from the seed bearing a tiny radicle on its tip. The radicle tip is covered with papillae which secrete a glue when they touch the surface of the host. The radicle then enlarges to form a cup-like sucker from which it penetrates the host tissue. Below the epidermis a branched green callus is formed. In many tropical mistletoes (Loranthaceae and Eremolepidaceae) such as *Plicosepalus*, epicortical roots are produced, which scramble along the branch, dodder-like, producing secondary haustoria wherever they touch. Some mistletoes have extra-ovular tissue (a pseudo-berry) on the seed in addition to chemical exudates for attachment (stickiness) and germination. The testa is highly reduced so that embryos are released in a sticky mass. Many mistletoes are pollinated by birds and have attractive flowers (including explosive corollas). The pseudo-berry is also disseminated by them.

The mistletoe families Loranthaceae and Viscaceae possibly evolved from non-parasitic groups via a hemi-parasitic root parasite. The mistletoe, *Anothofixus*, which is an obligate epiparasite taps into the phloem of its mistletoe host, *Amyema*, while the *Amyema* only taps the xylem of its own host, *Casuarina*. Often there is host specificity like *Arceuthobium* on *Pinus* or *Juniperus*.

One very unusual feature of tropical mistletoes (family Loranthaceae) is the way some seem to resemble the leaf shape of their host. *Amyema linophyllum* is a parasite of *Casuarina* and has leaves shaped like *Casuarina* branches. Other species resemble the phyllodes of *Acacia*. Over 75% of Australian mistletoes display these peculiar characteristics. *Dendrophatae shirleyi* may resemble three different kinds of hosts with either flat linear–lanceolate leaves, thick rounded leaves or linear compressed leaves.

6.4.6 Carnivorous plants

Carnivorous plants usually grow in acid bogs or sandy soils where nitrogen is limited or locked up due to low pH, but some prefer alkaline bogs or limestone. Therefore, carnivory probably arose as a result of selection pressure on plants living in nutrient deficient environments. The most important elements lacking or unobtainable are usually nitrogen and phosphorous. Carnivory can be regarded as only one extreme nutrient relationship, but most plants are capable of absorbing nutrients through the leaf surface to some degree, while at least half of the world's flowering plants have opted for a symbiotic relationship with fungal mycorrhizae as a means of obtaining scarce nutrients. Plant carnivory occurs in only about 0.2% of the world flora. Probably it is less cost-effective in energy terms than mycorrhizal systems. The carnivorous families are: Nepenthaceae, Sarraceniaceae, Dioncophyllaceae, Droseraceae, Cephalotaceae, Byblidaceae, and Lentibulariaceae, and belong to several unrelated orders. There are remarkable examples of convergence in those with pitfall traps, but there is considerable diversity in trap mechanisms. In the Caryophyllales it is the presence of a certain kind of gland that has permitted the evolution of insectivory. Here it is used for digestion, but elsewhere in the order it produces mucilage or excretes salt. The tank bromeliads (Bromeliaceae) absorb organic material via their leaf trichomes, a kind of incipient carnivory. There are some other

Table 6.2 Types of plant traps (number of species in each genus given in parentheses)

Passive traps	Species or Genus (number of species)	Order
1. Pitfall Traps	*Heliamphora* (6)	Ericales
	Sarracenia (10)	Ericales
	Darlingtonia californica	Ericales
	Nepenthes (c. 67)	Caryophyllales
	Cephalotus follicularis	Oxalidales
2. Lobster Pot Traps	*Sarracenia psittacina*	Ericales
	Genlisea (15)	Lamiales
3. Passive Flypaper Traps	*Byblis* (2)	Lamiales
	Roridula (2)	Lamiales
	Drosophyllum lusitanicum	Caryophyllales
	Triphyophyllum peltatum	Caryophyllales
Active traps		
4. Active Flypaper Traps	*Drosera* (100)	Caryophyllales
	Pinguicula (46)	Lamiales
5. Spring Traps	*Dionaea muscipula*	Caryophyllales
	Aldrovanda vesiculosa	Caryophyllales
6. Trapdoor Traps	*Utricularia* (250)	Lamiales
	Polypompholyx (2)	Lamiales

bizarre examples of carnivory. For example, Shepherd's Purse (*Capsella bursa-pastoris*) has seeds that are small and low on food reserves. They secrete a mucilage containing proteases and thus break down proteins present in the soil (which is generally poor in nutrients). This is a strategy that obviously has advantages for such a small annual.

Since the classic work of Darwin there have been very few studies of carnivorous plants as a whole. They are best discussed in terms of the types of traps they possess (Table 6.2). Venus's fly-trap (*Dionaea muscipula*) has rapid, repeatable, touch-sensitive movements. Movement is accomplished in seconds or fractions of a second. The two lobes on the trap will snap shut only when one or more of the trigger hairs are bent over. Movement of the Venus's fly-trap is always preceded by an action potential (at about 14 mm/s). Usually two hits are required to trigger trap movement, which begins precisely after the second action potential is received by the motor cells. If the second stimulus occurs within one second of the first, the trap fails to close. This is because of the so-called refraction period. The trigger hairs themselves create 'receptor' potentials which transduce the bending stress into an electrical code, which in turn fires the action potential. The actual movement is a growth movement, resulting from differential expansion of cells and rapid change in turgor on opposite sides of the motor organs. Stimulation of the trigger hairs also initiates secretion of digestive juices.

(a)

(b)

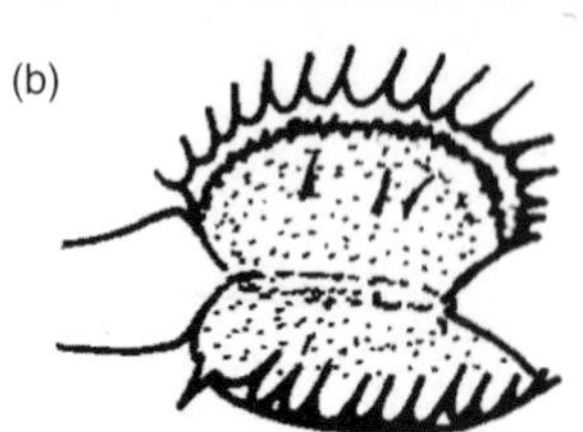

Figure 6.30. Venus's fly-trap (*Dionaea*: Droseraceae): (a) whole plant; (b) leaf showing trigger hairs.

Pitfall traps have evolved in three orders of plants. *Nepenthes* spp. are climbers, often found in bogs or in areas of poor soils such as peat-swamp forest. The pitcher develops from swelling at the end of a tendril. In *Darlingtonia* and *Sarracenia* it forms from a folded/urcelate leaf. The lid of the pitcher is a seductive device, has nectar secreting glands on its inner surface, and is a rain-protecting canopy. The rim is hard, round and glossy, with nectar-secreting glands within the angles, and downward-pointing hairs. The water in the pitcher may contain a wetting agent. Its pH is acid to neutral, but when food is added it becomes acidic for the digestive enzymes to work. After digestion it returns to neutral again. Digestive enzymes are ribonucleases, lipases, esterases, acid phosphatases, proteases and possibly chitinase. The fauna of the pitcher includes ants, spiders, protozoans, rotifers, crustaceans and fly larvae (including mosquitoes).

In *Genlisea*, which occurs on both palaeotropic and neotropic inselbergs, the prey is protozoa that are attracted by chemotaxis, the only known example of this phenomenon in the plant kingdom. In *Drosera*, the leaf surface has mucilaginous tentacles which are longest around the leaf margins. They are only able to bend towards the leaf centre. The outer ones respond quicker and move more quickly than inner ones. The glands are egg-shaped and reddish and have three functions:

(1) to secrete mucilage and catch prey
(2) to secrete enzymes (peroxidase, acid phosphatase, esterase, protease)
(3) to absorb the resultant fluid into the plant's system (assisted by microscopic hairs on leaves)

There is communication between the tentacles and insects are moved from the outer area to the centre. Leaves also infold as protection from rain. There are no nectar glands but the mucilage may mimic nectar.

Pinguicula catches prey by an active flypaper trap and secures victims with viscid glandular secretions. The leaves have a limited facility for in-rolling. Inflection, which brings more leaf area and more digestive glands into contact with prey, also prevents loss of nutritive digest from the margins. There are two kinds of glands on the leaf surface:

(1) stalked sticky glands which catch and detain prey and secondarily secrete globules of mucilaginous fluid
(2) stalkless glands with no mucilage, seated in a depression, digest the prey

The leaves have no nectar but possess a slight scent. The presence of an insect stimulates the glands, which are mildly acidic. The main supply of enzymes is from the stalkless glands. The enzymes are ribonuclease, esterase, acid phosphatase, amylase and protease. The secretion contains a mild bacteriocide to prevent infection from undigested prey.

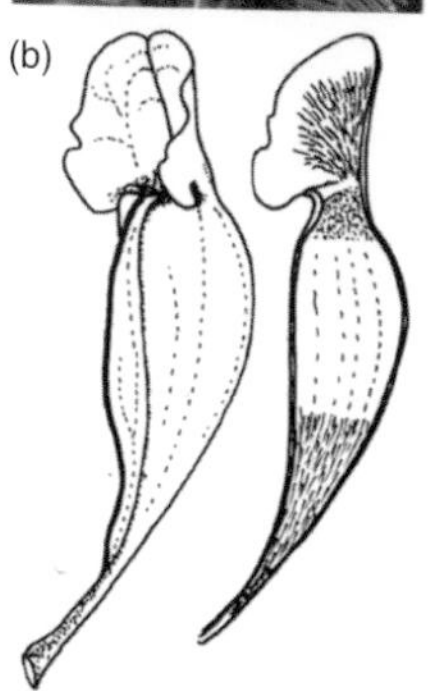

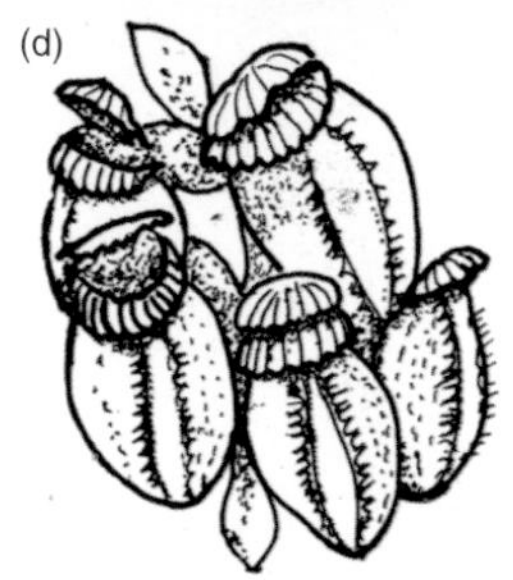

Figure 6.31. Pitcher plants showing convergence in different families: (a) Nepenthes (Nepenthaceae); (b) *Sarracenia* and (c) *Darlingtonia* (Sarracenniaceae); and (d) *Cephalotus* (Cephalotaceae).

Figure 6.32. *Pinguicula* (Lentibulariaceae).

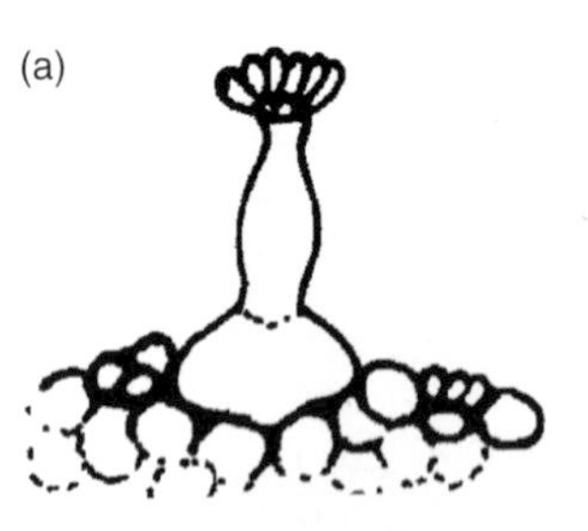

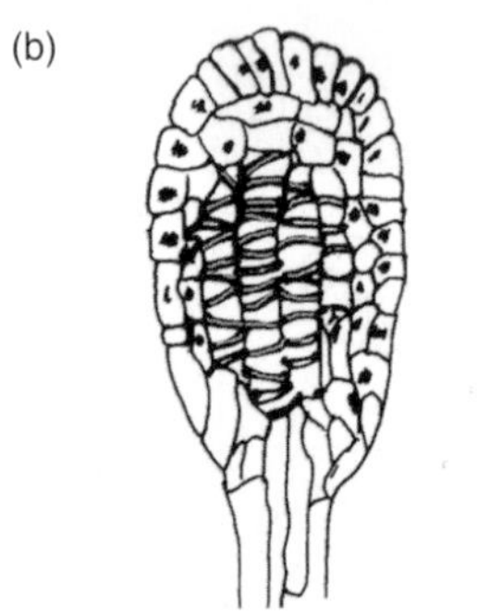

Figure 6.33. Digestive glands: (a) from *Pinguicula*; (b) from *Drosera*.

6.5 Plants of moist shady habitats (sciophytes)

Plants are sometimes arbitrarily divided into sun plants (heliophytes) and shade plants (sciophytes), because their abilities to absorb incident light differ, but it is clear that dividing plants into just two polarised categories does not exemplify the wide range of optimal light regimes required by plants from diverse habitats and at different stages in their life cycles. Heliophytes are plants that can tolerate high light intensities, and often high temperatures as well (epiphytes, desert plants, etc.) but, as could be expected, few such plants are found in deep forest. Heliophytes often possess crassulacean acid metabolism (CAM), or have a C4 photosynthetic pathway. C4 plants are scarce in shady habitats because they are less efficient at energy utilisation, and only at higher temperatures do C4 plants achieve a superior quantum yield to C3 plants. Epiphytic shade plants of tropical forests will be dealt with more fully in the section on epiphytes.

Sciophytes can be crudely described as 'shade-loving' if they are limited by an intolerance of high light saturation deficit, or as 'shade-enduring' if they are relatively tolerant of such a deficit. The level of light reaching forest floors is so low that shade plants quickly reach the compensation point where oxygen evolved through photosynthesis balances the uptake of aerobic respiration. Shade plants cannot obtain the higher rates of respiration found in non-shade plants, and prolonged exposure to light can even inhibit photosynthesis. For shade plants, it is not so much the time that their leaves are kept above or below the compensation point as the net energy balance for the year.

In woodlands, many plants are adapted to the buffered conditions of high moisture and shade and, although outwardly they do not look so different from non-woodland plants, they often possess many adaptations to their unique habitat, particularly physiological ones. In temperate and cold climates, forests provide relative warmth for many creatures, while, in the tropics, the coolness of the forest offers relief from searing temperatures. The ameliorating effect of forests and woodlands may also allow plants with a wide ecological amplitude to extend their ranges. For example *Primula vulgaris*, which is a plant of more open habitats in the moist western parts of the British Isles, is a woodland plant elsewhere.

The net input of radiant energy into a forest ecosystem may be low owing to seasonality and climatic effects. In temperate latitudes, particularly in areas influenced by oceanic climates, such as the Atlantic coasts of northern Europe, the Pacific northwest of North America, the Fuegian region of South America, and New Zealand, the amount of cloud cover throughout the year is great, and it would be expected that many of the plants of those regions are adapted to dull conditions rather than forest shade. In such areas it is difficult to distinguish between true shade plants and plants that are generally tolerant of cloudy conditions. For example, the majority of forest trees in the British Isles are able to survive with 20% or less of full sunlight through the major part of the growing season.

At high latitudes, flowering plants are close to the limits of their ability to obtain enough energy to maintain themselves throughout the year, and this is probably exacerbated by the overcast conditions, but the situation is complicated locally by the effects of elevation and exposure, and the number of frost-free days. Conifers and grasses may become increasingly dominant in these regions. A similar situation may prevail on tropical mountains, particularly those subjected to monsoon conditions or daily thunderstorms, such as the eastern Himalayas and many parts of southeast Asia from Burma to New Guinea, and from the tropical parts of the northern Andes to southern Central America.

In temperate and tropical forests, light is probably the most important microclimatic parameter. The shade plants of the understorey and ground layers are relatively small and slow growing. Low light levels appear to prevent herbaceous flowering plants from being dominant in the ground layers. Here, that role is often taken by bryophytes and ferns, and their allies such as club mosses (Lycopodiaceae), and *Selaginella* spp. (Selaginellaceae) although, in tropical rainforests, much of the ground layer comprises the seedlings of canopy trees. In addition to low light levels reducing the density of herbs, nutrients, water or oxygen may be reduced due to root competition by trees because the absorbing roots of forest trees are in the surface layers of the soil. Nutrient cycling is crucial to a relatively closed system such as a tropical rainforest, but the potential energy of nutrients may be locked up in the standing biomass. In very low light conditions, nutrient limitation, especially phosphorus, may be a critical factor in the acid infertile soils of high rainfall areas. The situation can vary greatly even at the microsite level. For example, ants and termites can produce patches of nutrient-enriched soils. Particulary in coniferous forests, acid conditions may prevent uptake of nutrients, and here most plants have a mycorrhizal association to help overcome this problem.

In tropical rainforests there is much competition at the ground level from palms and, additionally, in South and Central America from the panama-hat family, the Cyclanthaceae (Figure 6.36). On steeper slopes where tree density is less, and more light reaches the forest floor, herbaceous angiosperms may be more abundant. The largest herbs of the rainforest are shade-enduring monocots such as bananas, gingers, heliconias, and balsams. They tend to be social, grow in patches created by tree falls, or along streambanks, and reach maximum development under stronger illumination. The majority of herbs of the more open European deciduous forests fall into this open-area, patchwork category, for example foxgloves (*Digitalis purpurea*). Many of these social species, including the balsams (Balsaminaceae) have explosive capsules to aid the dissemination of their seeds. This is not surprising since there is very little wind at the forest floor.

Figure 6.34. Vernal herbs flowering in an English coppice.

Figure 6.35. Cloud forest rich in bryophytes, ferns and pteridophytes.

Figure 6.36. (a) *Cyclanthus*, the panama-hat plant; (b) fan palm grows up to 2 m in the shade of the canopy.

Figure 6.37. *Selaginella* is adapted for shady habitats.

6.5.1 The quality of light

Direct sunlight is a major component of the total radiation on the canopy, whereas, the total daily photosynthetically active radiation (PAR) within the forest is derived from two components: direct radiation in the form of sun-flecks, and indirect background light. Growth of understorey plants varies approximately linearly with radiation input to about 20% of full sunlight. Sun-flecks are thus of great significance because, in many forests, only about 2% of the PAR incident on the forest canopy reaches the floor and about 50%–70% of this is sun-fleck light. The few sun-flecks of longer than 10 min may contribute two thirds of the daily photon flux. It is common for 30%–60% of total carbon gain by forest floor plants to result from sun-flecks. Canopy gaps (chablis) are the main pathways for radiation below closed canopies, and where this reaches the forest floor there is a 'penumbral' influence around the periphery. In terms of the actual input of solar radiation, there is greater variability between small gaps than large gaps and, at temperate latitudes, the difference between north and south sides of a gap become considerable. Many shade plants are adapted to intercept sun-flecks, especially in the tropics. Short periods of photosynthesis provided by sun-flecks, or in the early part of the year in temperate forests, before the canopy closes, may be enough to provide the plant with a net energy surplus for the year.

Light reaching the forest floor is also different in spectral composition from that on the canopy and upper strata of the forest. In shade there is a general diminution of all wavelengths, but, in addition, in diffuse undergrowth light is depleted in red wavelengths owing to selective absorption of blue and red light by chlorophyll, resulting in a relative increase in green and far-red light, and a very low red:far-red ratio. In contrast sun-fleck light has a high red:far-red ratio, as does that of gaps, in contrast to the diffuse light of the forest floor. Shade-loving plants are adapted to light levels often as low as 2% of that of the canopy, but they vary in their tolerance of higher light levels that often accompany disturbance such as tree falls.

Shade plants also have a plasticity that allows them to maintain a constant relative growth rate over a range of light intensities. There seems to be greater plasticity of physiological response in light demanders although this is not always so. Canopy-top leaves usually have a higher rate of light-saturated photosynthesis, of dark respiration, and greater stomatal conductance than understorey leaves. When conditions change from sun to shade some species such as *Pentaclethra macroloba* seedlings will abort their sun leaves. The net effect of this is to increase the proportion of photosynthetically active leaf material. The higher the amount of chlorophyll, the greater the amount of light absorbed by the leaf.

Figure 6.38. Shady conditions of a *Sequoia* woodland in northern California.

6.5.2 The leaves of shade plants

The texture of leaves in shade plants is often striking. Velvety leaves occur in many genera such as *Neckia* (Ochnaceae) and *Kohleria* (Gesneriaceae), as well as in warm temperate species such as *Musschia wollastoni* (Campanulaceae). This surface texture may be acquired in different ways. There may be dense pubescence, or the cells of the upper epidermis may project as papillae. One easily observed effect of such leaves is that they encourage water to collect into a thin film, which is then shed, often from drip tips. Although they may also act as light traps, and increase light and heat absorption, velvety leaves are probably primarily devices for increasing rate of transpiration. Yet, shade plants seem unable to support a high rate of transpiration without setting up a permanent water deficit.

Even in per-humid climates with no regular or marked dry season, there are periods when evapotranspiration exceeds rainfall. Leaf cells of shade plants have the lowest suction pressures of any land plant. Some open their stomata at light intensities as low as 1/70–1/50 full daylight. Shade herbs often have thin leaves allowing greater transmission of light. Associated with this is a lower chlorophyll content per unit volume, and a lower photosynthetic rate when calculated on a leaf weight basis. This allows the plant to avoid harvesting excess light energy per unit cell volume that could drive potentially damaging reactions.

The metabolism of temperate shade plants is relatively insensitive to temperature, an adaptation which is thought to help prevent overactive catabolic activity during the heat of summer. In the tropics the situation is more complex. In shade-tolerant species there is an increasing sensitivity of photosynthesis to leaf temperature with increasing shade tolerance. *Neobalanocarpus heimii* showed the strongest reduction away from maximum (to 35%), and *Acacia auriculiformis*, a pioneer, was relatively insensitive (10% drop).

With longer wavelengths reaching the forest floor heat stress on ground herbs and tree seedlings may be a real problem. In some species an increased anthocyanin concentration in the epidermis enhances a greater reflectance of longer wavelengths, thus reducing heat load. The cut leaves of species such as *Monstera deliciosa* may be another strategy to encourage heat dissipation. Blue light is effective

Figure 6.39. Colour patterning of leaves of shade plants: (a) Acanthaceae; (b) *Maranta* (Marantaceae); (c) many plants of shady habitats have a purple underleaf that back scatters light into the leaf.

in inducing the opening of stomata, and there appears to be a separate receptor for the morphogenic effects of blue light, possibly a flavin.

The distribution of ground plants in forests is often correlated with subtle differences in microclimate, especially net annual light levels. Capture of radiant energy may be enhanced by simple movement of leaves (phototropism) or by increase in leaf size. Leaf mosaic strategy is often more evident among shade plants. Leaves may orientate themselves to lie at angles to the light rays striking them, in a position to intercept the maximum amount of light. This is usually accomplished by bending of the leaf stalk (e.g. *Fatsia japonica*), resulting in a leaf mosaic with minimum overlapping. Phototropic response is usually manifested by differential growth, although, in those species that have leaves that move relative to the position of the Sun, it is accomplished by turgor changes in the tissues at the base of the leaf stalk.

6.6 Epiphytes, hemi-epiphytes and vines

The name epiphyte is derived from Greek, the prefix 'epi-' means 'upon'. An epiphyte is a plant that grows upon another plant. Epiphytes are not parasites and do not directly obtain nutrition from the host tree upon which they grow, although they can be said to harm the host indirectly. They are not innocuous hitchhikers, and heavily infested trees often show signs of morbidity or injury. Trees in a senescent state tend to be prone to epiphyte infestation. Since epiphytes live in an environment dominated by fluctuating nutrient, moisture and light levels, they can be said to be adapted to withstand periodic stresses. The means of procuring mineral nutrients and moisture are crucial to an epiphyte and can impact on other organisms within its sphere of influence. In many mature forests epiphyte load is greater than that of understorey herbs and not infrequently the collective leaf surface areas of epiphytes exceeds that of the host tree.

Epiphytes are characteristic of tropical forests, oceanic islands, etc., where there is year-round high energy levels and high humidity. There is a decrease in vascular epiphytes with increasing latitude. They occur also in temperate forests of New Zealand, Tasmania, South America and the monsoon regions of the Himalayas where the macro-epiphytes are mostly ferns. This asymmetry of distribution is due to mesic conditions and a more or less oceanic climate in the Southern Hemisphere. Only four macro-epiphytes occur north of Florida in North America, and perhaps only the fern *Polypodium* in the British Isles. At high latitudes where winter light levels are low and where winter drought and frost, particularly air-frosts, are severe, flowering plants are at a disadvantage, especially for the critical stage of establishment. Frost allows survival of only micro-epiphytes such as bryophytes and lichens.

There are about 900 genera and almost 30 000 species of epiphytes in the world, but there are no totally epiphytic families. In most plant families epiphytes are insignificant, a spectacular exception being the Orchidaceae. There are between 20 000 and 25 000 orchids in the world and two out of three (70%) of them are epiphytic. Some 44% of all vascular plant orders and 16% (or about 65 families, 11 of which are ferns) of all vascular plant families have epiphytic species, but only 32 seed plant families have five or more. About 20% of the pteridophytes are epiphytic. There are about 143 species of *Lycopodium* that are epiphytic while only five species of *Selaginella* are epiphytic. Gymnosperms are rarely epiphytic and this is consistent with their slow maturation, massive axes, anemophily and heavy seeds.

There are slightly more families with epiphytes in the Palaeotropics than in the Neotropics (43:42), and there are six times more epiphytes in Central and South America than in Africa.

Speciation of epiphytes is greatest in the Neotropics; the numbers of cacti and bromeliads account for this. Africa, with about 2400 epiphytes, and only about 50% of the families found in other Palaeotropical areas is poorest. This was probably because of impoverishment during dry periods of the Pleistocene. Australasia is impoverished compared with the Americas (10 200 compared with 15 500). These distributions and diversities are the result of historical accident. Each continent has evolved its epiphytic flora independently from terrestrial relatives (sometmes several times over); for example most epiphytic Neotropical orchids belong to the subtribes Maxillarinae, Oncidinae and Pleurothallidinae, whereas the epiphytic Palaeotropical orchids belong to the subtribes Dendrobiinae and Bulbophyllinae. The Palaeotropics are richer in ferns, Araceae and Asclepiadaceae while Australasia is better represented by Rubiaceae (see Figure 6.41).

Figure 6.40. Epiphytic orchids: (a) pendulous orchid in flower; (b) pseudo-bulbs.

Figure 6.41. Epiphytes with a rosette form trap detritus and water: (a) epiphytic bromeliads in neotropical forest; (b) bird's nest fern (*Asplenium*).

Epiphytes may be classified in several different ways, for example by size as micro- or macro-epiphytes; by morphology as 'trash-basket epiphytes' or 'succulent epiphytes'; by ecology, physiology or behaviour as 'shade-tolerant or sciophytic epiphytes', ant-plants, 'sun-lovers or photophytic epiphytes', stranglers, bole-climbers, etc.; by dominant habit as proto-epiphytes (facultative and obligate); hemi-epiphytes (primary and secondary); and ultra- and hyper-epiphytes, etc.; or by a combination of many factors. The last method is probably the best because it takes into account specific adaptations and allows us to directly inter-relate the epiphyte with its immediate environment. However, there is a complete gradation from ill-adapted proto-epiphytes such as *Schefflera* (Araliaceae), *Episcia* (Gesneriaceae), through primary hemi-epiphytes such as stranglers (*Ficus* spp.: Moraceae) and secondary epiphytes such as climbers to the highly adapted hyper- and ultra-epiphytes epiphytes such as bromeliads and orchids. A rigid classification is not only difficult, it is inadvisable.

6.6.1 The herbaceous vines and woody climbers

Lianes (lianas) and climbers have been called proto-epiphytes probably because they begin life rooted in the soil. As they climb they establish connections with the host or with pockets of humus and become hemi-epiphytes. The contact with the soil may become insignificant or be lost altogether, so that at maturity they are holo-epiphytes. Almost all lianes are flowering plants. *Gnetum* is one exception. Ferns are unusual climbers. *Stenochlaena*, the vine fern, has slender green rhizomes. *Lygodium* has an indeterminate frond that produces pinnae continuously as it grows forward. The simplest climbers are those lianes that lean against or scramble over their supporting trees without any intimate connection. Others produce long arching stems that reach up to find support with hooks or thorns derived from leaves, petioles or lateral branches to aid their scrambling. The climbing palms (rattans), such as *Calamus*, are very common lianes in South East Asia. The distal pinnae of the pinnate leaves are backward pointing spines. The rattans grow very fast and have stems which may be well over 100 m long. The stems provide canes for the furniture, basket making and mat industries. Another interesting group are the climbing bamboos like *Dinochloa*, which has a zig-zag culm and roughened leaf sheaths to aid climbing.

Figure 6.42. An unknown climber clinging to the bark of a tree. Many bole climbers of the tropics attach by means of adventitious roots, and conserve moisture by closely hugging the trunk of the host tree.

A closer connection to the host is achieved by the vines with tendrils modified from leaf or stem, or which twine around the supporting tree. Unlike lianes, they may not conform to any simple architectural model. They show varying degrees of specialisation. An even closer connection is achieved by those climbers that produce adventitious roots. These may penetrate the bark, as in the climbing pandanus, *Freycinetia*. Others, such as the familiar houseplants *Monstera* and *Philodendron*, produce large corky aerial roots that take advantage of pockets of humus. These root climbers tend to be highly adapted epiphytes. They migrate up through the canopy with rounded leaves hugging the trunk in early stages and with large out-reaching

Figure 6.43. (a) Rattan palm *Calamus* with grapnel hooked rhachis; (b) *Monstera*.

compound leaves in later stages. Root climbers may obtain significant amounts of water through their adventitious roots, so that they can become holo-epiphytes relatively easily.

Some of the most important adaptations of lianes and climbers are in their vascular anatomy. The free hanging lianes must have a pliant stem able to withstand torsion movements. Parenchyma is abundant in the stems of lianes and vines. In part, this may be because fibres are not required the parenchyma may confer greater flexibility. The xylem and phloem have to remain functional at a great age because of the plant's restricted ability to replace them by secondary growth. There is a great variety of anatomical patterns, the result of differential activity of the cambium. Many have a ribbed xylem (lobed in transverse section) as a consequence of the cambium ceasing activity in places. The furrows between the arms of the xylem are filled with phloem (Bignoniaceae, Apocyanaceae, Acanthaceae). Some have only two lobes, giving a flat stem that is pressed against the supporting tree. In others, an interfascicular cambium does not develop except to produce extra separate bundles. In some lianes, for example in the Sapindaceae, and in *Gnetum*, successive cambia are produced in the cortex, so that they are polystelic. Some have intra-xylary phloem or bicollateral bundles.

Figure 6.44. *Freycinetia* (Pandanaceae) in New Guinea.

A very narrow stem supplies a profuse canopy with water, and conductive ability is maximised by having large diameter vessel elements, although this is hazardous because of the liability of cavitation, i.e. the water columns breaking. Conduction is maintained by also having narrow diameter vessel elements and/or vasicentric tracheids. The preponderance of parenchyma and the more even distribution of phloem through the stele that results from the irregular cambium may also confer the ability of the xylem to recover from cavitation. Photosynthates are distributed throughout the stem. The parenchyma provides sites for starch storage, since lianes have no other area where it might be stored, but in addition this source of soluble sugars may be important in the recovery of cavitated vessel elements. Sugars transferred into the vessels will increase osmotic

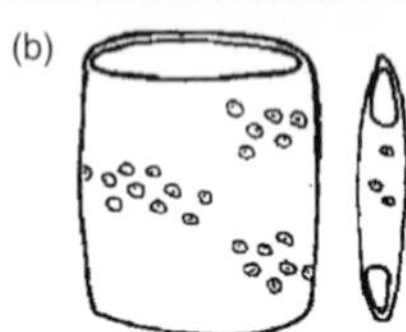

Figure 6.45. Woody lianes have a specialised xylem. (a) Crinkly lianes have flexibility; (b) dimorphic vessel elements, broad ones for water conduction, narrow ones for safety.

pressure thereby encouraging the flow of water back into them. The parenchyma also provides relatively unspecialised cells, which may allow regeneration of the vascular tissue through the formation of successive cambia, or after wounding.

There is distinct stratification among bole climbers. Top layers are distinctly photophytic, for example *Freycinetia* (Pandanaceae). Below this are mixed groups of aroids (Araceae), Gesneriaceae and Ericaceae, which are themselves stratified. Below this again are the sciophytic ferns. Most are shade-loving woody or suffrutescent perennials and don't display many obvious adaptations apart from aerial roots, but a few are succulent, for example Cactaceae and some Piperaceae. Many have increased amounts of chlorophyll or special pigmentation (see Section 6.3), and have varying degrees of dorsiventrality. Many start life in the soil but later lose contact with it as they root to the support tree. In areas with lots of sunlight (gaps, etc.) the climbers are often scrambling herbs.

6.6.2 Stranglers

Primary hemi-epiphytes such as stranglers (*Ficus*, *Schefflera*, *Fagraea*, etc.) start as holo-epiphytes in the crowns of young trees, and are carried upwards with the replacement canopy. They may have crowns larger than the host crown, which may show considerable loss of photosynthate through crown competition. Stranglers can maintain their large canopy because they send their roots, which are often free-hanging, to the soil. The roots increase in number and girth and eventually self-graft or anastomose so that they eventually encircle the host tree and 'strangle' it (Figure 6.46). Most grow in clearings, forest fringes and gaps.

Figure 6.46. A mature strangling *Ficus* that has completely surrounded its supporting tree.

6.6.3 Hyper-epiphytes and ultra-epiphytes

Hyper-epiphytes and ultra-epiphytes are photophytic epiphytes and include some trash-basket ferns (*Drynaria* spp.: Polypodiaceae), orchids, bromeliads and ant-plants. Most of them grow in the zone occupied by hemi-parasites such as mistletoes. Their adaptations reach an extreme in some tiny orchids which live at the tips of twigs in the canopy or on canopy leaves themselves, as epiphylls. There is a whole suite of adaptations to restrict water loss and allow them to live in the hot, dry, uppermost parts of the canopy. The adaptations here include reduction in surface area: volume ratio; aerial photosynthetic velaminous roots; pseudo-bulbs; succulence; stomatal sensitivity; loss of geotropism and polarisation (epiphyllous orchids); holdfasts; tanks and trichomes; farina; incipient carnivory; ant symbioses; and reversed myrmecotrophy (Piperaceae, Gesneriaceae, Orchidaceae), etc.

6.6.4 Adaptations of epiphytes

The tropical rainforest is heterogeneous in four dimensions. Epiphytes live along primary flux routes and, by virtue of their location and scavenging capacity, their strategy is to interrupt the nutrient/water cycle used by soil rooted perennials. In this way they retrieve nutrients lost from the above-ground parts (for example, through leaching and leaf drop), and intercept water and atmospheric inputs, which would otherwise be accessible to their host.

Their low productivity and substantial powers of nutrient-accretion increase their impact on biogeochemical cycling. They are thus major participants in the movement of mineral nutrients within tropical forests. Attempts to analyse the structure and function of tropical forest ecosystems cannot be wholly successful until epiphytes are given due consideration. Apart from anchorage in trees, there is no common factor of growth form, seed type, pollen vector, water/carbon balance, source of nutrient ions or resource procurement mode. Therefore, the life-form concept of 'epiphyte' must include a greater diversity of more subtle variation. Species inhabiting the same area of forest and the same tree crown may differ in their light and humidity requirements. Many orchids, including closely related species, form assemblages on the same host, preferring similar bark qualities, humidity and exposure.

Open-crowned, slow-growing trees with absorbent stable bark make the best hosts. Epiphytes are generally commoner where tree canopies are humid for most of the year, for example in swamp forests and other humid situations such as enclosed valleys. Moisture is probably the most important criterion of all. Temporal access to moisture, avoiding drought injury, is the most immediate challenge, and year-round, high atmospheric humidity rather than high total rainfall is most conducive. Epiphyte diversity is greatest in wet mid-montane forests, peaking at 1000–2000 m. Diversity diminishes at elevations above 2000 m and where there is increasing severity of the dry season. Cool montane cloud forest supports the most luxuriant epiphytic growth, with density peaking at 2000–2500 m where epiphytes make up to 30% of the foliar biomass. Often diversity is low and comprises only bryophytes, orchids and ericoids. They will be present in drier forest where dew or mist occurs but they are less diverse and abundant in areas with poor soils owing to their extreme vulnerability to disturbance. In such areas specialised epiphytes such as ant-plants or carnivorous plants such as *Nepenthes* spp. are more prevalent.

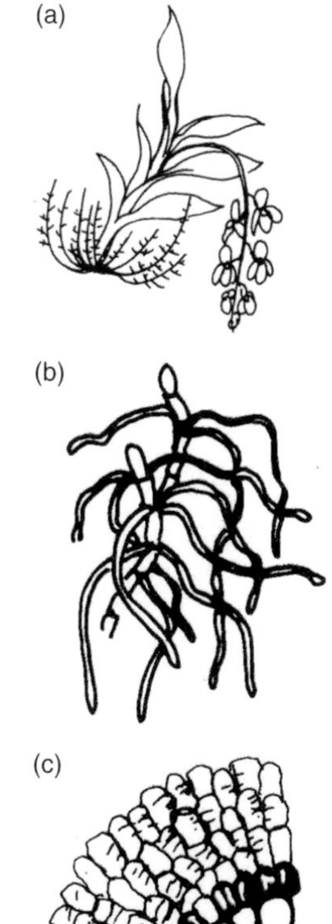

Figure 6.47. Epiphytic orchids: (a) with negatively gravitropic roots that trap leaf litter; (b) leafless epiphytic orchid; (c) velamen of an orchid root: the cells outside the endodermis die and form a sponge like layer.

Light levels and hence the leaf area index of the host are critical and many epiphytes appear to be more tolerant of low nutrients than heavy shade. The species of host tree is also important although, unlike parasites, most epiphytes usually have a broad host preference. Bark texture, stability and wettability are the most important physical determinants of seedling success. Genera with exfoliating bark such as *Eucalyptus*, *Syzigium* (Myrtaceae) are generally useless for epiphytes.

The susceptibility of the host to leaching and the nutritional quality of the canopy fluids is important. Different amounts of nitrogen and phosphorus can be extracted from the same kind of bark depending on nutritional status of the tree. Nutrient-charged water passes through a forest with some regularity but its movements are rapid and it leaves little residue. Stem flow and through-fall are usually dilute. Atmospheric inputs may be very uneven and almost every canopy may be characterised by frequent or prolonged intervals of extreme deprivation. Leachates may be important in breaking seed dormancy. Some orchids may be confined to trees that can support mycorrhizal fungi.

Perhaps the best way to begin analyses of epiphytes is to consider the forest to be analogous to the oceans. At the top there is a photosynthetic layer (euphotic zone) where most of the production occurs. This is the canopy. Below the canopy, in the shade zone (oligophotic zone), photosynthesis decreases along with diversity until we reach the bottom layers where there are only specialist scavengers that feed on the detritus falling down. The fluctuating boundary between the two is called the 'morphological inversion surface' or MIS. Above this there are air-movement, moisture and temperature fluctuations; below this there are stillness and uniformity. The MIS also effectively defines the holding level for understorey and juvenile trees.

Epiphytes may also be classified by their means of obtaining water, for example many that are 'continuously supplied' (CS) occur within the shade zone and are mostly ferns or aroids. Some have CAM (not aroids) while in others the velamen of the roots is not so developed. Many trap organic detritus by means of 'trash baskets', or have mycorrhizal associations and a prolonged life cycle. In contrast, other epiphytes which are 'pulse-supplied' (PS) are found in the sun zone. The PS epiphytes usually have CAM, a reduction in surface area:volume ratio, and a telescoping of parts (e.g. orchids, bromeliads and cacti). Many possess velamen on their roots, absorbing trichomes, etc., and absorb moisture from the atmosphere. The PS epiphytes are often ant-plants (myrmecotrophs) while many also have mycorrhizal associations.

Carbon fixation by means of crassulacean acid metabolism (CAM) is widespread in canopy epiphytes although this phenomenon is most frequently found in plants of arid climates. Canopy epiphytes usually also possess succulence, low surface area:volume ratios and low transpiration rates. They usually lack a well-developed palisade layer and most of the photosynthetic cells are spongy mesophyll. CAM is present in 26 flowering plant families. CAM plants, like all plants, must obtain water and CO_2 but if they fully open their stomata during daylight they transpire too much water. They therefore open their stomata at night only and fix CO_2 into malic acid by the enzyme PEP carboxylase. Malic acid is stored in the vacuole. Starch is degraded by glycolysis to PEP. HCO_3^- reacts with PEP to form oxaloacetate, which is then reduced to malic acid by the enzyme malate dehydrogenase. Malic acid disappears during the day. It diffuses out of the vacuole and is decarboxylated with the release of CO_2. This CO_2 is then utilised by the plant in daylight via the Calvin cycle of photosynthesis. Often CAM plants are facultative C3 and can switch to this mode in cloudy weather or following rainstorms.

Figure 6.48. Myrmecophytes provide hollow organs (domatia) as a home for ants: (a) *Dischidia* (Asclepiadaceae) has hollow leaves; (b) *Myrmecodia* (Rubiaceae) has a chambered tuberous stem.

Forest canopies are unusually hostile. There are many constraints or 'stresses' on epiphytes, particularly the true epiphytes of the forest canopy. Epiphytes ameliorate stress by several strategies. Tissue concentrations of nitrogen, phosphorus and potassium may be unusually low. In the tropical forest environment, because of the level of energy input, there is intense competition for living space. Selection pressure has led to the evolution of specialisation towards many available niches so that the pattern of epiphyte diversity within the canopy reflects microsite heterogeneity. Because of the intense selection pressure, there is rapid growth and rapid turnover of individuals. Disturbances such as bark exfoliation, tree movement, falling branches and tree death provide opportunities for individuals to establish. Most epiphytes are herbaceous perennials. Woodiness only occurs in regions of abundant moisture. They can establish in the canopy of forests without soil or an extensive root system, and thus have an economy which is very cost effective. In the majority the root system functions as a 'holdfast', analagous to that of seaweeds; for example, in bromeliads the roots are very reduced. Because the canopy has a very fluctuating water supply (mineral nutrients and moisture are intermittent rather than continuously available in all but the wettest climates) specialised means of nutrient and moisture procurement must be available. The water balance is assured by considerable mechanistic diversity, for example drought-sensitive roots of ferns, velamen of roots in orchids and aroids, ant domatia, and tanks and trichomes of bromeliads (Figures 6.48–49). Many have structures for impounding nutrients, water and other debris, for example tanks and trash-baskets. Carnivory is rare but myrmecotrophy is common. Most epiphytes have mycorrhizal associations and for some, such as orchids, this is vital for seedling establishment.

Epiphytes experience daily and seasonal drought. Subsequently they have xerophytic features such as economical water use, succulence and extensive water storage capacities, pseudo-bulbs, unusual osmotic qualities and stomatal sensitiveness. Deciduousness occurs where there is seasonal, but not too severe, drought. Some alternate

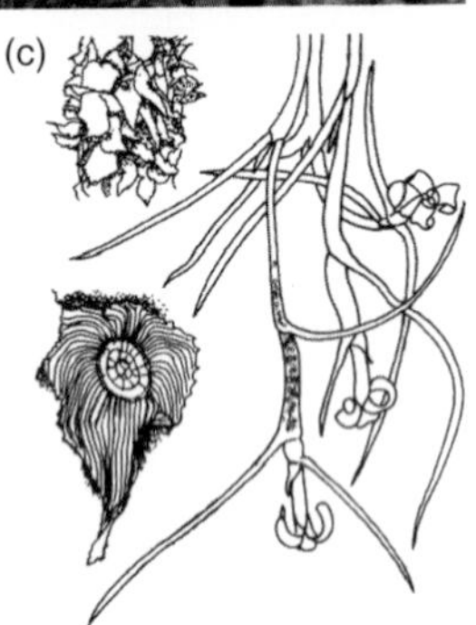

Figure 6.49. *Tillandsia* (Bromeliaceae), the genus of air-plants, has two main forms: (a) upright rosettes and (b) pendulous. Bromeliad leaves absorb water through scale-like trichomes, as shown in (c).

between 'wet active' and 'dry inactive' and a few are 'resurrection plants' with an ability to 'rebound' rapidly, although poikilohydry in epiphytes is found only in areas where moisture is abundant. Most epiphytes are homoiohydric (avoid dessication).

Because of the patchiness of suitable microsites (as a result of both the dispersion of host trees within the forest and the separation of their branches), there is often a scarcity of conspecifics and so aerial dispersal is the most frequent mode of spread (i.e. zoophilous pollination and wind-dispersed seeds). Usually the seeds are tiny and lack appendages (but see *Aeschynanthus* spp.: Gesnericeae).

Monocots (about 25% of all flowering plants) have five times the number of epiphytic species (especially in the families Orchidaceae, Bromeliaceae and Araceae) in comparison with the rest of the flowering plants, and twice the number with fern epiphytes. However, there is no common monocot adaptive theme, although they possess many features that appear to confer advantages as epiphytes. Most species are iteroparous, with a rhizomatous, sympodial habit, and serial perennation with determinate offshoots. Each 'phyton' is relatively autonomous with leaf, associated adventitious roots, buds and subtending stem segment. The meristems receive fixed carbon, mainly from nearby leaves (i.e. there is reduced translocation over the whole plant). The reticulate stele gives greater capacity for functional integration and extensive vegetative renewal with a minimum of tissue space. The meristematic regions remain as nutrient sinks whereas in times of stress dicots will self-prune by aborting leaves, branches and flowers.

Orchids have an affinity for acidic, humic, infertile soils (i.e. with reduced nitrogen), and utilise NH_4 rather than oxidised nitrogen, and this may have predisposed them to epiphytism since mycorrhizae mobilise nitrogen and phosphorus from sterile soil. Epiphytes, especially impounding ones, increase canopy humidity, which makes it more favourable for nitrogen-fixation. Many orchids have extensive nitrogen-fixing epiphyllae (which in turn have a symbiotic relationship with *Nostoc*, a blue-green alga). Epiphytes may be more important to the forest fauna than their numbers and biomass would suggest. Epiphytosis causes treefall and an increase in the physiognomic diversity of the forest, especially montane forest.

Bryophytes are most important in the water balance of tropical montane forests and the dynamics of their vascular plant associations (Figure 6.50). By intercepting more than 25% of precipitation they control and impede drainage, and can thus influence climate on a local scale. Many species, especially epiphylls in tropical forests, are associated with blue-green algae and fix atmospheric nitrogen. This is the main input for nitrogen in tropical rainforests. As with other epiphytes, bryophytes can profoundly alter the physiognomy of forests due to 'epiphyte load' and consequently affect biodiversity. When wet, mosses can be up to four times their live weight. The epiphyte biomass and interceptive capacity are proportional to annual rainfall where the monthly average is more than 100 mm, whereas

Figure 6.50. Epiphytic bryophytes: (a) coating a branch; (b) epiphyllous leafy liverworts and mosses. The leaves of many plants in the tropical forest have drip-tips to gather up moisture and shed it from the leaf to prevent the colonisation of bryophytes.

tropical macro-vegetation biomass (and diversity) increases only in areas of up to about 150 mm before tailing off.

6.7 Grasslands and savannas

The great grasslands occupy a climatic zone between forest and desert, in the North American prairies, the Asian steppes, the African savannas, and the South American pampas, but much of their recent distribution has been assisted by humans. Grasses (Poaceae or Gramineae) are one of the most familiar groups of flowering plants, yet their identification and biology remain problematical or mysterious. For much of the year they are seen in a non-flowering state and all appear to look alike, but they are of the utmost importance as a food source for humans and animals, and dominate much of the world's vegetation. There are about 651 genera with about 10 000 species of grasses, cosmopolitan in distribution, and forming one of the largest and most successful of flowering plant families. They occur in every kind of habitat, from mountains to the seashore, in forests and savanna, and in deserts, rivers and marshes, and are estimated to be the principal component of about 20% of the Earth's vegetation cover.

6.7.1 Grasses

The widespread occurrence and predominance of grasses in the various types of world vegetation results from:

- adaptation to a range of soil types
- adaptation to a diversity of climates and a broad ecological amplitude
- ability to compete successfully with other plant types
- ability to survive high levels of predation

This is brought about by the following adaptive features:

- unique morphology
- specialised physiology, especially connected with their modes of photosynthesis
- various strategies for vegetative reproduction
- specialised flowering mechanisms
- a diversity of breeding systems

All green plants utilise ribulose diphosphate for the initial capture of carbon dioxide from the atmosphere. Many grasses, however, possess an additional chemical pathway that utilises the three-carbon compound phosphoenol pyruvate (PEP). This is known as the Hatch/Slack or C4 pathway. This pathway is possible because of its spatial separation in the leaf-blade anatomy (Kranz anatomy) from the Calvin cycle. In grasses without this extra pathway (i.e. those with only a C3 pathway) the Calvin cycle alone, operates in the diffusely arranged cells of the chlorenchyma. The Hatch/Slack pathway operates in the radially arranged cells of the chlorenchyma and releases carbon dioxide into the outer bundle sheaths, where it is incorporated into the Calvin cycle. Plants with the Hatch/Slack pathway have a much higher rate of carbon dioxide uptake and higher growth rates than those in which it is absent. The C4 pathway reduces photorespiration. Because it operates most efficiently in high temperatures and high light intensities it occurs widely in tropical grasses such as *Andropogon*, *Panicum* and *Eragrostis*. Temperate grasses such as *Poa*, *Bromus* and *Festuca* gain no advantage from this extra pathway so they retain the Calvin cycle alone.

Figure 6.51. One of the more familiar grasses, *Cortaderia selloana* (pampas grass). Grasses are mostly small tufted plants, but occasionally may, like pampas grass, form large robust tussocks with a stout central axis composed of tightly packed dead leaf bases.

Most grasses have stems that are hollow, easily bent and yield to the wind. Stems do not increase their size by growing thicker at the sides and longer at the top. The thickness is fixed from the beginning, and the stem increases by the meristematic region (intercalary) at the base just above each node. The intercalary meristem is protected by sheathing leaf bases. Grasses do not develop taproots, and the adventitious roots are slender, relatively short, with infrequent branching. They originate in large numbers from the base of the plant (e.g. *Festuca*). Alternatively, the roots may be few and very long, extending deeply in the soil or remaining near the surface (e.g. *Aristida pungens*). Grasses with such roots are able to utilise all available surface water.

Many grasses, especially perennials, have horizontal underground rhizomes (e.g. *Agropyron*) or overground stolons (e.g. *Cynodon*), and produce new plants at intervals along their length. Rhizomes are often tough and serve to anchor the plant in the soil, as well as to colonise new ground (e.g. marram grass, *Ammophila arenaria*; rice grass, *Spartina townsendii*) (see Figure 6.52).

Grasses do not develop a permanent main stem with side branches (except bamboos). They grow from a basal rootstock, and the leaves, which are simple, often die back at the end of each season in perennial species. They are therefore mostly small tufted plants, but occasionally may be large, forming robust tussocks with a stout central

Figure 6.52. Marram grass (*Ammophila arenaria*) builds dune systems by trapping sand. It can grow rapidly, thus preventing burial by sand.

axis composed of tightly packed dead leaf bases (e.g. pampas grass, *Cortaderia selloana*). In tropical regions of seasonal or low rainfall there is often a danger of fire started by lightning or by humans. Grasses are able to survive because new growth is initiated at the rootstock below ground level. Tropical savanna grasses such as *Andropogon* and *Saccharum*, and the giant reed *Arundo donax*, may have stems exceeding 3 m. The leaves are borne in two ranks at intervals along the stem. They originate from nodes and comprise a basal sheath clasping the stem, and a blade which is usually narrow and flat, folded or rolled. Just above each node is the intercalary meristem. Differential growth of this meristem allows grasses to bend upright after trampling. Leaf blades also grow by a basal meristem situated at the junction with the sheath, permitting the blade to grow despite the removal of distal parts by grazing. At its upper end, the sheath passes into a parallel-veined blade. The blade is typically long and narrow but may be broad in shade-loving species. At the junction of sheath and blade is a short membranous rim, called a ligule, that may prevent rain entering the sheath.

Figure 6.53. Bamboo is rich in hard fibres, giving the the hollow stems great strength.

Grasses only need a piece of stem bearing leaves with node and internode in order to reproduce vegetatively. Adventitious roots readily grow from a node, while new shoots grow from buds in the axils of the leaves. This is the way that grass cover and sugarcane is established in horticulture and agriculture, respectively (i.e. by 'seeding'). In many tufted or tussock species vegetative growth is by tillering. New shoots grow out from the leaf axils at the base of the plant, to form a rosette or tussock (Figure 6.54). Other grasses may spread by stolons, stems that grow from the base of the mother plant and spread horizontally over the surface of the soil, producing a new plant at each node. This method is common in tropical grasses such as Bermuda grass (*Cynodon dactylon*). In temperate grasses a similar effect is achieved by rhizomes (under the soil surface). It has been estimated that a single plant of *Festuca rubra*, which spreads by rhizomes, may be some 250 m (>800 ft) in diameter and up to 400 years old,

Figure 6.54. Tussock grassland in New Zealand.

and that a large tussock of *Festuca ovina* (8 m or 26 ft across) could be 1000 years old. In a few grasses the base of the stems may become swollen to produce storage organs similar to bulbs or corms (e.g. False Oat grass, *Arrhenatherum elatius*; Bulbous meadow grass, *Poa bulbosa*), usually as a way of combating drought. The transformation of spikelets into bulbils with little leafy shoots is a regular means of propagation in some mountain or arctic species.

Figure 6.55. African savannas are grazed and browsed by diverse species of ungulates that partition the available vegetation. Such ecological separation is most marked in transition areas between woodland and grassland. For example, small antelopes such as Grant's Gazelle (shown here) favour more open areas, whereas the gerenuk is a browser of bushes. Smaller species of antelope such as dikdiks can graze under thorn bushes more easily.

Figure 6.56. Two related species of African acacia: *A. fustula* with a white bark has ant thorns, and *A. seyal* lacks protection from ants but has a red powdery bark rich in chemical defence compounds.

The growth form of grasses, having basal meristems and mechanisms of rapid lateral growth, allows them to survive herbivory while competitors are eliminated by the herbivores. The grass leaf can be strongly sclerified, and the epidermis may contain silica cells which make them less palatable. Some herbivores do show preferences for particular grasses. In grasslands in Britain, sheep prefer soft species such as *Lolium* to the hard-leaved *Nardus*. However, there is generally no close herbivore/grass relationship. Graminivorous herbivores such as cattle and horses are not very choosy, and no sophisticated chemical defences have evolved. In *Sorghum*, where some species are distasteful to locusts, the alkanes and acid esters in the wax are general repellents.

6.7.2 Adaptations of grassland and savanna plants

Savanna trees display a range of adaptations: for seasonal drought (sclerophylly or deciduousness, deep-rooting), for fire (protective bark) or against herbivores (latex, resin, chemical defences, thorns). In the savannas of America, Africa and Australia, thorn bushes and trees, especially of *Acacia*, are frequent (Figures 6.55 and 6.57).

Some *Acacia* species harbour ant colonies, supplying them with nectar from extra-floral nectaries, or with food-bodies produced at the tips of the leaves, and providing a home for them in swollen thorns. The ants attack herbivores and keep the surface of the plant clean.

Most other grassland plants exhibit physical or chemical defences against herbivory. Trichomes constitute a first line of defence often

Figure 6.57. The Reticulated Giraffe is a specialist browser of acacia trees and other thorny scrub (Kenya).

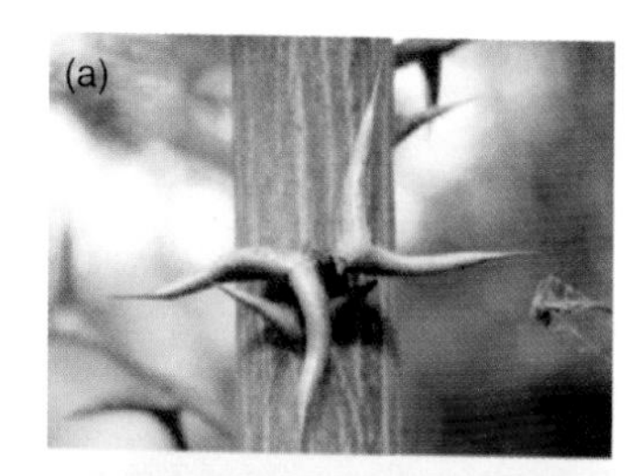

Figure 6.58. Ant thorns have evolved separately in African and American acacias: (a) from central America; (b) from Africa.

associated with chemical deterrence. A second line of defence is the leaf-surface waxes or resins, which are frequently mixed with toxic constituents. The evolution of more advanced compounds has restricted the range of possible herbivores and a close relationship with a particular herbivore has resulted. Co-evolution of plant and herbivore has resulted in greater rates of speciation.

6.8 Plants of cold or hot arid habitats

There are usually no sharp distinctions between arid and non-arid regions. These zones often blend into one another either on a global scale or locally in terms of variations in microclimatic or edaphic conditions. On a local scale this may be caused by rain shadow on the leeward sides of mountains or it may be the result of moisture being concentrated in valleys, canyons and gulleys. Even in extreme deserts, in pockets of moisture we may find some bizarre adaptations, for example poikylohydry or 'resurrection' phenomena. Mostly, such plants are bryophytes, *Selaginella* spp., or ferns such as *Notholaena parryi*. These plants usually survive in shade and absorb dew. Many epiphytes of tropical forests may be thought of as adapted to arid conditions. The intermittent moisture experienced by plants living on trees has led some workers to describe such conditions as desert-like. Therefore adaptations to aridity are not only the consequence of widespread geographic aridity but also very localised conditions within the immediate vicinity of the plant.

In many regions of the world there is seasonal aridity only, for example the Mediterranean regions. This aridity may not always be caused by lack of rainfall. In severe climates of mountains and at high latitude, water is unavailable due to freezing conditions and the

Figure 6.59. Diverse forms of plants of arid areas: (a) living stones *Lithops*; (b) barrel succulent Euphorbiaceae; (c) xerophytic pachycaul rosette tree: *Yucca brevifolia*, Joshua tree.

environment for plants is effectively arid. In addition, many plants of cold steppe regions of Asia, North America and the high altiplano of South America are actually halophytes growing in salty soils and may share many characteristics, of true desert plants, such as succulence. The majority of these halophytes belong to the Chenopodiaceae or Plumbaginaceae.

Plants of desert regions are often referred to as xerophytes, exemplified by members of the Cactaceae, Euphorbiaceae and other familiar succulents such as the Crassulaceae and Liliaceae *s.l.* (e.g. *Aloe* spp.). Species of more extreme desert conditions display the features we most commonly associate with xerophytism, for example thick cuticles, succulence, reduced surface area: volume ratio, sunken stomata, spines, CAM photosynthetic pathway, etc., but the diversity of xerophytes greatly exceeds this narrow selection. This is because such plants have diverse phylogenetic histories and are subject to tremendous permutations of rainfall patterns and other forms of precipitation with environmental parameters such as soil type, temperature, etc. For example, in the most arid areas of southeast Spain, Sicily and North Africa, and usually associated with salty soils, a few succulent asclepiads such as *Caralluma* spp. occur. We may be reluctant to treat such Mediterranean plants as desert plants, but should we call them halophytes, xerophytes, or both? It is hard to place such plants into rigid categories. In the adaptations to arid conditions there is every gradation in terms of morphology and physiology across many unrelated families and these patterns of variation are repeated in many different areas of the world. There are also some peculiar anomalies, for example, succulents are almost entirely absent from the arid regions of Australia.

Other herbaceous plants, especially monocots survive the dry season as bulbs or corms. These plants are called 'geophytes', for example *Crocus*, *Tulipa*, *Urginea*, *Asphodelus*, and many Orchidaceae, etc., plus dicots such as *Cyclamen*. There is considerable convergence of geophytic form among the herbaceous monocots. Many produce a flush of new leaves in the autumn and winter, and flower in the early spring or autumn. Some of the best-loved garden plants, such as daffodils, narcissi, tulips, crocuses and irises, belong to this group of monocots, while many of the South African geophytes, such as *Amaryllis*, *Clivia*, *Agapanthus*, *Zantedeschia*, *Crocosmia*, *Gladiolus* and *Watsonia*, have also become important for Northern gardens. There are about 1350 species of monocot geophytes alone. The geophyte flora also occurs in mountains (e.g. *Galanthus*, *Soldanella*, etc.) adjacent to Mediterranean ecosystems, but the species composition often differs, and many of the species are adapted more to the rigours of an alpine environment.

As temperatures drop below freezing, ice formation in the extra-cellular spaces of plants leads to a movement of unfrozen water from inside the cell to the extra-cellular spaces where it freezes, causing severe dehydration stress.

Numerous herbs lie dormant as seeds until the rains return and they can recommence their annual life cycle, for example *Roucela* spp. (Campanulaceae). In more extreme arid climates the annual life cycle is modified to become an ephemeral one, for example *Wahlenbergia campanuloides* (Campanulaceae), from the deserts of Namibia, NE Africa and Arabia, only flowers every other year.

6.8.1 Desert plants

Contrary to popular belief, cacti and succulents do not inhabit only conventional deserts. They have a rich range of habitats from snow-clad alpine slopes (up to 15 500 ft in South America) and arid plains to humid rainforest, but the characteristic of all these habitats is scarcity of water. It is the desert regions where such extreme xerophytes have greatest prominence, so these regions will be highlighted. The intensity of the climatic conditions under which xerophytes have evolved has meant that the life strategies and opportunities for evolutionary diversification of morphology has been limited. Thus we see among several unrelated families considerable convergent evolution, especially between the Cactaceae and the Euphorbiaceae. Also, different regions have had independent histories so it is not surprising that the composition of floras is different. The Cactaceae dominate the American arid lands while the succulent Euphorbiaceae are features of the Old World tropics. Other families show remarkable resemblances but are relatively isolated geographically; for example the ocotillos of the family Fouquieriaceae resemble the Didieriaceae from Madagascar, and both resemble some of the Madagascan Euphorbiaceae, such as *E. splendens*. Many of the Crassulaceae of the American tropics have morphological equivalents in Africa, while the succulent liliaceous genera of the American deserts such as *Yucca* and *Agave* have equivalents in the African genus *Aloe*. The Asclepiadaceae is another succulent family which is rich in species in the Old World tropics. Australia is exceedingly poor in succulents although it is rich in sclerophyllous and phreatophytic plant types. Many epiphytes are succulent, including cacti and bromeliads.

Figure 6.60. *Idria columnaris* *Fouquieriaceae*, Baja California.

Figure 6.61. Succulents in the Sudanese desert include succulent trees *Euphorbia abyssinica*, shrubby succulents *Carulluma* (Asclepiadaceae) and aloes (Amarallyidaceae) in flower.

Succulence, spines, a thickened cuticle and lack of branches and leaves make these two families superficially similar in appearance but one or two genera in the Asclepiadaceae also have this form. The flowers of each family are radically different. In addition the Euphorbiaceae have a toxic latex, which is a hazard for would-be herbivores. With experience one can recognise plants from each family by their overall appearance, even in the absence of flowers (Figure 6.62). The form of the succulent stem is highly adaptive and conforms to variations in surface area:volume relations and light intensity, as well as growing conditions, life-span, etc. From a theoretical point of view the shape which offers the least amount of surface area per unit volume is a sphere. This shape minimises the amount of surface through which transpiration can occur and which can absorb solar energy. Many cacti are in fact globular in shape and are to be found in the hottest of deserts, often on the ground or on exposed hillsides among rocks, for example the barrel cacti, *Echinocactus* and *Ferocactus*. Support for the succulent stem is provided by a number of vertical ribs, which may also give the cylindrical form greater surface area by created a series of fluted ribs. In this way the surface area may be increased without exposing the plant to the rigours of solar radiation, but these ribs also allow the plant to expand or contract.

The tallest cactus in the world is the giant saguaro *Carnegia gigantea*, which occurs in the Sonoran Desert of Arizona. When the saguaro is very young it requires shade protection from other species such as paloverde (*Cercidium*) or ironwood (*Olneya*).

Cacti have a slow rate of transpiration (about one thirtieth that of an average 'normal' plant) and the stomata are widely dispersed and sunken. The tissues contain large amounts of water-retaining mucilage and some species may contain up to 90% water in their stems. Many cacti have extensive fibrous or tuberous roots, located near the soil surface and these can absorb dew as well as rainwater. Most of the rainwater absorbed by cacti falls over a very short period of time so rapid uptake is vital. Some cacti have underground storage organs such as tap roots, whereas the columnar cacti utilise taproots as a stabiliser. They often have a dense covering of white hairs that probably aids in protecting them from intense sunlight as well as night-time cold, and almost all cacti are armed with spines that are really modified leaves. The spines form part of an absorbent structure called an areole, the base of which often has minuscule hairs called glochids. From the areole, two sets of buds develop, one for flowers, the other for spines and glochids.

Not all desert plants are succulents. Many have deep root systems and look relatively normal. Those that root deeply into the water table are called phreatophytes, for example *Acacia albida*, which, as a result, is also able to produce new leaves at the beginning of the dry season. Upon germination, the seedlings of these phreatophytes rapidly produce a deep tap-root. As in Mediterranean ecosystems there are sclerophyllous-leaved deciduous species, some of which may photosynthesise through their branches, for example, paloverde in Arizona

(a)

(b)

(c)

Figure 6.62. Different forms in the Cactaceae: (a) *Parodia* or *Mamillaria* type; (b) *Opuntia*; (c) *Carnegia gigantea*.

and the fever tree (*Acacia*) in Africa. Some plants with very succulent or swollen stems may have normal leaves; for example, *Pachypodium* (Apocynaceae) from Madagascar; *Adenium* (Apocynaceae) from northeast Africa, *Vitex* from South Africa, and even the baobabs and their relatives (Sterculiaceae) from Africa and Australia. The boojum (*Idria columnanis*: Fouquieriaceae) from Mexico is another such plant (Figure 6.61).

Figure 6.63. *Adansonia* baobab (Sterculiaceae) from Australia.

6.8.2 Arctic and alpine plants

The environmental conditions for Arctic, Antarctic and alpine plants are so varied worldwide that only an outline sketch can be given here. The tundras of the Arctic Regions, Antarctica and Greenland, and the treeless alpine areas of mountains occupy about 15% of the land surface yet contain relatively few plant species. In comparison with high latitudes, the alpine and tropicalpine regions are smaller in area, but their floras are of often richer and more diverse.

Mountain ranges provide unique conditions for the adaptation of plants and the isolation provided by mountains has led to the evolution of many bizarre species. Despite the severity of conditions experienced by alpine and high latitude plants, and the recognition of widespread convergence of form, there is nevetheless a large structural and functional diversity among them, especially those of high elevations. High mountains can fragment plant populations, leading to increasing rarity, or they can stimulate novel evolution. In addition mountains can act as corridors, enabling species to spread. Needless to say, many alpine plants are rare and in urgent need of conservation. It is this uniqueness that gives these plants their charm. The contrast between lowland and alpine plants is often manifested at the species or genus level, probably as a result of the intensity of selection pressures of extreme environments, and nowhere do we find steeper environmental gradients than on mountains.

Figure 6.64. Alpine scrub from Mount Ruapehu in New Zealand is rich in *Pimelea* and *Hebe* species.

Figure 6.65. The alpine zone at high latitudes starts at low altitudes. Here in the Cairngorms of Scotland it is at 1500 m. The vegetation on this plateau comprises mainly prostrate shrubs such as *Loiseleuria procumbens* (Ericaceae), which can withstand the combination of freezing temperatures and gale-force winds.

The delimitation of alpine zones on mountains is usually indicated by the natural tree line. The upper limit of tree growth is correlated with mean temperatures in warmest and coldest months, in addition to other factors such as exposure, avalanches and solifluction. This may be as high as 4000 m in the tropics, but varies considerably with region, climate and exposure, and at high latitudes it may even be at sea level. Then there is no effective distinction between alpine and high latitude plants. At these latitudes, relief becomes more important than altitude. A treeline may be absent altogether on dry mountains; for example, over much of the Mediterranean region it is lacking, and there is a zone of prickly 'hedgehog plants' at 1600–2000 m that may blend into a more typical alpine zone. Generally, where montane forests exist, the alpine zone encompasses low-stature vegetation above the natural tree line worldwide, and the zone between the closed upper montane forest and the uppermost limits of small individual trees is often termed 'subalpine'.

Figure 6.66. *Celmisia* (Asteraceae), an alpine plant from New Zealand is hairy and has a leaf rosette.

Environments at temperate and high latitudes are strongly seasonal. They have cool, short summers and cold long winters and, taken as a whole, are deficient in solar energy input. Temperature is a critical factor while water is often deficient or inaccessible. Water uptake may be severely impaired by low soil temperatures (0–5 °C). In Arctic regions physiological drought may extend throughout the summer, if roots are situated above permafrost. The ratio of above-ground biomass to below-ground biomass reflects the severity of the climate. Usually below-ground biomass is greater, as plants are generally of low stature and remain close to the ground, but there are exceptions; for example in many of the willows (*Salix* spp.) the stems are prostrate and extend considerable distances. Survival of both drought and freezing temperatures requires cell membranes that can tolerate dehydration. When plant tissues freeze, ice is first formed in gaps between cells, which draws water from protoplasts. A link between the ultrastructural and molecular basis for freezing tolerance and the evolution of dehydration tolerance of biomembranes has been suggested. Thus plant survival in hot and cold deserts may have common evolutionary roots.

Most arctic and alpine floras are dominated by chamaephytes (including cushion plants) and hemi-cryptophytes. With increased severity, cushion chamaephytes and mat hemi-cryptophytes increase, while tussock and rosette hemi-cryptophytes decrease. There is intense solar radiation at high elevations and latitudes, and plants in these areas have a greater photosynthetic capacity. Consequently the majority have some means of reducing transpiration, for example sclerophyllous leaves, sunken stomata, stomatal closure, succulence, and especially a wooly pubescence. Sun tracking is widespread, at least in the Arctic. *Soldanella* ssp. are able to open their flowers as they push through the melting snow.

In all tundras, perennials predominate, with very few annuals or biennials. Annuals usually require heat for development and seed-setting. The dominant forms are low shrubs and herbs, including

Figure 6.67. *Raoulia* here growing at near sea level. Some species have a white woolly pubescence and, as a result, have earned the nickname 'vegetable sheep'.

geophytes and cushion plants. Evergreen cushion plants and dwarf trailing shrubs, such as *Diapensia lapponica,* predominate on wind-swept ridges and plateaux. This is explained by the need for energy conservation. Cushion-form causes an increased resistance to CO_2, water vapour and heat fluxes. Herbs tend to occur on meadows with winter snow cover, although *Ranunculus glacialis* has been recorded at 2600 m. Upright shrubs such as *Saxifraga oppositifolia*, occur more on poorly drained sites or protected clefts. Arctic and alpine shrubs are generally found in less severe habitats than herbs but they may occur in polar deserts and at high elevations.

Figure 6.68. Hedgehog plants such as *Astragalus* shown here grow at sea level on exposed headlands in the Mediterranean region.

In the alpine regions of New Zealand and the more southern parts of the Andes, cushion plants occur which are unique. Genera such as *Azorella* (Apiaceae) are found in both regions and may reflect more ancient connections. There are many other families that also form cushions, some of them unexpected, for example Violaceae and Cruciferae. The most striking feature of these cushion plants (*Raoulia*, *Bolax*) is their large size (Figure 6.68). Many of the cushion plants of the Andes are actually cacti or bromeliads. In New Zealand, tussock grasses (*Chionophila*) are dominant in many areas. This genus has marcescent leaves around the pedestal, which is formed by a cylindrical mass of stems roots and leaf bases. This skirt buffers against diurnal extremes of temperature and humidity fluctuations.

Perhaps the simplest way to avoid winter conditions is either to survive the unfavourable period as seed (i.e. as annuals) or as cryptophytes. Cryptophytes winter underground whereas hemi-cryptophytes winter as a rosette or a bud at ground level. In plants that overwinter above ground (chamaephytes and phanerophytes), freezing avoidance may be achieved by several means. Mostly this involves protection of plant organs from freezing by physical insulation and/or by the prevention of ice formation by the production of certain chemical compounds or cryoprotectants. In the Norway Spruce (*Picea abies*), frost hardiness is characterised by a high proportion of unsaturated fatty acids in the membrane lipids and by a shift from

Figure 6.69. *Saxifraga oppositifolia* was a widespread and frequent component of the flora of north-western Europe during the dry cold glacial period of the Pleistocene because it could cope by growing up through wind-blown sand and silt.

photosynthetic starch formation to the production of sucrose and galactosides.

Most chamaephytes are evergreen and have lower photosynthetic and respiratory rates, which conserve energy because the plant does not need to replace leaves each year. Leaves may live for up to four years and act as storage organs for lipids and proteins. In winter, cuticular transpiration is high and drought resistance low in *Calluna vulgaris* and *Empetrum hermaphroditum*. This restricts them to sites of winter snow protection. In contrast *Loiseluria procumbens* grows on exposed sites, often with little snow cover because it has adventitious roots that enable it to absorb meltwater when the soil is frozen. In all the above types, growth occurs rapidly on the arrival of favourable conditions. Shoot growth is more correlated with soil temperature than with air temperature. Not all chamaephytes are evergreen, the dwarf willow, *Salix arctica*, is a notable exception.

6.8.3 Tropicalpine plants

Tropicalpine (tropical alpine) conditions are found in Hawaii, Borneo, New Guinea, East Africa and in northern South America where it is known as páramo. Above this zone is the Puna Zone or 'superpáramo', although various names are given, depending on the locality. The major influences are perpetually cool temperatures; frequent diurnal frosts throughout the year, little or no seasonality of temperature; some seasonality of precipitation; high UV and very variable levels of PAR in relation to diurnal cloud formation. Surface temperatures in páramo fluctuate from below freezing to 25–30 °C in the afternoon. Soil heaving by frost and needle ice discourages seedling establishment, whereas rocky areas retain heat, are buffered against extreme temperatures, and have higher species diversity. The fundamental difference between alpine regions of temperate latitudes and those of the tropics is that the onset of freezing conditions at high latitudes is gradual with the onset of the winter whereas in the tropics the environment is largely aseasonal. In tropicalpine areas it is 'winter every night and summer every day' with a high incidence of night-time radiation, frosts and high daytime insolation, especially in the morning. Often there is heavy cloud in the afternoon, especially in Borneo and New Guinea. In tropicalpine environments there is a higher above-ground biomass because of the perpetually frozen soil and nightly solifluction. The cold soil inhibits root growth. These profound differences have led to a very unique tropicalpine flora.

Tropicalpine plants fall into several categories, such as rosette plants, tussock grasses, acaulescent rosette plants, cushion plants and sclerophyllous shrubs (e.g. Hawaiian species of *Geranium*). Many of the sclerophyllous chamaephytes and cushion plants, as well as the annuals and tufted grasses are probably similar in their adaptations to those of other regions, although their phenology, being linked to aseasonality, undoubtedly differs. In tropicalpine regions, tussock grass is often dominant (e.g. *Festuca pilgeri* in Africa; *Deschampsia klossii* in New Guinea). Night frosts at the canopy level of tussock grasses on Mount Wilhelm occur on 88% of the nights at 4300 m.

Chief genera of rosette trees are *Espeletia*, *Dendrosenecio* and *Lobelia*. High-elevation lobelias are columnar, not arborescent. Columnar taxa include *Echium* in the Canary Islands and *Lupinus* in the Andes, while in Hawaii *Argyroxiphium* is characteristic. Growth is continuous (at least in *Argyroxiphium* spp.) and there is a slow accumulation of secondary xylem. Similar life forms can be seen in many other plants such as the 'pandani' (*Dracophyllum*) of Tasmania. Sometimes such plants are called 'megaherbs'. The woodiest species of *Lobelia* are found in mossy forest (e.g. *L. gibberoa* in East Africa). Most of these giant columnar species are, in reality, giant inflorescences. The giant senecios and lobelias of East Africa, the genera *Espeletia*, *Puya*, and *Lupinus* in the Andes and *Argyroxiphium* of Hawaii are subjected to temperatures of 10–12 °C during the day and −5 °C to −6 °C at night. The rosette leaves close at night forming a 'night bud' (nyctinasty) and this, together with the masses of marcescent leaves around the stem and a small reservoir of water within the rosette, provides insulation. (See also Section 6.9.4.)

The anatomy of these tropicalpine species shows characteristics of juvenilisation or paedomorphosis, and this has led workers such as Carlquist to hypothesise that the ancestors of these plants were herbaceous. *Espeletia*, like all these giant genera, has a wide succulent pith and a thick cortex or bark, and there is evidence of a water-storage function in the pith. Morning transpiration is compensated for by transfer of water from the pith, allowing early stomatal opening. Species with the largest relative water-storage capacity tend to be those that occur in the highest and coldest sites where the potential for physiological drought is the greatest. Therefore, height of giant rosette species usually increases with elevation. The inflorescence of *Lobelia telekii* contains about 3–5 litres of fluid. When air temperature is −6 °C the temperature of the fluid is +0.1 °C so there is thermal buffering. The fluid contains polysaccharides, mostly sucrose, which is a cryoprotectant. In giant Lobelia species, under normal freezing conditions, 80%–90% of the photosynthetic carbon gain is accumulated as sucrose, whereas, in non-freezing conditions, it accumulates as starch. Similarly in *Espeletia* and *Polylepis*, enhancement of osmotic potential and depression of the freezing point is related to soluble carbohydrate accumulation, so that metabolic processes may occur to −6 °C to −8 °C. Supercooling is beneficial under climatic conditions where only brief periods of mild nocturnal freezing occur, for example in páramo.

Shrubs, including the silverswords such as *Dubautia menziesii* and *Argyroxiphium sandwicense*, and the asteraceous *Tetramolopium humile* on the summit of Haleakala in the Hawaiian Islands, have a greater occurrence of vasicentric tracheids that offer subsidiary conductance and thus can help maintain water columns to leaves when the vessels embolise.

Leaf anatomy suggests adaptations for xeromorphy. Many, such as *Dubautia menziesii*, have thick and succulent leaves and some are clothed in dense hairs that also provide insulation and reduce transpiration, especially on young leaves. Leaf hairs do not lower UV absorption by much, unless extremely dense and felty but, in *Argyroxiphium sandwicense*, the highly reflective leaves are caused by trichomes that are flattened in their distal portions. Dense leaf hairs occur in many other species, including *Geranium tridens* from Hawaii. The leaves of *Espeletia* have areolar cavities (pockets) on the underside that may serve for water storage, whereas the leaves of *Argyroxiphium* have massive water-retaining gels in the inter-cellular spaces. These gels occur in other genera of the tribe Madiinae (*Blepharizonia*, *Hemizonia* and *Madia*). Old leaves fail to abscise, and they form a skirt around the stem, preventing freezing. In the higher-elevation *Argyroxiphium sandwicense* the trunks are constrained by frost, whereas in the lower-elevation *Argyroxiphium kauense* the trunks are taller.

6.9 Island floras

In the preceding sections we have given examples of form in the plant world and the astonishing diversity that has resulted from the interplay of plants with their environment through time. The focus has been on form, ecology and adaptation. In this final section we look directly at space and time and the relevance that these two interrelated aspects have for plant evolution. The diversity of islands and their degrees of isolation have given them a unique status as 'living laboratories' for the study of evolution, a fact that has been long known since the days of Alexander von Humboldt, Charles Darwin and Alfred Russel Wallace.

Islands have an importance for plant-life disproportionate to their geographical area. Although islands occupy only 3% of Earth's land area, the pooled number of species amounts to about 15% of all known plants. The importance of islands for biological research continues, although all too often such research is an attempt to minimise and perhaps occasionally to reverse the damage caused mostly by European exploration since the 16th century. The plight of so many island organisms, and indeed island ecosystems, now depends on an acceleration of research efforts to determine population numbers, genetic makeup and threats to their survival.

Many of the plants and animals from islands around the world have become extinct in the past few hundred years and many continue to teeter on the verge of extinction. Islands have a higher rate of natural extinctions, but human activities have accelerated that rate. In the past two hundred years the rates of extinction are 59%, 40%, and 79% for the Galapagos Islands, Hawaiian Islands, and the Juan Fernandez Islands, respectively. This is a damning indictment of how little we have cared for the other living organisms that we share the Earth with and depend upon, and how little we have understood the world in which they live. We now know that vulnerability is a major characteristic of island biota, a vulnerability which is the flip-side of a wonderful and unique diversity, which arose in ecosystems far removed from mainstream patterns of evolution on continents.

'Islands', from a biological perspective, include small offshore islands, continental islands and oceanic islands. It was Alfred Russel Wallace who first appreciated the differences between oceanic islands (mainly volcanic origin) and continental islands (fragments of continents) but there is every gradation from large continental islands such as Madagascar, New Guinea, New Zealand and New Caledonia to the most remote oceanic islands such as the Marquesas, the Tuamotou group or Easter Island in the South Pacific. Our description must also include islands in fresh-water lakes and islands of vegetation that are isolated either climatically or spatially on mountain tops.

The so-called 'islands in the sky' – the isolated mountains of south-eastern Arizona, the Sierra Madre ranges of Mexico and the tepuis of Guyana and Venezuela are good examples of such phenomena. Nunataks, glacial refugia and inselbergs are further examples of areas which display insular characteristics. Valleys, which are isolated from one another by mountain ranges, could also be regarded as 'islands'. The island syndrome is therefore a very broad and loose categorisation, in a sense, a fine-tuning of our interpretation of geographical and, to a lesser extent, ecological isolation. The latter almost always involves some measure of spatial isolation, and there are numerous examples, for example the distinctive flora of savanna termitaria, microsite species diversity among bryophytes and epiphytes, etc. In this account we will describe some of the unique features of true island plants and their adaptations.

Conifer and flowering plant generic diversity in the Pacific Islands

ordered by area emphasises the uniqueness of New Caledonia: total (endemic).

Solomon Islands	654 (3)
New Caledonia	655 (104)
Fiji Islands	476 (10)
New Hebrides	396 (0)
Samoa group	302 (1)
Society Islands	201 (2)
Tonga group	263 (0)
Cook Islands	126 (0)

6.9.1 The composition of floras on different islands

Islands near continents may have floras that are not very different from those of neighbouring mainlands. The flora of the British Isles is only a sample of the flora of continental Europe. A re-occurring feature of oceanic islands is that they are poor in families and genera, but rich in species, many of which are endemic, whereas old continental islands may have endemic families and even orders. New Caledonia, Fiji, Vanuatu, and the Solomons are part of Gondwana geologically, whereas the remainder of the Pacific islands such as Samoa are considered oceanic and, with the exception of New Caledonia, the land is relatively recent (Miocene for Hawaii and Melanesia; Pliocene for Polynesia). Distances across this huge region are great but endemism is high. In Hawaii alone, endemism is about 95%, while in New Caledonia overall endemism is 75% (in tree ferns it is about 90%). Fiji has about 66% endemism, including the relictual family Degeneriaceae. Many palms have a restricted distribution in the Pacific. There are 15 endemic genera in New Caledonia, and one in Fiji, Vanuatu and Samoa. A third group of palms is restricted to the Solomons, while the single palm genus in Hawaii has radiated into no less than

30 species. There are many unexplained anomalies; for example there are no epiphytic orchids in Hawaii. *Vaccinium* (Ericaceae) occurs in many high islands, but is absent from New Caledonia.

The flora of the Juan Fernandez Islands has developed in isolation, some 700 km west of central Chile, on three small islands of volcanic origin. The most ancient of these islands, Robinson Crusoe Island, is some 4 million years old and harbours plant communities with up to 70% species endemism. Of the plants found on the Galapagos Islands, 42% are endemic whereas 81% are endemic to New Zealand. Some 70% of the Hawaiian fern flora is endemic while the figure for flowering plants is 94%. The percentage of endemic taxa is more a reflection of the degree of isolation in time and space than the mode of origin of the island, but the actual situation is complicated by ecological factors.

Some species may become adapted to island conditions and, therefore, the present dispersibility of any given insular species cannot always be used as an estimate of dispersibility of its ancestral species in the past. Normal dispersal mechanisms may allow them to establish on other islands within an archipelago, even though distances between islands are often greater than to the mainland, for example *Galvezia speciosa* (Scrophulariaceae), *Crossosoma californicum* (Crossosomataceae), *Jepsonia malvifolia* (Saxifragaceae) and *Haplopappus canus* (Asteraceae) on the Channel Islands off the California coast. Their means of dispersal probably do not prevent them from getting back to the mainland, it is their changed ecological requirements that do. Sometimes species with broader tolerances will occasionally spread back to their ancestral home on the mainland, such as *Coreopsis gigantea* (Asteraceae) in California.

Most oceanic islands are volcanic and many may be too young, too small, i.e. ecologically too poor, and too remote to have autochthonous groups. The floras of the raised coral limestone islands and atolls of the Pacific are generally very impoverished. The combination of limestone and salty conditions suggests high selectivity. On the Iles Loyauté the forest is only about 15–25 m tall with a restricted diversity. In the lower strata there are few terrestrial ferns. In western Micronesia the dominant species on raised coral limestone are in the Moraceae (*Artocarpus*, *Ficus*), Myrtaceae, Sapotaceae, *Elaeocarpus* and *Hernandia*. Even a young tropical archipelago, the Galapagos Islands, which is relatively close to the continent is relatively poor ecologically. The Hawaiian flora, which is predominantly Indo-Malesian (Indo-Malayan: 40%; American: 12%) is exceptionally diverse for an oceanic archipelago.

The present Hawaiian islands originate at least from the Miocene, which is moderately young geologically, but other factors such as their remoteness, topography and size must have contributed to their greater floristic balance and richness. Also, in the early geological evolution of the Hawaiian islands, they may have been closer to rich floras of continental land masses than at present, and therefore have been more receptive to colonisation by normal species expansion.

In addition there has been substantial adaptive radiation within the islands. The alpine flora of the volcanic uplands has a higher percentage (91%) of endemic species compared with the archipelago as a whole (20%).

The more isolated these ecosystems are, the more fragile they are. Western Micronesia is influenced by the Asian monsoon. Otherwise, trade winds dominate the remainder of the Pacific Islands (from the southeast in the Southern Hemisphere, and from the northeast in the Northern Hemisphere). Most Islands have ample precipitation and, with few periods of drought, have a climate compatible with evergreen forest vegetation. On the leeward sides of high islands, rain shadow areas occur creating different climatic regimes (e.g. New Caledonia, Vanuatu and Hawaii). Cyclonic disturbances at the end of the warm season affect mainly the western Pacific, and may strongly affect forest dynamics. *Araucaria* spp., with their deep tap roots, are at an advantage. On Isles Loyauté, *Araucaria columnaris* grows on sublittoral cliffs in the prevailing wind. Its peculiar stem morphology may help it survive strong winds. Post-cyclone forests occur in the Solomons and these depauperate forests comprise mainly *Campnosperma brevipetiolatum*, *Endospermum medullosum*, *Terminalia calamansanai*.

There is a tremendous variation in these montane forests throughout the vast Pacific region, but generally with a decreasing richness the further east one travels. Each island has its own peculiarities, particularly New Caledonia and Hawaii. In the Solomon Islands the montane forests are especially rich in Myrtaceae, and there are four epiphytic species of *Rhododendron*. In Fiji, the tallest forests have two species of *Podocarpus*. The Podocarpaceae are represented throughout Melanesia and in Tonga, but only in Fiji and in New Caledonia do they play an important role. *Agathis macrophylla* is the only araucarioid in Fiji, Vanuatu and Santa Cruz. In Fiji and Vanuatu there are mixed communities of hard and softwoods (Araucariaceae and Podcarpaceae) from sea level to 1200 m.

The floras of older continental islands such as Tasmania, New Caledonia, New Zealand and New Guinea have a more balanced (harmonic) composition, including ancient families such as the Araucariaceae, Austrobaileyaceae and Nothofagaceae, which are not only notable for their poor dispersability and slow growth, but also for their inability to establish easily. With the exception of New Guinea, these islands, together with Fiji, have been isolated since the Cretaceous, and are part of an ancient island arc that connected South America to Antarctica during those times. New Guinea, which is currently isolated from Australia by a shallow sea, was formed in the Miocene by the collision of the Australian plate with island arcs of an ancient Melanesian foreland in the west Pacific region. Although geologically quite young, the flora of New Guinea is a mix of relict taxa from Australia and those which have dispersed from the Indo-Malaysian and Philippine regions subsequent to the Miocene collision.

Figure 6.70. *Dracaena Draco*, which is now confined to the Canary Islands, is a remnant of the Tertiary laurel forests once widespread in North Africa.

6.9.2 Relict floras and extreme disjunction

Many island plants would appear to be relicts, survivors of ancient plant groups that were once much more widespread and which have disappeared from continental source areas. In the Southern Hemisphere, *Lactoris fernandeziana* (Lactoridaceae) is found nowhere else outside the Juan Fernandez group and, as with *Amborella* (Amborellaceae) on New Caledonia and *Degeneria* (Degeneriaceae) on Fiji, it may be the sole survivor of a unique family of plants once widespread in the Southern Hemisphere. *Medusagyne* (Medusagynaceae) and *Lodoicea* (Arecaceae) on the Seychelles in the Indian Ocean, and several small vesselless families of the 'magnolioid' complex (Winteraceae, Himantandraceae, Eupomatiaceae and Austrobaileyaceae) are other prime examples. In the Canary Islands (Macaronesia), the laurel forests are relicts of an extensive flora that extended over much of North Africa and Europe until the late Tertiary (Miocene and Pliocene) but has now all but disappeared. Conifers are numerous in the forests of New Caledonia and Fiji but are not found further east than Tonga. In eastern Polynesia Hawaiian and Austral elements occur; for example, *Astelia* reaches many islands.

New Caledonia is the most distinct of the Pacific islands, and has a unique flora that is totally distinct from other forests of Melanesia. The 75% endemism of the forest flora is the result of the long isolation and great age of the island, as well as the general prevalence of ultrabasic rocks. These factors have influenced the montane flora of New Caledonia no less than the lowland vegetation. The commonest tree ferns on ultrabasic soils are *Dicksonia* spp., whereas *Cyathea* spp. are on schists. Endemic families of the New Caledonian montane forests include Paracryphiaceae (*Paracryphia*), and Strasburgeriaceae (*Strasburgeria*). Unique endemic genera include *Apiopetalum* (Araliaceae), and *Canacomyrica* (Myricaceae). Among the conifers there are five endemic species of *Agathis* and twelve endemic species of *Araucaria*, plus the endemic genera *Neocallitropsis* and *Libocedrus* (Cupressaceae), and *Austrotaxus* (Taxaceae). The latter is an isolated member of the essentially Northern Hemisphere yew family. It reaches heights of about 25 m, but it is rather rare, growing at relatively low elevations from 600 to 800 m.

If we look at the extinction and localisation of floras on large land masses such as North America or Australia we see that it is precisely older elements such as conifers and woody flowering plants that are also restricted to narrow and declining ranges, for example *Agathis*, *Araucaria* (Araucariaceae), *Nothofagus* (Fagaceae), *Fremontodendron* (Sterculiaceae), *Franklinia* (Theaceae), and *Lyonothamnus* (Rosaceae). *Lyonothamnus*, now confined to offshore islands of California, is very closely related to the genus *Vauquelinia* that has about nine species extant in the Sierra Madre ranges of Mexico where one species extends to the Big Bend region of Texas. The recent discovery of a new genus (*Wollemia*) of the Araucariaceae in a relatively well-explored part of Australia clearly shows how radically different the floras of the past must have been.

Groups that show extreme disjunction, on the Mascarene Islands and Hawaii include:

Astelia, *Dianella* (Liliaceae s.l.), *Cordyline* (Lomandraceae), *Pipturus* (Urticaceae), *Peperomia* (Peperomiaceae), *Diospyros* (Ebenaceae), *Myrsine* (Myrsinaceae), *Canthium*, *Psychotria* (Rubiaceae), *Pittosporum* (Pittosporaceae), *Elaeocarpus* (Elaeocarpaceae), *Zanthoxylum* (Rutaceae).

6.9.3 Adaptive radiation

Morphological and molecular evidence suggests that rapid evolution has been a characteristic of many island taxa. The species involved need not be from the most advanced or labile groups. The sources for re-colonisation could also be older floristic associations; for example the two species of the genus *Musschia* (Campanulaceae) on Madeira, although relatively young in absolute terms, have probably evolved from very old elements within the family that were part of the Tertiary flora of North Africa and Europe. They thus display features of both relictualism and modernity.

Examples of relictualism and modernity:

Nesocodon on Mauritius; *Heterochaenia* on Reunion; *Argyroxiphium* (Asteraceae) on Hawaii; *Aeonium* (Crassulaceae), *Echium* (Boraginaceae), *Isoplexis* (Scrophulariaceae), *Sideritis* (Lamiaceae), *Argyranthemum* and *Sonchus* (Asteraceae) in Macaronesia.

Two of the best examples of adaptive radiation on islands are *Aeschynanthus* and *Cyrtandra* (Gesneriaceae). *Aeschynanthus* is widely distributed from the Himalayas, through South East Asia and Indonesia, to New Guinea and has radiated in forest and mountain habitats on many of the islands to produce a welter of species, most of which are epiphytic. This has been paralleled in other genera of the Gesneriaceae in the Old World such as *Agalmyla*, but the genus *Cyrtandra* probably holds the all-time record for adaptive radiation. It is

Figure 6.71. *Aeschynanthus.*

Figure 6.72. *Hebe.*

Hyper-pachycaul forms are found in the following genera (on high mountains):

Lupinus (Papilionaceae),
Espeletia, Senecio
Argyroxiphium (Asteraceae),
Lobelia (Lobeliaceae),
Puya (Bromeliaceae),
Yucca (Agavaceae),
Aloe (Aloeaceae),
Eryngium (Apiaceae),
Echium (Boraginaceae).

distributed from the Nicobar Islands to the Marquesas and is the most widely distributed genus of its family. Across its range it has speciated greatly on many islands such as Borneo and New Guinea, but it is on Hawaii that it has undergone the most explosive speciation. *Metrosideros* (Myrtaceae) is another genus with species that are found on all the major islands of the Pacific, while *Hebe* (Scrophulariaceae), in New Zealand is another classical example.

Many island plants have evolved in isolation from a single invasion event. The gene pool of the original coloniser was only a small fraction of the total gene pool of the species. The effect of selection on this limited gene pool is known as the founder effect and is the mechanism postulated for the evolution of bizarre giant herbs of many islands. We expect rapid evolution on islands to be aided by lack of predators and competitors, a wide spectrum of ecological opportunities and isolating mechanisms, but the ability to evolve also depends on the genetic makeup of each particular plant group. The effects of small mutations on developmental programmes, and of pleiotropic genes, can cause massive shifts in morphological and ecological parameters. Polynesian lowland forests contain woody groups that are herbaceous on continental mainlands, for example Lobeliaceae, Asteraceae, Amaranthaceae. Not all plant families have developed arborescence. For example the genus *Viola* (Violaceae) on Hawaii is herbaceous or woody but does not form 'trees'. This is in marked contrast to the lobelioids (Lobeliaceae) and silversword alliance (Asteraceae) that have radiated into diverse arborescent forms.

6.9.4 Characteristics of island plants

Convergent evolution is displayed most clearly among island plants. For example, different lineages of the normally herbaceous or shrubby Asteraceae have diversified and become morphologically similar on the Hawaiian Islands (*Argyroxiphium* tribe Heliantheae) and on the Juan Fernandez Islands (*Robinsonia*, in the Senecioneae). Although the history of each island is unique, and no two islands are identical, there are sufficient similarities among their plants to recognise a 'syndrome' and to consider them in much the same way as other ecological groups such as desert plants, marshland plants.

'Insular' woodiness, the pachycaul habit, is also characteristic of continental areas, for example tropicalpine zones of South America and Africa. Generally such similar plants from continental regions and 'islands' have been ignored, or their form explained as being the result of their own unique ecological relationships, rather than the retention of an ancient form as championed by Corner and Mabberley. The trend is towards increased stature, principally through an increase in woodiness, either from herbs to shrubs, herbs or shrubs to pachycaul arborescent shrubs, or shrubs to true trees. There is an ecological advantage in becoming tree-like and long-lived, since the competitive ability for the long-term occupation of a particular site increases considerably. Hyper-pachycaul forms have evolved on high mountains (see Section 6.8.3).

The reproductive structures of island plants often show evidence of evolutionary change, either as a reduction in conspicuousness, or as a shift from insect- to wind-pollination; for example the ancestor of the wind-pollinated Kerguelen Cabbage (*Pringlea antiscorbutica*: Brassicaceae) almost certainly was insect-pollinated. Anemophilous species that are frequent on islands include *Rhetinodendron* (Asteraceae); *Plantago* (Plantaginaceae); and *Coprosma* (Rubiaceae). There is paucity of conspicuous flowers and a greater frequency of white flowers on many islands such as the Galapagos and in New Zealand. Hawaiian forests have an abundance of species with small green or whitish flowers that are also poor in scent. Self-compatibility and the ability to hybridise would be advantageous for establishment on islands. Evidence of hybridisation can be found in numerous genera. Having lost self-incompatibility, the predominant mechanism preventing self-pollination is dioecism, which is particularly common in some island floras.

Figure 6.73. *Echium*, one of a number of large shrubby echiums in Macaronesia.

Flowers on oceanic islands are frequently more promiscuous in terms of their pollinators, for example *Azorina* (Campanulaceae) of the Azores. In other instances there may have been a shift from bird-pollination to other animals. Although purely conjectural, the Canary Islands may once have been populated by sunbirds of the family Nectariniidae, which are confined nowadays to regions further south in Africa. Several Canarian genera of plants may have been pollinated in the past by these birds, for example *Canarina canariensis* (Campanulaceae) and *Lotus berthelotii* (Papilionaceae). In Madeira the genus *Musschia* (Campanulaceae) has copious nectar, unusual coloration and is frequently visited by lacertid lizards. It too, may have been regularly pollinated by birds in the past.

Endangered island genera

Delissea,
Brighamia,
Rollandia,
Cyanea,
Clermontia (Lobeliaceae)
Hesperomannia,
Remya (Asteraceae)
Hibiscadelphus,
Kokia (Malvaceae) *Trochetiopsis* (Sterculiaceae),
Pteralyxia (Apocynaceae).

There is a marked increase in the size of fruits among forest trees on islands, while many also display massed flowers, especially insular species. More conventional trees such as the relict Channel Islands Ironwood (*Lyonothamnus floribundus*: Rosaceae) also have massed flowers in large inflorescences.

The dispersal ability of island plants often alters through time and may be lost altogether, presumably because it is selectively advantageous in an out-crossing species with tiny populations; for example several species of the genus *Campanula* (*C. incurva*, *C. sartorii*) in the Aegean region of Greece have independently lost capsular dehiscence or at best only dehisce tardily. Such ecological shifts may be irreversible. In seeds and fruits that have modified dispersal mechanisms there often appear to be malformations that do not occur in mainland relatives, and which have no selective advantage, for example in *Yunquea*, *Dendroseris*, and *Bidens* (Asteraceae).

Many endemic genera that have apparently irreversible adaptations to highly specialised locations are now severely endangered owing to competition from alien plants such as *Psidium* (Myrtaceae) or *Schinus* (Anacardiaceae). Their ability to compete is weak and they seem to be unable to re-establish after disturbance. The cause appears to be genetic. In cultivation many of them appear to have a 'death wish' and prove difficult to maintain for any length of time. In the wild a paucity of individuals also plays a key role in loss of genetic variability. When loss of variability occurs, potential for change is halted. Some of these vanishing genera that show this unfortunate tendency include *Nesocodon*, *Heterochaenia* and *Musschia* (Campanulaceae); *Delissea*, *Brighamia*, *Rollandia*, *Cyanea* and *Clermontia* (Lobeliaceae); *Hesperomannia* and *Romya* (Asteraceae); *Hibiscadelphus* and *Kokia* (Malvaceae); *Trochetiopsis* (Sterculiaceae); and *Pteralyxia* (Apocynaceae).

Further reading for Chapter 6

Arber, A. R. *Water Plants: A Study of Aquatic Angiosperms* (Cambridge: Cambridge University Press, 1920).

Benzing, D. H. *Vascular Epiphytes: General Biology and Related Biota* (Cambridge: Cambridge University Press, 1990).

Bramwell, D. (ed.) *Plants and Islands* (London: Academic Press, 1979).

Breckle, S.-W. *Walter's Vegetation of the Earth* (Berlin: Springer, 2002).

Burrows, C. J. *Processes of Vegetation Change* (London: Hyman, 1990).

Carlquist, S. *Island Biology* (New York: Columbia University Press, 1974).

Chapman, V. J. *Ecosystems of the World, vol. 1. Wet Coastal Ecosystems* (Amsterdam: Elsevier Scientific, 1977).

Crawford, R. M. M. (ed.) *Plant Life in Aquatic and Amphibious Habitats* (Oxford: Blackwell Scientific Publications, 1987).

Grant, P. R. (ed.) *Evolution on Islands* (Oxford: Oxford University Press, 1998).

Kuijt, J. *The Biology of Parasitic Flowering Plants* (Berkeley: University of California Press, 1969).

Lloyd, F. E. *The Carnivorous Plants* (New York: Dover Publications Inc., 1942, reissued 1976).

McArthur, R. H. and Wilson, E. O. *The Theory of Island Biogeography* (New Jersey: Princeton University Press, 1967).

Mühlberg, H. *The Complete Guide to Water Plants* (Yorkshire: EP Publishing Limited, 1980).

Chapter 7

The fruits of the Earth

But lo! men have become the tools of their tools. The man who independently plucked the fruits when he was hungry is become a farmer; and he stood under a tree for shelter, a housekeeper.

Henry David Thoreau, 1817–1862

7.1 Exploiting plants

Human beings use thousands of species of plants for food, either as food or flavourings, for fuel, for construction materials and as sources of chemicals (oils, resins, gums, dyes, medicines and poisons). Almost all our calories and protein come either directly from plants or indirectly from plants used as food for our domesticated animals (the remainder comes from algae and fungi). All parts of plants have been directly exploited. Food has been obtained from the root (root-tubers, tap-roots), stem (tubers, rhizomes, and canes), leaf, flower (nectar and pollen in honey), seed and fruit. Wood, timber, fibres and other materials such as resins and latex have been obtained from roots, stems and leaves.

Humanity has always exploited plants but perhaps only for the past 10 000 years or 1% of human history have we cultivated them. Almost uniquely, human kind is a gardener, a cultivator. The first gardeners, in hunter–gatherer societies, were likely to have been women. They selected favoured plants, helped their cultivation and, in doing so, unconsciously changed the plants. The first cultivation of plants may have occurred in the Mesolithic period some time after 15 000 years ago. It was a time when the climate fluctuated rapidly. Human populations turned increasingly to plants to provide an assured supply of food. Many of the plants chosen were those growing in dense stands at the margins of human habitations. By about 10 000 years ago it was the availability of domesticated plants and their cultivation that enabled civilisation of human kind, the living in permanent settlements tending the surrounding gardens and fields. The availability of domesticated crops and a settled life-style permitted populations to grow, continuing the pressure to continue to rely on crops.

Some plants are still exploited directly from nature and are unchanged from wild plants but many of our chosen plants have been radically changed by their relationship with us. Once chosen they came under intense evolutionary pressure; selected for favourable traits they evolved rapidly. Early selection of improved varieties may have been accidental but it soon became more conscious as particularly favoured plants were tended and propagated. Some seed or roots from a particularly palatable plant or one particularly easy to harvest were saved and sown to provide more. Generation after generation of this kind of selection transformed the plant. A few like maize (*Zea mays*) are so transformed that they are quite unlike any possible wild ancestor. Most crop plants could now not survive without the hand of humanity to cultivate them. In addition their diversity has been exploited and their genetic pool fragmented into thousands of local cultivars or land races.

Only a tiny handful of plant species have been domesticated. There are about 2000–3000 species of cultivated plants, but less than 100 important ones, from a possible 250 000 ($<0.04\%$). Others are collected from natural populations especially as a source of timber or grazing. However, of the thousands of species utilised by humanity only a relative few have entered world commerce. Fewer than 20 species provide most of the world's food and just three, the main cereal crops of wheat, rice and maize, account for about 60% of our calorie intake and directly more than half of our protein. Along with potatoes they dominate world agricultural production.

Why have these species become so important? Primitive human cultures are much more sophisticated and exploit a much wider range of wild species. Obviously the chosen crops had some features that favoured them. However, there must have been a large element of chance in their selection and it is startling to think we might have ended up with a very different array of crops.

Crop plants share many features that differentiate them from their wild ancestors. Increased yield has been achieved in some cases by selecting a more favourable partition towards the utilised part of the plant; the amount and size of grain produced relative to the foliage in cereals for example. In this way smaller plants have been favoured. A whole set of changes has improved the quality of the crop especially as a food; reduced spines and toxic constituents, increased sugar or starch content, increased attractiveness. The length and strength or flexibility of fibres in fibre-plants has been improved. Important adaptations have been related to agricultural practice. An inflorescence that does not break apart and so allows easier harvesting is one example. One of the most important changes in this century has been the selection of varieties that are able to make use of the high levels of fertilisers. Cultivars have become adapted to different climatic regimes and different photoperiods (that is, day lengths in different latitudes).

Crop plants have faced less competition, while irrigation and the the application of fertilisers have minimised stress. A rapid or annual life cycle has been selected. Seeds and tubers lacking dormancy and with rapid and uniform germination, or clonal reproduction at the expense of sexual reproduction have been selected. Synchronous development and ripening has been selected to enable more efficient harvesting.

Plants are a source of food and pleasure and material for construction. The same species can provide all three. Many crop species were initially domesticated for one purpose and then a different part of the plant was exploited. The brassicas are an outstanding example of a kind of plant in which almost all parts of the plant have been exploited in one way or another. *Brassica oleracea*, *B. campestris*, and *B. napus* are closely related species that exist as an astonishing range of crops.

Table 7.1 Brassicaceae cultivated for food

Cabbage
Cauliflower
Kales
Mustards
Chinese brassicas
wong nga baak
baak choi
hoi sum, etc.
Broccolis
Turnips
Swedes
Kohlrabi
Sarson
Oilseed rapes

Two particular kinds of plants, the bamboos and the palms, are the supreme champions for the number of different uses they have. There are supposed to be over 1000 uses of bamboo. There are more than 40 different genera of bamboo and hundreds of species. Many grow to great heights; up to 37 m has been recorded even though bamboo is an herbaceous species. There is a branched rhizome from which the culms expand very rapidly like the sections of a telescope; growth of nearly 1 m per day has been recorded. So important is bamboo that it holds a central place in Chinese and Japanese art and mythology. The first monograph on any plant group was *ChubPhu* written by Tai Khai-Chih about AD 460. Bamboo is said to combine the strength of steel with great lightness. The strength and lightness comes from the hollow tubular structure with fibres around its margin. The hollowness has been utilised as pipes even for piping natural gas. The elasticity of the fibres is a great advantage in regions of hurricanes. The exterior has a thick waxy and silicon-impregnated coat that is highly resistant to the weather and to pests and diseases.

Palms do not just provide coconuts. There are nearly 3000 species of palm. They provide drink, seed, fruit, timber, and fibre. Even just from the coconut (*Cocos nucifera*) alone there is drink, a food and a fibre. The borassus palm (*Borassus flabellifer*) (Figure 7.1) has 801 uses according to a Tamil song. Palm wine is produced by tapping the immature male inflorescence in several other species including *Elaeis guineensis* (Figure 7.2) and *Nypa fruiticans*, the mangrove plam.

Date palms, *Phoenix dactylifera*, provide large quantities of fruit with yields often in excess of 30 kg per tree. Favoured trees have been propagated from vegetative suckers for millennia. *Phoenix dactylifera* is dioecious and wind-pollinated but it was early understood that the amount of fruit set could be enhanced by pollination by hand. Today

Figure 7.1. Borassus palm (*Borassus flabellifer*) grown on the margin of paddy fields: it provides timber, leaves for thatch and making baskets and mats, fibres for paper, seedlings are ground for flour, fruits roasted to eat and the inflorescence tapped to make palm wine.

in date palm plantations a ratio of 1 to 25 or less male to female trees is maintained to provide pollen. An interesting phenomenon is that the male pollen directly affects the quality of the developing maternal fruit tissues (metaxenia). The betel palm, *Areca catechu*, provides betel nuts that are scraped and chewed along with betel leaves (*Piper chavica*, *P. betel*) and also, sometimes, nutmeg, cardamom, and lime, as a panacea and stimulant. The lime helps salivation and together the lime and saliva encourage the release of the stimulant arecoline that increases respiration. The leaf stains the mouth deep red. Chewed betel is the stimulant of choice for millions of people from India through to the Pacific Islands.

Figure 7.2. Oil-palm (*Elaeis guineensis*) plantation.

Oil-palm, *Elaeis guineensis*, is one of the most productive of all crops. Bunches of fruits weigh about 25 kg and the fleshy fruit has up to 70% oil content. Palm oil is used in wide variety of foods and in the manufacture of margarines, soap and cosmetics. The centre of origin of oil palm is in the wetter parts of West Africa but there is now a very substantial plantation-based industry established in South East Asia. Palm oil has a markedly different fatty-acid spectrum from other vegetable oils; 44% palmitic acid (a C16 saturated fatty acid) but with a roughly equal content of unsaturated fatty acids. Other oil-palms are exploited on a more limited scale, such as the babassu palm *Orbignya phalerata* in Brazil.

The stem of the palm is utilised in several ways. The sago palm, *Metroxylon sagu*, and moriche, *Mauritia flexuosa*, provide starch from their pith. Palm hearts, used in oriental cuisine, are the centre of the apical buds.

Palm leaves are used for thatch and weaving. Rattan palms are climbing palms that provide the long flexible stems extensively used in furniture. Many species also provide fibre or timber from the trunk, while fuel comes from all parts of the plant. Palms are also extraordinarily beautiful plants, widely planted for their ornamental value. Many tropical gardens are adorned with stately avenues of palms such as the royal palm.

7.2 Plants for food

Some families have provided a large proportion of species exploited for food by humankind. In the forefront are the grasses (Poaceae) that provide the world with most of its carbohydrate either directly from the grain of the cereal crops (wheat, rice, maize, barley, oats, sorghum and millet) or as feed for animals. Cereals occupy over 70% of the world's croplands. Other important families are the Fabaceae (legumes and beans of all sorts), the Brassicaceae (leaf, root and oil-seed crops) and the Solanaceae (fruits and the potato). Human civilisation and cereal cultivation go hand in hand. The Russian botanist Vavilov proposed eight 'hearth' areas where most plants were domesticated but this may be an oversimplification. Certainly there are three major areas for the domestication of plants, each associated especially with the domestication of one of the three major cereals, as well as a range of other crops:

- the Near East, the so-called fertile crescent running from Egypt to Mesopotamia (wheat)
- Central America (maize)
- Southern China and South East Asia (rice)

There are some shared features of these cradles of domesticated plants. They are regions with well-marked wet and dry seasons. Plants in these areas have to store nutrients either in dormant roots and tubers or in seeds for survival through the dry season. This was the source exploited by Neolithic people.

The folklore around particular crops is extensive. Hindus believe that the Lord Vishnu caused the Earth to give birth to rice, and the God Indra taught the people how to raise it. In China there is a proverb: 'the precious things are not pearls and jade but the five grains'. Rice is the foremost of these. In Shinto belief, the Emperor of Japan is the living embodiment of Ninigo-no-mikoto, the god of the ripened rice plant.

Table 7.2 Main cultivated cereals listed in order of area cultivated

Wheat *Triticum*
Rice (Paddy) *Oryza*
Maize *Zea*
Barley *Hordeum*
Sorghum *Sorghum*
Millet (*Pennisetum*, *Eleusine*)
Oats *Avena*
Rye *Secale*
Triticale
Fonio (Acha) *Digitaria*
Canary Seed *Phalaris*

7.2.1 Cereal plants

The earliest archaeological evidence for the domestication of a cereal is of wheat. Charred remains of domesticated einkorn and emmer wheat, and a little later of barley, have been identified from the remains of Neolithic villages of about 11 000 years ago in the area of western Turkey, Iraq, Syria and Israel. Two major changes occurred in the domestication of wheat from its wild ancestors (Figure 7.3):

- a non-brittle rachis permitting the harvesting of whole wheat ears
- free-threshing grain permitting the separation of the grain from its surrounding chaff (glumes, lemma and palea)

Subsequent changes to wheat cultivars occurred as Neolithic agriculture spread. The spread of Neolithic agriculture into northwestern

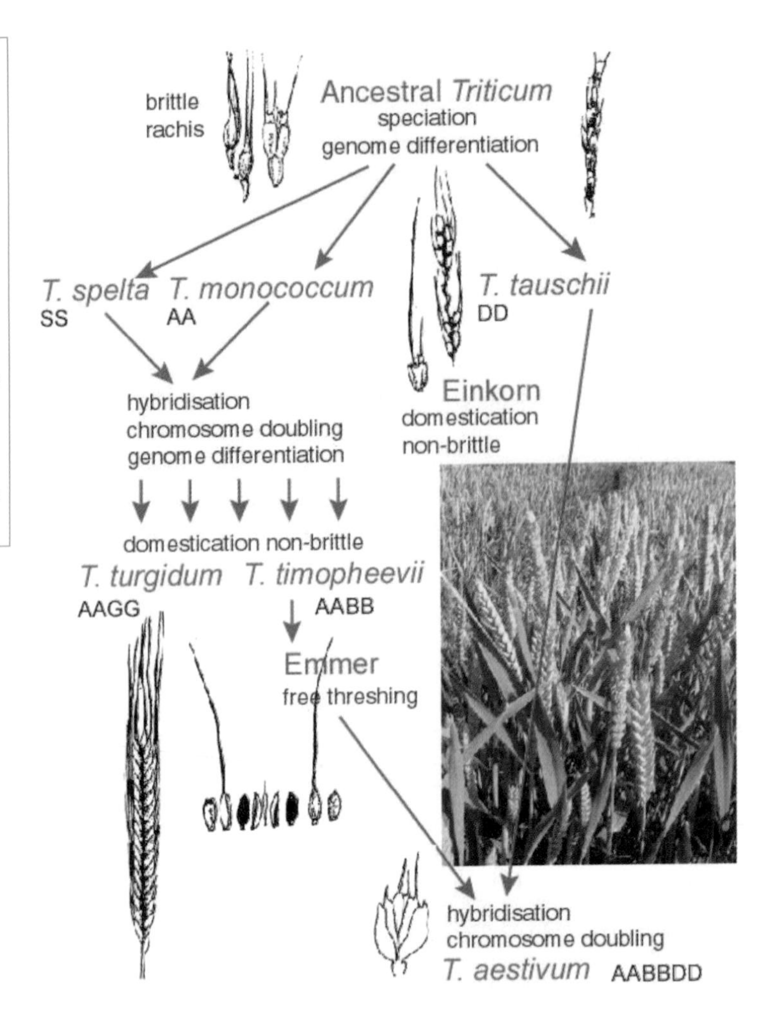

Figure 7.3. The evolutionary history of domesticated wheat: it involves hybridisation between cultivated species, chromosome doubling and selection. Today four related species of wheat are cultivated but there are thousands of cultivars. *Triticum monococcum*, closest to the ancestral domesticate, and *T. timopheevii* have a very minor importance. A variety of *T. turgidum* (var. *durum*) is widely grown in drier areas and provides macaroni or semolina flour. It is *T. aestivum*, the bread wheat, that is cultivated worldwide on a vast scale.

Europe has been mapped, reaching Britain 5000 years ago. Cultivars arose that were adapted to different daylength (or were day-neutral), seasonality, rainfall and temperature and different soils.

The first domestication of rice occurred a little later than wheat. Pottery shards bearing the imprint of both grains and husks of *Oryza sativa* were discovered at Non Nok Tha in the Korat area of Thailand dating from 6000 years ago. The spread of rice from its proposed area of domestication in northern Thailand, Myanmar (Burma), Assam or southwestern China northwards into China depended to some extent on changes in agricultural practice. Puddling of soils extended soil moisture availability. Transplanting of seedlings 1–6 weeks old in paddies gave rice plants a start over other plants and increased yield. Elsewhere rice remained an upland dryland crop. Only later was wetland cultivation spread by the migration of peoples to other areas in southeastern Asia. Wetland cultivation is so extensive now that methane-producing bacteria in ricefields are suspected of putting 115 Mt of methane, equivalent to all the world's natural swamplands, into the atmosphere each year adding to global warming.

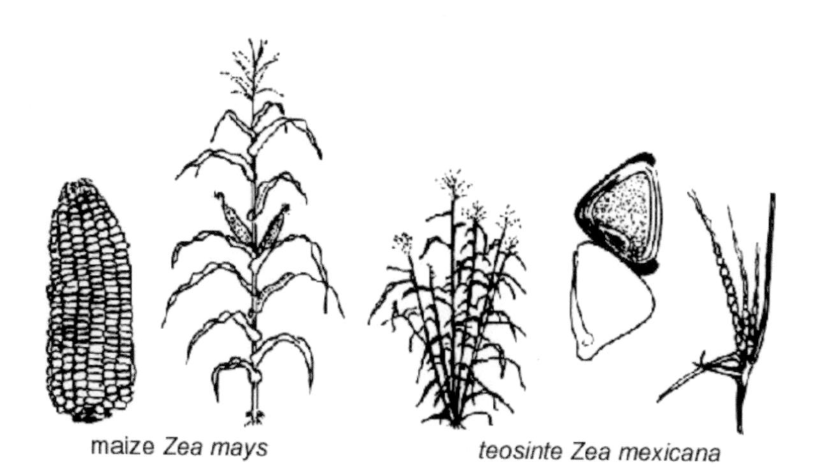

Figure 7.4. Cultivated maize and its wild progenitor teosinte. Teosinte has a series of lateral branches each with a terminal male tassel and lateral female spikes, compared with maize which has a strong main stem with a terminal male tassel and lateral female cobs. In teosinte the female spike is slender with 6–12 grains arranged alternately. Each grain is surrounded by a hard case. In maize, in the cob, there are several rows of grain each in a shallow soft cupule. However, teosinte and maize hybridise easily and produce an intermediate hybrid plant.

It was the introduction of new cultivars of rice that was one of the key aspects of the so-called green revolution of the 1960s. The new varieties had a shorter stature with a greater allocation of energy to the grain rather than leaf, and higher responsiveness to nitrogen-based fertilisers. The new varieties (one particularly successful one was called IR8), had the potential for much greater yield and were day-neutral, so they could be grown at any latitude. At the same time new varieties of wheat enabled the enhanced cultivation of wheat in drier areas such as Pakistan and Mexico.

However, there were negative consequences. Some of the new varieties were disliked because of aspects that had not been identified in the breeding programme, for example IR8 was disliked for its cooking quality. Some new wheat varieties were red rather than amber in colour. Perhaps, more importantly, there was a loss of genetic diversity as long-cultivated land races were abandoned. The new varieties were actually part of a package that included new methods of cultivation. Although yields were greater so were inputs, especially in fertiliser.

The early history of maize (*Zea mays*) is obscure. It is related to teosinte *Zea mexicana*, a weedy annual. Teosinte looks, at least superficially, very different, so different that for many years it was placed in its own genus called *Euchlaena*.

Maize and teosinte are diclinous, having separate male and female inflorescences on the same plant. Maize and teosinte differ especially in the form of the female inflorescence (Figure 7.4). The origin of maize may have been by transmogrification of the corn so that each lateral branch was shortened and the terminal male tassel feminised. In maize, like rice and wheat, there has been very extensive diversification of cultivars. Three major types of maize are grown; grain or field corn, sweetcorn and popcorn. Grain or field corn is by far the main kind; 70% is fed to animals and more used as silage. It is classified into four main types; Dent, Flint, Flour and Waxy, which differ in the distribution and type of starch in the kernel.

Table 7.3 Maize varieties differ in the composition of their endosperm

Field	Dent	Hard starch at the sides and a soft type in the centre of kernel, but shrinking at apex on drying to produce a dent (starch = 30% amylose, 70% amylopectin)
	Flint	Hard starch layer entirely surrounding the outer part of the kernel (30% amylose, 70% amylopectin
	Flour	Very thin layer of hard starch and almost entirely soft starch (30% amylose, 70% amylopectin)
	Waxy	Starch consists almost entirely of amylopectin
Sweetcorn		Contains a high proportion of sucrose in the kernel
Popcorn		High proportion of hard starch, pops as water in the soft interior expands

Nearly all the maize grown in the USA and Canada is hybrid corn produced by crossing inbred lines. Crosses are ensured by detasseling (i.e. emasculating one line) or by the manipulation of male sterility genes. The utilisation of a particular sterility genotype exposed maize to attack by the mildew *Helmonthosporum maydis*.

Within the regions once covered by savanna grasslands a range of other cereals are grown; sorghum and various kinds of millets. Sorghum is the staple food for 500 million people in 30 countries in Africa, India and China. It matures in as little as 75 days and is highly photosynthetically efficient. It can grow in both temperate and tropical conditions but is especially favoured in tropical conditions where maize cultivation is marginal because of drought, salt and waterlogging. It is utilised in various ways; boiled like rice, cracked like oats for porridge, malted like barley for beer, baked into flatbreads, popped for snacks, or eaten green like sweetcorn. However, it has a high tannin content that depresses nutrient absorption and a large quantity of protein content is the poorly absorbed prolamine. There are several different types of *Sorghum arundinaceum* including durra (the main grain in Africa), milo (the main grain in Central America), kaolang (grown in China) and feterita, which has large red-yellow or white grains.

Pearl millet (*Pennisetum glaucum*) has a relatively low yield but is more adapted to heat and drought than maize, and it is more nutritious than sorghum. Steam cooked, it is known as couscous, but it is also used to make breads, fermented foods and porridges. Finger millet (*Eleusine coracana*) is a useful grain because it is rich in the amino acid methionine lacking in the other tropical staples such as cassava and plantain. It is a demanding crop requiring intensive cultivation at weeding and harvesting stages. Fonio (Acha or white fonio) *Digitaria exilis* and *Digitaria iburua* (black fonio) are two more kinds of millet grown in the West African savanna region. Fonio is favoured because of its taste and is rich in methionine and cysteine, normally deficient in grains. It matures very rapidly, in as little as 6–8 weeks after sowing.

Grasslands are also very important as pasture for domesticated animals, especially cattle and sheep, and are also cultivated as fodder crops or for silage. Some pure grasslands of maize are cultivated but mixed grasslands, generally dominated by *Festuca* (Fescues) or *Lolium* (perennial ryegrass), are the most important sources of fodder. Usually they are improved by mixture with legumes, like alfalfa and clover, which, because of their nitrogen fixing-abilities, produce a fodder or silage with higher protein content. Legumes are sometimes sown separately for fodder and various kinds of leafy vegetables (beet, swedes, carrots) are also sometimes used. In drier areas grassy sorghums, including Sudan grass (*Sorghum sudanense*), are widely grown as fodder and pasture.

There are a few other species that have been harvested as grains but are not grasses, especially from the family Polygonaceae. They were important in Neolithic cultures and have been important famine food. Buckwheat (*Fagopyrum esculentum* and *F. tataricum*) is the most, important, with a reported harvested area of 2 582 589 ha worldwide. Used mainly as food for poultry, buckwheat also provides a kind of flour useful for gluten-free diets. It is also used in the Far East to make noodles. Bees make a dark, highly flavoured honey from the pollen and nectar. Quinoa (*Chenopodium quinoa*) is a similar 'grain' popular especially with health-food enthusiats.

7.2.2 Pulses

The pulses are all legumes from the family Fabaceae (or Leguminosae). There are many different species. They are especially important as food not only because they are two to three times richer in protein than cereals, but also because the spectrum of proteins they contain is different from cereals. They are especially important as a source of protein for poorer people in the tropics. Soybeans (*Glycine max*) are said to contain all the essential amino acids in a quantity sufficient for an adult. Legumes are generally rich in lysine, which is in low concentration in cereal protein, but they are generally low in methionine and cysteine.

The range of proteins in pulses is very great though this can cause dietary problems. Peanut allergy, mainly to a number of low molecular weight proteins in peanuts, has become a serious problem in the west in recent decades. The potential dangers for some people of eating faba or broad beans (*Vicia faba*) have been known for many millennia in the Mediterranean region; some classical authors advised against eating them. A large minority of people in this region have favism, an allergic sensitivity to eating the beans or even inhaling their pollen, causing haemolytic anaemia. People with favism have a form of genetic deficiency in an enzyme called glucose-6-phosphate dehydrogenase (G-6-PD). This is the commonest human enzyme deficiency and its distribution is associated with the distribution of malaria in past times, because it confers some protection against malaria.

Lentils, peas, vetches and beans were some of the earliest crop plants to be cultivated. In lupins it has been possible to observe the process of domestication because it has taken place recently. *Lupinus albus* were eaten by the Romans, but were mainly used as fodder.

Table 7.4 Legume crops

Soybeans *Glycine max*
Beans *Phaseolus vulgaris*
Groundnuts *Arachis hypogea*
Chick-Peas *Cicer arietinum*
Cow Peas *Vigna unguiculata*
Peas *Pisum sativum*
Pigeon Peas *Cajanus cajan*
Lentils *Lens culinaris*
Broad Beans *Vicia faba*
Vetches *Vicia* species
Lupins *Lupinus albus*, *L. angustifolius*
String Beans *Phaseolus coccineus*
Carobs *Ceratonia siliqua*

However, in the twentieth century more determined efforts were made to domesticate other species of lupins. Mutants with lower alkaloid content (more sweet), reduced pod shattering, a more permeable seed coat, early flowering and disease resistance were all relatively quickly selected.

Many pulses have multiple uses. For example soya bean provides oil, fodder, flour used in confectionary, in biscuits soya-milk and ice-cream, and, as a fermented form in soy sauce, and curd (tofu). Groundnuts (peanuts) are another important source of oil.

Legumes are also an important source of dietary fibre. Legumes differ in their digestibility, which, in some cases, is related to the tannin content. People vary in their ability to digest beans, as we all know, causing terrible flatulence in some cases.

Many legumes are also important fodder crops and green vegetables and leguminous trees are important sources of timber and fuel in the tropics.

Figure 7.5. Oilseed rape (*Brassica oleracea* or *B. napus*).

7.2.3 Oilseed crops

Oilseed crops are derived from a remarkably wide range of plant families. Rape seed has been cultivated in Northern Europe since the Middle Ages, mainly as animal feed, and was used, for example, in reclamation of polders, fenland and salt-marshes. *Brassica napus* and *B. rapa* (turnip) are relatives of the cabbage but have been bred for high oil content in their seed. Many different varieties are grown for human consumption, in cooking oil and margarine, animal feed, or industrial use (Figure 7.5). Varieties cultivated for consumption have low erucic acid (<1%) and sulphur-containing glucosinolate content. Both these compounds reduce the digestibility and palatability of the crop. One disadvantage that low-glucosinolate varieties have is that they are more susceptible to pests. Double low varieties are also marketed under the name 'Canola'. Alternatively HEAR varieties have high erucic acid content (50%–60% of the oil). The erucic acid is converted to erucamide, which is used in the process of manufacturing polythene, to reduce friction and prevent films sticking together.

Chemical manipulation of the fatty-acid spectrum and content provides oil for a multiplicity of purposes, from detergents to pharmaceutical manufacture. For example a reduction in longer-chain fatty acids such as linolenic acid content increases shelf-life for food-oils. Some manufacturing processes prefer high oleic acid content, and this is also favoured in the diet for the prevention of coronary heart disease. The potential market for rape seed oil is great. Already it is incorporated into lubricants for two-stroke petrol engines, but derived methyl esters could be used as a diesel substitute when mineral-derived diesel oil becomes more expensive.

Related to oilseed rape are the mustard oils *B. juncea* and *Sinapis alba* cultivated in India and elsewhere for their oils but also for the production of condiment. For this purpose it is the high glucosinolate content that provides the taste of mustard. *B. juncea* contributes volatile, or nasal, pungency, and *S. alba* contributes heat and

Table 7.5 Families of oil seed crops

Fabaceae
Soybeans *Glycine max*
Brassicaceae
Rapeseed and Mustards, *Brassica, Sinapis*
Asteraceae
Sunflower *Helianthemum*
Safflower *Carthamnus*
Arecaceae
Oil Palm Fruit *Elaeis*
Oleacaeae
Olives *Olea*
Pedialaceae
Sesame Seed *Sesamum*
Linaceae
Linseed *Linum*
Euphorbiaceae
Castor Beans *Ricinis,*
Tallowtree Seeds *Sapium*
Tung nuts *Aleurites*
Cucurbitaceae
Melonseed *Cucumis*
Sapotaceae
Karite Nuts (Sheanuts) *Vitellaria*
Papaveraceae
Poppy Seed *Papaver*
Cannabidaceae
Hempseed *Cannabis*
Simmondsiaceae
Jojoba Seeds *Simmondsia*

sweetness in the mouth. *B. nigra*, black mustard, is one of the oldest recorded spices domesticated in Asia Minor or Iran, but in recent decades has been largely replaced by *B. juncea*, which has better harvesting qualities (a non-dehiscent silique).

In lower latitudes, where oilseed rape is not normally cultivated, sunflower (*Helianthus annua*) becomes the most significant oilseed crop. The oil has over 90% of oleic and linoleic acids in roughly equal proportions. As well as providing a cooking oil valued for the high linoleic acid content, it is useful in manufacture as a 'drying oil' in paints and varnishes. It also does not yellow over time, in contrast to the nearly 50% linolenic acid provided by linseed. It is the linolenic acid in some oils that leaves a pervasive odour when it is used for frying. Safflower (*Carthamnus tinctorius*) is another oilseed from the Asteraceae.

Castor oil, *Ricinis communis*, has little use in the diet. Its value as a purgative is well known. However, it has extensive industrial use as a non-drying oil and was used in lamps by the Egyptians more than 4000 years ago. Flax provides linseed, an unsaturated oil used in the paint industry. Frederick Walton first produced linoleum in 1863 from oxidised linseed oil mixed with cork and pigments and pressed onto a jute or hemp backing.

Tallowtree seeds, Karite nuts and Tung nuts are all valuable sources of oils and fats. Tallowtree (*Sapium sebiferum*), cultivated in China, provides fats for candlewax and soap. *Aleurites* species (Tung nuts), cultivated in South East Asia, provide drying oils for paints and varnishes, and candle-nut oil, and are also used in curries. *Vitellona* (*Butyrospermum*) (Karite Nuts) is primarily grown in the semi-arid Sahel belt of Africa. A solid fat (butter or stearin) and the liquid oil (olein) have a wide range of uses, especially as a substitute for cocoa butter.

7.2.4 Root crops

The four major root crops are from four different families (Table 7.6). The potato is the king of vegetables. *Solanum tuberosum* is only one of several hundred species in its genus. The tomato (*Lycopersicon*) is also closely related. An important step in the domestication of potato was the selection of mutants with alkaloid free tubers. By the time it was being cultivated by the Incas many varieties already existed. By the year 1600 some had been introduced to Europe. 'Andigena' potatoes introduced from the Andes remained a horticultural curiosity in Europe. They were adapted to the short days of their origin. However, selection gave rise to clones adapted to long days and cultivation became established in northern Europe. New varieties had shorter stalks and stolons, fewer flowers, larger leaves and fewer larger tubers. Potato remained for a long time a garden crop or cattle food in most of Europe but very swiftly became a staple crop in Ireland where it flourished in the wet cool climate. Its needs were small. Cultivation of a small patch on raised and manured lazybeds could easily supply a family. The population of Ireland rose

Table 7.6 The world's most important root crops ranked by area harvested

Solanaceae
Potatoes *Solanum*
Euphorbiaceae
Cassava *Manihot*
Convolvulaceae
Sweet Potatoes *Ipomoea*
Dioscoreaceae
Yams (Cocoyam)
mainly Taro *Colocasia*
and Yautia *Xanthosoma*

Figure 7.6. Taro (*Colocasia esculenta*: Araceae).

from 1.5 million in 1760 to 9 million in 1840. Over-reliance on the potato was fatal. Potato blight caused by the fungus *Phytophthora infestans* became prevalent, causing a full-scale epidemic in the early 1840s. By death from famine and emigration the population of Ireland fell by 2.5 million. The pattern for emigration was set and by World War I a further 5.5 million people emigrated to Britain and America.

Sweet potato, *Ipomoea batatas* (family Convolvulaceae), was one of the crops Columbus brought back to Europe, and from there was introduced throughout the Old World by the Portuguese and Spanish. There is some linguistic evidence for an earlier introduction to Polynesia from South America, because sweet potato is called cumara in Peru and kumara in Polynesia. It was one of the staples in Polynesia by the time of European discovery. Thor Heyerdahl showed that an introduction to Polynesia from the east was possible by his voyage in the balsa (*Ochroma lagopus*) raft, the Kon-Tiki. It was kumara that permitted the colonisation of Aotearoa (New Zealand). The colonists also brought breadfruit and coconuts, which did not flourish in the cooler New Zealand climate although kumara did.

Cassava or manioc, *Manihot esculenta* (family Euphorbiaceae), is the most important tropical root crop and provides more than half of the energy requirement for over 420 million people in tropical countries. There are two forms distinguished by the content and distribution of hydrocyanic acid (HCN) in the tubers. Sweet cassava has low HCN that is confined to the tuber bark. Bitter cassava has high CN distributed throughout the tuber. Sweet cassava was cultivated and, perhaps, first domesticated by the Mayans. Bitter cassava may have an origin further south. Fresh cassava has a very limited shelf-life but is readily converted into 'flour'.

Different species of yam were domesticated in Asia (*Dioscorea alata*), Africa (*Dioscorea rotundata*) and America (*Dioscorea trifida*). Later, yams were used as ship's victuals. They are rich in vitamin C and combatted scurvy. They also have a long shelf-life unlike cassava and became widely distributed along trade routes.

7.2.5 Vegetables

Tomatoes come sixteenth in the top 30 world crops. Although treated as a vegetable they are, of course, a fruit. In some dictionary definitions a vegetable is a plant eaten with an entrée or main course or in a salad but not as a dessert or table fruit. In fact, in 1893 the United States Supreme Court ruled that a tomato was legally a vegetable rather than a fruit. By this definition a vegetable may originate from any part of the plant, including the flower (e.g. globe artichoke). Tomato was widely cultivated in the Americas before the arrival of Columbus. Aubergines (eggplant), chillies and green peppers are also fruit-vegetables from the potato family Solanaceae.

Figure 7.7. Vegetable fruits: (a) Breadfruit; (b) Avocado or alligator pear.

Other families important as sources of vegetables are the Alliaceae (onions, leeks, garlic, shallots), Brassicaceae (cabbage, broccoli, cauliflower, etc.), Leguminosae (peas, green beans, runner beans, carob) and Cucurbitaceae (cucumber, gherkins, squash, marrow and pumpkin). Hybridisation and selection have resulted in an extraordinary proliferation of diverse vegetables. The Brassicaceae provides a remarkable example of a group of closely related species providing root, leaf and seed crops.

Artocarpus altilis (Moraceae) or breadfruit, is another fruit treated as a vegetable, which has a strong regional and historical importance in Polynesia, even though it is not important in world trade (Figure 7.7). It had the great advantage of preservability. The starchy fruit was stored in large air-tight pits. A sour semi-fermented pudding produced from the dough and called Ma was a staple of Polynesia. Of course it was an attempt by the British to transport such an efficacious plant to the West Indies that spurred the mutiny on the *Bounty*.

Table 7.7 Vegetables by family

Solanaceae
Tomatoes
Chillies and green peppers
Eggplants
Alliaceae
Onions and shallots
Garlic
Leeks and other Alliaceae
Brassicaceae
Cabbages
Cauliflower
Cucurbitaceae
Cucumbers and gherkins
Pumpkins, squash, gourds
Asteraceae
Lettuce
Artichokes
Chicory roots
Asparagaceae
Asparagus
Apiaceae
Carrots
Malvaceae
Okra
Chenopodiaceae
Spinach

7.2.6 Fruits and nuts

The range of fruits eaten mainly for pleasure rather than as a source of energy is very diverse (Table 7.8). However, it is remarkable that most of the commercially-important ones come from only two families, the citrus family Rutaceae and the rose family Rosaceae. Fruits of the citrus family (oranges, lemons, limes, grapefruit, pomelo, tangerine, satsuma, mandarines) all share a characteristic form, a kind of berry called a hesperidium with a succulent glandular hairy endocarp. In contrast the fruits of Rosaceae are very diverse including pomes (apple, pear, medlar, quince), drupes (peach, apricot, plum and cherry), drupecetum (= an aggregate of drupelets) (blackberry and raspberry) and pseudocarp (strawberry). Compared with their wild ancestors, cultivated fruits are fleshier, sweeter, with fewer tannins and glycosides, and frequently less seedy, but the relationship to wild crops is often close. For example in Turkestan there is a range of wild varieties of the apple (*Malus pumila*) that are large and sweet, bridging the gap between the crab apple (*M. sylvestris*) and the domesticated varieties (*M. domestica*).

Table 7.8 Fruits and nuts

Fruits and nuts by family	Fruit type
Rosaceae	
Apples, Pears, Quince	Pomes
Plums, Peaches, Apricots, Cherries, Almond	Drupe
Strawberries	Pseudocarp
Raspberries	Drupecetum
Arecaceae Coconuts, dates, Areca Nuts (Betel) (see also oil-palm)	Drupe, Berry
Musaceae Bananas including Plantains	Berry
Vitaceae Grapes	Berry
Rutaceae Citrus fruits	Hesperidium
Anacardiaceae Mangoes, Cashew nut and apple, Pistachio	Drupe, aril
Cucurbitaceae Melons, Cantaloupes, etc.	Berry
Bromeliaceae Pineapples	Aggregate fruit
Juglandaceae Walnut	Drupe
Betulaceae Hazelnuts	Nut
Moraceae Figs	Syconium
Sterculiaceae Kola nuts	Berry?
Lauraceae Avocados	Berry
Caricaceae Papayas	Berry
Ebenaceae Persimmons	Berry
Fagaceae Chestnuts	Berry
Grossulariaceae Gooseberries and Currants	Berry
Ericaceae Blueberries, Cranberries	Drupe
Actinidiceae Kiwi Fruit	Berry

Figure 7.8. Banana (*Musa acuminata*).

There are about 35 species of *Musa* (banana). The wild ancestors are the species *Musa acuminata* and *M. balbisiana*. Wild plants have a pulp that does not develop unless seeds are present but mutant varieties have a pulp that starts to develop even when the flower is not pollinated. Further mutation of these parthenocarpic clones gave rise to seed-sterile parthenocarpic clones. This may have occurred independently in *M. maclayi* to give rise to the Fe'i bananas of New Guinea. Hybridisation occurred between *M. acuminata* clones and between *M. acuminata* and *M. balbisiana*. This latter cross enabled bananas to be cultivated outside the moist tropics of their origin in South East Asia and in areas of seasonal drought. The hybrids are triploid or tetraploid and are more vigorous. Cultivation spread thoughout the Old World tropics by the end of the sixteenth century when they were then introduced to Central America. Cultivation of the banana became an expression of imperialism as U.S. fruit companies came to dominate the trade. Relatively few clones are grown, exposing bananas to a significant risk of disease from banana wilt (Panama disease), or leaf spot, the pathogenic fungi adapted to attack individual banana genotypes that are grown on a large scale.

Another fruit crop with a narrow genetic base is the pineapple. It also sets fruit parthenocarpically. Cultivars had already been selected by native Americans before the arrival of the Spanish and Portuguese. Pineapple was used to provision ships, and because the crown of the fruit can be used to establish a new plant it was readily established around the world. However, many cultivars originate from just a few plants taken initially to France where a fad for growing 'pineries' took root.

As well as soft fruits there are many species harvested for nuts. In fact some of the most popular nuts like Brazil nuts and almonds are not nuts. Coconuts, almonds, and walnuts are the inner part of a drupe. Brazil nuts and kola nuts are seeds. The cashew nut is produced on the end of a fleshy aril (cashew-apple) that is also eaten and in Brazil made into a drink called cajuado. The nut is very hard and has to be cracked open by large lever-like pliers (Figure 7.10).

Figure 7.9. Pineapple (*Ananas comosus*).

7.2.7 Sugar crops

Sugar-cane has had an influence on human history that rivals that of the cereal crops. Sugar-cane probably originated as a crop in New Guinea. Its cultivation spread north and west so that even by 3000 years ago it was chewed for its sweetness in India and China. A variety called puri was first extracted to provide refined sugar in India. It spread westward, notably following the spread of Islamic culture, and was cultivated along the North African coast and in Spain by the Moors. In the early fifteenth century the colonies of the Portuguese in Madeira and the Azores, and the Spanish in the Canaries, were cultivating it. From there it was a short step to its introduction into the West Indies and the Americas, encouraged by the loss of other sources in North Africa and the Near East because of the expansion of the Ottoman empire.

Figure 7.10. Cashew nut (*Anacardium occidentale*) factory in Malaysia. The edible seed must be prised from its hard case with a pin, a process made more difficult by the acrid liquid that surrounds it.

Refined cane-sugar, sucrose, was a valuable commodity; in the fourteenth century it was worth ten times more than honey. Its exploitation required intensive labour at least twice a year, at planting and harvest. The native Carib people had been decimated by disease and could not provide this labour, and in the early sixteenth century slaves from West Africa were brought to the Caribbean. The Middle Passage from West Africa was to become notorious for its bestial treatment of its human cargo. In the Caribbean, slaves were traded for sugar and rum, which were then traded in Europe. The third leg back to Africa was with firearms, cloth, salt, and trinkets to buy slaves. On each leg of the triangular trade, profits could be made. In the late eighteenth century about 25% of British maritime effort was involved in the trade and a quarter of a million workers in Britain supplied it. The value of a slave was low, about half a ton of sugar in 1700, and one ton might represent the life-time production of a slave. But the mark-up was very great. A slave bought for £3.00 in Africa could sell for £25 in the West Indies. Many died on the passage.

The naval blockades of continental Europe by Great Britain in the late eighteenth and early nineteenth centuries spurred the search for an alternative source of sugar. Following the discovery in 1747 by the German chemist Marggraf of 4% sucrose in beet, Achard selected a high yielding variety called 'White Silesian Beet' with 6% sucrose. With sponsorship from the King of Prussia beet-sugar refining was established by 1802. In 1811 Napoleon established schools for the study of beet and required it to be grown. Over subsequent decades sugar-beet breeding raised the sugar content to over 20%.

Other plants also provide syrup, and not just in their fruits. Sweet sorghums (sorgo) for example, produce sweet pith and are cultivated for syrup.

Figure 7.11. An early variety of sugar-cane (*Saccharum officinarum*) in cultivation in Madeira from whence it was introduced to the Americas along with slavery.

7.3 Plants for craft and fuel

Throughout human history plants have provided a major source of material for construction and craft. Wood and timber was the main constructional and craft material before the industrial revolution. A medium-sized timber framed farm house from Suffolk built in 1500 used 330 trees. Half of these were less than nine inches in diameter and only three exceeded 18 inches in diameter. Most trees were felled after 25–75 years. Trees older than 75 years were difficult to fell, transport and cut up into usable timber. In 1483 the first statutes in England were introduced to protect woodland. Woodland had to contain a minimum number of standard trees. It is also impossible to over-estimate the importance of plants as a source of fuel until the last 200 hundred years.

7.3.1 Timber

The timber from trees has enabled the construction of extraordinary buildings like the cathedrals of Europe, but there is no better example of the value of timber in construction than the tremendous wooden ships that have been built in past ages. In the British Navy, oak (*Quercus*) was the favoured species but pine was also important. The craft that permitted first colonisation and then trade across the vast distances of Polynesia were constructed of many timbers but at their heart were vesi logs (*Intsia bijuga*) from Fiji, hollowed because of their great density. The maintenance of timber supplies for the British Navy moulded aspects of foreign policy throughout the eighteenth and early nineteenth centuries; keeping the Baltic open for trade in the Napoleonic era, gaining Minorca in the Mediterranean and so forth.

Some species have been particularly favoured because of their rot-resistant qualities. Teak (*Tectona grandis*) was used for the keel of HMS *Victory*, although elm (*Ulmus*) was more commonly used for this purpose. Teak is native to India and Malaysia. Natural forests have been extensively exploited and plantations have been established in the Philippines and Indonesia. It has been highly favoured because of its durability, resistance to insect attack and the weather. Satinay

(*Syncarpia hillii*) from Northeast Australia is resistant both to white ants and marine borers and was utilised in the construction of the Suez Canal. Similarly, greenheart (*Ocotea rodiaei*), a relative of camphorwood, was used in the construction of the Panama Canal because of its resistance to termites and other borers.

Various kinds of hardwoods, often called mahogany, were favoured for superior furniture and joinery. The mahoganies *Swietenia* from the Americas and *Khaya* from Africa are two of the most used. Mahogany wood is heavy and strong, resists rot and termites and has been favoured for joinery because it is easily worked. Rosewood *Dalbergia* was another important species. Various species of ebony (*Diospyros*) were favoured for fancy veneers and beading.

Hardwoods vary in their density. The so-called ironwoods sink in water. Several different genera provide ironwoods. Lignum vitae (*Guaicum officinale*) is the hardest of the commercial timbers with a specific gravity of 1.37. The genus *Tabebuia* from South America with either yellow or pink trumpet-like flowers also provides ironwoods. The wood is the most durable of any American timber and is reputed to have been used as the propeller shaft bearings of submarines (*Tabebuia serratifolia* has a specific gravity of 1.20). Ebony (*Diospyros ebenum*) has a specific gravity of 1.12. These compare with the lightest hardwood, American balsa (*Ochroma pyramidale*), with a specific gravity of only 0.17. Teak (*Tectona grandis*) has a specific gravity of 0.63. The density and high resin content of Lignum vitae or 'wood of life' (*Guaiacum officinale*), make it resistant to friction and abrasion. It is self-lubricating and under certain conditions the wood wears better than iron. Because of this, the wood has been highly valued for pulley sheaves, bearings, casters, food-handling machinery, and for end grain thrust blocks, which line the propeller shafts of steamships. The sweet-smelling resin contains 15% vanillin and was highly sought after as a cure-all. Peroxidase enzymes in blood cells oxidise chemicals in the resin resulting in a characteristic blue-green colour change. Actually the heaviest plant tissue comes not from wood but from vegetable ivory, the seed of various species of palm.

Hardwoods have a mixture of cell types in the wood but always have a high proportion of tightly-packed thick-walled lignified fibres. Lignified cell-wall material is about 1.5 times as heavy as water. Bouyancy is conferred by the air in the spaces in and between cells. The softwoods, from conifers such as pine (*Pinus*), spruce (*Picea*) and fir (*Abies*), are so-called because they are easy to work in carpentry; they have a homogeneous tissue composed mainly of tracheids. They are also favoured as a source of wood pulp that can be turned into a variety of materials, not just paper. Some softwoods are actually harder and heavier than most hardwoods. Juniper (*Juniperus*) is one. It has small densely packed tracheids. Most of these timbers have been, and are being exploited relentlessly from the wild. Many of the most favoured have become very rare as a result. Afromosia (*Pericopsis elata*), which was more expensive than African Mahogany, was made nearly extinct in West Africa.

Modern forestry in Britain can perhaps be dated back to the introduction of the *Larix decidua* (European Larch) in the early 1600s from the Alps, the first of many species of conifer to be introduced. In the tropics and warm temperate regions such as the Mediterranean basin, the genus *Eucalyptus* has been favoured to provide shelter belts as well as a fast-growing timber tree. In many areas they have become naturalised and pose a threat to native vegetation. The red gum *Eucalyptus camaldulensis* is a prolific seeder and, initially lacking any native pests or diseases, became a weed in California until in the late 1990s, when an Australian defoliating psyllid pest called *Glycaspis brimblecombei* became established. There are about 450 species of eucalypts native to Australia. The wood of different species varies from relatively soft and light to hard and dense. Karri (*E. diversicolor*), spotted gum (*E. maculata*), blackbutt (*E. pilularis*), and jarrah (*E. marginata*) are favoured for furniture and wood carving. Jarrah is stronger and more durable than oak and resistant to termites and marine borers and has a reputation for use as marine piles and planking.

7.3.2 Fibres

Figure 7.12. Ramie (*Boehmeria nivea*) provides the longest, and most absorbent and silkiest of bast fibres, but its fibres are difficult to free from the stem. The fibres have great strength if not bent and twisted.

Fibres from plants have been woven into fabrics for millennia. A large proportion of harvested trees are not utilised for timber but as a source of fibre as wood pulp for paper or the production of boards. It is the cellulose that is most wanted; the lignin that is present can cause discoloration and make a fibre more brittle, though it is resistant to decay. Flax and ramie (Figure 7.12) are mainly cellulose and are white. They provide a fine fibre. Jute and hemp are coarse and brownish, and contain 10%–15% lignin.

Fibres from stems

There are three main stem fibres (bast) sources: flax, jute and kenaf. Flax (*Linum usitatissimum*) is the source of linen yarn. Linen is one of the oldest of all the plant-based fabrics. The fibre is obtained by subjecting the stalks to soaking in water (retting), drying, crushing and beating. The fibre strands measure about 30–75 cm. The yarn has strength, durability, is resistant to attack by microorganisms, and has a smooth surface that repels dirt. It also has a beautiful lustre. It absorbs and releases moisture quickly and so is comfortable to wear. Jute (*Corchorus*) provides a coarser fibre used in sacking (burlap) and carpeting. It has been grown in the Bengal area of India and Bangladesh from ancient times. Manufacture in the west began in the 1790s centred on Dundee in Scotland. The two species (*C. capsularis*, or white jute, and *C. olitorius*) grown for jute fibre are similar and differ only in the shape of their seed pods, growth habit, and fibre characteristics. The fibres are held together by gummy materials that are softened by retting for 10–30 days during which time bacteria break them down. The fibres run the length of the stem, and are loosened and jerked out of the stem. Kenaf (*Hibiscus cannabinus*), is used as an alternative to jute. The plant grows to a height of 5 m and provides fibres up to about 1 m long.

Of much less significance today are the fibres from barks, often beaten into sheet-like cloths directly. The most famous is the tapa cloth of Polynesia and New Guinea obtained by soaking and pounding with mallets the bark of the paper mulberry (*Broussonetia papyrifera*). Cork used in bottling wine comes from the cork oak *Quercus suber*, which is widely cultivated in the Mediterranean region.

Papyrus was an early form of paper made from the sedge *Cyperus papyrus*. Rice paper of the Orient was made by pounding sheets spirally cut from the pith of the rice paper plant, *Tetrapanax papyrifera* (Araliaceae). The Aztecs and Mayans used the bark of a *Ficus* species to make a kind of paper not unlike papyrus.

Paper making was first perfected by the Chinese about 2000 years ago. Using the stem fibres of paper mulberry separated in water and floated, they were allowed to settle on a mesh, before drying. Most paper today is made from the fibres of woody plants. The first step is to turn the wood into a pulpy mass. Several methods are used. In one process wood chips are cooked with bisulphites and then digested with strong acids. The softened fibres are then blown to separate them. Paper produced by this process has a tendency to discolour and become brittle with age. The alternative process is alkaline sulphate production using sodium sulphate, sodium sulphide and sodium hydroxide, a process that also removes resins from conifer wood. Acid-free paper is normally used for books.

Purified wood cellulose is used in the manufacture of cellophane or rayon depending on the way in which it is extruded in manufacture.

Fibres from leaves

Fibres are obtained from leaves with numerous vascular bundles in the veins. Monocot leaves are the most useful because of the long unbranched veins.

Sisal (*Agave sisalana*) of the agave family (Agavaceae) is the most important source. It is native to Central America, and has been used since pre-Columbian times. The fleshy and spiny lance-shaped leaves grow out in a dense rosette. Outer leaves are cut off close to the stalk as they reach their full length. The fibre is usually obtained by crushing the leaf between rollers and scraping the resulting pulp from the fibres. Sisal fibre is coarse and inflexible but valued for ropes and twine because of its strength, durability, ability to stretch and resistance to deterioration in salt water.

Pineapple is another source of fibre but Manila hemp or abaca provides the best quality ropes. It comes from the leaves of a banana species (*Musa textilis*) native to the Philippines.

Fibres from seeds

Cotton is the king of plant fibres. Seed is embedded in a mass of white unicellular hairs (trichomes) forming the boll inside the capsule (see Kapok, Figure 7.13). The hairs (lint) are twisted into usable thread which is tough and strong. Each hair may be up to 50 mm long.

Figure 7.13. Kapok (*Ceiba pentandra*): hairs are produced on the inner surface of the seed capsule of the kapok tree, a tall buttressed tropical tree. Kapok hairs are difficult to spin and hence it is largely used as packing for mattresses and pillows as well as life-savers.

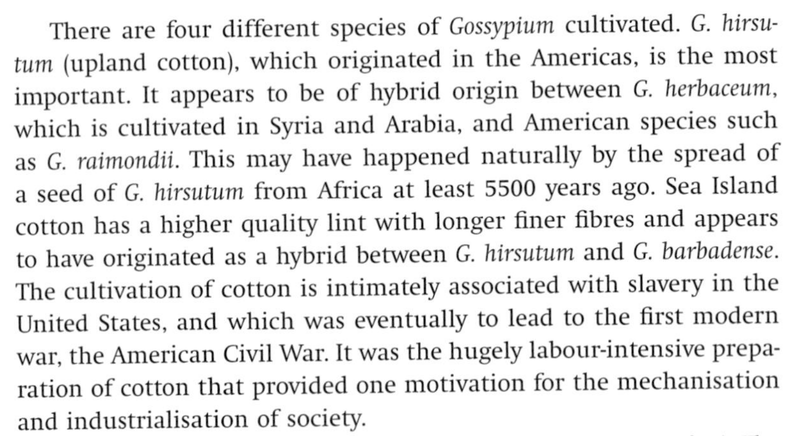

There are four different species of *Gossypium* cultivated. *G. hirsutum* (upland cotton), which originated in the Americas, is the most important. It appears to be of hybrid origin between *G. herbaceum*, which is cultivated in Syria and Arabia, and American species such as *G. raimondii*. This may have happened naturally by the spread of a seed of *G. hirsutum* from Africa at least 5500 years ago. Sea Island cotton has a higher quality lint with longer finer fibres and appears to have originated as a hybrid between *G. hirsutum* and *G. barbadense*. The cultivation of cotton is intimately associated with slavery in the United States, and which was eventually to lead to the first modern war, the American Civil War. It was the hugely labour-intensive preparation of cotton that provided one motivation for the mechanisation and industrialisation of society.

Coir comes from the outer husk of the coconut (*Cocos nucifera*). The processed fibres, made up of smaller lignified threads, each of which are only up to 1 mm long, are up to 30 cm long, light, elastic and highly resistant to abrasion, are used to make brushes and matting as well as cordage.

7.3.3 Assorted materials

Plants are also a source of many assorted materials.

Turpentines, gums, industrial chemicals

Turpentines are a mixture of volatile oils and non-volatile resins (rosin) extracted from different species of trees. Pines (*Pinus sylvestris*, *P. pinaster*, *P. palustris*, *P. caribea*) are the main sources. The volatile part (turpentine) is used as a thinner for paints and varnishes and as a synthetic substrate for the production of a wide range of materials. Turpentine is only one of the range of products of pine used in wooden shipbuilding manufacture ('naval stores'). Canada balsam is the turpentine extracted from *Abies balsamea*. Balsams are aromatic resins extracted from a range of plants, some of which have a medicinal value. *Myroxylon pereirae* provides balsam of Peru. Balm of Gilead and myrrrh, mentioned in the Bible, are sweet smelling resins used as incense and in cosmetics, and come from small trees in the genus *Commiphora*, for example *C. gileadensis*, which is native to the areas around the Red Sea.

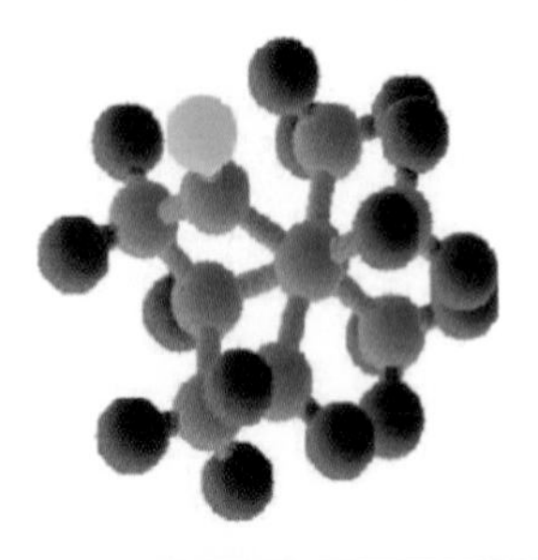

Figure 7.14. Camphor comes from *Ocotea usambarensis* from East Africa, and a number of other species.

'Gum trees' are *Eucalyptus* species. Oil of eucalyptus (eucalyptol) is a volatile terpene compound distilled from the leaves. It is used for flavourings, cough drops, and for the synthesis of menthol. Another volatile terpene with a lemony fragrance that is supposed to prevent insects biting, comes from the leaves of *E. citriodora*. Eucalypts are rich in kinotannic acid and are used as a source of tannins to convert animal hide into leather. One of the main Australian sources of kino is the common red gum (*Eucalyptus camaldulensis*).

True polysaccharide gums, come from sources such as the carob tree (*Ceratonia siliqua*).

Rubber

Rubber, *Hevea brasiliensis*, was exploited exclusively from the wild forests of South America until it was introduced to South East Asia. The first plantation was established in Sri Lankla in 1872. A few plants sent to Singapore, Malaysia and Indonesia were used to establish commercial plantations there. Several other species have been used to provide latex for rubber but none has really become commercially viable except when rubber has been in short supply. Gutta percha (*Palaquium gutta*) provides a rubber that is more brittle. *Manilkara bidentata* provides balata rubber, which is also non-elastic and used to make machine belting. It is also sometimes used as a substitute for chicle, a terpene gum from the latex of the sapodilla tree (*Manilkara zapota*), the chewing gum of the Aztecs.

Figure 7.15. Castor oil from *Ricinis communis* is used in the production of nylon and other synthetic resins and fibres. It is first converted to sebacic acid.

Dyes

When westerners travelled to the Pacific Islands they were impressed by the tattooing they observed, produced from a blue pigment prepared from the baked nuts of the candlenut tree *Aleurites moluccana*. The fashion was adopted by sailors and exported to the west. Indigo, a deep-blue and fast dye, from woad (*Isatis tinctoria*) was used to paint the body in Ancient Britain. Described by Caesar when he invaded Britain, it may even have given Britain its name from the Celtic word ‘Brith’ meaning paint or mottled. Indigo was also used as a dye of fibres, even used to dye the first blue uniforms of British policemen. Indigo is also obtained from various legumes (*Indigofera tinctoria, I. suffruticosa*).

Henna (*Lawsonia inermis*) is the orange-red counterpart of woad but is still used to dye the body and hair, especially in the region from North Africa to South Asia. Its main dye component is lawsone, a quinine. It is a very fast dye and can also be utilised to dye fabrics and leather. Juglone obtained from a diversity of walnuts (*Juglans*), is also a quinine. Confederate grey is obtained from butternuts (*Juglans cinerea*). Bloodroot (*Sanguinaria canadensis*) provides another red dye.

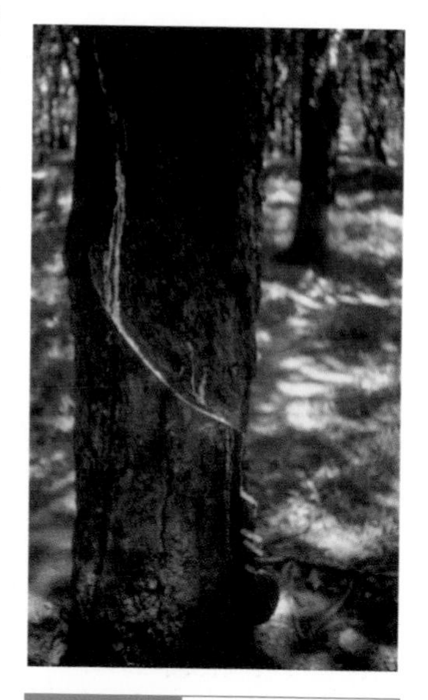

Figure 7.16. Rubber tree (*Hevea brasiliensis*) cut to release the latex in a plantation in Malaysia.

Many plant dyes have been supplanted by chemically produced aniline dyes. Such is the case with the red dye (alizarin) produced by various species of madder (*Rubia cordifolia, R. peregrina, R. tinctorum*). However concern for the use of natural products as food additives has maintained the use of foodstuff dyes. *Bixa orellana* (Figure 7.17) is the source of annatto (mainly a carotenoid called bixin) the yellow dye utilised in many food products such as margarine. Crocetin is a related carotenoid; the main dye in the stigmas of saffron (*Crocus sativus*) is used as a spice as well as a dye. It has an ancient use and importance, hence the saffron-coloured robes of members of several religious orders. The spice turmeric (*Curcuma longa*) has increasing use as a yellow food colourant.

7.3.4 Fuel

Throughout human history wood has provided the primary source of fuel. It is only in the past few hundred years that fossil fuels have

Figure 7.17. *Bixa orellana* the source of the yellow food dye annatto.

Welcome the Creations Guest,
Lord of Earth, and Heavens Heir.
Lay aside that Warlike Crest,
And of Nature's banquet share:
Where the Souls of fruits and flow'rs Stand prepar'd to heighten yours.
Andrew Marvell, 1621–1678

provided an alternative. The industrial-scale use of wood for fuel came with the great civilisations. Wood is still a vital fuel in many parts of the developing world. In the semi-arid areas of Africa species of acacia provide a major source. Unfortunateley, these trees and bushes also have an important place in the delicate ecology of areas sensitive to desertification.

7.4 Plants for the soul

There is a close connection between the use of plants for health, for recreation and pleasure, and for spiritual purposes. They range from tea, to the sugary sweet, the flavoursome spice, and to the hallucinatory drug. The search for and the use of plants that supply these needs has had a profound influence on human history. Many of these pleasure plants contain compounds with a physiological effect on humans, some are psychoactive and provide poisons to panaceas, or both depending upon the concentration. In 1924, a German toxicologist called Lewin published a system for classifying narcotic and stimulating drugs:

- euphorica – sedatives (morphine, heroin)
- phantastica – hallucinogens (mescaline)
- inebriantia – excitation followed by depression (alcohol)
- hypnotica – inducing sleep (kavaine)
- excitantia – stimulants (caffeine, nicotine)

There is a deep-seated and ancient ambivalence about the use of plants for the soul rather than for anything other than the primary ones, of food or craft. This ambivalence has included not just the extraordinary hallucinogens but also the mundane coffee or tobacco. There is less ambivalence about the display of plants for their beauty to assuage the trials of the soul.

7.4.1 Herbs and spices

The history of spices, tea and opium are inexorably linked. Spices and herbs characterise many different cuisines: the lemon grass (*Cymbopogon citrates*) of Thai food is just one.

The use of herbs to flavour foods is an ancient practice. The Romans used local herbs and also imported spices: lovage and fennel for sauces; savory for bean dishes; nutmeg and aniseed to preserve meat, and cumin to flavour pastry. They also flavoured some dishes with silphium asafetida (*Ferula foetida*), called either the food of the gods or the devil's dung depending on whether you like its very strong garlicky smell and flavour. It is still used in India and Persia as a condiment and digestive remedy. It flavours Worcester Sauce. *Ferula* belongs to the umbel family (Apiaceae), one of the just three families that provide the majority of herbs (Table 7.9). The others are the labiates (Lamiaceae), and composites (Asteraceae). These are three very distinct families and three of the first families to be

Table 7.9 Culinary herbs from the three most important families as sources

Lamiaceae (labiates)
Anise hyssop - *Agastache foeniculum*
Basil - *Ocimum basilicum*
Dittany - *Origanum dictamnus*
Horehound - *Marrubium vulgare*
Hyssop - *Hyssopus officinalis*
Mints - *Mentha* spp.
Monarda, bergamot - *Monarda didyma*
Lavender - *Lavandula angustifolia*
Lemon balm - *Melissa officinalis*
Oregano or marjoram - *Origanum* spp.
Patchouli – *Pogostemon* sp.
Rosemary - *Rosmarinus officinalis*
Sage - *Salvia officinalis*
Savory: *Satureja hortensis, S. montana*
Shiso - *Perilla frutescens*
Thyme - *Thymus vulgaris*

Apiaceae (umbellifers)
Angelica - *Angelica archangelica*
Anise - *Pimpinella anisum*
Caraway - *Carum carvi*
Coriander - *Coriandrum sativum*
Cumin - *Cuminum cyminum*
Dill - *Anethum graveolens*
Fennel - *Foeniculum vulgare*
Horseradish - *Armoracia rusticana*
Lovage – *Ligusticum officinale*
Parsley - *Petroselinum crispum*
Silphium asafetida - *Ferula foetida*
Sweet Cicely - *Myrrhis odorata*

Asteraceae (composites)
Chamomile - *Matricaria recutita*
Curry plant - *Helichrysum angustifolium*
Feverfew - *Tanacetum parthenium*
Mexican tarragon - *Tagetes lucida*
Tarragon - *Artemisia dracunculus* var. *sativa*

taxonomically recognised. They are also relatively evolutionarily-advanced. Their importance as herbs may be related to their adaptive radiation in the production of insect deterrents. The strong-tasting compounds discourage insect predators but add savour to our cooking.

The essential oils and other compounds that give the smell and flavours are anti-feedants. Some species have been favoured in particular as insect repellants, such as Citronella grass, *Cymbopogon nardus*, or natural insecticides, such as Pyrethrum, *Tanacetum cinerariifolium*. Others provide the base scents of many perfumes (Lavender, Patchouli).

The mints (*Mentha*, Labiatae) are interesting because of the way a few species and their hybrids provide a number of different flavours and smells: Peppermint – *Mentha* × *piperita* (*M. aquatica* × *M. spicata*), Eau de Cologne mint – *M.* × *piperata* var *citrata*, Spearmint – *Mentha* × *spicata* (*M. longifolia* × *M. suaveolens*), Apple mint – *M. villosa*, Pennyroyal – *Mentha pulegium*, Round-leaved mint – *M. suaveolens*.

Many of the most important spices come from relatively primitive families, particularly from South East Asia. Trade in these spices by boat and overland caravan was established to the Near East and China even before the Roman era. Pliny's *History* includes fascinating entries about the source of cinnamon and cardamon. Pliny thought they came from Arabia and Medea, although in fact they came from South East Asia, and only travelled through these lands on their way to Rome. In later Roman times the trade came to be centred on

Figure 7.18. Part of the spice market in Istanbul.

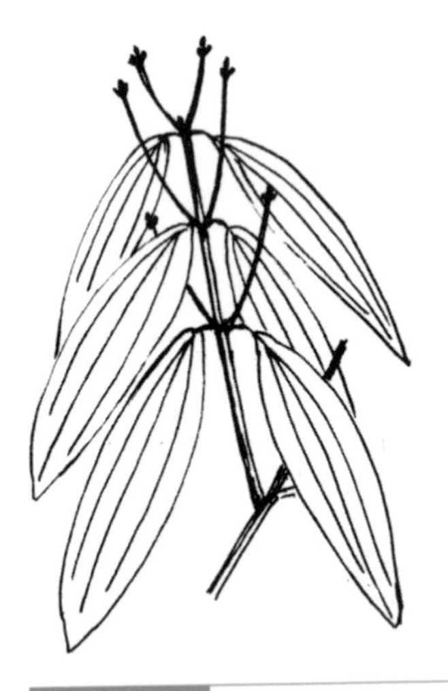

Figure 7.19. Cinnamon (*Cinnamomum zeylanicum*).

Figure 7.20. *Myristica fragrans*, source of nutmeg and mace.

Constantinople (See Figure 7.18). In the Middle Ages Venice and Genoa controlled the trade with the rest of Europe.

There was pepper (*Piper nigrum*: Piperaceae) from southwestern India, cardamon (*Elettaria cardomomum*: Zingiberaceae) from India to western South East Asia, cinnamon (*Cinnamomum zeylanicum*: Lauraceae) from Sri Lanka, Cassia bark or Chinese cinnamon (*C. aromaticum* from further east), and ginger and turmeric (*Zingiber officinale* and *Curcuma longa* both Zingiberaceae) from Malaysia. From the furthest east and the most valuable of all were cloves (*Eugenia caryophyllus*: Myrtaceae) from the South Moluccas and centred on Amboyna, and nutmeg and mace (*Myristica fragrans*: Myristicacae) from the Banda Islands just west of New Guinea. Nutmeg is the seed, mace is the aril that surrounds it. It originates from the 40 square miles of the Banda Islands far to the east of the East Indies. Nutmeg was used as flavouring, for example in ale, but also gained a high value as a prophylactic against plague.

One aspect of the westernisation of the world was the attempt to gain and control the trade in cloves and nutmeg. In the Middle Ages, with the spread of the Ottoman Empire, the age-old route of supply from the Spice Islands via Arabia to Egypt and thence to Venice was disrupted. Now the maritime western European nations, first Portugal and Spain, then Holland and England, sought supplies directly from the Indies and so began their conquest of the world. The competition between Portugal, Holland and England for a part of the trade was intense and often brutal. The Dutch East Indies company V.O.C. (Vereenigde Oost-Indische Compagnie) and the East India Company (John Company) vied for command. After the Amboyna massacre in 1629 the Dutch gained unchallenged mastery from Portugal and England in the East Indies and a monopoly of trade there that was to last for a century. It was finally destroyed when, in 1770, a French botanist called Pierre Poivre smuggled nutmeg and clove seeds out. In 1796 under the pretext of the Napoleonic wars the English invaded Banda and transported nutmeg plants to British colonies. Soon an alternative industry was established in the West Indies, especially Grenada. Subsidiary trades were established in Asia, initially to finance the spice trade, that in time became more important than the spice trade itself: cloves for salt in the Persian Gulf, cloves for gold or opium in India, opium for gold, silver, silk and tea in China. Cotton fabrics and china porcelain were carried as well.

Flavourings from the Americas include chilli peppers (*Capsicum*), vanilla from the orchid *Vanilla planifolia* (Figure 7.21) and cocoa from *Theobroma cacoa*.

It is a remarkable fact that these three flavourings were combined with honey in the drink 'Chocolatl' that was consumed in large quantities by the Aztecs, in a custom they probably learnt from the south. They believed that the cocoa tree was of divine origin. They obtained cocoa from Central America by trade. Indeed cocoa beans were even a form of currency. Cortés brought cocoa beans back to Spain in

Figure 7.21. Vanilla plant growing in the wild in Mexico.

1528, and gradually the custom of drinking chocolate spread, especially when chilli was left out, sugar, nutmeg and cinnamon added, and the drink served hot. By the mid seventeenth century the drink, favoured in part because of supposed medicinal qualities, had reached England. Both the chocolate makers, Fry's of Bristol and Terry's of York, started as apothecaries. In London, around 1700, someone thought to add milk and then in early Victorian times methods to solidify chocolate were perfected.

The Spanish monopoly on the trade in cocoa beans was broken by the Dutch when they captured Curaçao, and later when the French captured Cuba and Haiti. Today Ghana is the major producer although cocoa was first planted there only in 1879. Chocolate contains caffeine, like tea and coffee, but also theobromine. There are three main types of cocoa, called Forastero and Crillo, and their

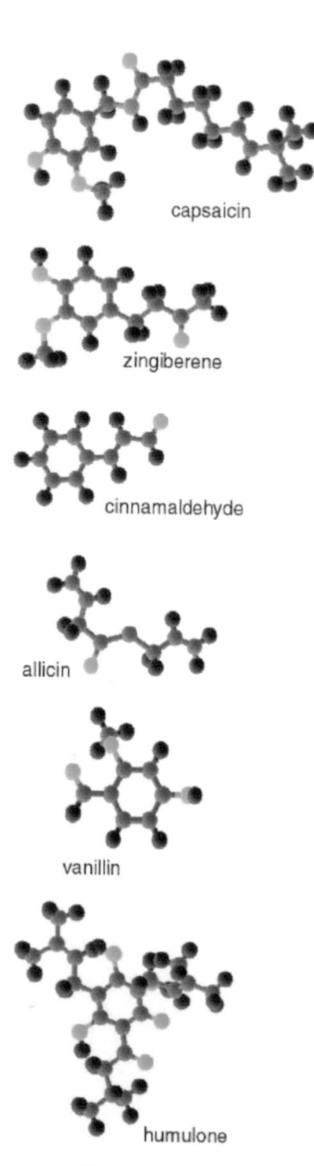

Figure 7.22. Chemical composition of flavourings. Capsaicin is the hotness of chilli peppers; zingiberene is ginger; cinnamaldehyde is the flavour of cinnamon; allicin is the flavour of garlic; vanillin is the flavour of vanilla; and humulone is the flavour of hops.

hybrid Trinitario. Forastero is the most important, producing strong flavoured beans. The beans are extracted from the pods and cleaned before being allowed to ferment and dry.

7.3.2 Drugs as medicines and remedies

Knowledge of the use of plants as a source of drugs was mystical, magical and powerful. On Egyptian papyri from 3500 years ago there are some of the earliest surviving lists of medicinal plants. The power was jealously protected – different kinds of plants were given secret sacred names. *Ambrosia maritima*, a ragweed, now used to flavour liqueurs, was called vulture's heart. Theophrastus makes fun of the superstitions of some herb gatherers who advocated, for example, the collection of peony roots at night lest their eyes were pecked out by woodpeckers. Another superstition was that the only way to pull mandrake from the ground was to tie a dog to it and from a distance call the dog with a horn. Pulled from the ground the shriek of the mandrake killed the dog but out of earshot the herbalist was spared. However, Theophrastus seems to have respected, and quotes, the Athenian herbalist Diocles who had collected together the existing knowledge of medicinal herbs. Only fragments of Diocles' written work survive, but this work probably provided the foundations of all later works by herbalists like Crateus, doctor of King Mithradates of Pontus, and Sextus Nigereach, both of whom made their own original additions.

The culmination was the herbal produced by Dioscorides in about AD 60. The *Materia Medica* was the foundation of botanical knowledge for a millennium and a half even though its descriptions of plants were poor. Widely read, and widely travelled, probably while serving as a doctor in the armies of the Emperor Nero, Dioscorides was able to incorporate his own good sense and experience in the work. It included nearly 600 plants.

Figure 7.23. An image from the *Codex Vindobonensis*.

Luckily a beautifully illustrated version of the *Materia Medica*, called the *Codex Vindobonensis*, has survived. Produced in AD 512 for Juliana Anicia, who was the daughter of Flavius Anicius Olybrius, the Emperor in the West, the *Codex* has a chequered history. It first turned up in 1406 in a monastery in Constantinople. After 1453, with the conquest of the city, it was in the hands of the Turks. The Jewish doctor of Suleiman the Magnificent seems to have purloined it. Busbecq, that adventurous ambassador of the Holy Roman Emperor, saw it and managed to get some drawings for Mattioli. Seven years later a sale had been negotiated and the *Codex* arrived in the Imperial Library in Vienna. The importance of the *Codex* is that there is good reason to believe that many of the drawings, which are very naturalistic in style, are derived from earlier ones drawn from nature, perhaps even by Crateus himself. The *Codex Vindobonensis* represents a peak of botanical knowledge and observation. For the millennium after it was produced, there was a sad decline in the quality of copies of Herbals and scarcely any new observations were made.

The collapse of the Roman Empire almost extinguished botanical knowledge in the west. Only Christian monks kept a flicker of the classical expertise alive. There was a more ancient herbal tradition in the east: the Ayurveda, from the Sanskrit *ayur* (life) and *veda* (knowledge) that probably dates back 5000 years. By 800 BC different herbals mention 500 and 760 medicinal plants. *Manushi* is the use of plants in treatment. Herbal remedies are substances (*dravyas*) that work in the body by their properties (*guna*) such as ushnatva (hotness), *ruksha* (dryness) and *pichhilatva* (sliminess). Herbs are classified according to their habitat and their actions on *dosha* or body type. The medicinal properties can be increased by various treatments (*sanskar*). Their purpose is to promote the body's own healing processes. The combination of ginger (*Zingiber officinalis*), turmeric (*Curcumin longa*), frankincense (*Boswellia serrata*) and ashwagandha (*Withania sominiferum*) has been shown in a couple of studies to reduce swelling in rheumatoid arthritis and osteoarthritis. Triphala ('three fruits') is an Ayurvedic combination of amalaki or Indian gooseberry (*Phyllanthus embilici*), bibhitaki (*Terminalia belerica*) and haritaki or herda (*Terminalia chebula*) that provides a general health tonic promoting good digestion, increasing red blood cells and removing undesirable fat. *Phyllanthus embilici* is also the main component of Chyavanprash, a general tonic containing 40 or more different herbs provided as a jam. It has 30 times more vitamin C than oranges. Other components are ashwagandha, haritaki, cinnamon, shtavari (asparagus), bamboo, clove, cardamom, pippali or long pepper (*Piper longum*) and Bilva or Bael Tree (*Aegle marmelos*). These ingredients have a variety of activities; most have multiple activities. *Aegle marmelos* is astringent and has antiviral antihelmintic, anti-inflamatory and antimicrobial (against *Vibrio cholera* and *Salmonella*) properties. Another Ayurvedic herb, kutki or katuka (*Picrorhiza kuroa*), stimulates the immune system, is anti-allergic and preliminary clinical trials showed it to be an anti-asthmatic.

Traditional Chinese medicine is also part of an old and surviving tradition of herbalism in Asia. The Chinese emperor Shen Nung compiled the herbal Pen Tsao more than 2500 years before Dioscorides. Traditional Chinese medicine is becoming increasingly popular and there are said to be 400–500 Chinese medicinal herbs now available in the west. Medicines are available for any illness or disease. Efficacy relies on a complex mixture of species so that there has been a problem, first in identifying the active components, and second in ensuring their safety and freedom from adulteration. One notable case occurred in Belgium in 1991–1992: slimming capsules that contained two Chinese herbal medicines had the climbing herb *Stephania tetrandra* along with the superficially similar looking *Aristolochia fangchi*, which contains kidney toxins and carcinogens. Over 100 cases of acute renal failure were reported.

Herbalism has, at times, been overcome by strange ideas. Plants with red organs, like the red seeds of peony, were prescribed for menstrual problems or bleeding. The dark-purple loosestrife, *Lysimachia atropurpurea* (Primulaceae), was used to stop bleeding. The hardness of the seeds of gromwell, *Lithospermum officinale*, meant it should be used to break up kidney stones. The quackery present in many works

Figure 7.24. Herbal remedies. (a) *Hyoscyamus niger* (henbane) is a source of alkaloids with a hypnotic and narcotic effect. Gerard called it 'English tobacco', and he recommended it use for cuts and bruises but cautioned against smoking it. (b) *Aristolochia* (birthwort). By the *Doctrine of Signatures* it was recommended for female complaints because the flower resembled the shape of the womb.

had a long history. Pliny had reported the use of the ashes of a rose gall, looking like a ball of down, to be mixed with honey and applied to a bald head, to make hair grow. Man's vanity and credulity has not changed much in two thousand years. Some remedies, by happy accident, were effective (Figure 7.24). The root of mandrake, *Mandragora officinalis*, shaped like a man, was especially powerful. It was used to put people to sleep before surgery, and used along with henbane and opium, but if its strength was miscalculated the anaesthesia became permanent! Similarly the efficacy of Ginseng root, *Panax ginseng* was advertised in part because of the similarity of the root to the human body. It is perhaps the most famous of all Chinese medicines, but paradoxically its efficacy against anything much has been doubted. The root contains a wide range of constituents, 2%–3% of which are 'ginsenosides', triterpenes and saponins.

Figure 7.25. *Catharanthus roseus* (Apocyanaceae), a source of vincristine, a drug that counters childhood leukaemia.

Plants provide one quarter of all prescribed medicines. An A–Z of a few are listed in Table 7.10. Many others have been recommended for one kind of ailment or another. Not all perhaps, are very efficacious. However, some plants do provide drugs that have a profound effect on human physiology. These are the poisons. In different parts of the world different plants have provided arrow or fish poisons. Most are sources of alkaloids. A different species of *Stephania* (*S. hernadifolia*) is used as a fish poison in Australia. Aconitine from the monk's hood *Aconitum* (Ranunculaceae) was used in various parts of the Northern Hemisphere as an arrow poison. *Strophanthus* and *Strychnos* in Africa and Asia, and the latter also in South America, were other sources. *Chondrodendron* and *Curarea* (both Menispermaceae) and *Hura* (Euphorbiaceae) provided other poisons of remarkable efficacy. *Hura* provides a poison half a million times more toxic than potassium cyanide.

Poisons have also been used to kill people. Socrates, convicted of not recognising the gods of the state, introducing new divine

Table 7.10 An A–Z of notable medicines derived from plants to illustrate the range of medicinal plants. There are hundreds of others

Species (Family)	Vernacular name or drug name	Ailment or action
Agave sisalina (Agavaceae)	Steroids	Hormones
Berberis vulgaris (Berberidaceae)	Berberine	Eye disease
Catharanthus roseus (Apocyanaceae)	Vincristine	Leukaemia and Hodgkin's disease
Dioscorea sp (Dioscoreaceae)	Diosgenin	Contraceptive
Ephedra sinica (Ephedraceae)	Ephedrine, Pseudoephedrine	Bronchodilation, rhinitis
Filipendula ulmaria (Rosaceae)	Aspirin	Analgesic
Glycyrrhiza glabra (Fabaceae)	Liquorice	Cough mixture, Constipation
Hydnocarpus kurzii (Flacourtiaceae)	Chaulmoogra oil	Leprosy
Inula helenium, I. racemosa (Asteraceae)	Radix helenii, absinthe	Skin and chest disease
Juniperus communis (Cupressaceae)	Volatile oil, tannins	Cystitis and rheumatism
Karwinskia humboldtiana (Rhamnaceae)		Tetanus
Lavandula angustifolia, L. latifolia (Lamiaceae)	Lavender	Antiseptic, insecticide
Melilotus officinalis (Fabaceae)	Dicoumarol	Anticoagulant
Nepeta cataria (Lamiaceae)	Catnip	Colds and colic
Podophyllum peltatum (Berberidaceae)	Etoposide, podophyllotoxin	Warts, cancer
Quassia cedron (Simaroubaceae)	Cedron	Vermifuge, fly-papers
Rauvolfia serpentina (Apocyanaceae)	Reserpine (alkaloid)	Hypertension
Syzygium aromaticum (Myrtaceae)	Eugenol	Toothache
Taxus brevifolia, T. baccata (Taxaceae)	Taxol	Cancer
Uvaria chamae (Annonaceae)	Finger-root	Eyewash
Veratrum album (Liliaceae)	Protoveratrine	Hypertension
Withania somnifera (Solanaceae)	Ashwagandha	Narcotic, diuretic
Xanthocephalum sarothrae (Asteraceae)	Snakeweed	Fever, infection, anti-tumour
Yucca sp. (Agavaceae)	Yucca root	Anti-inflammatory
Zanthoxylum americanum, Z. clava-herculis (Rutaceae)	Prickly ash	Febrifuge, toothache

things, and corrupting the youth, was sentenced to drink an infusion of hemlock (*Conium maculatum*), which, unusually for the umbels (Apiaceae), is rich in alkaloids. The bean of Calabar (*Physostigma venenosum*) was used in a trial by poison by the Efik tribe in Nigeria. If the accused vomited up the poisonous potion they were innocent. But such powerful drugs have also been used as medicines. Today *Physostigma* is the source of physostigmine which is used in the treatment of glaucoma.

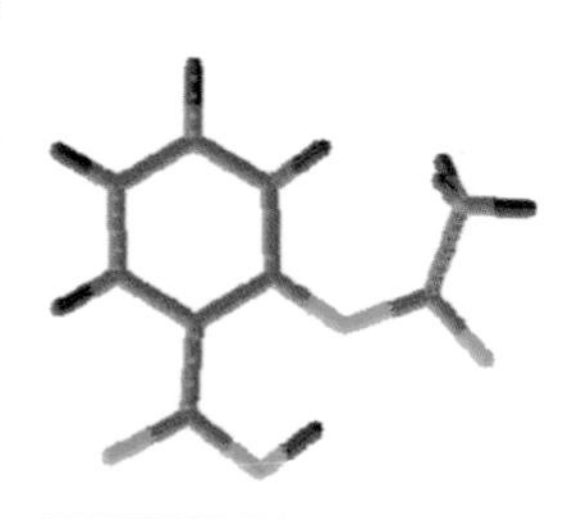

Figure 7.26. Aspirin.

Perhaps the widest used medicine of all is aspirin, first marketed by the drug firm Bayer in 1899 as 'a' 'spirin', a synthetic acetylsalicylic acid that had previously been obtained from *Filipendula ulmaria*, then called *Spiraea ulmaria*. Gerard had recommended it boiled in wine to provide a remedy for pains of the bladder. The Greeks of the classical world and the native North Americans had obtained their own acetylsalicylic acid from willow bark (*Salix*).

One overriding danger for the explorer in the tropics was catching malaria. There are various legends and stories associated with the discovery that the bark of a tree native to South America provided a remedy. In 1633 a Jesuit priest called Father Calancha in Peru had noted its efficacy. In 1638 the wife of the Count of Cinchon, the Viceroy of Peru, lay very ill with malaria. The desperate physician successfully tried 'Peruvian bark' and a remedy was borne. Peruvian bark was exported by the Jesuits to Rome, at that time a highly malarial location. Its use spread throughout Europe.

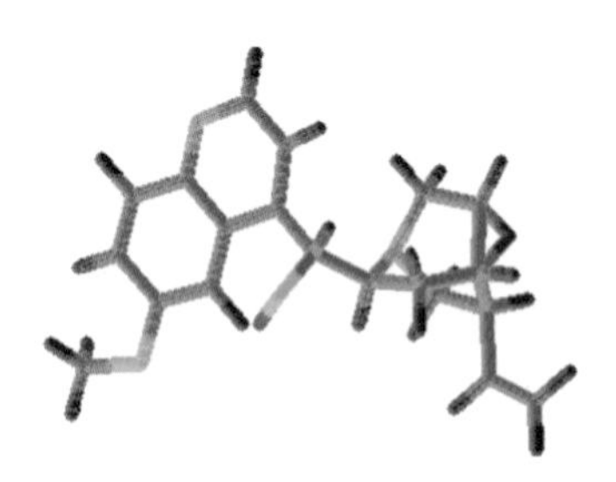

Figure 7.27. Quinine.

In the eighteenth century Joseph de Jussieu described the tree. Linnaeus later named it *Cinchona* after its first famous recipient. In 1820 the French chemists Pelletier and Caventou isolated the active component of the bark, the alkaloid quinine. The value of the bark was so great that the Dutch and the British attempted to break the South American monopoly. Joseph Banks was the first to suggest the collection of the species of cinchona from the Andes. Clement Markham, who had retired at the age of twenty-one from the Royal Navy to become an explorer, managed to persuade the Indian Office and Hooker to fund an expedition to collect *Cinchona* trees and transport them to India for cultivation, some via Kew, some directly to Calcutta. The transfer, made between 1852 and 1854, was made possible only by the invention of the Wardian Case, a kind of transportable mini-glass-house. Meanwhile, in 1852, Hasskarl the Dutch director of the botanical garden in Java entered South America under a false name and bribed an official with a bag of gold for some *Cinchona* seeds. The results of these competing efforts were a relative failure. The plants gained were low-yielding varieties. An alternative source was from an Australian called Ledger who persuaded an Aymara Indian called Incra to smuggle seeds of high yielding plants out of Bolivia. Incra was successful but was later discovered and tortured to death. Ledger failed to sell his seed to Britain. They were wary because plants previously supplied by Ledger were low yielding. The Dutch government took the chance and thereby a multi-million dollar industry was established in Java based on high-yielding *C. ledgeriana*. At one time it produced 97% of the world's quinine, which was traded through Amsterdam. The South American industry based on wild trees was destroyed. However, the capture of Java by the Japanese, and Amsterdam by the Germans in the Second World War, gravely threatened the allies ability to fight in the tropics. This re-ignited the South American industry and also the search for synthetic substitutes that, after the war, eclipsed the use of quinine.

Figure 7.28. Neem (*Azadirachta indica*). Chewed neem twigs keep teeth healthy and neem extract, containing salannin, is now included in some toothpastes. Neem oil repels insects of many sorts including mosquitos, chiggers and ticks, lice and scabies and is also fungicidal, bactericidal and antiviral. It has also been used as a contraceptive douche.

Although several effective drugs, like aspirin and quinine that started out as natural plant products, have now been replaced by a synthetic product or a synthetic analogue, plants are still a store of huge potential. There is no better example than the Neem tree (*Azadirachta indica*). It has been used for hundreds of years in India and Burma and is a veritable cure-all.

7.4.3 Beverages

Plants have also provided recreation. Teas of various sorts made from infusing leaves or roots with boiling water, have been used for millennia. As well as extracting the chemicals from the leaves, boiling water had the benefit of sterilisation when water sources were likely to be polluted. Mint (*Mentha*) tea is popular in Islamic cultures. The mints contain the essential oil menthol, menthone and menthyl acetate, flavoured with many other minor components good for indigestion and colic. Valerian tea (*Valeriana officinalis*) is noted for its sedative effects, probably induced by the iridoid alkaloids called valepotriates. Pacific Islanders prepare Kava from the rhizomes of *Piper methysticum*. A mixture of different lactones produces a feeling of brotherhood and tranquility when administered in the formality of the Kava ceremony. Like the ritualistic drinking of wine, a few drops are allowed to fall to the floor as a libation to the ancient gods. The three most important components are kavain, dihydrokavain and dihydromethysticin. The first anaesthetises the mouth like cocaine but the second is more readily absorbed and metabolised.

The most widely used teas are stimulants and most, including 'tea' itself, contain caffeine. The caffeine content of the raw product does not really indicate the caffeine 'hit' that a beverage provides because of the different ways they are prepared and the strength of the brew. Typically, a cup of instant coffee can have more than twice as much caffeine as a cup of tea brewed for 1 minute. Ground coffee has much more but so has stewed tea. Cocoa has a quarter the caffeine of tea, but chocolate has more. Theobromine is a weaker stimulant than caffeine, and is also found in tea and maté. Colas of various sorts rival instant coffee for their caffeine content. In recent years guarana (*Paullinia cupana*), used with cassava to make an alcoholic beverage in the Matto Grosso, has been widely added to alcopops to make stimulant or reviving drinks because of its high caffeine and theobromine content.

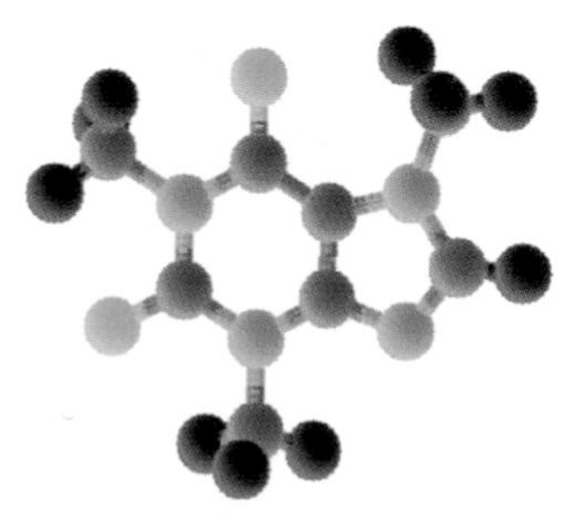

Figure 7.29. Caffeine.

The Chinese emperor Shen Nung discovered green tea in 2737 BCE, from *Camellia sinensis*, when some leaves fell into his drinking bowl. The Portuguese probably introduced tea to Europe from the Far East. By the early nineteenth century a huge trade through the port of Canton, probably half to Britain, was in place. It was the tax on tea re-exported to America that was used as one excuse for the rebellion that led to the War of Independence of the United States: 'no taxation without representation'. Chinese civilisation was brought low by other westerners by the sale of opium, traded for tea, among other items. Introduction of tea plants from China to India had failed, but then Wallich, once director of the botanic gardens in Calcutta, identified a native Indian variety of tea plant in the hills of Assam that became the foundation of the Indian tea industry. Today tea plantations are found throughout the highlands of Asia and Africa (Figure 7.31). For a long time Indian and Sri Lankan tea was regarded as inferior in quality to Chinese tea. Green Chinese tea has the tannin epigallocatechin as its main flavour component. The tannins in black fermented tea are oxidised and very complex.

Figure 7.30. Cocoa tree (*Theobroma cacoa*) with cauliflorous fruits.

Figure 7.31. Tea plantation.

Like many plants used for pleasure, the drinking of tea became the centre of a ritual in both Britain and Japan. Different varieties, conditions of growth, and treatment after harvest, result in the subtle differences in taste and colour: Assam, deep bronze and malty; Ceylon, pale gold and delicate; Darjeeling, light and fragrant; Kenyan, coppery and strong; Lapsang Souchong, smoky. It is curious that some of the more specialist teas like Earl Grey, with its citrus and bergamot, arose firstly as teas adulterated with other plant material to make up bulk and weight before they were traded.

Coffee is another Old World species. It probably originated as a crop in Ethiopia. Its oldest recorded use is as the leaves or beans chewed to relieve fatigue and hunger. Coffee drinking originated in Arabia in the fifteenth or sixteenth century and spread via Turkey to the rest of the world. Coffee cultivation was introduced to the east from plantations in Yemen, especially by the Dutch who established plantations on Java by the late seventeenth century. There were two main original introductions to the New World. One was via the Amsterdam botanical garden of a single plant sent from Java. This was the source of a few seedlings sent to the Paris Jardin des Plantes that were then sent to Martinique. Another introduction was from Yemen via Réunion and then to the Caribbean. As a result the genetic base of New World coffee is very narrow and it is very vulnerable to leaf rust (*Hemileia vastratix*).

Maté (Brazilian or Paraguay tea) or *Ilex paraguarensis*, provides a caffeine-rich tea that is popular in South America. Wild trees are also harvested, along with a few related species to supplement the plantation supply.

Kola tree (*Cola nitida* and *C. acuminata*) originated in West Africa. The nuts that provide the caffeine rich flavouring of many soft drinks are actually the embryos after the fleshy seed coat is removed. Kola is a relatively recent domesticate, within the past 1000 years or so.

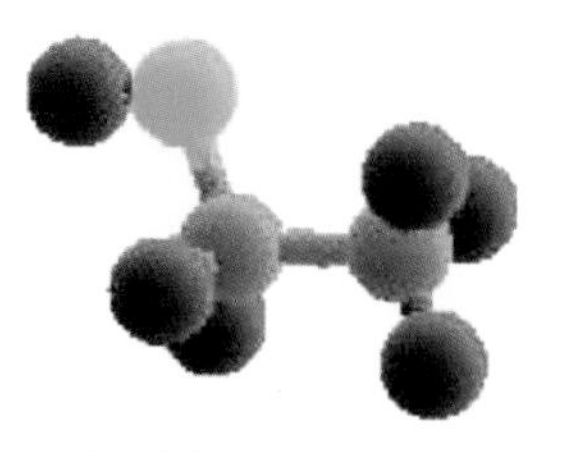

Figure 7.32. Ethanol.

7.4.4 Alcohols from plants

Almost every crop has been utilised to produce alcohol by the fermentation of sugar, or starch first converted to sugar. The process is as varied as the kinds of drinks created. South American women chew maize kernels and spit it into a pot before leaving it to ferment. The Sumerians made bread with wheat and barley and then fermented bread crumbs to create an alcoholic beverage called Sikaru sucked through reeds as straws to avoid the debris.

Alcohol has also a long history of medicinal use. The strong beer produced in classical times called Zythum was a component of many medicinal remedies. Medicines were dissolved in it, and in previous centuries it was given in great quantities to the ill. At least, if drunk in large enough quantity, it kept the patient quiet. It was not until the late Victorian era that hospitals were founded in Britain on the basis of strict abstinence. But it is not so long since stout was marketed as being 'good for you'. Although today, advertisers might be wary of claiming medicinal value for alcoholic beverages, who can doubt the profound pleasure and soothing of the soul that alcohol can bring and not just from intoxication?

Figure 7.33. Barley (*Hordeum vulgare*).

The most important alcoholic drinks are made from grain or from grapes. Barley is the grain used most often to make beer and spirits. It is first soaked and then allowed to germinate so that the starch is turned into sugar. Germination is brought to a halt by baking. The temperature and length of time baking helps to determine the flavour. Bourbon is distilled from a mash of grain containing, according to US Federal Law, not less than 51% corn, along with barley and either wheat or rye. Each distillery has its own unique blend of grain and some of the mash recipes are generations-old family formulas jealously guarded. In Africa, sorghum and millet are fermented to make alcoholic porridge and beer. Some notable spirits have other sources of carbohydrate than grain. Vodka is made from grain or from potatoes. Cassava is harvested in Brazil for industrial alcohol production. Tequila is usually clear in colour and unaged and is distilled from the fermented juice of the Mexican *Agave* plant, specifically several varieties of *Agave tequileana* (Figure 7.34). Mescal, a similar beverage to Tequila, is less expensive and stronger in flavour, and is made from an agave plant that grows wild in the Oaxaca region.

As important as the source of carbohydrate for fermentation are the many plants used to give flavour to alcoholic drinks. Honey and dates were used to flavour Sumerian Sikaru. The Ancient Egyptians used juniper, ginger and saffron to flavour their own strong beer called *heqet* or, in Greek, *zythum*. Distilled alcohol is a pretty flavourless substance. Other plants add flavour, whether it is the oak of sherry barrels used to age whisky or the wood charcoals used to filter bourbon. Bombay sapphire gin has ten different 'botanicals', the

Figure 7.34. *Agave tequileana* plantation in Oaxaca province, Mexico.

most important being juniper berries and coriander. Gin gets its name from the Dutch word *genever*, meaning juniper. The reputation of gin to bring on a miscarriage may be related to the effects of juniper. Beer gains its bitterness and aroma from hops *Humulus lupulus* which also prolongs the life of beer. Hopped beers were established in Medieval Europe and when they spread from Holland to England in the Tudor period the native brewers of unhopped ale complained and tried to prevent it.

Figure 7.35. Hop (*Humulus lupulus*) is a climber related to *Cannabis sativa*. It is dioecious; the female flowers have bitter alkaloids.

Absinthe has a fascinating history. In little more than a century, between the writing of the original recipe by Dr Pierre Ordinaire in 1792, and the founding by Henri-Louis Pernod of the most important absinthe distillery in France in the early 1800s, to its banning in France in 1915, it came to be favoured in the bohemian circles of France for its ability to stimulate creative activity. Its users and abusers included the poet Rimbaud, the writer Baudelaire and the painter Van Gogh, among many others. Absinthe is an emerald green alcohol, as a result of the presence of chlorophyll, with added herbs (aniseed, fennel, hyssop, lemonbalm, angelica, star anise, dittany, juniper, nutmeg and veronica), the most important being the bitter wormwood (*Artemisia absinthium*). The name wormwood denotes its former use to counteract parasitic worms. Vermouth, made from the flower heads of wormwood, gets its name from the German for wormwood (*wermuth*). Up to 90% of wormwood oil is thujone, also isolated from *Thuja occidentalis* and other plants. The psychoactive role of thujone is uncertain, but the related species *Artemisia nilagirica* was smoked in West Bengal for its psychoactive effects, and *Artemisia caruthii* was inhaled by the Zuni native Americans as an analgesic. Calamus (*Acorus calamus*) and nutmeg (*Myristica fragrans*) were also sometimes used in making absinthe and may have enhanced the psychoactive effect. Absinthe was diluted with cold water poured over a perforated spoonful of sugar, turning the shot of absinthe milky white as the essential oils were precipitated out of the alcohol. Much of the effect of absinthe was purely alcoholic. Undiluted absinthe had 60%–85% alcohol so a large part of its peril was associated with alcohol abuse.

Wormwood is used to flavour the Swedish brannvin made from potatoes. Chartreuse, made by Carthusian monks supposedly following an ancient recipe called the 'Elixir of Life', contains 130 herbs and spices, including small amounts of thujone. So also does Benedictine, which is made by Benedictine monks. Other drinks no longer use wormwood. Herb Sainte, from New Orleans, and Pernod are wormwood-free absinthes but contain star-anise (*Illicium verum*) for flavour. Pastis is a similar liqueur to absinthe and was also originally made with wormwood, but is now flavoured with liquorice (*Glycyrrhiza glabra*). Other essential oils are used to flavour many other drinks like Ouzo and Jägermeister.

Wine is one alcoholic drink that requires no adulteration but it too provides an example of the ambivalence there is about the use of plant products for pleasure. Dionysus, the god of wine, was the son of Zeus and a mortal woman. He taught the art of vine cultivation and gave the gift of wine, but he has two natures, bringing joy, health and divine ecstasy, or brutality and unthinking rage. He was accompanied by the Maenads, drunken women bearing rods tipped by pine cones (a reference, perhaps, to the use of resin to preserve wine), and who might go mad and rip apart and eat animals raw. Dionysus was associated with rebirth after death, like the vine growing back after it is pruned. Wine conferred a feeling of power as if by drinking it the drinker gained part of the divinity of Dionysus himself. There are echoes here in the use of wine in the Eucharist as a celebration of the Christian faith.

Wine was also associated with creativity. Most of the Greek plays were first performed at the feast of Dionysus. Nietzsche contrasted the creativism of Apollo with that of Dionysus: the first cool, structured, full of meaning and controlled; the second unpredictable, instinctual, wild, ecstatic, pleasurable and emerging from uncontrolled Nature. The Dionysian cult magnified the access to the natural world by use of fennel *Foeniculum vulgare* (Apiaceae).

The range of grape varieties allows wines to be savoured of such subtle difference and speciality (Table 7.11). Grapes (*Vitis vinifera* and other species) are widespread throughout the Northern Hemisphere. The grape vine was probably domesticated in South West Asia, spreading from there to reach Greece about 3000 years ago. When the Asian grape was introduced to North America it hybridised to native grape species especially *V. labrusca*. American grapes are resistant to the root louse *Phylloxera* and after an epidemic in 1867 they were widely introduced to Europe for use as rootstocks. The dried fruit, raisins, sultanas (seedless) and currants (from Corinth), are particularly rich in anti-oxidant compounds, especially the phenolic compound resveratrol.

Alcohol has fuelled the work of many artists and writers and destroyed the lives of not a few. Unfortunately the excessive use of alcohol does not guarantee the production of great artistic work.

Table 7.11 Famous varieties of wine

Cabernet Sauvignon	Deep red	Complex, depth of fruit flavours, blackcurrant, blackberry, long maturing
Merlot	Red	More acid than above and faster maturing
Shiraz, Syrah	Red	Luscious, silky, spicy
Grenache (Garnacha)	Red or Rosé	Sweet, fruity, low tannin
Pinot Noir	Red	Delicate fruits and flowers to rotting vegetables, perhaps the oldest variety
Gamay	Red	Beaujolais, acid but low in tannin
Chardonnay	White	Light and subtle but often heavily oaked
Sémillon	White	Dry or sweet, yellow, with hints of citrus
Sauvignon Blanc	White	Light with a touch of dryness, enjoyed young or slightly aged, made more robust by fermenting in oak
Riesling	White	German, sweet and fragrant with a touch of spice
Muscadet	White	Extremely dry, and light
Muscat	Red or White	Provides fruit as well as wine

7.4.5 Smokes, snuff and chews

Smoking has an age-old history. Pliny records the smoking of several materials including dried dung from an ox fed on grass. It was the Scythians who are first recorded as using cannabis for recreational purpose. They took steam baths placing cannabis seeds on heated stones. Mayan pottery figures record smoking, probably of tobacco, although *Nicotiana tabacum* is only one of about 60 species in the tobacco genus and other species were utilised, chewed, smoked or sniffed as snuff for the alkaloid nicotine they contain.

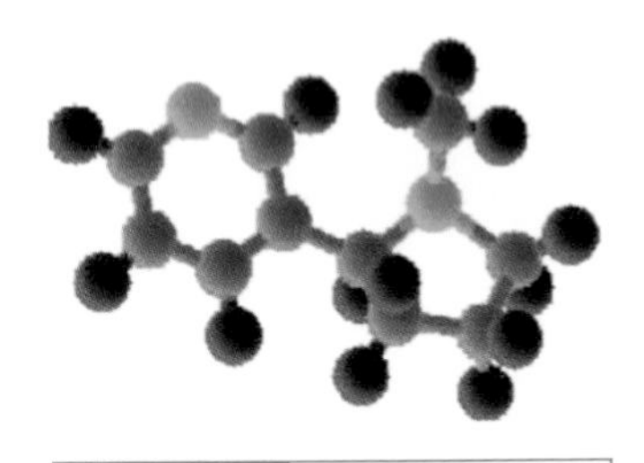

Figure 7.36. Nicotine.

Like other Caribbean tribes the Taino of Cuba used tobacco in rituals, recreation and medicine. They smoked tobacco in pipes and cigars but also made cigarettes wrapped in palm leaves, corn husk or bark. They sniffed a mixture of tobacco and coca-leaf dust to relieve fatigue and chewed tobacco as a stimulant. The first European to try tobacco was Rodrigo de Jerez, whom Columbus sent to explore inland Cuba. He liked it so much that he took the habit home to Spain but was imprisoned by the Inquisition for being in league with the devil for puffing away. It was the Portuguese who first cultivated tobacco outside the Americas from 1512 onwards, and snuff was being sold in the markets of Lisbon by 1558. In 1559, a French envoy sent to Portugal to negotiate the marriage of the King of Portugal to the daughter of the French King introduced the habit into northern Europe.

The taking of snuff was already an established habit in Europe. Shakespeare records aromatic powders of orris root (sweet flag iris, *Iris florentina*), camomile (*Chamaelium nobile*) and white pellitory (*Anacyclus pyrethrum*) being passed around banquets. Other plants used were alehoof (*Glechora hederacea*) and sneezewort (*Achillea* sp.). Tobacco snuff was combined with orris, cinnamon, cloves, fennel, sage, bergamot and lavender. Sir Walter Raleigh may have first popularised the pipe-smoking of tobacco. He brought the habit back from the Virginia colonies where it was popular among the native Americans. By 1596 a German visitor to London noted the English passion for smoking. Like nutmeg, tobacco became especially praised as a prophylactic against plague. The habit was widespread in men, women and children. At Eton College the boys were beaten if they forgot to smoke their pipe of tobacco in the morning. The first successful commercial crop was cultivated in Virginia in 1612 by John Rolfe and within seven years it was the colony's largest export. The cultivation of tobacco increased the demand for slave labour in North America. Cigars and cigarettes did not become popular until the nineteenth century. The name is said to come either from the rustling sound of cigars, like cicadas (cigarra), or from the little garden (cigarral) where tobacco was first cultivated in Portugal. In 1839 'Bright' tobacco, or golden Virginia tobacco, was discovered accidentally by quickly drying the leaf. The later introduction of 'White Burley' tobacco leaf and the introduction of the first practical cigarette-making machine in the late 1880s spurred the wider smoking of cigarettes. It was another 50 years before the first health concerns were raised, and these were not related to the insecticidal properties of nicotine, but the correlation between the incidence of cancer and smoking first noticed by researchers in Cologne in 1930.

The smoking of cannabis has a parallel history with tobacco. Cannabis has been utilised as a medicinal remedy for many millennia. The name has an ancient Sumerian root. Cloth made from cannabis fibre has been dated at 9000–10 000 years, but the first pharmacological use is recorded as such in Chinese herbals dated to the Emperor Shen Nung more than 4000 years ago. Zoroaster placed cannabis at the top of his list of medicinal herbs in his Zend-Avesta written in 550 BCE. Leaves, fruits and also the resin obtained from female plants are utilised. Cannabis smoking became established in the Middle East. Notoriously, the Muslim sect, the hasshasshin, established by Al-Hasan, who lived by robbing passing trade caravans, used a cannabis beverage to induce a state of bliss. The use of cannabis in Europe is indicated by the ban placed on it by Pope Innocent VIII in 1484. Nevertheless, for centuries after this, the growing of cannabis was officially encouraged in Elizabethan England and throughout the Spanish Empire, and this was not just for the fibre. The Napoleonic adventure in Egypt at the end of the eighteenth century encouraged the use of cannabis in Europe, especially the resin, as part of a fad for everything oriental. Le Club des Haschischins was founded in Paris in 1844 for monthly meetings where the members could enjoy the hallucinogenic effects of the resin in convivial surroundings. By the late nineteenth century hashish smoking parlours were to be found in every American city, with 500 in New York City alone. However, the tide had turned, perhaps partly because, following the Spanish–American war, Mexicans were caricatured in Randolph Hearst's newspapers as pot-smoking layabouts. In 1914 the US Congress passed the Harrison Narcotics Act to control recreational use of drugs. Hearst's

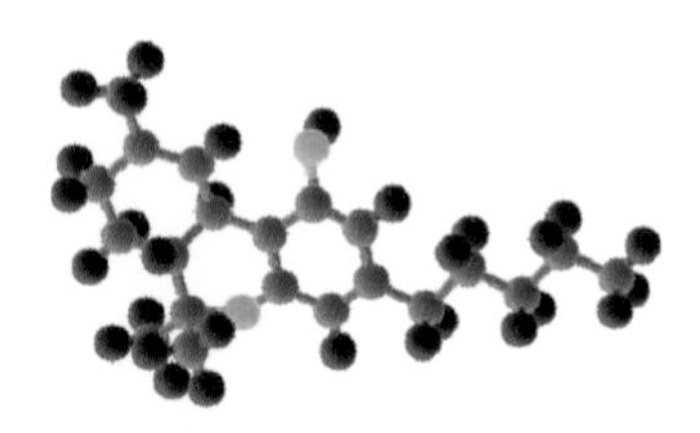

Figure 7.37. THC, the psychoactive component of cannabis.

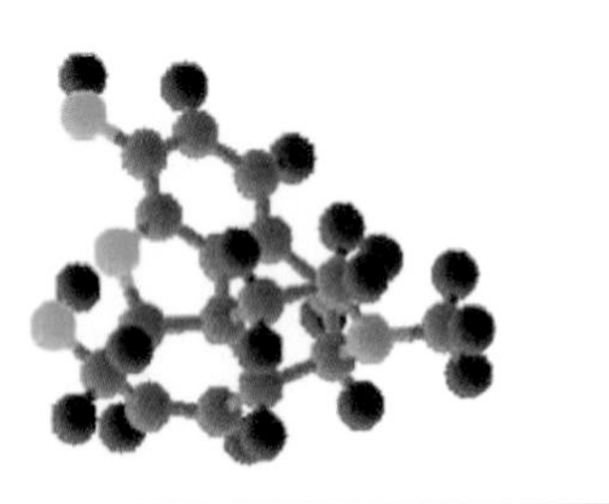

Figure 7.38. Morphine.

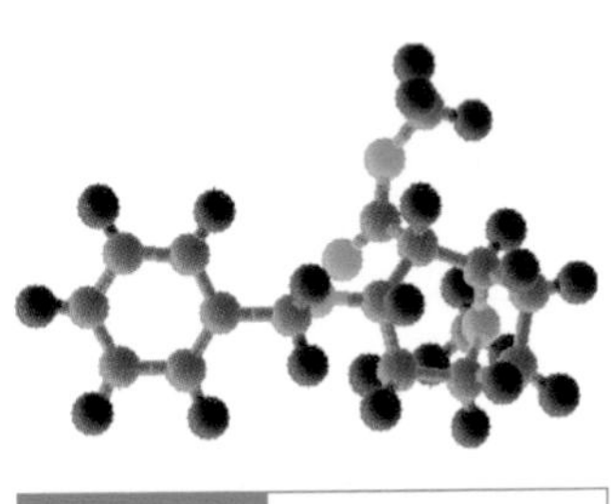

Figure 7.39. Cocaine.

tabloid campaign continued with articles having such headlines as 'killer weed from Mexico' and 'Marihuana Makes Fiends of Boys in 30 Days', and telling tales of 'marijuana-crazed negroes' raping white women. In 1937 The Marijuana Tax Act prohibited cannabis production, including for fibre.

Opium from the opium poppy (*Papaver somniferum*) now has a terrible reputation but it was not always like this. It had long history as a remedy for diarrhoea in China. Paregoric tincture still provides this remedy. However, in the seventeenth century the recreational use of the drug became more important. Opium contains more than 30 alkaloids including codeine, morphine, and papaverine harvested from the latex of the fruit capsule. Heroin is a synthetic derivative. Trade in opium helped to pay for tea. New sources from India from the eighteenth century onwards traded by the East Indies Company made it more widely available in China and they exported 1500 tons of opium each year, worth a billion dollars in today's values. One ton of opium could pay for nearly 40 tons of tea. Its import into China was banned in 1729 and 1800 but the ban was widely flouted by the western trading nations. In 1839 the Chinese authorities confiscated and burned a year's supply, precipitating the First Opium War between Great Britain and China. The result was that Britain was ceded Hong Kong and granted favoured trading status along with France and the USA. Trade conflict continued and, after the Second Opium War in 1856, France and Britain were given further rights. Chinese culture, shielded for millennia, was now exposed to western influences, including Christian proselytising. In addition, secrets of the tea industry were revealed.

Coca eases the headache and nausea associated with high altitudes. Coca leaves (*Erythroxylon coca*), like betel, are often chewed with lime to help the mouth absorb the alkaloids present. Mariani tonic wine enriched with an extract of coca leaves gained a cachet in nineteenth-century Europe. Extract of coca with caffeine from cola nuts was used in the early recipes for Coca Cola. In 1904 a law forced the removal of the cocaine. Now coca leaves are used to manufacture cocaine paste and powder. The mark-up from the unprocessed leaves to the powder sold illegally on the street is enormous, perhaps more than 1500 times. Not since nutmeg has such a mark-up been available for a plant product.

Psychoactive drugs, used medicinally, can also be used or abused recreationally. Opium, cocaine, cannabis have all been mentioned already but today perhaps the greatest fear and odium is applied to the use of hallucinogenic drugs even though their use has always played an important part in human spirituality and they have normally been administered in a highly ritualistic manner. The Priestess of the Oracle at Delphi uttered her prophesies while drunk on the vapours that issued from a cleft in the rocks beneath her feet, likely from some burning plant material. Psychoactive drugs also have had a contradictory fame/infamy for releasing the creative instincts of some artists. William Burroughs and Philip K. Dick are only two of many writers who have produced work of startling vision. The influence on artists is as profound. One wonders what drug Breughel was on to produce his startling visions of heaven and hell. It has even been suggested that the gentle poetry 'Leaves of Grass' of Walt Whitman may have been influenced by the psychoactive effect of the grass-like *Acorus calamus*.

Ebena snuff, used in shamanistic rituals by Amazonian natives, is made from three main plants: *Virola theiodora* (Myristicaceae), *Justicia pectoralis* (Acanthaceae) and *Elizabetha princeps* (Leguminosae). It is the *Virola* species that contain the rich mixture of psychoactive tryptamines and beta-carbolamines. The other species, *E. princeps*, used as ash, and *J. pectoralis*, used as calcium carbonate crystals, aid the extraction of these psychoactive components. Similar compounds are found in the legume, *Anadenanthera peregrina* (niopo or yopo). Another South American hallucinogenic plant is called ayahuasca, the 'vine of the soul', *Banisteriopsis caapi* (Malpighiaceae) and *B. inebrians*. Different varieties of the former give visions of different colours and content. Other important American hallucinogens come from two related families: *Turbinia corymbosa* and *Ipomoea violacea* (both Convolvulaceae), and *Datura* and *Brugmansia* (Solanaceae). *Datura* (Jimsonweed) has also been used in Africa and Asia for its hallucinogenic properties. In Europe the use of Belladonna (*Atropa belladonna*, Solanaceae) was associated with witchcraft. *Tabernanthe iboga* (Apocyanaceae) is used in West Africa.

Peyote (*Lophophora williamsii*), the sacred 'mushroom' of the Aztecs, is actually a cactus. There are several other cacti that have been utilised by native peoples for their psychoactive effects. The dried peyote is sliced into buttons for consumption. It is rich in alkaloids; the principal one (30% of total alkaloid content) is mescaline. Three hours after consumption, visual, auditory, olfactory and tactile hallucinations begin. They last for three days. Mescaline, like many of the other hallucinogenic compounds, mimics the chemical structure of the brain messenger compound serotonin. Other psychoactive drugs mimic, and gain their activity by mimicking, different compounds involved in neurotransmitter pathways. This is their strength and danger because many are highly addictive. A single exposure to cocaine produces long-term changes to brain dopamine cells.

(a)

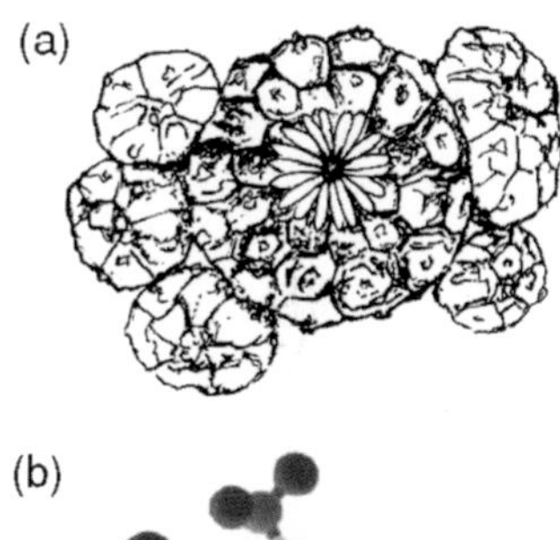

(b)

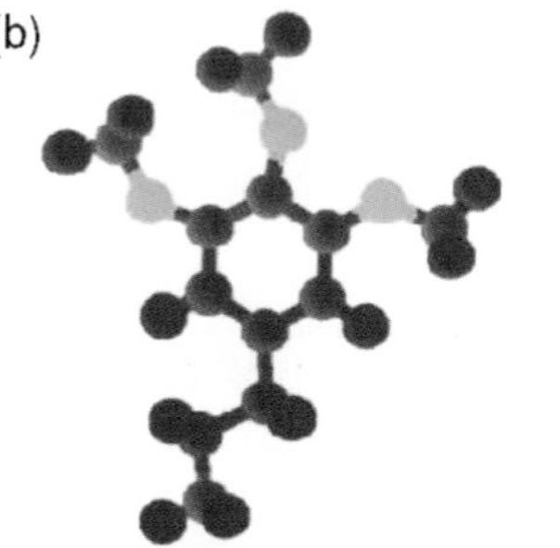

Figure 7.40. Peyote: (a) the button cactus; (b) mescaline.

Other dangers of abusing hallucinogenic plants are aptly illustrated by the abuse of *Salvia divinorum* from Oaxaca in Mexico that has become a fad in recent years. Known as Yerba de María ('Herb of María'), hojas de la Pastora ('leaves of the Shepherdess'), and Hierba de la Virgen ('Herb of the Virgin'), it contains a compound Salvinorin, which is said to be the most potent naturally occurring hallucinogen so far isolated. When the herb *Salvia divinorum* is consumed, either by smoking the dried leaf or chewing the fresh leaves, the effects are usually (but not always) pleasant and interesting because the amount absorbed is very small, but when vaporised and inhaled the smoker loses awareness and control over his or her body and may inflict self-injury.

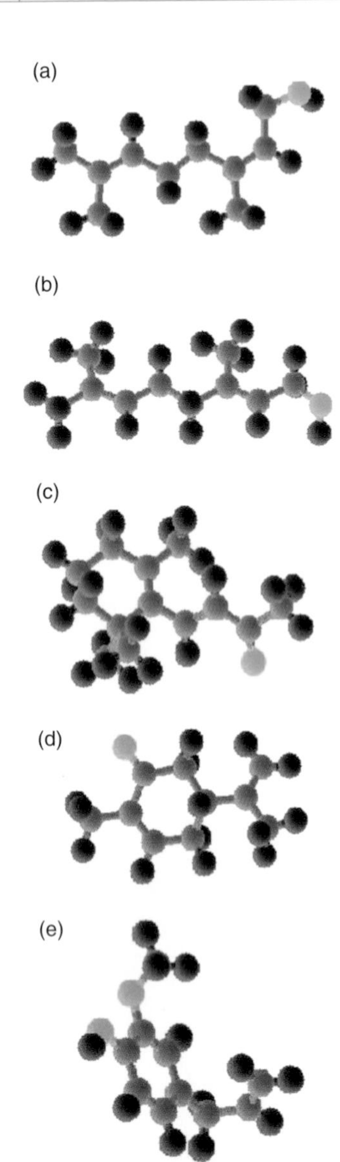

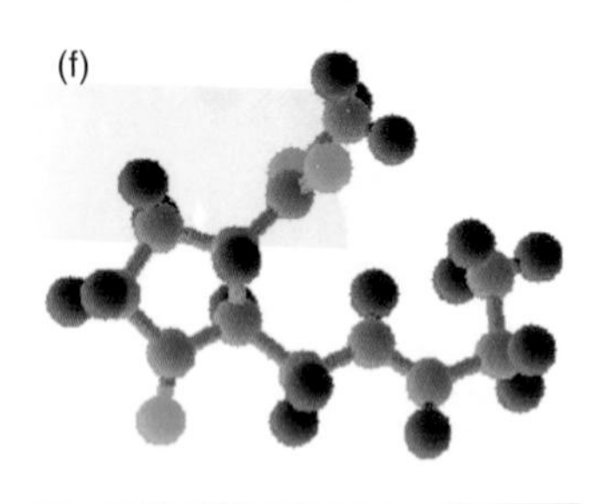

Figure 7.41. Scents from plants; (a) nerol; (b) geraniol; (c) ionone; (d) D-carvone; (e) eugenol; (f) jasmone.

Not all plant products used for pleasure have a psychoactive component. Chicle collected from the latex of *Manilkara zapota* trees in Central America provides the base for chewing gum. It contains no active components itself, though mint and other flavourings are normally added, but the process of chewing encourages the release of natural endorphins.

7.4.6 Scents and perfumes

The word 'perfume' comes from the Latin *per fumen* (through smoke), a reference to the use of incense in religious ceremonies. The same scented flowers cultivated by the Ptolomaic Egyptians are favoured in gardens today: jasmine, rose, lily-of-the-valley and stock. Flowers were grown in gardens in the Roman Empire for garlands and for perfume, and there was an extensive luxury trade in flowers. Egypt exported fresh flowers to Rome! The technique of distillation was established by Islamic chemists. Centres such as Damascus were famous for the perfumes they produced. The distilled scent of Damask rose produced a lasting perfume that was mainly the result of two components, geraniol and 2-phenylethanol.

Distillation involves the passing of steam through the flowers so that it becomes laden with volatile oils, and then allowing it to condense. The insoluble oil is then drawn off. Around 2000 roses yield just 1 g of the attar of roses. Alcoholic extracts of perfumes were first produced in Europe in the Middle Ages after contact with Islamic science. Various parts of the plant were used. The process today involves either cold-pressed extraction, extraction with ether, or steam distillation, and then washing and dilution with alcohol. Roses, violets, jasmine, tuberose, mimosa, jonquil, orange-blossom and bergamot from the Mediterranean region, plus lavender and peppermint from more northern regions are the most important sources of scented essential oils. Alcoholic extracts are complex mixtures. Bergamot from *Citrus bergamia* contains α-pinene, β-pinene, myrcene, limonene, α-bergaptene, β-bisabolene, linalool, linalyl acetate, nerol, nery acetate, geraniol, gerianiol actetate and α-terpineol.

7.5 The scientific improvement of plants

Food crops of diverse sorts have undergone similar changes as they have been domesticated: to extend their geographical area and season of growth, and to increase yield, palatability or attractiveness. These changes to crops have often been accompanied by radical changes that make them reliant on human intervention to be propagated. Crops grown for their vegetative parts, and propagated vegetatively, like cassava, sweet potato, yam and sugar-cane have reduced flowering, and flowers are sometimes sterile or partially sterile. These crops often have highly distorted chromosome complements and are evenly or unevenly polyploid, and aneuploid. Crops grown for their fleshy fruit and propagated vegetatively normally have maintained seed fertility, although seed yield may be reduced, but those propagated vegetatively, like bananas, may have become parthenocarpic. The scientific improvement of plants can perhaps be dated to the effort used to improve the sugar content of beet in the late eighteenth and nineteenth centuries.

7.5.1 Plant breeding

The nature of plant breeding is related to the biology of the crop; whether it is normally outbreeding, inbreeding or propagated clonally. Four distinct patterns can be identified. There are two distinct stages; first, the creation or discovery of new variation; second, the selection of a new variety with favourable characteristics. New variation can be discovered in existing land races or wild relatives of the crop and introduced into the crop by crossing. Alternatively, crossing different established varieties will produce an array of genetically distinct individuals among which the parent of a new variety may be sought. A first-generation cross produces a first generation, or F_1, of more or less identical siblings if the parents are highly homogeneous from being inbred for many generations; but between heterozygous clones or outbreeding plants a variable progeny is produced that may be selected. These F_1 plants are highly heterozygous and if crossed among themselves produce a second generation (F_2) that is highly variable. A slightly different strategy is adopted to introduce a particular gene, such as a disease-resistance gene, into a crop cultivar from a related cultivar or wild species. A hybrid is created between the cultivar and the plant bearing the wanted gene. Among the hybrid progeny those plants carrying the new gene are selected, for example by exposing the F_1 to the disease. However, these hybrid plants are usually intermediate between the crop and the source plant and lack many of the favourable characteristics of the original crop. These characteristics are re-established by repeated backcrossing to the original cultivar.

An important aspect of modern plant breeding is that the commercial variety should be relatively uniform. Selection increases uniformity by narrowing the genetic base of a crop, but potentially the initial crossing between different plants may widen it. This is not a problem with clonally propagated crops or in crops that are inbred, but, in plants that normally outcross, several generations of selection may be required to stabilise and make the crop variety uniform. An alternative is to produce a hybrid variety by crossing two known parents. Various techniques have been used to enable this, including the manipulation of male sterility either by mechanical emasculation or by the use of male-sterility genes. The seed source is grown as alternate rows of hermaphrodite and female plants and hybrid seed is collected from the female rows only.

7.5.2 Crop plants and disease

A major target of crop plant improvement has been to develop disease-resistant varieties. The modern practice of cultivation of highly genetically homogeneous cultivars makes crops especially vulnerable to pathogens adapted to that cultivar. Examples include potato blight in the mid nineteenth century, the vulnerability of European grape clones to American *Phylloxera* root aphid in the 1860s, the vulnerability of American banana clones to Panama banana wilt disease and leafspot (*Mycosphaerella*) root from 1933, and the vulnerability of Arabica coffees in Latin America to *Hemileia* leaf rust from 1970. The progress of the disease is exacerbated by the growing of the crop as a monoculture in large fields or plantations.

These diseases exhibit a characteristic boom and bust cycle that is related to the pattern of coevolution of the diseases and crop. The presence of new virulence in the pathogen selects for new resistance genes in the host to overcome it, and when these evolve the pathogen becomes less successful. It is then the turn of the host resistance genes to select for new virulence genes in the pathogen, thus repeating the cycle. The history of late potato blight, caused by *Phytophthora infestans*, provides a good example of this pattern, indeed it was first described by Van der Plank in a variety of potato called *vertifolia* and is sometimes called the *vertifolia*-effect. The disease resistance is called vertical resistance because it normally relies on one or few resistance genes. It is a strong but short-lived resistance. An alternative resistance, horizontal resistance, relies on the weak effect of many genes. It takes longer to evolve but is also more difficult for the pathogen to overcome. After the mid nineteenth-century potato blight, it took about 40 years of breeding to develop a useful level of horizontal resistance to blight in northern temperate potato varieties. Along with what has been called integrated pest management, the use of a combination of fungicides like metalaxyl, and new ways of cultivation and harvesting, blight ceased to be a serious problem in northern temperate fields. However, in the 1940s potato blight, which originated in Mexico's Toluca Valley, spread to South America. Here, in the centre of potato genetic diversity, horizontal resistance developed relatively rapidly, probably by crossing within and among genetically variable cultivars and perhaps even by natural crosses with wild relatives.

The story has a remarkable twist in its tail. *Phytophthora infestans* blight exists as two mating types in Mexico. Each can reproduce asexually but they can also cross sexually. The nineteenth-century epidemic in northern temperate potato clones was caused by just one mating type that was introduced by chance, probably on a single plant. It reproduced asexually and evolved relatively slowly by mutation. However, in the 1970s the other mating type was imported into Europe on potatoes, then sold on around the world. By the early 1980s there was a sharp increase in the prevalence of aggressive new variants of blight, and now blight is very difficult to control. Now that there are both mating types present *Phytophthora infestans* can reproduce sexually and a vast new genetic diversity has been released. It will be very difficult to develop any lasting disease-resistance, either horizontal or vertical.

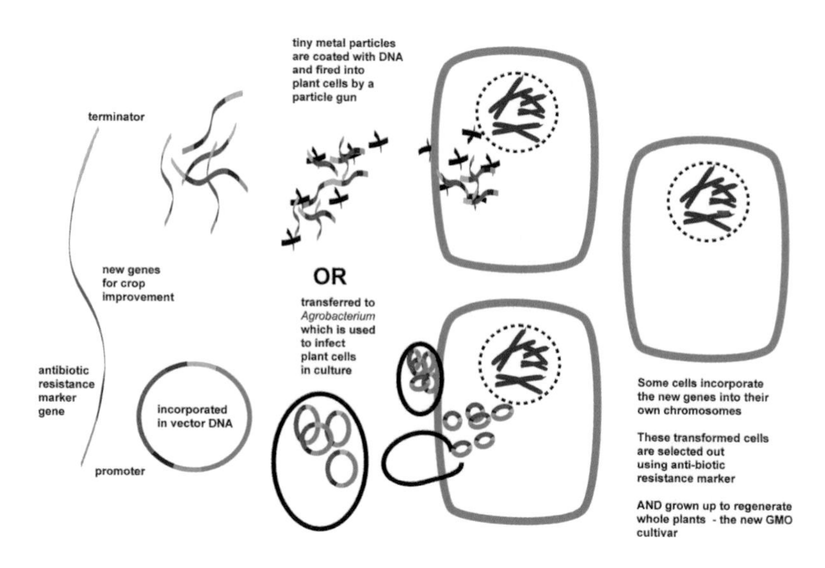

Figure 7.42. The production of GM crops.

7.5.3 GM (genetically manipulated) crops

The 'genetic engineering' of plants is often confused with plant-breeding, which has been going on for thousands of years through the process of hybridisation and selection, first by farmers and then by plant breeders. What has become possible more recently is the insertion into plants of completely foreign, even animal, genes, which are then expressed (Figure 7.42). This new technology may seem to offer exciting prospects for some, but it also harbours uncertain dangers.

Arising from the new technology is the ability to identify and target particular genes. A gene from an exotic species can be introduced into an existing crop along with the necessary DNA sequences that control its expression. In this way the crop can be made to produce novel products. These products may be new food compounds or may allow a plant to grow in new conditions. For example, the Flavr Savr tomato has the gene that causes fruit softening inserted in reverse and hence the ripening process has been slowed down. Rice has been transformed to improve its nutritional quality.

There have been two major uses so far. First, resistance to a proprietary herbicide has been introduced into a crop so that the herbicide can be used more effectively to make a weed-free field. Second, crops have been manipulated to produce a crystalline protein normally produced by the bacterium *Bacillus thuringiensis* (Bt). The protein prevents caterpillars and beetles feeding on the crop. There has subsequently been a marked drop in pesticide use in areas where Bt cotton is grown.

Genetic manipulation will also permit the introduction of disease-resistance genes from either related or unrelated species into a crop. For example, in 1977 genetic resistance to rice blight caused by *Xanthosoma oryzae* pv. *oryzae* was identified in a wild rice from Mali called *Oryza longistaminata*. The introduction of this resistance into cultivated rice by traditional plant-breeding would have taken many years of crossing and backcrossing. Instead the gene was isolated and cloned in a microbe to multiply it up. It was then introduced into crop rice by coating gold particles with the gene and firing them into a rice cell culture using a helium powered gun. Alternative methods of introducing exotic genes include the use of genetically modified plant viruses like the tobacco mosaic virus as the transporter.

The second stage in GM technology is selection of the transformed cells/plants. In the rice blight example this was done by raising plants from the cultured rice cells and exposing those plants to the disease. Those carrying and expressing the resistance genes remained healthy. Untransformed plants became diseased. However, more normally a controversial technique is to include, tightly linked to the introduced gene, a marker gene such as one for antibiotic resistance co-expressed with the desired product. This can be used to select the transformed plant cells. Kanamycin resistance was used in the development of Flavr Savr tomatoes. However, there is a serious concern that such marker genes may jump species in the gut of humans and make human gut microbes antibiotic resistant. There is also the worry that unplanned gene expression may take place. Ironically, a new variety of celery produced by traditional plant-breeding skills illustrates this point; it proved to be highly allergenic to some people.

Environmental worries focus on three areas. The first is that species such as butterflies and birds that rely on insects will suffer because of the increased industrialisation of the agricultural system, and the combined use of chemical herbicides and insecticides that many GM crops require. Such data as exist indicate that there are considerable marginal effects but this is unlikely to sway growers desperate to find a cost-effective way to combat pests and disease. The second worry is that, by crossing with wild relatives, super-weeds will arise. It has been shown that crossing does occur; the environmental impact of this is uncertain but the potential is great. Some of the most persistent weeds of grain crops are wild oats species. The third concern is that any gains will be short-lived because the very strong selection applied to pests and weeds means that new virulent pests and herbicide-resistant weeds will soon evolve.

Soya beans (58%), maize (23%), cotton (12%) and oilseed rape (6%) were the most important GM crops in 2000, covering 16% of the total acreage mainly in the USA, Argentina and Canada. But the cat is out of the bag and China is likely to become another significant grower of GM crops very soon.

Nevertheless the idea that plants are factories where gene machines can be housed to make any product we desire is dangerously attractive. The production of phaseolin seed-storage protein from French beans in sunflower cells was an early achievement. Another early example was the production of the component of plastics polyhydroxybutyrate (PHB) in *Arabidopsis*. Attempts have been made to improve jojoba for production of wax esters and guayule for latex production and other species for technically useful oils. However, these initiatives have so far had only limited success because of lack of productivity and quality.

The genetic engineering of plants for the production of novel materials requires a deeper understanding of plants as a whole: an understanding that humans lack. Just because we know the complete genetic sequence of *Arabidopsis* doesn't mean we know what the 24 000 genes do, nor crucially do we understand how all the products of these genes integrate with each other and the environment. There is a growing number of people who find the whole idea of GM crops deeply offensive, and who advocate a more enlightened approach using organic methods of food production. For many, GM crops are simply a means for globalised corporations to entrench their control and dominance of the world's food supplies.

7.6 The flowering of civilisation

Flowers hold a central place in human history. The use of the word 'flower' for 'the best', 'the most attractive part', 'the essence', and the way the word 'flowering' is used to describe a new cultural development, are a measure of the central place of flowers in our culture. Agriculture had separate origins in Western Asia, Central America and China. The earliest traces provide abundant evidence for the botanical knowledge that gave rise to civilisation. Even the monuments are marked with patterns derived from plants.

7.6.1 Symbolic flowers

Tomb paintings and carvings in Egypt, dating back 5000 years, show gardens; a formal pool containing lotus, fringed with papyrus, and shaded by palms and figs. One painting records an expedition in 1480 BCE to the south to collect perfume providing shrubs for Queen Hatshepsut's garden at Thebes. The parks and gardens of Mesopotamia are not just legendary, they may have provided the inspiration for the description of the Garden of Eden. The Hanging Gardens of Babylon, constructed in the reign of Nebuchadnezzar II in the sixth century BCE, were artificial, terraced and irrigated hills. The gardening tradition lived on. The word for paradise comes from the Persian word *pairadaeza* for a walled park or garden. Epicurus (341–270 BCE) who advocated the seeking of pleasure as the supreme goal in life, also made the first garden in the city of Athens. His students were known as 'the philosophers of the garden' because he instructed them there. Lotus buds, rosettes and palmettes are recurrent decorative themes both in Egypt and in Mesopotamia. In Egypt a heraldic lily, representing the south, was produced prolifically on sculptures from Old Kingdom times. Papyrus represented the north. Later *Acanthus* leaves provided the inspiration for the carved capitals of columns in Greece.

Flowers provided inspiration for styles of personal adornment, like the beautiful necklaces of blue and white lotus, daisies and cornflowers from Egypt, and the Sumerian caps, crowned with gold daisies, that can be seen in the British Museum.

Plants have had an important symbolic significance. Olive foliage was the sign of goodwill (the olive branch). Cherry laurel (*Prunus laurocerasus*) was used to crown the Caesars, perhaps quite appropriate considering their later murderous behaviour, since the leaves contain cyanide that is released when they are damaged. The rose, sweet scented and full of soft petals, was associated with luxury, with Venus, with love and spring, but also with death. Garlands of roses were placed as wreaths at tombs. This was justified in early Christian times because, with its thorns, it also came to symbolise Christ's agony. Flowers as garlands or chaplets were incorporated into many religious ceremonies. The cult of the Virgin Mary resulted in many flowers being baptised with Christian names. *Calendula* (Marigold), *Alchemilla* (Our Lady's mantle), *Spiranthes* (Our Lady's tresses), *Cardamine* (Our Lady's smock), *Anthyllis* (Our Lady's fingers) are just some of them. Flowers also entered heraldic imagery. Broom (*Genista*) was the symbol of the Plantagenet dynasty (planta-genista).

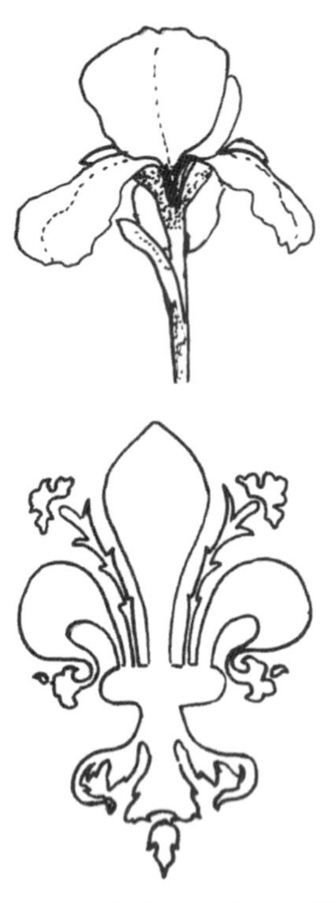

Figure 7.43. The derivation of the fleur de lis symbol. Some writers identify the Fleur de Lis (flower of Louis) as a lily but it is probably an *Iris florentina* (top), the source of the violet-scented insecticidal and medicinal orris root.

This kind of floral imagery is very ancient and widespread. Perhaps the Indus Valley civilisation of 5000 years ago provided the precursor for the strong identification with nature and the sacredness of plants in Hinduism. The lotus blossom had a paramount significance but many other plants like the bright yellow and orange flower garlands of marigolds were favoured in ceremonials and also in private life. Garlands were a bed-time adornment of men as well as women. Jasmine flowers were used to dress the hair.

'Say it with flowers' was a slogan coined for the American Society of Florists, but the *Language of Flowers* was quite literally an attempt to systematise the symbolic character of flowers. In a book published in Paris in 1819, Charlotte de Latour codified the language of flowers by providing a dictionary of meanings. It is all very amorous and ridiculously detailed. A rosebud with its thorns and leaves signified 'I fear, but I am in hope', while without its thorns it meant 'there is everything to hope for'. Without its leaves meant 'there is everything to fear', and when upright but upside down meant 'One mustn't fear or hope'. A marigold on the head meant trouble to the mind, on the heart, the pain of love, and on the breast, boredom. The whole nonsense, which became extremely popular, and was elaborated by many other authors, was given the gloss of being an ancient science practised in classical times and the Orient.

There is rosemary, that's for remembrance . . . and there is pansies, that's for thoughts . . .
Shakespeare

A different symbolic meaning has been given to the same flowers over the ages. In Shakespeare's *Hamlet* the deranged Ophelia gives fennel to the King signifying flattery and columbines to the Queen signifying adultery. The symbolism is poetic and complex. Three flowers are supreme in their symbolic weight: the lotus, the rose and the lily. The lotus (*Nymphaea coerulea*) is the blue lotus (actually, a water-lily) that was sacred in Ancient Egypt. Its flowers open at dawn and close at night and so it became associated with Ra the Sun god. The bud clearly had a phallic significance too. The flowers are fragrant. There is also a white water-lily, the true or sacred, *N. lotus* that opens at night and also grows in the Nile delta. The Indian Lotus (*Nelumbo nucifera*), was introduced in the Persian period. The Madonna lily (*Lilium candidum*) figured on Cretan frescos from 5000 years ago. It is native to the eastern Mediterranean region but has long been cultivated for its white flowers, used for making scent. White lilies are mentioned in the Bible, but some have suggested that the Rose of Sharon may be *L. candidum*. Another suggestion is that this is *Pancratium maritimum* (or Sea Daffodil), while the Lily of the Field is *Narcissus tazetta* or *Hyacinthus orientalis*. The sweet-smelling lily was identified with springtime and rebirth in the Roman period. In the early Christian period it was rejected for being a symbol of luxury and idolatry but later it came to symbolise purity and was associated with chastity and the Virgin Mary.

There was also a cult of flowers and gardening in China, even a peony mania in the early ninth century. Peach, chrysanthemum, lotus, peony, *Cymbidium* (orchid) and *Lilium lancifolium* (tiger-lily) were all cultivated for their beauty from 1000 BCE. From about 600 AD many others like camellia and magnolia were also cultivated. Gardens provided a quiet retreat from the world where the beauty of nature could be contemplated. The arts of poetry and painting were intimately connected with the study of flowers. (See, for example, the poem describing the peony, on this page.) The four noble plants were the plum, bamboo, orchid and *Chrysanthemum*. *Chrysanthemum*, the flower of autumn, was the symbol of joviality and contentment. Bamboo, for spring, was the symbol of the courteous gentleman, and of companionship and modesty. The orchid, for summer, was the symbol of the refinement of beautiful women and great men. The plum, symbol of winter, was for chastity and feminine beauty and also had a complex number symbolism: the blossoms are Yang, Heaven, the branches, Yin, Earth, the pedicel is the Ridgepole of the Universe, the three sepals are Heaven, Earth and Man, the petals are the five elements and the stamens are the seven planets.

The deep green foliage is quiet and reposeful.
The petals are clad in various shades of red.
The pistil drops with melancholy.
Wondering if spring knows her intimate thoughts.

Chinese poem about the peony

7.6.2 The plant collectors

In Europe from the sixteenth century there was a renewed appreciation of the beauty of flowers. Plants were no longer recorded and collected primarily for their uses but out of intrinsic interest and for their beauty. Francis Bacon wrote that the creation of a beautiful garden was the highest form of artistic achievement. The publication in 1629 of John Parkinson's *Paradisi in Sole Paradisus Terrestris* was influential in encouraging the placing of plants in a garden landscape. Botanical gardens were being established across Europe to grow exotics and rarities. Royal and aristocratic enthusiasts employed a succession of famous gardeners to create an earthly paradise for them. They introduced exotic plants from abroad and encouraged the cultivation of rarities from home.

Busbecq brought tulip bulbs back to the Imperial Garden in Vienna in 1554. There, between 1573 and 1589, Clusius experimented growing new cultivars. Tulips were not popularised until about 1608 in France but soon a craze started which spread north to Holland. By 1629 Parkinson, in perhaps the first great garden book, the *Paradisi in Sole*, numbered 140 different varieties in English gardens. Huge sums were risked on the newest varieties. One bulb cost a carriage and pair; another cost twelve acres of land. In Haarlem one merchant bullied a rival into selling a duplicate bulb for 1500 guilders and then stamped on it to preserve the value of his one. A seaman accidentally mistook another bulb worth 3000 guilders for an onion and ate it. He ended up in jail for six months. Soon 'paper' tulips, futures on potential tulips, were being traded, but in 1637 the bubble burst, ruining many speculators. However, the enthusiasm for flowers did not disappear.

The opening up of the New World provided an immense source of new plants – many of which could grow very well in the gardens of north-west Europe. Aristocratic collectors competed by financing collectors to improve their stock. John Tradescant laid out a garden for Robert Cecil, Earl of Salisbury, at Hatfield House and later served the first Duke of Buckingham and Charles I. His own garden in Lambeth became a treasury of exotic plants from abroad. His son John Tradescant the Younger continued the tradition; he travelled to the north coast of Africa, and made three trips to Virginia. Many North American flowers were introduced to Europe through the Tradescant garden. They include the tulip tree, cornflowers, Michaelmas Daisies, Virginia Creeper and, of course, *Tradescantia*. In France another father and son team, Jean and Vespasian Robin, were as influential. Their Paris garden was later to transmogrify, much enlarged, into the Jardin des Plantes. The black locust introduced from North America, but probably first grown in the Tradescants' garden, was given the name *Robinia pseudoacacia* in their honour. The close connection with North America continued throughout the seventeenth and eighteenth centuries. Philip Miller, at Chelsea Physic Garden in London, was eager to receive seeds from the Quaker farmer John Bartram. Botany was no insipid past-time but a vigorous and exciting adventure at home and abroad. Rumbustious botanical excursions to the Physic Garden caused a public scandal.

It is only possible to mention here a few of the most significant and intrepid plant collectors. Francis Masson introduced many heathers, pelargoniums and other plants from South Africa, but after surviving capture by French privateers and hostile natives, Boers and escaped convicts, he too died young, frozen to death in North America at the age of 33.

The story of Philibert Commerson, on the rival French expedition to Cook's by Bougainville, appeals because it includes more of the kind of lunacy and obsession that accompanies many of the adventures of these botanical maniacs. Philibert Commerson (1727–1773) was trained at Montpellier where he got into trouble for stealing plants for his herbarium from the gardens of residents and even from the botanical garden. He made his name as a heroic plant collector, escaping one mountain avalanche by rolling down ahead of it like a ball. Falling into a mountain ravine, on another occasion, he was trapped by his hair in a bush. Cutting himself free meant that he fell into a raging torrent and nearly drowned. Commerson seems to have taken up the chance to travel with Bougainville's expedition with alacrity, as if he was desperate to get away from home. Accompanied by a faithful assistant, a young fresh-faced lad called Jean Baret, he travelled halfway around the world collecting specimens and making notes. Things went awry in the New Hebrides when a chieftain took a fancy to Jean who was subsequently revealed to be a Jeanne, Commerson's housekeeper from home. She had chased Commerson to his ship where she disguised herself in order to be with him. Perhaps it is too much to suggest that Commerson refused to go back to France in embarrassment. He ended his days botanising in Madagascar.

Another botanical eccertric was David Douglas credited with introducing 200 species to Europe from North America including the giant conifers such as the one that bears his name the Douglas Fir (*Pseudotsuga menziesii* – the specific epithet recording the efforts of Archibald Menzies before him. In the third decade of the nineteenth century, Douglas survived a bolting horse that understood commands only in French, having all his belongings stolen while he was climbing a tree to collect seeds, encounters with bears, falling down ravines, suffering cold and starvation, as well as several episodes of capsize, only to die trampled to death in a pit constructed to capture feral cattle on Hawaii.

Not all collectors suffered such hardships all the time. Joseph Dalton Hooker, collecting in Sikkim in 1848, was accompanied at first by a retinue of 56 porters and servants.

7.6.3 Plantscapes

Plants in their natural setting have come to be regarded as 'heritage' and as an amenity resource. Floras have been transformed and enriched by the introduction of exotic species. A cult of trees has never been far from the surface. Trees were associated with national strength and defence. In the seventeenth century, in his best-selling book called *Silva*, a report on tree planting to the Royal Society, Evelyn noted the loss of yew: 'Since the use of bows is laid aside amongst us the propagation of the yew-tree is quite forborn'. He praised oak because 'ships of oak become our wooden walls'. Evelyn was a champion of native trees and the planting of trees not just for their uses but for their aesthetic beauty.

Towards the end of the seventeenth century and in the eighteenth century there was a change of sensibility. There was a shift away from the emphasis on plants for food and for survival to plants for expression. The landscape artists, either the idealised landscapes of Lorraine or the romanticism of Friedrich and Constable, and in Japan the woodcuts of Hiroshige and Hokusai, have all trained our eye. The clipped geometrically shaped symmetrical gardens of the Restoration, with their symmetric parterres gave way to rococo asymmetry, which at its margins blurred the distinction between garden and Nature, and

Figure 7.44. The arcadian landscape of Stourhead.

then there was the exuberance, the wildness, of baroque. Nature and the countryside were colonised again, but now by the intellect. Just as towns and cities were starting to burgeon like great excressences, gardens were now to be constructed to recall an Arcadian landscape when life was pure and simple. They were to be artfully constructed to present a picture of an idyllic landscape. The landscaped parks of William Kent, Charles Bridgeman, Lancelot 'Capability' Brown and Humphrey Repton marked the final phase of taking possession of the landscape. Now even the wild or semi-natural was to become reformed in an ideal image as part of the estate of the landowner. Improvement could be theatrical and painterly as at Stourhead, involving large-scale works like the diverting of rivers and streams, the damming of lakes and the moving of mature trees (Figure 7.44).

Gardening on the grand scale can be seen at Blenheim and many other grand country houses (Figure 7.45). It might also include more subtle changes of landscape. Views were very important. Through the views the park seemed to include the countryside around. This could include, not just scenes of wilderness, with natural features incorporated or enhanced, but the productive countryside of fields and woods. Ha-has, hidden boundary banks and ditches were constructed so that the riff-raff were excluded, but the eye of his lordship was unimpeded.

In part, agricultural development funded the development of landscaped parks but in its turn the timber of the landscape parks was an important source of income. Humphry Repton railed against the nouveau riche who had only a commercial interest, but in time the improvement came to be seen more pragmatically as necessary investment for commercial return.

Grand landscape design was in decline by the beginning of Victoria's reign, but by this time the Romantic Movement had taken psychological possession of the landscape and its flora. The countryside and Nature became now as much a mental landscape as a real one. At

Figure 7.45. Formal garden at Hampton Court.

its most shallow, this was expressed as a search, not just for beauty, but for the 'picturesque', a scene capable of being painted. This was a way in which beauty could be circumscribed, described, possessed. It has provided the vocabulary of our appreciation of the countryside and Nature. Jane Austin poked fun at the cult of the 'picturesque' in her novel *Sense and Sensibility* published in 1797.

In the works of the Romantic Poets something more sophisticated and much more important was re-created: an emotional relationship with Nature. For Wordsworth, 'Nature' became a medium through which the most profound, even religious, thoughts were conveyed. In his long autobiographical poem *The Prelude*, Nature could speak through the medium of a 'rugged' landscape, an individual primrose growing from a rock-face, or the wind blowing through some trees. The romantic poets were writing in the context of the beginnings of industrialisation and a world that seemed more and more mechanistic.

In the same way the long tradition of gardens has led to the development of a spiritual space seen to its perfection in the highly artificial Japanese gardens that still manage to personify Nature (Figure 7.46). It is gardeners and horticulturalists, not botanical scientists, who now know plants and appreciate their endless forms.

Figure 7.46. Japanese garden.

Plants are now manipulated to create artificial habitats that are pleasing to the eye. Plants are reduced to being tools of the architect. Street trees soften the line of roads. Parks, squares and gardens provide a 'green lung' to city dwellers. No new office block is complete without its arrangement of glossy 'houseplants' to welcome the visitor in the lobby. London has six million trees in parks, gardens and streets. Favoured street trees are those that grow upright and do not produce slimy honey-dew or leaf litter. It is extraordinary to see *Ginkgo biloba* fringing a polluted city street, like the last descendant of a royal lineage, 200 million years old, reduced to the status of a street cleaner. But we are thankful for it!

During the nineteenth century the growing appreciation of green places for the physical health of the populace led to the establishment

Figure 7.47. Yosemite.

of city parks and other preserved spaces. In London, the great nineteenth century metropolis, it was the business elders of the City of London who protected Hampstead Heath and Epping Forest. The romantic landscape was in some ways a reaction against the narrowing spirituality of the day. In the same way plants have become more important to us in our dry sterile city lives, reminders of both Nature and another world. It was John Muir who emphasised the value of wildness. His efforts led to the preserving of areas of wilderness in the USA, starting with Yosemite in 1890, leading the way for the rest of the world.

Walk away quietly in any direction and taste the freedom of the mountaineer. Camp out among the grasses and gentians of glacial meadows, in craggy garden nooks full of Nature's darlings. Climb the mountains and get their good tidings, Nature's peace flow into you as sunshine flows into trees. The winds will blow their own freshness into you and the storms their energy, while cares will drip off like autumn leaves. As age comes on, one source of enjoyment after another is closed, but nature's sources never fail.

John Muir

It has also become so much more urgent for us to protect the alternative natural plantscape of trees and flowers and rocks and rivers, an other-world that is closer to the infinite and to the mystic. The need for a countryside that we can access is strong. Rambling is perhaps the biggest leisure activity in Britain . . . after gardening. Today the spiritual and recreational value of plants rival their importance economically and commercially, but ecotourism is taking an increasing share of the tourism industry. The United Nations designated the year 2002 as the International Year of Ecotourism (IYE) to promote the three basic goals of its Convention on Biological Diversity. It is noteworthy though that only one of these is to conserve biological (and cultural) diversity. The other two were economic and commercial.

7.6.4 Future plantscapes

There can be little doubt that plants are going through more rapid changes now than they have ever done before. It is astonishing to think how rapidly the landscape has been changed by human beings, starting with Neolithic culture perhaps a little over 10 000 years ago, but mainly in the past two centuries. Within a few thousand years huge areas of the world's vegetation has become dominated by the influence of human beings. First, organised pastoralism, and then civilisation with its reliance on agriculture, have created new plant communities and even new kinds of plants. Forests have been replaced by patchworks of field and grassland. Marshes have been drained. The wild plants have been pushed to the margins.

The most diverse and ancient vegetations, the tropical forests, managed to escape very much destruction until the twentieth century. They have taken many millions of years to evolve and we have scarcely begun to understand them, and yet almost incredibly they are likely to be lost or changed out of all recognition within the space of our own lifetime. Even the tropical forests have come to be regarded primarily as a resource to be exploited for the benefit of mankind. One sees this, even in arguments put forward for the preservation of the forests. A mere 10 km^2 of Amazonian forest can contain 2200 different species of plants; they might be useful as new types of crops or for providing new drugs. This may be true, but how depressing it is to see the forests only as a possible commodity to be exploited for profit Botanical research is becoming a tool of the market place; it is to be concentrated only on the exploitable, the potentially profitable. What a poverty of the imagination and spirit this represents!

7.6.5 Threatened plant species and vegetation

Exploitation and loss of habits has either caused the extinction of, or brought to the edge of extinction, many plants. For example, California has 20% of plant species at risk. The threats are several. Wild tree species with particularly favoured timber have been particular endangered by over-logging. The most important danger is habitat loss or degradation either for urban development or for agriculture and pasture. Climate change, in part brought on by human activity, may make extinct small populations at the limits of the current tolerance. One of the most damaging results of human activities has been the introduction of plants to new areas. Humans have introduced exotic species everywhere they have colonised and in some places these have become a significant threat to the native flora. Sometimes concern about introduced plants is exaggerated. The flora of Britain is an entirely immigrant one with very low levels of endemism. However, in the wet woodlands of western Scotland that are a hotspot for bryophyte diversity, *Rhododendron ponticum*, once native but re-introduced as a garden plant, poses a significant threat to biodiversity.

Regions that have been isolated for a long time and have a high proportion of endemic plants are particularly threatened by introductions. Hawaii has 1200 endemic plant species, 90% of the total native flora. A third of these are rare, and about 150 have fewer than 50 individuals. In New Zealand 88 species are specifically listed as pest plants, including *Clematis vitalba* (old man's beard), wild ginger, purple pampas grass (*Cortaderia jubata*) and *Pinus contorta*. New Zealand's worst weed, the gorse *Ulex europaeus*, has had millions of dollars and much effort expended on it to try to control its spread, but paradoxically, if left, it acts as a nurse plant for native bush to protect native plants from grazing while they establish. In many areas aquatic and riverine environments are especially vulnerable because of the rapidity with which exotic weeds can spread along waterways.

7.6.6 *In situ* and *ex situ* conservation

The most important and valuable form of conservation has been the establishment of *in situ* nature reserves and wildlife parks. The number of protected areas has grown rapidly, but there are signs that this period of growth is ending because the size of the areas becoming designated as protected is declining. The majority are less than 100 ha. According to a UN sponsored survey, 8% of the world's remaining forests are in protected areas but unfortunately this does not always necessarily protect some of them from logging or encroachment for other uses, especially in South East Asia where human populations are growing rapidly. And other kinds of vegetation are relatively unprotected. Concerns about the fate of the world's plants have encouraged the establishment of seed banks to preserve samples of seed, especially for crop plants and their relatives. The Millennium Seed Bank Project at the Royal Botanic Gardens at Kew aims to collect and conserve 10% of the world's flora. Plants from dry lands have been targeted especially, in part because they are regarded as especially vulnerable from climate change and over-grazing but also because seeds from the wet tropics often prove to be rather difficult to store.

Conservationists have resorted to economic arguments in order to promote conservation. Ecotourism has been seen as the saviour of our biodiversity, but this is a dangerous mistake because it exposes our surviving natural plantscapes to the vagaries of the marketplace. Rather we must promote the conservation of biodiversity, of which we are part, as one of the fundamental human values. Plants have created the terrestrial environment we live in and make it habitable for us. All the myriad kinds of terrestrial animals rely upon them too. Each plant species represents a unique point, the result of a unique evolutionary dance, in space and time. Isn't that reason enough?

Further reading for Chapter 7

Balick, M. J. and Cox, P. A. *Plants, People, and Culture. The Science of Ethnobotany* (New York: Scientific American Library, 1997).

Frankel, O. H., Brown, A. H. D. and Burdon, J. J. *The Conservation of Plant Biodiversity* (Cambridge: Cambridge University Press, 1995).

Hobhouse, H. *Seeds of Change. Five Plants that Changed Mankind* (New York: Harper and Row, 1987).

Lewington, A. *Plants for People* (New York: Oxford University Press, 1990).

Simmonds, N. W. 1977. *Evolution of Crop Plants* (London and New York: Longman, 1997).

Tivy, J. *Agricultural Ecology* (London: Longman, 1990).

Chapter 8

Knowing plants

Knowledge is 'seeing' this vital meaning behind the appearance of things. It is penetrating the mystery of life. Thus, it is only through this process of learning 'to see' that we come to know ourselves.

Socrates, 469–399 BCE

8.1 The emergence of scientific botany

The study of plants must be one of the oldest occupations of humans who, even in their most primitive state, required a wide knowledge of the plants that provided food or remedies for illness. By trial and error they knew which plants were poisonous and which were edible. This expertise led to the first sowing of wild seeds, the start of agriculture and therefore the beginning of civilisation. The earliest classification systems were utilitarian 'common-sense' classifications but could be extremely sophisticated. The Mayan folk classification of plants, for example, is no less systematic than the latest scientific classifications based largely on analyses of DNA sequences.

The long history of botany is a record of our attempts to describe and understand plants. This is not as straightforward as it might seem. Even a simple term such as 'leaf' can be interpreted in several ways and its meaning depends upon the context of its use. A concept such as species is more complex.

Every description exists on a background of biological theory, to which it is intimately related – whether this relationship is expressed or merely understood.
Agnes Arber, 1954

It is common for new botany students to complain about the number of terms, names and concepts they have to learn. Botany uses language in which the things are, in a sense, 'created' by the words we use to describe them. Students who look down a microscope at a botanical specimen for the first time, a section of a leaf perhaps, and are told simply to draw what they see, often have difficulties. They ask, 'what am I looking at?'. Without a conceptual framework they cannot see. Conceptualisation is fundamental to science but, because they are abstractions, concepts do not necessarily correspond to our individual experiences of reality. Epistemology is the study of how we know things. To understand how botany has developed as a science in western cultures, and the epistemological difficulties we currently face, we need to go back to its beginnings in Ancient Greece and start with the world of the Pre-Socratic philosophers. This may not at first seem directly relevant to the practice of modern botany, or indeed other descriptive sciences, but we believe strongly that it is.

8.1.1 The legacy of the Ancient Greeks

The Greeks were the first to say that the world was knowable, because they believed in the human power of reason. Through understanding the nature of the Universe and the nature of humans, they had the key to understanding their place in the scheme of things. For the Ancient Greeks the Universe and the world were harmonious and unchanging; they were to be understood and admired without any attempt to change them. They did not see the Universe or the world as having directionality in time (i.e. a history).

There were basically two major schools of Pre-Socratic philosophers, the Ionian School and the Pythagoreans. Both schools made a distinction between form and matter. In thinking of form and matter, we must not confuse form simply with the shape of an object, or matter with the stuff from which the object is made. This would be to misunderstand the problems the Greek philosophers were trying to solve. The Greek word for form comes from a verb 'idein' which can mean both to see and to know. The form of anything was that which was knowable about it, but no object is the same as its definition. Therefore, from early on, the Pre-Socratic philosophers had a deep mistrust of the senses, and the idea that only thought processes can give us information about the nature of reality predominated.

In classical times the Ionian School was famous as one which sought scientific answers to questions about nature. Because the school was mainly concerned with observing nature, its followers were called natural philosophers. In contrast, the Pythagoreans saw number or form as the first principle. Heraclitus, who was part of the Ionian School, was born about 535 BCE in Ephesos, the second great Greek Ionian city, and he was probably the most significant philosopher of ancient Greece until Socrates, Plato and Aristotle. From a modern perspective, some of his ideas would seem to parallel those of Lao Tzu of ancient China. Heraclitus was the philosopher of eternal change, the doctrine that everything is in a state of flux. His most famous opinion, and the one most emphasised by his disciples, as described in Plato's *Theaetetus*, is: 'You cannot step twice into the same river, for fresh waters are forever flowing in upon you'. He focused on the internal rhythm of Nature, the Logos (Rule or Way), which moves and regulates things. Fire ($\approx$ energy), being the most fluctuating of all things, is therefore the essential reality of the Universe, and the only material source of natural substances. Heraclitus saw the harmony of the world as the resolution of many diverse forces. The essence of Logos creates an infinite and uncorrupted world, without beginning, and converts this world into various shapes as a harmony of the opposites, the composition of which sustains everything in Nature. This concept of unity in diversity, of the 'One as Many', is Heraclitus's most significant contribution to philosophy. He was one of the first philosophers to suggest that we cannot rely wholly on our powers of observation, a suggestion that would have profound implications for western philosophy for the next 2000 years.

The school of Pythagoras greatly influenced the course of Greek philosophy, and particularly Plato. Pythagoras, who was born on the island of Samos, founded an ethical, religious, and mystical system of teaching about 530 BCE in Croton in southern Italy. The Pythagoreans believed that knowledge was inherently mystical. For them, numbers were the ultimate elements of the Universe. This may seem a rather bizarre idea to us but the Pythagoreans thought of numbers spatially. To say that all things are numbers is another way of saying that everything that exists consists of points in space, which taken together make a number. There is a link to modern physics here with the use of string theory to provide a grand unifying theory of everything. In making number the first substance of the world, the Pythagoreans transferred these mathematical concepts to material reality. Since the fundamental realities of the world were structural and mathematical, those which display greater simplicity, regularity, and coherence in mathematical proportions or parameters are aesthetically more beautiful and 'better'. The correct proportion between the whole and its parts was the cause of beauty in the object, and was called harmony, meaning perfect arrangement. The similarity between the whole and its parts can be expressed in terms of some proportion supposed to exist between them. Pythagorean interests in proportion extended to music and they devised a musical scale based on vibrating strings of different lengths with a constant ratio of 3:2. This mathematically elegant system was extended to the Universe since the five planets known at that time had movements with similar ratios. The Pythagoreans imagined 'a music of the spheres' that was created by the Universe, a wonderful idea that has inspired many composers in the modern era. As we saw in Chapter 2, the influences of mathematics, particularly the Fibonacci Series and the Golden Mean, had powerful impacts on plant morphology and development in the nineteenth and twentieth centuries.

The identity of harmony with good order is considered by some to have been the main contribution to Greek philosophy by the Pythagoreans. The study of mathematics was indispensible for intellectual and spiritual progress, while medicine was the science that brought harmony to the body. Ultimately Pythagoreanism has been a dynamic force in western culture, influencing not only philosophers, theologians, mathematicians and astronomers (notably Copernicus, Kepler, Descartes and Newton) but also musicians and poets.

The Greek mistrust of the senses found its greatest expression in Parmenides (*c.* 540–470 BCE) who was the first to make the distinction between truth and appearance. He thought that the senses were deceptive, and that knowledge of the truth could only be gained through 'pure thought'. Anything rationally conceivable must exist. Reality is not primarily what can be experienced by the senses, but is what can be expressed in language. The sensible world for Parmenides was not the real world, with the result that he created a dichotomy, the sensible world and the world of thought. What is meant by 'objective' is what exists independently of any particular mind or viewpoint. Subjective existence that is dependent or a human viewpoint (i.e. via the senses) is considered to be an invalid method of inquiry. This way of obtaining knowledge was to have a profound influence on Plato, and far-reaching effects for western science.

The changes that were taking place in the Mediterranean world during the fifth and fourth centuries BCE, particularly in trade and agriculture, provided a tremendous stimulus for the study of animals and plants that was to reach in zenith in Aristotle and Theophrastus. The most pervasive influence was Plato (427–347 BCE) who separated the immaterial world or the world of Ideas (i.e. heaven, the eternal world of inner spiritual perception) from the material world or the transitory world of phenomena (i.e. earth, the world of outer sense perception). In this theory, the originality of the knowledge of an object does not reside in any phenomenal reality of the object itself but in the universal Idea of the object. All worldly things are recognised as imperfect replicas of its perfect form. They are only the individuations of the indestructible essential forms or Ideas that are outside of the space-time continuum, and therefore are infinite.

Aristotle (384–322 BCE), whose philosophy and work are often still misunderstood today, is usually regarded as the 'father of scientific methodology' since the scientific study of organisms in the west is first recorded in his works, much of which surpasses that of later Middle Age scholars. Aristotle came from Stagira in Macedonia; he was a tutor of Alexander the Great and a member of Plato's Academy in Athens. His conception of the close relationship of all living organisms based on a comparative approach was derived from the Ionian philosophers. The influence of Plato is clear, but Aristotle attempted to overcome the dualism of Plato by seeing nature as a unity. The archetypes of Aristotle are arrived at by *empirical means* and are fundamentally different from those of Plato. Plato's Ideas are sometimes thought of as archetypes, and thus much confusion still exists with the use of this term. The affirmation of Plato's 'real' world was not to be sought by empirical means but by what subsequently become known as 'cataphatic theology', whereas Aristotle's science was based on the rational observation, comparison and interpretation of observed phenomena rather than divine revelation. For Aristotle, the originality of the

Figure 8.1. Aristotle.

knowledge of an object *resides in the phenomenal reality of the object itself, not in the universal Idea of the object.*

Aristotle was really concerned with fundamental questions on the essential nature of plants and animals. Despite his awareness of the dynamic nature of the world, he believed in the eternal order of the Universe, and was not an evolutionist in a Darwinian sense. Consequently he sought clear distinctions and boundaries between phenomena. Organisms were regarded as separated, immutable entities, which were characterised by their 'essences'. He did not classify organisms into rigid categories but, rather, he adopted broader categories of morphological and ecological similarity. In his philosophy, individuals alone are real entities. Species and ideas are not realities, but ways of understanding reality and, as such, they exist only in the intellect. In his *Doctrine of Entelechy*, organisms were created for a purpose and the world is a hierarchy of entelechies or purposeful arrangements leading upwards from the lowest to the highest. For Aristotle, the Platonic Idea (Form) is the original purpose which actualises the developmental potential of the organic world in the image of itself.

According to Linnaeus, Theophrastus (*c.* 372–287 BCE) was 'the father of scientific botany'. He was born on Lesvos where Aristotle had studied plants as a young man, and later became a student of Aristotle as well as his friend and collaborator. He took over the Peripatetic school of the Lyceum after Aristotle retired. It was Theophrastus's *Enquiry into Plants* and *Causes of Plants* that were the first botanical texts. They were the notes of some of his hugely popular lectures that he presented for 35 years to students at the Lyceum. At one time he had more than 2000 students. Theophrastus's botanical lectures have survived almost complete. For the first time botany appeared as a distinct science with a comprehensive and clearly defined field of enquiry. About 550 different plant species are mentioned.

Figure 8.2. Theophrastos.

As the boundary of the Greek world expanded with Alexander's military conquests, Theophrastus incorporated knowledge gleaned from reports of travellers and merchants who had followed in Alexander's wake. In the *Enquiry* there is an emphasis on the uses of plants and in the *Causes* there is an emphasis on their growth and reproduction. It is Theophrastus's keen powers of observation, especially his ability to make sharp comparisons, that mark his work. For example, he recognised that grasses have flowers like those of more showy plants even though they are small and inconspicuous. He rejected some aspects of Aristotle, such as his method of classification, and he held the view that the essence of plants consists of their parts. Of great importance is the fact that he was aware that the categories in his scheme of classification intergrade and that the distinction between them is one of convenience.

Theophrastus represented the culmination of Greek inquiry into plants but, with the destruction of Greek rationalism, botany declined, especially after the second century AD. Diocles of Carystos was a herbalist and contemporary of Theophrastus and it was largely his writings that became the foundation for later herbals such as those of Dioscorides and Herophilus of Alexandria. Herbalism was just one important branch of botanical knowledge that was taught in the Lyceum of Athens, but it is virtually the only one that survived until the renaissance of botany in the second millenium, marked by the reprinting of Theophrastus's works in the late fifteenth century. Dioscorides thus represents the final chapter in Greek herbalism. His *Materia Medica*, which was written about AD 60 largely replaced the *Rhizotomikon* of Crateuas (120–60 BCE) and became the basis of Arabic pharmacopeia, and was often accompanied by beautiful illustrations. Theophrastus's work thus remained influential in the Islamic world. Nestorian Christians driven from the Eastern Roman Empire had carried the European classical tradition from Syria to Persia. At Jundeshapur, a university was established about AD 500 that kept the tradition alive. Arab herbalists both preserved the classical herbalism of the west and added to it. For example, Avicenna (ibn Sina) (AD 980–1037), the foremost Islamic philosopher who came from Iran, produced the *Canon of Medicine*, which included many plants unknown to Dioscorides. The spread of Islam, and the trade and commercial prosperity that accompanied it, provided cultural links between Persian learning, and that of India and China. In the Islamic centres of Sicily and Toledo, Jewish scholars translated the classical works and thus helped to spread botanical knowledge back to the west, where, after a long period of stagnation lasting almost 1800 years, it was rediscovered. Meanwhile, in China, botany was developing in a uniquely independent way.

8.1.2 Botany in China

The development of botany in China began at roughly the same time as in Ancient Greece. The Chinese ideas embodied in concepts such as the interplay of yin and yang were a remarkable echo of Heraclitus and the Ionian philosophers, while the Chinese concept of Tao (The Way) can be compared with the Greek idea of Logos. With its emphasis on herbalism, early Chinese botany paralleled much of that in the west, and this strong herbal tradition was encouraged by bureaucrats and emperors. Organised herbal medical knowledge has remained influential in Chinese culture until the present day. By the fourth century BCE, botanical knowledge had increased to a descriptive level and included sophisticated plant terminology. In the *Erh Ya*, over 330 plants are mentioned, some of which were illustrated. From this period onwards, for over a millennium, Chinese botany steadily progressed, particularly in descriptive aspects and in herbalism until, in the Middle Ages, it far surpassed western botany. Chinese botanical nomenclature was stable and had a binomial structure somewhat analogous to that developed by Theophrastus.

Over the centuries in China many different kinds of plants were included in herbals, and monographic treatment of certain groups of plants reached a sophisticated level. The earliest monograph on chrysanthemums dates from the beginning of the twelfth century AD.

It lists 25 different cultivars. In contrast to the west, the tradition was for a high degree of accuracy in the description and figuring of plants. Apart from the utilitarian aspects of plants, China failed to develop a theoretical knowledge of, or a natural classification of plants as a whole, and lacked the driving curiosity and exploitation characteristic of western science that developed in Europe after the Renaissance. Perhaps the reason for this was that, in the religious traditions of Ancient China (and of India), there was no sense of separation between humans and Nature. Nature was not something to be conquered and therefore science, in the western sense, never developed. There was no need for it when the goal was to conquer the self and transcend appearances. The Ancient Chinese view of the world, which has valuable lessons for conservation today, was more akin to the descriptive and contemplative approach that was characteristic of Goethe and later phenomenologists.

8.1.3 Botany in the Renaissance

The rediscovery of classical Greek science and philosophy in Europe, partly influenced by Islam, ushered in the Renaissance in the west. Pope Nicholas V instigated a new translation of Theophrastus's *Historia Plantarum* and *Causae Plantarum* by Theodore Gaza (1400–1475) from copies discovered in the Vatican library. A new translation of Dioscorides's *Materia Medica* was also made by Pietro d'Abano in 1478. Most of the manuscripts on botany at this time were hand-written copies created by monks in monasteries, but Italian manuscript herbals of the fifteenth century showed a renewed appreciation of the value of accurate drawings of plants. A realistic depiction of nature is seen to perfection in masterly drawings by Da Vinci, Dürer, and in the early sixteenth century detailed and beautiful woodcuts were being produced by a pupil of Dürer called Weiditz.

Figure 8.3. Weiditz woodcut.

At the same time, while the works of Theophrastus and Dioscorides were becoming more available, there was a realisation of the limitations of the classical authors. The prevailing philosophies of the Middle Ages remained Plato's doctrine of Idealism and Aristotle's Essentialism, but by the mid fourteenth century the new philosophy of Nominalism was developing. Nominalism was a doctrine that considered the Platonic universals and concepts as mere necessities of thought which have no real existence except as names. Individual phenomena, as revealed by experience, are the primary reality. This inversion of Platonic doctrine had the most profound effect on the subsequent development of science. Nominalism flourished as a dominant doctrine through the Middle Ages thanks to the writings of people like the Franciscan, William of Ockham (1285?–1349?).

The widespread adoption of critical observation weakened the prevailing Idealistic dogma and had a catalytic effect on the development of a positivist institutional science. Throughout the sixteenth century universities were being established in many cities of Europe, particularly in Italy, and there was a great revival of interest in botany, although it was still largely from a medical perspective. One inevitable spin-off of this development was a proliferation of botanic gardens. An ever-widening diversity of flowers collected from around the world was being introduced to newly founded botanic gardens. Both Pisa and Padova claim to be the first 'scientific' botanic garden. Luca Ghini was perhaps most influential as an enthusiastic teacher and correspondent with other botanists throughout Europe, first in Bologna and then Pisa. One simple but lasting influence was his popularisation of the collection and exchange of pressed dried plants, herbarium specimens. The herbarium and the botanic garden became the two pillars of systematic botanical research and are still essential today.

Cesalpino (1519–1603) was a true follower of Aristotle, and was a student of Ghini. Through Cesalpino's work these two great botanists gained a lasting influence. Cesalpino, taught at the university of Pisa for nearly 40 years. He did not endear himself to the Inquisition with his original views but nevertheless eventually he became the physician to the Pope. He took up Ghini's focus on the recognition of species within genera, and made many other advances in botany, human anatomy and medicine. An interesting aspect of Cesalpino, and perhaps a forerunner of Ernst Mayr's *Biological Species Concept*, is that he thought a species' defining characteristic was its biological continuity, that is, its ability to reproduce its own kind. He was the first botanist to achieve the first working comprehensive classification of plants since Theophrastus with his *De Plantis* (1583).

Remarkably, Cesalpino attempted a natural classification and rejected the utilitarian or the alphabetical arrangements favoured by the herbalists. One of his most significant contributions was his classification of non-flowering plants. Natural classification became feasible because the description of plants had become more structured. The use of morphological characters dominated plant taxonomy from Cesalpino onwards, but the dominant method of classification remained downward by logical division until the time of Linnaeus. Cesalpino had an Aristotelian (typological or essentialistic) world view and believed that nutrition and growth were the highest reflection of a plant's essence. However, he made a clear distinction between essential or fundamental characters and accidental or superficial characters that are likely to change depending upon the climate or soil. The plant was dissected into its parts (or characters), which could then be compared, but not all characters were necessarily given equal weight.

Gesner was an outstanding botanical collector. He was also a remarkable botanical illustrator, although he never saw his work published. He produced 1500 drawings, which far surpassed the achievements of others, because they were drawn from living material, and included separate detailed diagrams of flowers, fruits and seeds. They were converted to woodcuts at his own expense. A few were utilised by other authors, without acknowledgement, but the bulk were not published until 1751. Nevertheless, Gesner had great influence on the development of botany through his contacts with other botanists. It was this sharing of knowledge through publication and by correspondence that greatly increased the sum total of botanical knowledge. Part of Gesner's correspondence was published in 1591 by Jean Bauhin, with whom he botanised in Switzerland. Set against a time of religious troubles, several other important botanists also travelled widely in Europe, sometimes to escape persecution, but by visiting each other there developed a new botany.

8.1.4 Botany in the seventeenth century

In the early seventeenth century the vastly increasing botanical knowledge was becoming a brake on further understanding. Some system had to be created out of the chaos of names. An important step was provided by Gaspard Bauhin, nearly 20 years younger than his brother Jean who, in his *Pinax* (1623), provided comprehensive references to previous works and a list of names thought to be synonymous. In addition, probably influenced by Gesner, he recognised genera, each of which could be split into a number of species. Bauhin frequently used a binomial name for each species, using a generic part and a specific descriptor, and it is thought that this had some influence on Linnaeus's development of a binomial system.

The Renaissance was a time of profound change in the way the world was observed and contemplated. In an increasingly mechanised world, new technology was developing, particularly the development of the microscope, which had the most significant impact on botany. René Descartes (1596–1650) developed the metaphor of the machine to an absurd level in his *Discourse on Method*. He was clearly influenced by Nominalism, but much of his reasoning was influenced by a deductive process that recalls Plato. He divided the world of Nature into the two fundamental realms of mind and matter. In his metaphor of the world as a machine, Nature was to be understood in terms of logical laws, as linear series of cause and effect. The inevitable result of a mechanical understanding of Nature was that whole organisms could be understood by the study of their parts. He had no insights into emergent phenomena. His claims that organisms were merely automata were considered offensive by many biologists. In addition, he believed that Nature was the result of accident, a view that conflicted with those who held the view that Nature was orderly and created by design. This led him into difficulty with regard to the place of humans in the grand scheme of things, and subsequently to develop his notion of the uniqueness of humans in their possession of soul, a privilege not accorded to animals. In the centuries that followed, together with the subsequent triumph of Isaac Newton (1642–1727), Descartes influenced the way positivist science functioned. If the metaphor of the machine was not taken literally, it still strongly influenced the belief that biological phenomena could be understood by reduction to chemical and physical processes.

John Locke (1632–1704), who acknowledged his indebtedness to Descartes, was an empirical philosopher and a contemporary of Newton, held a belief in the primacy of information derived from the senses, and was influential in the promotion of the experimental method. An important contribution of Locke to biological enquiry was his advocacy of comparative methods to substantiate the limits of genera founded largely on intuition. One of the most remarkable anti-reductionist statements made by Locke is his description of the contrast between a plant and mechanism. Locke saw that plants have bodily coherence and 'internal self-motion', and with their parts united to a 'common vital activity'.

Nevertheless the principal of analytical enquiry through reductionist empirical methods became firmly established among seventeenth-century biologists, and to this day, these methods have provided remarkable advances in understanding. Some of the earliest advances at that time were in an understanding of the nature of plant nutrition, thanks to the increasing use of experiments (and probably to the alchemists as well).

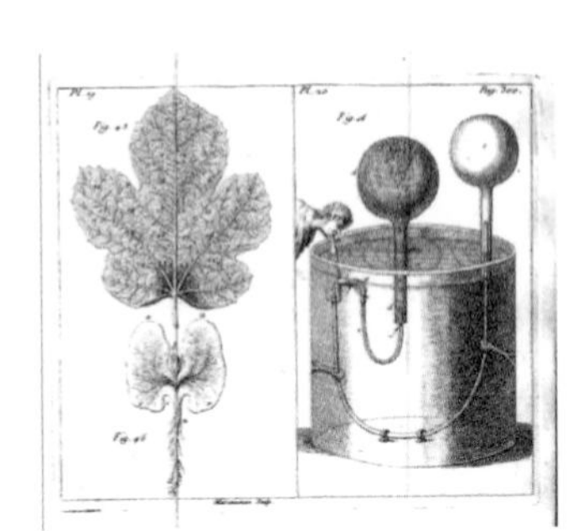

Figure 8.4. The start of experimental botany: one of Hales' experiments.

Hales (1677–1761) carried out further experiments and observed, in his *Vegetable Staticks* of 1727, that plants imbibed much greater quantities of water than animals and lost most of it by what he called perspiration. This observation was to lead to our modern understanding of transpiration as the motivating force in water movement in plants. Meanwhile, experimental botany was developing. For example, Richard Bradley reported the first experimental hybridisation, carried out by the London nurseryman called Thomas Fairchild in 1717, who had crossed two different species of *Dianthus*, a carnation and sweet william, to create the first artificial hybrid.

Plant morphology also received a considerable boost thanks to the systematic analysis of plants by Joachim Jung. He introduced a precise terminology for the parts of plants and the spatial and developmental relations between them, including phyllotaxis. His systematic analysis of plant form had a lasting influence on descriptive botany. His posthumous publications, *De Plantis Doxoscopiae Physicae Minores* (1662) and *Isagoge Phytoscopica* (1679) were apparently known to Ray, who adopted his methods of rigorous morphological analysis. Interestingly, Jung rejected the dichotomous division of plants into trees and herbs that can be traced back to Theophrastus, and which was to resurface again in the twentieth century in the works of John Hutchinson.

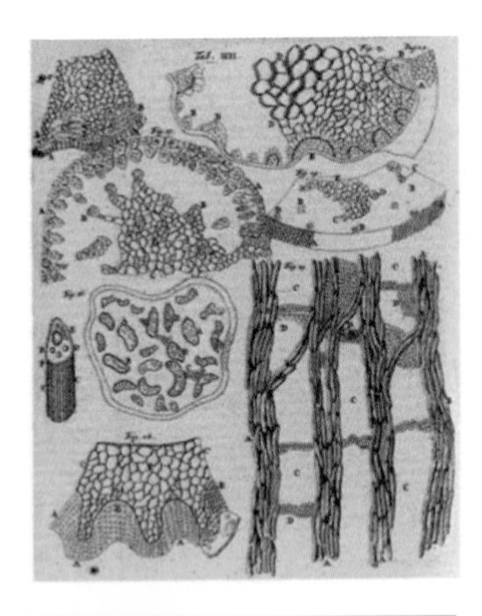

Figure 8.5. An example of Malpighi's plant anatomy drawings.

In 1671 two great anatomists, Nehemia Grew in England and Marcello Malpighi in Italy, working separately, published their accounts. Grew's *The Anatomy of Plants Begun*, first presented in manuscript in the Spring at the Royal Society, was presented again in December, now in printed form, at the same meeting as Malpighi's *Anatome Plantarum Idea* that had just been published in Bologna. By coincidence, at the same meeting, John Ray was made a Fellow of the Royal Society. It was Malpighi who firmly established the tradition of the use of Latin in botanical descriptions.

Ray's statement that 'pollen is the equivalent of the sperm of animals' may have come from Grew, or Bobart, who seems to have proposed it about 1682. They were far-sighted in their appreciation that morphology embraced the whole plant and not just finer aspects of anatomy. In addition they realised that a fuller understanding of plant morphology could only be gained by study of coordinated development in ontogeny. The microscopic observations made by Grew and Malpighi of plant and flower structure were not surpassed until there were improvements in microscope technology at the beginning of the nineteenth century.

John Ray (1623–1705) is considered to have influenced the theory and practice of botany more decisively than any other person in the second half of the seventeenth century. As well as writing the first proper Flora of the British Isles, it was Ray's intensely practical approach that allowed him to reach new conclusions from his own observations. His most original contribution was to the development of a plant classification, although he is also considered to be the founder of plant physiology. Ray developed a system of natural classification that grouped together plants on the basis of their natural affinity. Like Cesalpino, he rejected accidental characters and his works *Methodus Nova* (1682), *Historia Plantarum* (1686) and *Methodus Emendata* (1703) were remarkable for their sophisticated analysis of variation in plants. Especially influential was his establishment of the major groups of plants such as the monocotyledons and dicotyledons for plants with either one or two seedling leaves, or the enangiosperms (angiosperms) and gymnosperms for plants with and without enclosed seeds. However, he was confused about flowering plants with one-seeded fruits, which he put in the gymnosperms. He coined or popularised several terms, including petal, cotyledon and pollen. He was a detailed systematic collector and strongly advocated field knowledge of plants. At about the same time Rudolf Jacob Camerarius, the director of Tübingen Botanic Garden, was making experimental manipulations of pollination. He reported the first detailed observations of pollination in his *De sexu plantarum epistola* of 1705.

Steps were made towards establishing a taxonomic hierarchy, an effective way of representing the pattern of relationships between species by a series of less and less inclusive categories. Families and genera became more established as categories. Magnol (1638–1715), Professor of Medicine and later director of Montpellier Botanic Garden, defined a series of families, including the Ranunculaceae, Papaveraceae, Papilionaceae and Malvaceae, which are recognised today. He had met Ray in 1664 and was greatly influenced by him. Tournefort (1656–1708) also made a valuable contribution to botany at this time by providing brief descriptions of genera. In this way the genus became firmly established as a rank in the taxonomic hierarchy.

8.1.5 Botany in the Age of Enlightenment

The rationalism and empiricism of the seventeenth century gradually weakened religious dogma and tradition, and ushered in the Age of Enlightenment. There were great attempts at this time to explain all aspects of human endeavour with the certainty of mathematics and Newtonian mechanics, and there were discussions about human progress that inevitably fed into the first ideas of evolution in nature.

The concept of the *Scala Naturae*, or 'great chain of being', which can be traced back to Aristotle, took on a dynamic dimension in the seventeenth century. Baron Gottfried Wilhelm von Leibniz (1646–1716) was a German philosopher and contemporary of Newton, and is still considered to be one of the great thinkers of the seventeenth century. He developed a rationalist philosophy that attempted to reconcile the material world with the existence of God. Influenced by Pythagoreanism and Platonism, he believed in a 'pre-established harmony' or divine order (as shown by the *Scala Naturae*) created by God. In his *Monadologie* (1714) Leibniz maintained that this divine order was composed of autonomous units called monads that had no physical influence over each other, but that they all worked in harmony to the Creator's plan. Because it was created by God, the world of monads was the best possible one. Leibniz was later caricatured by Voltaire as Dr Pangloss in *Candide* (1759) for this apparently irrational optimism. However, Leibniz's ideas of development, unlimited potentiality, plenitude and progress were in direct conflict with Cartesian ideas of uniformity and mathematical constancy, and hinted at the evolutionary thinking that was to take root during The Enlightenment and later in the nineteenth century.

The idea of unlimited progress probably had considerable influence, particularly in France where it was enthusiastically adopted by the naturalist Charles Bonnet (1720–1793). However, Bonnet and other naturalists of that period such as Buffon (1707–1788) were not evolutionists in a modern sense because they saw evolution as the mere unfolding of immanent potential. Significantly, for the subsequent developments in the botany of Adanson and the De Jussieus, Buffon recommended that classification should be based on the totality of all characters and not on the arbitrary selection of a few.

The construction of taxonomic hierarchies did not resolve the problem of formal scientific names in the seventeenth century being confusing and unstable. Species names were really a means of identification. Some names consisted of a whole sentence of descriptive terms and were very cumbersome. Even Bauhin's binomials were no good because they were constantly superseded as new species were discovered and new characters had to be used to distinguish the known species. The Swede Carl Linnaeus had the lucky thought that, with a workable classification for identification, the second name of a binomial could act merely as a trivial label fixed for all time to that species. For example, the species called *Geranium columbinum majus dissectis foliis* by Ray became *Geranium molle*. This introduced a stability into naming species that was vital for scientific communication. His *Genera Plantarum* of 1737 and *Species Plantarum* of 1753 are taken as the start of modern plant taxonomy. Partly as a result of this and because his Sexual System of classification provided a ready, if artificial, means of identification, Linnaeus is, perhaps unfairly, regarded as the father of systematics (Figure 8.6).

Figure 8.6. Linnaeus' Sexual System of classification of plants. Linnaeus believed that the essence of a plant consisted of its sexual parts, and he provided a ready means by which all known plants could be identified by observing and counting the male and female parts. There were 24 classes based on the number, relative length and degree of fusion of stamens. And each class was subdivided into orders that differed in the number of pistils (illustration by Georg Dionysius Ehret from http://www.linnean.org/).

The Sexual System placed the grass *Anthoxanthum* in a different class from all other grasses because it had two stamens, but in the same class as the sage *Salvia*. It clearly did not reflect Nature and on several occasions Linnaeus published fragments of a more natural classification. However, it was in France, where Linnaeus' artificial classification did not become fashionable, that serious progress towards this goal was made.

The de Jussieu family, several generations of botanists to the French king, proposed a rival system that superseded the Sexual System. Antoine de Jussieu was taught by Magnol and succeeded Tournefort as the Professor of Botany at the Jardin du Roi. His approach is epitomised by his recognition of a class of plants he called the Fungosae, which included the fungi and lichens based on their total affinity. It was this latter point, the necessity of encompassing multiple affinities to make a natural classification, that was most important.

It was Antoine's younger brother Bernard de Jussieu, at first the assistant demonstrator at the Jardin du Roi, and then, from 1759,

This long-desired arrangement, far superior to all others, alone truly uniform and simple, always in conformity with the laws of affinities, is so-called natural method, *which links all kinds of plants by an unbroken bond, and proceeds step by step from simple to composite, from the smallest to the largest in a continuous series, as a chain whose links represent so many species or groups of species, or like a geographical map on which species, like districts, are distributed by territories and provinces and kingdoms.*

From *The Introduction to the Genera Plantarum 1789* (translated by Susan Rosa). In Stevens, P. F. *The Development of Biological Systematics* (New York: Columbia University Press, 1994).

supervisor of the royal garden at Versailles, who established a natural classification of flowering plants. He laid out a garden at Trianon as a living demonstration of his system.

A pupil, and then a friend and colleague, of Bernard de Jussieu, was Michel Adanson. In 1748 he went to live in Senegal for six years. Contact with the exuberant tropical flora had a profound influence on him. The existing classifications of Tournefort and Linnaeus proved woefully inadequate for identifying species and could not be easily modified to encompass the new species he encountered. On his return from Africa Adanson lived with Bernard de Jussieu, published the *Familles des Plantes* (1763), in which families are described in detail, and arranged them in a natural sequence that represents their relationships. He did not believe in the use of *a priori* characters, but thought that all characters should be given equal consideration as a first step. Adanson came close to understanding that a natural classification showed the genetic or familial relationship between plants. In 1789 Antoine-Laurent de Jussieu, nephew of Bernard, published his *Genera Plantarum*. The beauty of his work was that he adopted the Adansonian method of classification but used Linnaeus's system of naming species. One hundred families of plants were established, 94 of which were flowering plants. However, Antoine-Laurent failed to achieve a natural system of plants above the family level; not surprisingly since this task remained extremely challenging right until the end of the twentieth century. Instead, he resorted to a more artificial system based on what were thought to be a few important characters. Perhaps he was influenced in this by the utility of Ray's use of the number of cotyledons to separate the monocots and dicots, a single character which actually worked to separate two natural groups.

Important advances were also made in the study of bryophytes. J. Hedwig accurately described the life cycles of bryophytes for the first time and produced the first classification of them, his *Species Muscorum Frondosorum*, which was published posthumously in 1801. This work is now considered to be the valid starting point for the nomenclature of mosses, in the same way that Linnaeus's *Species Plantarum* is the starting point for flowering plants. The study of algae unfortunately did not reach such a level, although there were important contributions by S. G. Gmelin (*Historia Fucorum*), and by Joseph Gaertner on *Spirogyra* in 1788. The gradual elucidation of the life cycles in plants was to reach a crowning achievement half a century later in the seminal work of Wilhelm Hofmeister (1824–1877).

Meanwhile many of the most important aspects of flower sexuality and pollination were being established. Koelreuter, the son of a Tübingen apothecary, was inspired by Camerarius, and between 1761 and 1766 published the results of a series of experimental pollinations, including inter-specific hybridisations, which demonstrated plant sexuality and the process of pollination, thus firmly establishing the importance of insects in transferring pollen between plants.

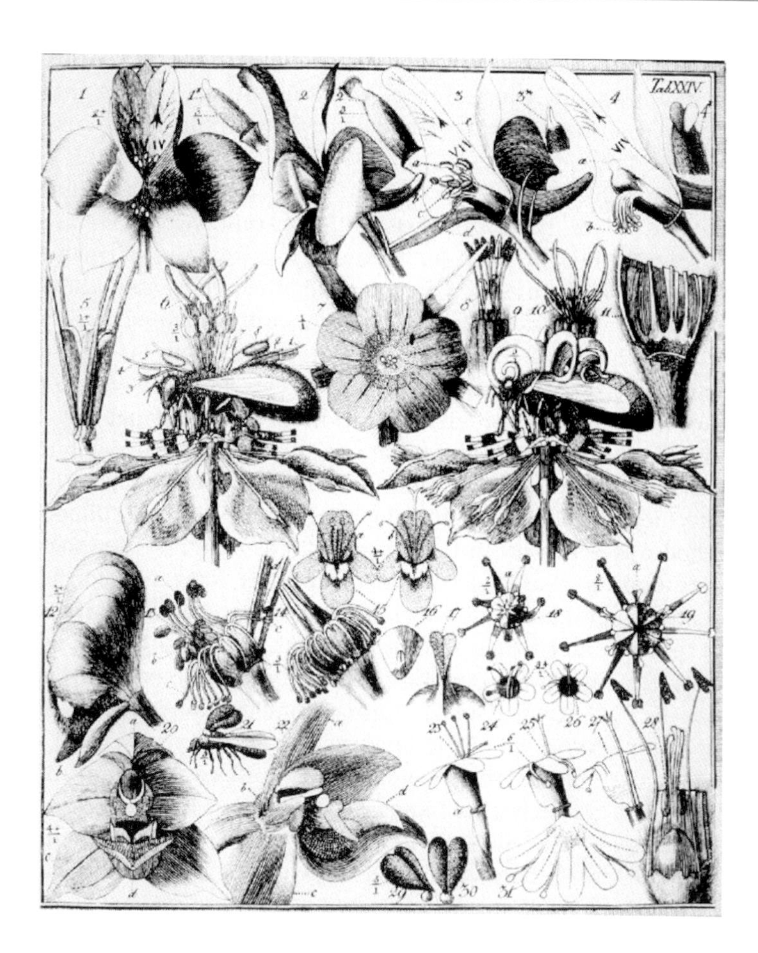

Figure 8.7. A page of illustrations from Sprengel's book.

He even recorded different kinds of pollination mechanisms and also noted that hybrids are rare in Nature because hybridisation normally can only occur between closely related species that are usually geographically isolated from each other. Noting the intermediacy of hybrid morphology between the parents, he formulated a 'proto-theory' of genetics. Sprengel brought the observation of flowers right up to date when, in Berlin in 1793, he published a book entitled *Revelation of the secret of nature in the construction and fertilisation of the flower* (Figure 8.7). Unfortunately this ground-breaking book, which established the relationship of flower shape to mode of pollination, was at first largely ignored.

8.1.6 Botany in the Age of Romanticism

The Romantic Movement was a revolt in the late eighteenth and early nineteenth centuries against the artistic, political and philosophical ideas associated with the neo-classicism of the mid-seventeenth to mid-eighteenth centuries. The mechanistic Newtonian world view was increasingly regarded as dry and sterile. At its root was the perceived loss of the spiritual dimension of Nature and humans. It was characterised in literature, music and painting by freedom of form, and creative imagination. Although it was largely eclipsed by the rise of modern science from about the middle of the nineteenth century onwards, the sensuous expression of the Romantics was not anti-science. On the contrary they have had a pervasive influence right up to the present day.

In Germany, mainly through the influence of Immanuel Kant (1724–1804) and K. G. J. Jacobi (1804–1851), the influence of empiricism waned, paving the way for German Idealism. In Germany, botanical research was more focused towards theoretical and philosophical aspects, whereas in Britain and France it remained largely empirical. Kant held that the content of knowledge comes *a posteriori* from sense perception but that its form is determined by *a priori* concepts of the mind (i.e. that all observations are theory dependent). He accepted Locke's view of the primacy of empirical knowledge but he insisted on being critical, by the use of theories that confirm or falsify, and the discarding of doubtful facts. He believed that a dialectical method was necessary in the process of reasoning, an idea that was later to find its greatest development in Hegel. Kant defined the attitudes and general methods which became the accepted norm for scientific research and which were established in botany largely by Schleiden. German universities were increasing in number at this time, and German science was to gain ascendancy and dominate botany in Europe for much of the nineteenth century.

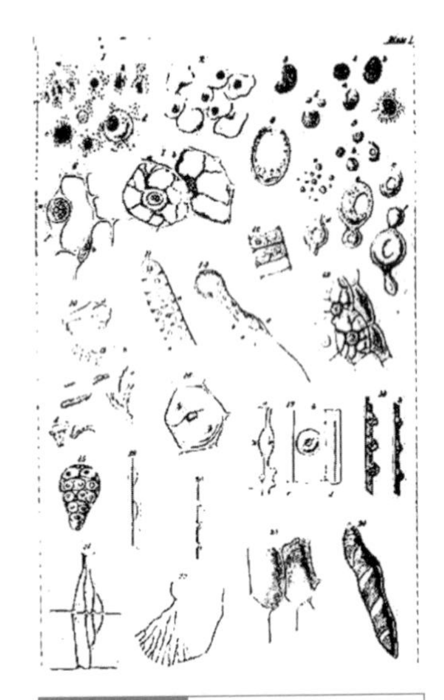

Figure 8.8. Schleiden's illustrations of cell development (Plate I in *Principles of Scientific Botany*).

If a man walks in the woods for love of them half of each day, he is in danger of being regarded as a loafer. But if he spends his days as a speculator, shearing off those woods and making the earth bald before her time, he is deemed an industrious and enterprising citizen.
Henry David Thoreau

From a modern perspective, one of the strangest developments of the Romantic era was the school known as the 'Naturphilosophie' or 'Nature-philosophy'. The Nature-philosophers knew that Newtonian physics must be wrong, because humans have feelings, consciousness and volition, which could not be explained by classical means. Their solution was to permeate the spiritual dimension through everything, thus unifying the Universe instead of perpetuating the division between humans and Creation. Despite much criticism from biologists such as Mayr, the Naturphilosophie was inspirational.

In the mid nineteenth century, the transcendentalist movement, in North America, was influenced by Lorenz Oken, and had its finest flowering in the writings of Henry David Thoreau (1817–1862), Ralph Waldo Emerson (1803–1882), Emily Dickinson (1830–1886), Walt Whitman (1819–1892) and John Muir (1838–1914). Thoreau in particular has had a lasting impact, and has been the inspiration of thousands of aspiring naturalists, particularly in North America but, as a champion for the conservation of nature, there could have been no finer example than John Muir. Other biologists associated with the Naturphilosophie movement include Schelling, Schimper, Hegel and Ernst von Baer, of whom the latter made some profound impacts on biology.

Georg Hegel (1770–1831) was interested in 'established' science and he opposed the reductionism of the mechanistic world view. Hegel's argument against reductionism was that trying to apply ideas from one level of the hierarchy to another would lead to confusion. His philosophy of the dialectic, the interplay or argument between alternative states, with its emphasis on historical contingency and conditional determination of phenomena, is relevant to the course of plant development and environmental interactions and subsequently on the study of evolution.

The genius of Johann Wolfgang von Goethe (1749–1832) claims a place here, although in contrast to the idealism of the Nature-philosophers, Goethe's method was essentially one of empiricism and inductivism. Like Kant, he was well aware that all observations are theory-laden, but his approach was to make the subjective aspect of observations an integral part of his empirical method since he believed that it led to deeper insights. This he did by varying the conditions under which the phenomena were observed, by studying plants throughout their ontogeny. He also paid particular attention to malformed plants. His insights were obtained through highly perceptive observation and intense visual impressions. He was a 'picture-thinker' (equivalent to the 'intuitive Anschauung' of Troll), which also has been described by Agnes Arber as 'thinking with the mind's eye'. He believed in open-mindedness, rejected dogma and took great care not to replace observations with abstractions. Although he believed in the harmony of the Universe, he sought revelation of it from the world of nature through phenomena.

A few minutes ago every tree was excited, bowing to the roaring storm, waving, swirling, tossing their branches in glorious enthusiasm like worship. But though to the outer ear these trees are now silent, their songs never cease. Every hidden cell is throbbing with music and life, every fiber thrilling like harp strings, while incense is ever flowing from the balsam bells and leaves. No wonder the hills and groves were God's first temples, and the more they are cut down and hewn into cathedrals and churches, the farther off and dimmer seems the Lord himself.

John Muir

For Goethe, science was not the search for abstract truth, but for synthetic and dynamic archetypal phenomena (Urphänomen), which, to him, were the highest levels of experience attainable. His 'delicate empiricism', which revealed his distrust of reason, was the approach that led him to formulate the Urphänomen. Unfortunately there has been much confusion over Goethe's concept of Urphänomen. It would be a mistake to attribute a Platonic Idealism to Goethe, for Plato's universals are not derived by empirical means. Goethe's Urpflanze has been erroneously interpreted as equivalent to an actual ancestral plant but, as Goethe wrote in 1827, 'the archetypal phenomenon . . . is to be seen as a fundamental appearance within which the manifold is to be held'. The Urpflanze was therefore a conjectural concept which allowed hypothetical situations to be visualised.

If we are to describe what a body is, the whole cycle of its alternations must be studied; for the true individuality of body does not exist in a single state but is exhausted and displayed only in this cycle of states.

Hegel

Goethe's botanical studies began some time during his first years in Weimar in the late 1770s, and in 1790 he published his *Versuch die Metamorphose der Pflanzen zu erklären*. This was his attempt to provide a theory of plant morphology, and it was based upon his observations of the serial homology of cotyledons, leaves and floral parts. Although other botanists at the time were thinking along similar lines, it was Goethe who clearly recognised that homology involves developmental processes. His views on developmental integration and its potential for evolutionary studies far exceeded that of Darwin although, owing to the magnitude of the task, he was never able to apply his methodology to a wider plant world. Goethe was also the first to coin the term 'morphology'. He can rightly be described as the originator of a phenomenological approach to botany and the father of comparative morphology of plants. He subsequently influenced the development of Gestalt Theory and theories of 'self-organisation', which are of relevance to plant systematics and plant development respectively.

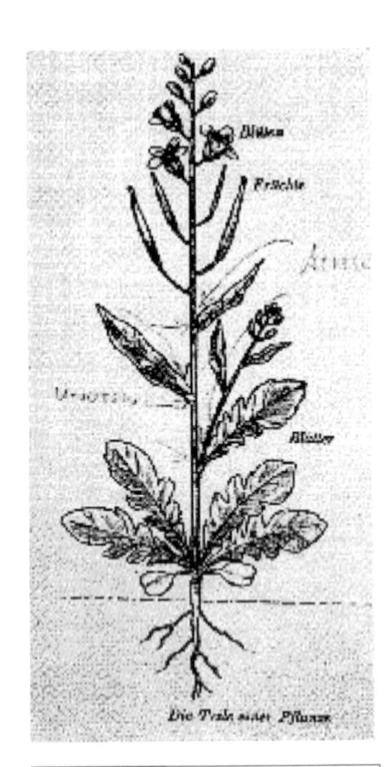

Figure 8.9. Illustration from Goethe's *Metamorphose der Pflanzen* (from www: odysseetheater.com/Goethe/ Goethe_20.htm).

> . . . the aims I strive for are an understanding of nature as a whole, proof of the working together of all the forces of nature.
>
> Humboldt

Many of Goethe's ideas resonate clearly in the writings of later continental philosophers such as Husserl, Merleau-Ponty and Heidegger, and continue to filter into European botany today. Unfortunately they remain largely ignored or derided by British and American schools of thought.

8.1.7 Voyages of discovery

In Britain, from the mid eighteenth century to the early nineteenth century, botanical studies were largely concentrated on the description and classification of new discoveries brought home from the expanding British Empire. Much of this activity was for economic or imperial gain, and was coordinated from the Royal Botanical Garden at Kew; and led ultimately to the introduction throughout the Empire of seven major exotic crops that were to entrench the British even more firmly as a world power. These were tea, coffee, rubber, opium, sugar, tobacco and cotton.

The eighteenth and nineteenth centuries were the times of the great exploratory voyages from England of Captain James Cook and botanists such as Joseph Banks and Robert Brown, and from Germany, of Alexander von Humboldt. Although the development of plant geography really got going in the nineteenth century with the researches of Alphonse de Candolle (1855), it had its beginnings in the work of K. L. Willdenow's *Grundriss der Krauterkunde* (1792). This was already prefigured by the work of Alexander von Humboldt who engaged on a five-year expedition (1799–1804) to northern Latin America and the Caribbean. Based on observations during his expedition, Humboldt was the first to propose formally the fundamental principles underlying the distribution of plant species. According to Ernst Mayr, Alexander von Humboldt was the father of ecological plant geography. Through the universality of his outlook, Humboldt anticipated the modern study of ecosystems. For him, detailed research was not an end in itself. Humboldt's book, *Personal Narrative of Travels to the Equinoctal Regions of the New Continent During the Years 1799–1804* (7 volumes, 1814–1829) and his *Essai sur la Géographie des Plantes* (1807) (with Bonpland) were later to influence both Darwin and J. D. Hooker, but it was in France that evolutionary studies really germinated.

8.2 Evolutionary botany

8.2.1 From Revolution to evolution

In France in the late eighteenth century there developed ideas of change and progress in nature. Foremost in the expression of the revolutionary changes in biology was J. B. Lamarck (1744–1829) who was the first to articulate fully theories of biological evolution. Although it was in his zoological work of 1809 that his mature ideas were fully expressed, Lamarck had already presented his basic findings in the first four volumes of his *Encyclopédie Méthodique* (1783–1793), which dealt with plants. His ideas on evolution can be traced to

Leibniz, Adanson and Buffon rather than to Darwin's grandfather Erasmus Darwin who published an outline sketch of evolution in 1794 (*Zoonomia*).

Let me repeat that the richer our collections grow, the more proofs do we find that everything is more or less merged into every thing else, that noticeable differences disappear, and that nature usually leaves us nothing but minute, nay puerile, details on which to found our distinctions.

Lamarck (1809) quoted by P. F. Stevens in *Why do we name organisms: some reminders from the past*, http://www.botany.wisc.edu/courses/botany 940/papers/Stevens2002.pdf

It is unfortunate that so great a botanist as Lamarck is often derided by modern biologists on account of his evolutionary theories. Much of this criticism stems from zoologists who ignore the context of time and place in Lamarck's ideas. For animals, he stressed their increasing complexity from simple to complex rather than their adaptation and ecology, but, as far as plants were concerned, Lamarck did make some very useful observations about the relationship of plants to their environment. He stressed the capacity of plants to alter their form through phenotypic plasticity, and the importance of growth factors in the evolution of both plants and animals. He was aware of time as a component of evolution and the long periods required for change, i.e. he was a gradualist. In some respects his integrative views of evolution are far superior to the naïve mechanistic mutation/selection theories of early Neo-Darwinists. One of the reasons why Lamarck was so dismissed, particularly in the first few decades of the twentieth century, was the recognition that the experience of an organism cannot be transmitted to the genes, and that a barrier exists between the reproductive and body cells in higher animals (the Weismann Barrier). However, the germ cells of all plants develop from somatic cells and, theoretically, somatic mutations can be inherited. There is now a considerable body of research on somatic mutations and the influence of the cytoplasm on the genome, but old habits die hard and Lamarckian inheritance is still spoken in hushed whispers in the hallowed halls of biology.

Evolution in various guises, influenced by various interpretations of the *Scala Naturae*, had been in the air since the time of Leibniz. By the early nineteenth century, evidence for it came from the study of geology and fossils. Lamarck probably developed his mature evolutionary ideas in the late 1790s as a result of his studies of fossil molluscs at the Paris Museum. Evolutionary interpretations of the fossil strata were also made for plants by two of the main founders of palaeobotany, Alexandre Brongniart (1770–1847) and his son Adolphe-Théodore (1801–1876) who studied Cretaceous fossil plant beds in the vicinity of Paris. They demonstrated that form had changed over time and they also developed a classification system for plant fossils. Adolphe-Théodore was able to recognise three major horizons of fossil plants: the Carboniferous, which contained ferns and their allies; the Mesozoic, which was dominated by gymnosperms; and the Tertiary, which had an increasing importance for the angiosperms. In 1828 he published his *Histoire des Végétaux Fossiles* in which he attempted to integrate knowledge of fossil plants with hypothesised palaeoclimates.

Figure 8.10. Drawing of fossil (Asterophyllites) by Brongniart (from mgs.md.gov/esic/brochures/fossils/carper.html)

In France towards the close of the eighteenth century, plant classification systems followed the work of Adanson and A.-L. de Jussieu but they were eventually superseded by the publications of the Swiss botanist, Augustin Pyrame de Candolle (1778–1841), whose *Théorie*

Élémentaire de Botanique appeared in 1813, and by the commencement of his *Prodromus Systematis regni Naturalis* (1824–1873), which was one of the most comprehensive systematic works on seed plants ever published. De Candolle's classification system was a natural one (in contrast to a purportedly phylogenetic one) and his criteria for the recognition of natural groups were highly sophisticated for its time. He made significant contributions to floral morphology, particularly in his recognition of symmetrical relations. Such quality of observation and taxonomic judgement was continued by his son, Alphonse de Candolle, whose publication in 1830 of *Monographie des Campanulées* was a model for monographic work.

The researches of the Scottish botanist Robert Brown (1773–1858), whom Goethe called 'this acknowledged greatest of botanists', were inspirational. His most significant contribution to botanical theory in his later years was his interpretation of floral development and his use of the vascular system to elucidate homology. As well as recording 'Brownian motion', the apparent random movement of dust like particles, first observed with the pollen of *Clarkia*, Brown made clear the profound difference between gymnosperms and angiosperms. It was also he who made the fundamental connection between pollination on the stigma, followed by growth of the pollen tube, to fertilisation in the ovule. In addition, he first recorded the nucleus as an essential organelle present in all cells and observed cytoplasmic streaming in the staminal hairs of *Tradescantia*.

M. J. Schleiden expressed the opinion that Robert Brown was the first to see that the history of development is the leading principle in understanding the morphology and nature of plants, but it was the Germans who dominated developmental morphology and plant anatomy during the nineteenth century. This was largely because of the high profile that botany was given in German universities and by the continued improvement of microscopes, many of which were manufactured in Germany. The astonishing breadth and detail of the German work is still evident today. In a negative way, the British contributed to this domination by their almost total preoccupation with systematic and floristic studies, although in an ironical twist, this same preoccupation was responsible in some measure for Darwin's epic voyage on the *Beagle*.

Several German individuals may be singled out as having made outstanding contributions, the most significant being M. J. Schleiden who published his ground-breaking book *Grundzüge der wissenschaftlichen Botanik* in 1842 (English translation: *Principles of Scientific Botany*, 1849), which was the foundation for modern botanical studies. Influenced by Kant, Schleiden's methodology was one of critical inductivism, and he was dismissive of intuitive methods, particularly some of the high-flown speculations of the Nature-philosophers. The most important outcome of Schleiden's researches is the theory of the dual role of the cell as the basic structural and developmental unit of plants.

Towards the middle of the nineteenth century there were several other German plant anatomists working on developmental problems, including C. Nägeli, F. Unger and W. Hofmeister. In 1844, Nägeli, who had studied under Hegel and tried to apply his thoughts to scientific methodology, did extensive investigation of cells. He showed that the nucleus was present in all cells of all the major groups of plants, and that division of the nucleus was a prerequisite to cell division. At about the same time, Franz Unger (1800–1870), working with *Tradescantia* (Commelinaceae) discounted earlier observations on plant apices by Schleiden, and produced firm evidence that new cells arise by division. He coined the term 'meristematic' for the formation of new cells at the tips of growing stems. Incidentally, in 1852, Unger published his *Attempt of a History of the Plant World* in which he outlined an evolutionary theory of plants that was a forerunner of Darwin, and alluded to the mechanism of variation that leads to diversity in the plant world as being internal. His use of the term Urpflanze differs somewhat from that of Goethe and this may have caused some misinterpretation of the latter. Gregor Mendel, who was one of Unger's students, was motivated by Unger's enquiries into the origin of plant species to undertake experimental work.

Although growth by division of single apical cell had been described in mosses and algae by Nägeli, it was Wilhelm Hofmeister (1824–1877) who demonstrated multicellular meristems in higher plants. It was Hofmeister's researches on the life cycle and reproduction of the cryptogams and the homologies of their reproductive structures that profoundly influenced the development of botany. His investigations clearly established that a relatively uniform plan of organisation ran through the entire plant kingdom, although this idea is implicit also in the botanical work of Goethe. More specifically, Hofmeister discovered regular alternation of two generations in the complete life cycle of a fern, and in bryophytes, gymnosperms and angiosperms. This clarified the previously puzzling relationships between non-flowering and flowering plants. One of his most profound insights was that the reproduction of heterosporic pteridophytes such as *Pilularia*, *Salvinia* and *Selaginella* were a key to the understanding of reproduction in higher plants. Lower plants thus became a vehicle for the study of higher-plant evolution.

In 1844 the British establishment was stunned by the anonymous publication of *Vestiges of the Natural History of Creation* by Robert Chambers, in which it was proposed that spontaneous generation and recapitulation were evolutionary mechanisms. Although the *Vestiges* was a timely catalyst for people such as Alfred Russel Wallace, and probably Darwin also, it offered nothing significantly new in terms of the actual causes of evolution. Nevetheless, the questions posed by Chambers festered in Wallace, culminating in his famous 'Law' essay of 1855. As The Age of Romanticism gradually drew to a close the milieu was just right for Charles Darwin to enter the stage. Darwin, since the return of the *Beagle* in October 1836, had been quietly gathering evidence for his ideas on evolution. In this task he was influenced by the works of Malthus, Humboldt, the geologist Lyell, and particularly by contemporary botanists such as Joseph Dalton Hooker who were ambivalent about many aspects of plant origins and distribution.

8.2.2 Darwin and Wallace as botanists

Both Charles Darwin and Alfred Russel Wallace gathered much of their evidence for the theory of natural selection and the origin of species from the diversity and geographical distribution of animals. However, despite his assertions to the contrary, Darwin was also an accomplished botanist, although, prior to the publication of *On the Origin of Species*, he was an observer of plants rather than an experimenter. Like many naturalists who are first drawn to the study of animals, Darwin came to appreciate the world of plants in his more mature years, and especially from his time at Cambridge where he was guided by his botanical mentor J. S. Henslow, Professor of Botany. The success of the flowering plants is largely the result of their interactions with pollinators, especially insects, and in his later years Darwin turned his attention to this fascinating and most striking aspect of the plant kingdom. Darwin's many scientific observations and results of experiments published between 1865 and 1880 took up from where Sprengel left off. His interest in plants was more than a vehicle for recreation and relaxation. From the attention and devotion he gave to plants it can readily be concluded that the interests of this self-styled 'incorrigible loafer' were highly contemplative.

Major botanical works by Darwin

The Various Contrivances by Which Orchids Are Fertilised By Insects (1862); *Climbing Plants* (1865); *Insectivorous Plants* (1875); *Effects of Cross and Self-Fertilisation in the Vegetable Kingdom* (1876); *Different Forms of Flowers on Plants of the Same Species* (1877); *The Power of Movement in Plants* (with F. Darwin, 1880)

Alfred Russel Wallace, co-discoverer of the theory of natural selection and key player in the development of biogeography, first began to appreciate the world of plants as a young man when he roamed the moors and mountains of South Wales, and especially the Vale of Neath with its delightful waterfalls. The year was 1841 and at that time he was working with his brother as a surveyor. He took a great delight in being able to identify the commoner species of wildflowers and developed an early feel for the orderliness of Nature. One of the first botanical books he bought was Lindley's *Elements of Botany*, which, despite it being an initial disappointment as a guide to British plants, nevertheless proved useful as an introduction to plant classification. Wallace visited flower shows and became aquainted with tropical species such as the orchid *Epidendrum fragrans*, which gave him a tremendous thrill and enjoyment. It was to these early experiences of exotic plants in such incongruous places that Wallace attributed his desire to visit the tropics although, like Darwin, he was also strongly influenced by the writings of Alexander von Humboldt. Self-taught, he started his own herbarium, taking care to collect good specimens and devote careful attention to their preservation. This early interest in plants was to be the turning point in Wallace's life.

Together with the entomologist Henry Walter Bates, Wallace's first adventure as a professional collector in the tropics took him to the Amazon region of Brazil in 1848 where he spent four years before returning to England on the ill-fated *Helen*, which caught fire and sank. His ideas on the origins of plants and animals were first aired at this time in conversations and correspondence with Bates. During his time in Brazil Wallace also met and teamed up with the botanist Richard Spruce. Wallace collected and described many new species of plants from Brazil, including three species of palms but unfortunately all of the specimens went up in flames along with the *Helen*. He began work on the origin of species during this period of the 1850s while in the field, publishing little-noticed papers that argued for the fact of evolution on the basis of geographical distributions. On his return to London in 1852 Wallace read a paper at the Zoological Society of London in which he stressed geographical distribution and in which it was clear that the seeds of his later, more mature ideas on evolution were sown. Like Darwin, the analysis of geographical distribution was, for Wallace, an important method in understanding biological form and crucial in an understanding of organic evolution.

8.2.3 The plant geography of Darwin and Wallace

During the voyage of the *Beagle*, Darwin's attention was drawn understandably to the geographic distribution of plants and it is these aspects of botany that are highlighted in *On the Origin of Species*, two entire chapters of which are devoted to the subject. It was from biogeography that Darwin gathered most of his evidence for evolution. But the researches of both Darwin and Wallace ultimately provided another context for understanding plants. Plants were adapted to their environment, and their geographical distribution could be explained by natural selection acting on their means of dispersal. When Darwin set out on his historic voyage on the *Beagle*, he took with him a translation of Humboldt's *Personal Narrative*, from which he memorised entire passages by heart.

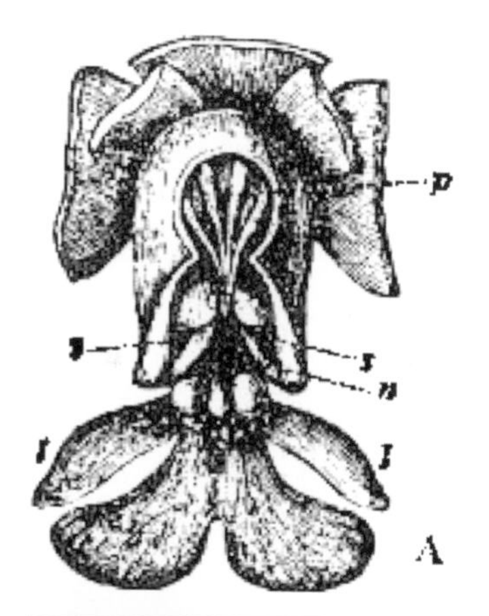

Figure 8.11. Part of a diagram Darwin used to illustrate the pollination of orchids.

Wallace pondered at great length on the diverse geographical and climatical boundaries which separated closely related species, even on continental land masses such as South America. In his *Journal* he wrote: 'During my residence in the Amazon district I took every opportunity of determining the limits of species, and I soon found that the Amazon, the Rio Negro and the Madeira formed the limits beyond which certain species never passed'. In his essay of 1855 (*On the Law Which has Regulated the Introduction of New Species*), which Darwin largely ignored, Wallace showed that the spatial and temporal connection between closely related species can be explained by common descent. He asserted that 'The most closely allied species are found in the same locality or in closely adjoining localities . . .' and 'every species has come into existence coincident both in space and time with a pre-existing closely allied species'. In 1858 he suddenly intuited the selection theory without realising that Darwin had already done so, and ironically wrote to him for help in getting his ideas published. This resulted in the joint paper read before the Linnean Society and published that year. For the remainder of his life Wallace generously credited much to Darwin, but his contributions were highly significant in their own right, and had an originality that is now being increasingly realised.

The analysis of patterns of geographic distribution were central to Wallace's ideas on evolutionary theory, much of which was brought to focus during his eight years in the Malay Archipelago (1854–62), but it is not clear as to the importance he attributed to sympatric speciation which is implicit in the above quotations from his 1855 essay. The insular nature of the Malay Archipelago was to impress on Wallace even more the factors that determine the boundaries of a species' range. This was particularly the case for the fauna of the region, which show affinities with both Asia and Australia. Wallace drew attention to the sharp faunal boundaries between Asia and Australia and the demarcation line separating the two great faunal regions has come to be known as Wallace's Line. We now know that this is the result of the tectonic history of the region and that Wallace's Line coincides roughly with the position of the continental shelf of Sundaland.

Although various ideas about movement of the continents had existed since the time of Francis Bacon, during the mid nineteenth century the continents were generally thought of as being static and fixed. A well-formulated theory of continental drift was almost half a century in the future. This led to concepts of floristic regions being largely coincident with the major continental areas of the world, and, of course, the corollary of static continents is dispersalism.

Darwin's main postulates concerning plant geography were that plants have achieved their present distributions through dispersal from a centre of origin. These postulates also apply to species with vicariant or discontinuous distribution patterns where the populations of a species are no longer in contact. Such dispersal could be achieved by gradual spread through contiguous land masses or by chance migration across natural barriers such as mountains or oceans. Apart from the possiblity of polytopic origin, particularly as a result of polyploidy, these postulates seemed reasonable and have generally been accepted by the botanical community, but confusion still remains about the proximate ability of a plant to disperse locally and its ultimate ability to colonise globally. Many botanists believe that, in orthodox biogeographic theory, there is a strong but naive bias towards the immediate means of dispersal rather than the ability of a plant to be part of a dynamic plant community.

Sir Joseph Dalton Hooker, who was a friend of Darwin, was highly influential in British botany from the mid nineteenth century onward. He originally favoured a land-bridge hypothesis to explain the distribution of plants in the Southern Hemisphere but later, although swayed by dispersalist hypotheses, he became sceptical about the truth of both explanations. Darwin was extremely suspicious of land-bridge theories and continental movements. He believed that the explanation of a plant's distribution was to be sought particularly in its past history with respect to climatic change and the plant's powers of migration and survival, that is, in its ecology.

There is no doubt that the growing influence of Darwin's causal analyses of evolution and the developing ecological approaches in biogeography led to a more dynamic concept of regional biogeography and one that could eventually be more comfortably accommodated within the theory of continental drift. Darwin's dispersalist ideas have found favour in theories such as the *Theory of Island Biogeography* (1967) by MacArthur and Wilson. Despite much criticism, the greatest virtue of this theory lies in its contribution to our understanding of the dynamics of island populations, and its application to the conservation of fragmented habitats such as rainforests.

Arguably, the most radical rethink of biogeographic theory in the twentieth century came as a result of the unique contributions of Leon Croizat. He was particularly impressed by the correlation between current distributions of plants and past tectonic movements. His most famous dicta are that 'the earth and life evolve together' and 'dispersal forever repeats'. Despite being accused of eccentricity, Croizat's wordy and somewhat idiosyncratic writings remain immensely important to phytogeographers and are a goldmine of information. Because he painstakingly plotted the distributions of numerous organisms using his track method in order to detect similar patterns, biogeography in his hands became 'panbiogeography'.

. . . it is certainly true that the variation elaborates the stuff on which Natural Selection comes to work; it is on the other hand radically false that variation but casually proceeds ('it is random in direction'). This cannot be because the variation is subjected to 'oriented evolution' and 'type of organization' . . .

Leon Croizat (translated by M. Heads in *Tuatara*, 1984)

8.2.4 The beginnings of ecology

Although ecology is prominent in the writings of Buffon, Linnaeus and Humboldt, the actual term 'ecology' was coined in 1866 by Haeckel as the science dealing with 'the household of nature'. In the second half of the nineteenth century orthodox biogeographic theory was given impetus by the works of distinguished botanists such as Asa Gray and Sir Joseph Dalton Hooker and by the more ecological approaches of Alphonse de Candolle (1855) and A. Grisebach (1872). The first major treatises on plant ecology and soil science were those of E. Warming (*Plantesamfund*, 1895) and N. N. Dokuchaev (*Principles of Soil Science*, 1890–1895), respectively. These were followed quickly by A. F. W. Schimper's *Pflanzengeographie auf physiologischer Grundlage* (1898) and C. Raunkiaer's classification of plant life-forms in his seminal *Videnskabernes Selskabs Oversigt* (1903). However, ecology remained largely descriptive of plant associations until it embraced population dynamics, and the mathematical methods for its analyses, in addition to the quantitative analysis of matter and energy flow in ecosystems.

The discreteness of certain species was well-known to naturalists such as John Ray, Linnaeus and Gilbert White. The variety was the only subdivision of the species recognised by Linnaeus in his *Philosophica Botanica* (1751) but he used the term to cover both geographical and individual variation. Buffon was one of the earliest biologists to remark on the geographic variation between similar animals of North America and Europe but he, like many other naturalists of his time, treated such populations as species. In 1825 Leopold von Buch stated clearly the principle of geographic speciation; that species can evolve by fragmentation of formerly contiguous populations. Observations were made by naturalists such as Pallas who suggested that vicariant populations of European and Siberian mammals were merely 'varieties' of the same species, while Gloger (1833) recommended that such geographical variants should be called 'races' or 'varieties'. Ultimately, the geographical 'variety' was subsequently designated 'subspecies', particularly if it differed morphologically, and was accorded a trinomial, while the term variety became restricted to individual variation. The first author to use trinomials regularly was Schlegel, as early as 1844. From the mid nineteenth century onwards the acceptance of the mutability of species and the range of variation presented within the boundaries of species initiated an intensified study of geographical variation in the new light of evolutionary theory. The strength of Darwin's adherence to geographical isolation as an important component of speciation was, in part, probably the result of his increasing study of plants from 1844 till 1859, and the writings of botanists such as Herbert. Darwin noted that oceanic islands were particularly favourable for the evolution of (new) endemic species although, by 1859, he was ready to accept sympatric speciation for many continental species owing to ecological isolation.

8.2.5 Adaptation and the theory natural selection

The origin of species was not a problem before the eighteenth century but it became one after Ray and Linnaeus insisted that the diversity of nature consisted of sharply demarcated and fixed entities (species). Their origin now had to be explained. Lyell saw the species as the unit of evolution and, by asking questions about the mutability of species, their origin and extinction, was pivotal in the subsequent preoccupation of Darwin with the species question. Darwin took a copy of Lyell's *Principles of Geology* with him on the *Beagle*.

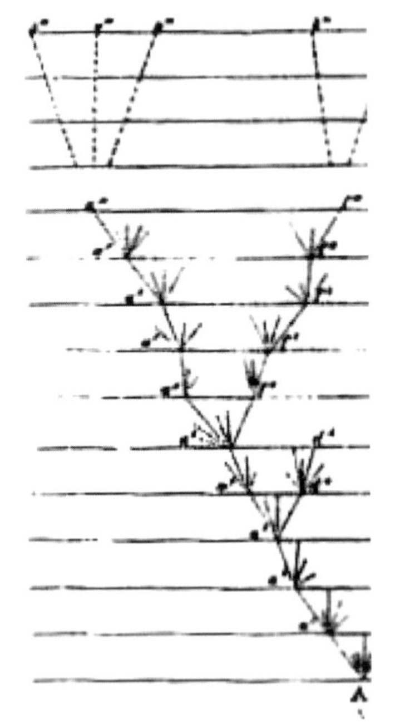

Figure 8.12. Fragment of Darwin's diagram showing the 'principle of divergence' from *On the Origin of Species*.

Darwin's publication of *On the Origin of Species* in 1859 did not remain obscure but shook the foundations of science and society. It provided a framework in which the differences between organisms could be understood. The theory of natural selection, which he had first formulated in 1838, provided a way in which biological diversity might have arisen by a process of adaptations becoming established. It was the insight of Darwin and Wallace to see that inherited differences between individuals could, by natural selection, lead to species that are better adapted to their environment. If different populations experienced different circumstances, different sets of adaptations would be selected. Thus, directionality was applied by nature to bring about the harmonisation of an organism with its environment. The process postulated by Darwin was entirely mechanistic: the organism proposes and the environment disposes. If this process of selection is combined with reproductive isolation between the differently selected populations then a new species can arise.

What may often be overlooked is that Darwin's empirical approach, although on par with the highest standards of his day, nevertheless brought his own subjectivity to his seemingly objective study of Nature. Darwin interpreted the phenomena that he observed by the concepts that he used to frame his observations. His use of the terms *heredity* and *inheritance* was based on a metaphorical extension of the inheritance of property in human society, and therefore misrepresents biological reproduction. His concept of homology is derived from the perspective of 'property' (the parts of an organism) being inherited. Homology was to be determined through descent from a common ancestor. Therefore, it is not surprising that some of his most compelling arguments come from animal behaviour such as competition for food and mates, territorial defence and from sexual selection.

Competition between plants is not as obvious as between animals and there is no real equivalent to animal behaviour. The analogue of behaviour in the plant world is growth. The majority of plants display some degree of phenotypic plasticity, which tends to be optimal for any given microhabitat. Phenotypic plasticity is the change in the expressed phenotype of a genotype as a function of the environment. However, there are many plants that have a relatively fixed developmental pattern and do not exhibit much plasticity. In such cases the norm of reaction is not expressed so much by the individual as by the genome and so selection on plasticity can occur only through selection in structured populations.

This highlights the great difference between natural selection in the animal world where phenotypic plasticity is manifested more in behaviour than in morphology. Had Darwin focused on plants earlier he might have given additional weight to the reciprocal relationship with the environment and its harmonising effects on the whole plant. However, in the absence of a theory of heredity, such a course would have brought him dangerously close to Lamarckism and, in any case, plants would have provided little in the way of popular appeal for his theory of natural selection.

Very few unequivocal examples of natural selection have been observed in nature, most of the evidence for natural selection being inferential, based of comparative data. It is difficult to perform evolutionary experiments because of the timescales involved between generations and that is why rapidly reproducing organisms such as bacteria, *Drosophila* and *Arabidopsis* have been chosen by researchers. But in most instances evolution must be inferred from observations and such inferences must be tested against subsequent observations. The study of adaptation is equally difficult, because the unit of selection is not a single character but the whole individual, the sum of all its adaptations. Adaptation is therefore relative. If it were not so, then organisms could not evolve or become extinct. It is a dynamic concept stressing the changing environmental parameters and a species response to them.

There are problems in the use of the words 'adaptation' and some see mechanistic connotations in its use. Sometimes adaptation is used in a physiological or sensory context when it then has a different meaning. In an evolutionary sense, as it is used throughout this book, many writers have noted its ambiguity. It is said to be non-scientific because the advantage characters confer is not measurable. Attempts to make the study of adaptation scientific, to measure relative adaptation, have centred on measuring a surrogate called 'fitness', i.e. reproductive success, as a kind of summation of the effectiveness of all the adaptations of an organism. However, even this measurement is extraordinarily difficult. Other measurements of adaptation have centred on establishing 'optimality models', assuming adaptations maximise the efficiency of some function of the plant. Most so-called adaptations are merely speculations, untested hypotheses, and if it is suspected that a structure has a particular function, then that hypothesis must be tested. In fact only a tiny proportion of variations, supposed adaptations, have actually been tested in this way. However, that does not mean that the concept is a faulty one, but it makes much more sense to consider organisms and their environment as two interacting components of a whole and the term 'adaptive' then becomes purely descriptive of their relationship or harmony.

A fundamental criticism of adaptationist hypotheses is that they do not allow the possibility that many variations are not adaptive in a Darwinian sense. The adaptationist approach is essentially teleological, seeking to find the purpose of every biological structure. This is particularly true when one studies arguments from design. Many biological structures seem so beautifully adapted to their function that an intelligent designer has often been invoked. Indeed the main resistance to the acceptance of evolutionary theory was from this particular viewpoint. But morphology may have been shaped by other forces: such as intrinsic processes and constraints that are developmental, and by environmental modifications of a plastic form, or it may have arisen by evolutionary accidents, such as genetic drift in small populations. It is not surprising if patterns of variation appear complex because we do not understand all the processes at work. Even attempts simply to describe natural patterns of variation in the ordered hierarchy of a classification are bound to have their limitations.

8.2.6 Historical contingency versus rational morphology

The conceptions of history, and the directionality of time, have had a very tortuous passage in biology. Although Aristotle had a concept of directionality (or polarisation) among living organisms, from 'lower' to 'higher', the legacy of 2000 years of essentialism was a concept of a harmonious if discontinuous universe. This was the view accepted and promoted by the religious establishment up to, and for long after, the Renaissance. The idea that there is a time dimension giving continuity to Nature, that progress and change are possible, became current in the seventeenth century thanks to philosophers such as Leibniz, although Aristotle had seen that organisms are joined by a 'unity of plan' and assumed that all organisms sharing such a plan share a unity of homologous parts. However, Aristotle's primary interest in function led him to mistake homologous parts for analogous parts and thus he united very dissimilar organisms under the same 'unity of plan', for example the fins of whales and fishes. During the first half of the nineteenth century the analogous similarities of organisms, such as the zygomorphic flowers of mints and foxgloves, were regarded as being the result of similarity of function alone, and not of historical contingency or convergent evolution. Homologous similarities were part of the plan of creation and were to be found also in the correlation of parts, and this blinded Cuvier to the possibility of evolutionary change. In the great debates of Cuvier and Geoffroy Saint-Hilaire it is clear that neither were really asking the right question; that is, what is the reciprocal relationship of the organism and its environment? As a result Cuvier insisted that function determines structure whereas for Geoffroy it was the other way round.

Darwin simply side-stepped the difficulty of interpretation by declaring that all similarity is the result of proximity of descent from the nearest common ancestor, and that analogous similarity is caused by modification over time due to similarity of function. The time dimension, plus, of course, the mechanism of natural selection, was the crucial component missing in the theological viewpoint. Thus, history was firmly re-instated as a fundamental factor of enquiry into Nature.

During the nineteenth century the study of embryogenesis and development was mostly focused on animals and this, in turn influenced their classification. The classification of plants was probably less influenced by evolutionary theory than that of animals, although the maturation of an organism at different stages its development (ontogeny) has been an important process in the evolution of plants. When a plant reproduces at an earlier, more immature stage, it is an example of paedomorphosis. Delayed reproduction or retardation is less common in plants. Both phenomena have not been given the prominence they deserve. According to Schleiden, Robert Brown was the first to see that the history of development is the leading principle in understanding the morphology and nature of plants. Others who contributed greatly to the comparative embryology of plants within an evolutionary context include the great plant anatomists of the late nineteenth century such as Hofmeister, Strasburger, Goebel and Bower.

The first comprehensive theory to explain differences in development among organisms was the idea of 'recapitulation', which was first proposed by the German Nature-philosophers. This idea had its roots in the *Scala Naturae* of previous centuries, and its basic tenet is that nature proceeds in steps from the simple to the complex, with each organism passing through ancestral stages during its development. The now (in)famous 'biogenetic law', that ontogeny recapitulates phylogeny, was formerly proposed by Ernst Haeckel in 1866 and who modified the original recapitulation idea by suggesting that the whole process involved changes in developmental timing, or heterochrony. It was an appealing, if deceptively parsimonious hypothesis. Some 40 years earlier, the theory of recapitulation had, in fact, been challenged by K. E. von Baer (1792–1876), who argued that each stage in an organism's developmental trajectory posed special challenges for that organism and should not be viewed as a simple recapitulation of ancestral form. The so-called laws of development or 'laws of growth' never really became part of the phylogenetic programme.

On my theory, unity of type is explained by unity of descent.
Darwin

. . .the chief part of the organisation of every being is simply due to inheritance . . .
Darwin

With the hegemony of Darwinism, developmental 'laws' were soon supplanted by historical narratives that would explain similarity by descent. According to Darwin, every structure could now be explained in terms of adaptation and selective advantage, but he did not say how different sets of adaptations actually arose in the first place. This is sometimes called Darwin's 'Black Box'. For Darwin, the morphological expression and 'traits' of organisms could be explained by, as yet unknown, 'factors of heredity'. His equally unknown 'Laws of Growth' were no longer an issue but, throughout his life, he maintained an intuitive, if unsure, belief in their importance. From this, it would appear that Darwin did have some difficulty with the ability of chance mutations and natural selection actually to produce the intricate and bewildering designs of the organic world. Since then many people also have had problems with this very question.

Life is simply the reification of the process of living.
Ernst Mayr

Like Richard Owen (1804–1892), Darwin rejected the idea that many constant features of organisms should be understood in functional terms. The explanation usually offered is that these archetypal features were of adaptive importance to ancestral forms in the past, most of which are now extinct. Ernst Mayr has explained the unity of the archetype by emphasising that the genetic programme for development consists of a set of such complex interactions that it can be modified only very slowly. By changing the focus from an organismic-centred research to one that was based on phylogeny or 'descent with modification', Darwin unwittingly (?) introduced a programme of research which explained everything by historical contingency and which was ultimately based on his ideas of genealogy. Whereas Linnaeus regarded his taxonomy as a preliminary statement for an understanding of the principles of biological creation, Darwin took the view that the reconstruction of actual, contingent genealogical succession was the procedure that would reveal the plan of creation. That so much of the biological realm is still unintelligible to modern phylogenetic analyses is the high price paid for this view. However, the question of a rational as well as an historical unification still remains on the agenda in biology.

The harmony of the world is made manifest in Form and Number, and the heart and soul and all the poetry of Natural Philosophy are embodied in the concept of mathematical beauty.
D'Arcy Thompson
(http://www-history.mcs.st-andrews.ac.uk/Quotations/Thompson D'Arcy.html)

Biologists in the late nineteenth and early twentieth centuries, such as Bateson, Driesch and D'Arcy Thompson, attempted to use structuralist concepts to counter the challenge of Darwinism with its preoccupation with functional adaptation, but the major difficulty for developmental research in the nineteenth century was that, unlike Darwin's evolutionary approach, there were no well-formulated theories within which it could be framed; there were no laws which could explain transformation of form. Goethe's revolutionary approach to understanding phenomena did not explain transformation of form. Bateson (1894) questioned whether the discontinuity between species is the consequence of natural selection acting on a formerly continuous variability of form or whether it is due to the manifestation of an intrinsic discontinuity. He believed that form in organisms could be understood in terms of relatively discontinuous and alternative stable states. Weak perturbations allow a particular state to re-establish itself but stronger ones result in transformation to an alternative determinate state. Driesch also viewed the organisms as a self-regulating system with an autonomous ordering principle. From a modern perspective, Bateson's ideas (which possibly were influenced by Galton) seem strikingly similar in some respects to those of Waddington, and the autopoiesis of Maturana and Varela.

Much of the effort by morphologists in the late nineteenth and early twentieth centuries was an attempt to refute one of the central tenets of Darwin's theory, that is that organisms are aggregates of independent parts constrained only by external functional requirements. Just as these ideas were starting to bear fruit Weismann proposed his theory of inheritance at about the same time as Mendel's 'Laws of Heredity' were being rediscovered. Darwinism almost totally eclipsed any alternative explanation for organic form and practically destroyed the science of rational morphology.

8.2.7 Classification and evolutionary theory

In the art and science of plant classification there is no agreement on what a classification is for. The philosophy behind classification is that different plants can be assigned to recognisable groups or classes if they possess the characteristics of the class. In other words, they constitute natural kinds and can be classified accordingly. Each class can then be incorporated into more inclusive groups based on more widely shared attributes. In this way, the natural classification is orderly, hierarchical, predictable and compatible with evolutionary theory. In other words, it has, itself, the structure of a theory. There are many different types of classification depending on the purpose of the classifier. In botany different classifications usually reflect different taxonomic viewpoints. Some workers regard a classification primarily as an information retrieval system, like a Linnaean system, which may also, almost as a subsidiary feature, indicate evolutionary relationship, whereas others, at the outset, aim to produce a classification that reflects evolutionary relationship. If the evolutionary relationship has been correctly determined the classification may provide a useful information retrieval system. There is currently a crisis in Systematic Biology with a major split between those who wish to retain the traditional Linnaean system and those who advocate a phylogenetic approach to classification. Pre-Darwinian classifications were empirical processes based on natural kinds, as indicated above. They were not based on evolutionary theory, phylogeny or genealogy, and were thus radically different in nature from modern phylogenetic classifications. The Linnaean system was never intended to reflect phylogeny. Despite its warts, it has proved remarkably workable as a general system for more than 200 years and does not seriously compromise phylogenetic efforts at classification.

Natural classifications of plants dominated the nineteenth century. The de Candolle family, at the Conservatoire et Jardin Botaniques in Geneva, influenced by the natural system of A.-L. de Jussieu (*Genera plantarum*, 1789), produced what remains for some flowering plant groups the latest world monograph. Auguste de Candolle started the *Prodromus Systematis Naturalis regni Vegetabili* in 1824, and it was continued after his death by his son Alphonse. Seventeen volumes covering 58 975 species were eventually published by 1873.

Darwin saw that genealogy could be the basis for understanding similarity of form and so the study of natural kinds gave way to phylogeny as the basis for constructing classifications. The problem remained how to represent that continuum in a classification, for Darwin gave no indication as to how organisms were to be ranked. Despite Hooker's championing of Darwin, Bentham and Hooker did not change the sequence of families in their *Genera Plantarum*, the first volume of which was being drafted as *On the Origin of Species* came out. Bentham and Hooker followed the de Candolles rather closely although their natural arrangement was substantially adopted by many later workers and given an evolutionary twist. *Genera Plantarum* was the last major work largely uninfluenced by Darwin's theory of evolution. Its value lies in the accurate descriptions of genera and families. Despite the practical difficulties of utilising phylogenetic data for classifications, Darwin's theory of evolution provided an explanation for the great diversity of nature. Since the mid-nineteenth century biologists have used putative phylogenetic relationships as the best means of establishing a classification. This was supposedly a major revolution in botany for it marked the end of natural classification systems. In practice the word natural changed its meaning so that it came to be equated with phylogenetic.

The most complete early phylogenetic system, a development of an even earlier one by Eichler, was that of another German, called Engler, who published prolifically with a number of other authors over a period of 40 years. His most important work, with Prantl, *Die Naturlichen Pflanzenfamilien* was published between 1887–1908. Many herbaria today are still organised according to its sequence of families. In many respects it follows de Jussieu rather than de Candolle; for example, the gymnosperms are excluded from the angiosperms. Engler divided dicots into the Archychlamydeae and the Sympetalae. The evolutionary sequence goes from the simplest, apetalous, unisexual flowers to those which are very complex, starting with families such as the Piperaceae and Chloranthaceae, followed by the catkin-bearing trees, a group he called the Amentiferae.

In America, Bessey (1915) laid down guidelines for the construction of phylogenies such as the recognition of evolutionary trends, and he proposed an alternative system that was also purportedly phylogenetic (Figure 8.13). This was closer to the system of A. P. de Candolle, and starts with 'perfect' bisexual flowers such as the buttercups, while other major lineages are derived by changes such as the fusion

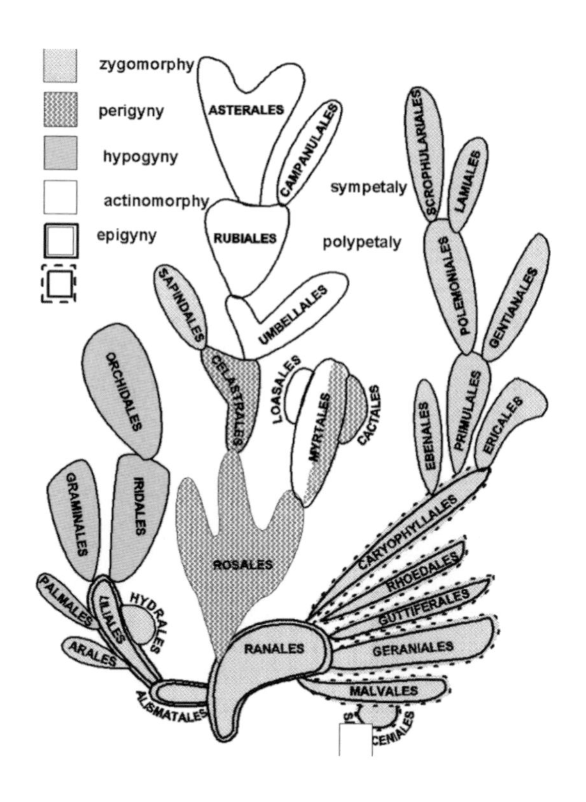

Figure 8.13. In America, Bessey used a cactus-like diagram to show the phylogenetic relationships of flowering plant groups. He laid down guidelines for the construction of phylogenies such as the recognition of evolutionary trends, and he proposed an alternative system which was purportedly phylogenetic. This was closer to the system of A. P. de Candolle and starts with 'perfect' bisexual flowers such as the buttercups, while other major lineages are derived by changes such as the fusion of floral parts.

of floral parts. Twentieth-century classifications have largely reflected these two early phylogenetic systems.

The search for an agreed natural phylogenetic classification of plants has been beset with problems of intelligibility. Darwin had eliminated a structuralist approach to understanding homology so that the homology of similar features of the plants could only be determined by genealogy. Many arguments in support of phylogenetic classifications were circular. Phylogeny was extrapolated from the classification and then was used as evidence for the classification. This naïve approach to classification largely dominated plant systematics for much of the twentieth century and is the basis of what subsequently became known as 'evolutionary systematics'. The literature of phylogenetic systematics has been burdened by voluminous writings on homology and its central importance in systematics. Many features of plants have evolved either in parallel or convergently in different lineages and it is consequently difficult to separate homologous characters from analogous ones.

8.3 Phylogeny, genetics and the New Systematics

8.3.1 Variation and the transformation of form

Despite the moral outrage, Darwin and Wallace's revolutionary insights swept away the old paradigm of the natural theologists, based

Alphonse de Candolle and others have shown that plants which have very wide ranges generally present varieties; and this might have been expected, as they are exposed to diverse physical conditions, and as they come into competition (which, as we shall hereafter see, is an equally or more important circumstance) with different sets of organic beings.

Charles Darwin, *On the Origin of Species*

on the *Scala Naturae*, that organisms were adapted to their particular environment by the utility of Divine design. Nature was now seen to be dynamic, and evolving by a materialistic process (natural selection) that could be understood in everyday terms. Revolutions, by their very nature, usually call for a total overthrow of the prevailing conditions or modes of thought, and this was no less true for Darwinism. But it would be unrealistic to believe that everything pre-revolution was outmoded or that the new paradigm would be a cure-all for archaic ways of thinking and doing science. Lamarckism still held sway for many people, and Darwin himself was still partly inclined towards the view that inheritance of acquired characters played a supplementary role in evolution. Plant classification in the immediate post-Darwinian period did not alter profoundly, due to the stability of natural systems such as those of A. P. de Candolle, Bentham and Hooker, and others.

As the furore mellowed, the gradualist evolution hypothesis of Darwin, was accepted by many influential botanists such as Sir Joseph Dalton Hooker and Asa Gray, although 'Darwin's bulldog' T. H. Huxley, and Darwin's cousin Francis Galton, believed that evolution occurs by discrete leaps or saltations. Darwin's explanation for the discontinuity among species and taxa was to postulate extinct hypothetical ancestors or missing links. The lack of fossil angiosperms led him to talk of their origin as an 'abominable mystery'. Even today, the discrete macro-evolutionary gaps between taxa are commonly regarded as entirely caused by the extinction of intermediate forms. But many of the leading geneticists of the day adhered to a saltationist view of evolution through mutations (the term 'mutation' was actually introduced by Waagen in 1869) in the genetic material, especially de Vries, who dismissed all alternative views, yet maintained that his theory was a modification of Darwin's. William Bateson (1861–1926), who strongly influenced many of his contemporaries, was opposed to gradualism, and denied that natural selection played a major role in evolution. Because species are discontinuous, he hypothesised that variation is also discontinuous, and has its origins, not in the environment nor in adaptation, but in the intrinsic nature of organisms themselves (heterogenesis). Bateson gathered such voluminous data on morphogenesis and variation (published in his *Materials for the Study of Variation* in 1894) that it potentially provided the foundation for a well-formulated science of rational morphology.

One of the areas that Bateson researched was teratology because, like Goethe before him, he saw such aberrant forms as providing additional clues for the transformation of form. It is noteworthy that this same approach has been used in the study of the genetic control of floral development through the study of such mutants as *apetala* (see Chapter 2). What the opponents of structuralism (who were mostly advocates of a narrow interpretation of Darwin's theory) failed to consider was development. They hypothesised about mutations, selection and gradual evolution, but they did not consider the coordination of development during ontogeny, which was one area where morphologists were qualified to contribute. In the absence of a theory of transformation, they defaulted to gradualism, even to the extent of rejecting Mendelian inheritance.

As the nineteenth century drew to a close, the momentum of the new paradigm faltered for want of a mechanism of heredity. In 1896, James Mark Baldwin published a paper that, while claiming that natural selection was sufficient to explain evolutionary change, intuitively hypothesised a 'new factor in evolution'. His hypothesis, subsequently known as the 'Baldwin Effect', states that those individuals which are more adaptable in their phenotypic response, and can accommodate to the vicissitudes of their environment, are more likely to survive and leave more progeny. This phenomenon may be accompanied or followed by a genotypically controlled response which has been coined 'genetic-assimilation' by Waddington. Waddington termed this superficially neo-Lamarckian response 'canalisation', a homeostatic process that favours a particular ontogenetic trajectory. Thus, plasticity is selected for, and builds 'an epigenetic landscape, which in turn guides the phenotypic effects of the mutations available'. It was a neat re-interpretation of the Darwinian model. Following on from this idea was the proposal by Woltereck of the concept of the reaction norm, of which W. Johannsen drew attention to its close similarity with that of the genotype. Johannsen, who actually criticised Woltereck for misunderstanding the significance of environmental context, was, at that time, already developing the concepts of genotype and phenotype. A century later these ideas seem particularly relevant but the rediscovery of Mendel's work pushed them into the shadows.

8.3.2 Gregor Mendel and the rise of genetics

Darwin's theory of evolution was flawed because it lacked a theory of inheritance of adaptive traits. He never understood how heritable changes occurred or resulted in variation. Although he made some suggestions about how this might happen, the explanations were not satisfactory. He thought that the features of the parents were blended together in their progeny. Unfortunately this meant that any evolutionary novelty was likely to be dissipated like a drop of ink in a bucket of water and unavailable for selection. It was the crossing experiments of Gregor Mendel (a correspondent of Nägeli), carried out in his monastery garden at just about the same time as *On the Origin of Species* was published that gave Darwinism credibility. Unknown to Darwin, Mendel had established the mechanism of inheritance, thereby founding the science of genetics (Figure 8.14). He made his discovery by a study of inheritance of characters such as seed colour and texture and flower colour in peas. Most importantly he showed that differences were not blended in the progeny but might be expressed in later generations. Mendel concentrated on elucidating traits in organisms and his research on peas and other plants eventually allowed him to hypothesise the existence of heredity

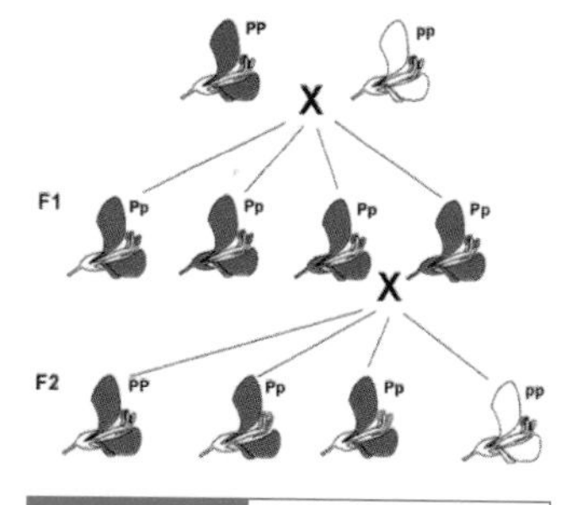

Figure 8.14. Gregor Mendel established the mechanism of inheritance.

'factors', which we now know to be genes. Mendel's work, first published in 1866, remained unappreciated until it was rediscovered in 1900. It was history repeating itself because, in a similar way, the earlier researches of Camerarius, upon which Mendel's own work relied, had remained neglected, until Koelreuter took them up half a century after their first publication in 1705.

8.3.3 Genetics and Neo-Darwinism

In an ironic twist, when Mendel's work was 'rediscovered' by de Vries, Tschermak and Correns in 1900, it was interpreted as supporting a discontinuous theory of variation and was thus cited to discredit Darwin. Incidentally, long before this, Mendel's work was known to Nägeli, who actually published Mendel's results in a book entitled *The Mechanical-Physiological Theory of Evolution* (1884). Darwin certainly knew of Mendel and, although he unfortunately never read his classic paper of 1866, he had studied a paper by Focke that repeatedly cited Mendel. Mendel, who was even mentioned in *Encyclopaedia Britannica*, also possessed a copy of *On the Origin of Species*. Meanwhile, in Cambridge, Edith Saunders (1865–1945), in collaboration with William Bateson, independently rediscovered some, at least, of Mendel's laws before his work became known to them.

With the realisation of the significance of Mendel's work, and the role of the nucleus as the carrier of heredity being recognised in the 1870s by Oskar Hertwig and others, and the researches of August Weismann (1834–1919), the science of genetics was born. Perhaps the greatest contribution of Weismann was his clarification of chromosomal rearrangements or 'crossing-over', and its importance for evolution. When Weismann, in 1883–1884, proposed the complete and permanent separation of soma and germ plasm, the so-called 'Weismann Barrier', the idea of the inheritance of acquired characters lost all credibility. By then, however, Darwin, who saw natural selection as only part of the evolutionary story, with inheritance as possibly dual in nature, was already dead. Inheritance of acquired characters was now firmly rejected by leading exponents of selection and gradualism, and was not to reappear until the somatic selection hypothesis of Steele *et al.* in the 1980s reopened old wounds.

Weismann's theory of the continuity of the germ-plasm had become the cornerstone of early twentieth-century evolutionary biology. The efficacy of the Weismann Barrier for animal evolution was assumed to apply equally to plants, if it was ever considered in the context of plants at all. The Weismann Barrier is hardly applicable to plants, which have several permanently embryonic regions (the meristems) derived from somatic tissue. Weismann's insistence of the overriding importance of selection led Romanes, in 1896, to coin the term 'Neo-Darwinism' which is sometimes erroneously equated with the 'synthetic theory of evolution' or 'modern synthesis'. From the turn of the century until about the 1930s, the focus of evolutionary research centred mainly on the cell and the mechanism of inheritance, as the new science of genetics gained ascendancy. Simultaneously, research in ecology was developing, and eventually it all culminated in the so-called 'synthetic theory', which became the dominant paradigm.

8.3.4 The 'modern synthesis' – new orthodoxy

In the west, during the 1930s, evolution was conceptualised as the result of natural selection acting on variation within populations rather than between individuals, and the scientific orthodoxy channelled research towards the study of changing gene frequencies in populations. The Hardy–Weinberg equation demonstrated that gene frequencies will remain unchanged in a large cross-breeding population unless there is a perturbation such as selection. Thanks to the brilliant applications of mathematics by R. A. Fisher, Sewall Wright and J. B. S. Haldane, this particular focus led to the development of mathematical population genetics. Fisher had a conception of genetic architecture based on genes as independent factors – their effects could be added together to produce the phenotype and even continuous traits. In contrast, Sewall Wright was impressed by the pleiotropic effects of single genes and genetic drift, changing gene frequencies as a result of random sampling in small populations.

During this period there were developments in ecology, spearheaded by G. Turesson, E. B. Ford, S. S. Tchetverikov, Timofeef-Ressovsky, and others, which were essentially field-based, an approach that became known as ecological genetics or genecology (also known as biosystematics). Tchetverikov rejected the idea of traits being determined by independent genes, but accepted the hypothesis that each trait is determined by a whole complex of interacting genes. Thus, the combination of ecology with genetics, and a powerful mathematical foundation, firmly established the importance of selection and provided the bases for the new evolutionary synthesis. Paradoxically, adaptation became a problem because individuals in populations rested on adaptive peaks, and a transition to a different adaptive peak required some individuals to cross valleys where they had less fitness. It was Sewall Wright who introduced the metaphor of the 'adaptive landscape', which was later to become associated with the ideas of Waddington.

The 'modern synthesis' was largely the outcome of the Jesup Lectures at Columbia University, and was completely biased towards zoology until Stebbins published his *Variation and Evolution in Plants* in 1950, apart from contributions by Gilmour (1940). Despite the fact that pioneering efforts had been already developed in the field of experimental genetics by Bauer, in phytosociology by Braun-Blanquet, and in the demography of populations by several ecologists, the efforts of plant biologists did not seem to contribute significantly to the new synthesis, at least in the early stages. In part this was because of the complicated nature of plant genetic systems, and the inability of botanists at the time to formulate a uniform species concept applicable to plants. Population genetics demonstrated that only small amounts of migration between populations could prevent their divergence re-emphasing the importance of reproductive isolation in evolution.

8.3.5 Speciation

Speciation, from Darwin onwards, was one of the most fundamental problems of evolution. Although Darwin and Wallace never actually subscribed to the idea that geographical isolation was a precursor to reproductive isolation and speciation, Mayr declared that the 'modern synthesis' was clearly only the maturation of Darwin's theory of evolution. They obviously saw the importance of islands in the speciation process, but did not go as far as Mayr in elevating geographical isolation as the major determining factor. By 1859 Darwin was ready to accept sympatric speciation for many continental species due to *ecological* isolation. Reproductive isolation is achieved in plants by many different mechanisms. Mayr clearly recognised that the actual mechanism of speciation ultimately resided in the genes and chromosomes. Nowadays, the role played by internal factors such as polyploidy and genetic turnover has to be addressed in any theory of plant evolution. Polyploidy, for example, is widespread in the angiosperms, and current research is revealing that the plant genome is a dynamic evolving system (see Figure 8.15). Several plant families such as the Campanulaceae have massively rearranged chloroplast genomes in all the major genera, the significance of which has yet to be realised.

Figure 8.15. An example of sympatric speciation following hybridisation and polyploidy is the evolution of *Senecio cambrensis* as an allopolyploid derivative of a hybrid between *S. squalidus* and *S. vulgaris*.

Mayr's Biological Species Concept (BSC) emphasises the reproductive isolation of species. The compelling attraction of the BSC is the fact that it provides biological criteria for the recognition of some of the discontinuity that exists in nature. The BSC places the species as the basic unit of evolutionary biology, and as more natural than higher categories. However, if time were somehow speeded up, we might observe species becoming and disappearing in rapid succession, dissolving into one another, and into genera, just as we observe the transient passage of individuals within generations. In a sense, over evolutionary time, the species is no more real than any other taxonomic category, and one can understand why the Nominalistic Species Concept remained favourable among some botanists. However, functionally, the species has an importance not found in collective higher categories.

Mayr applied his BSC concept to a local flora in northeast United States and concluded that an overwhelming majority of the species could be embraced perfectly adequately within this concept. However, in botany, the BSC has never been popular, and for most botanists, particularly herbarium workers, a taxonomic (morphological or phenetic) species concept was preferred. This has had at least one unfortunate consequence for botany: diverse phenotypes, often of a single species or ploidy level, and often based on a single collection, have been given specific status, thus burdening the nomenclature with superfluous names, and hindering subsequent investigation.

However, the most obvious reasons for the preference for a taxonomic species concept in botany are as follows.

- Plants display a wide amplitude of variation with respect to their environment
- Reproductive isolation in plant species is initially harder to prove than in animals because plants lack behavioural traits which could indicate some sort of reproductive barrier
- Plants are usually collected for study rather than observed in the field and, therefore, morphological criteria are used to define species boundaries
- Plants have several alternative reproductive strategies such as vegetative reproduction, and apomixis, which bypass sexual reproduction altogether

In practice, reproductive isolation in plants is frequently not absolute. Even though they may differ morphologically and genetically, hybrids may differ in degree of sterility, or be perfectly fertile. In addition variation is often reticulate and multi-dimensional, and is not amenable to discrete recognition. Turesson (1922) saw the Linnaean nomenclatural system as limiting in its ability to conceptualise the variation in plants in Nature, and therefore he developed his genecological terminology. He coined the term *ecospecies* for the Linnaean species from an ecological perspective, and *ecotype* for the total phenotypic expression of a particular ecospecies within a more localised habitat. For the complete aggregate of populations or indeed, species, which are capable of hybridisation, Turesson used the term *coenospecies*. In complex cases the genecological approach is inadequate and the 'deme' terminology was devised as a solution. It avoids formal names. The core of this terminology is the neutral suffix '-deme', to which is attached one or more prefixes that imply restricted applications of the complete term. The prefixes used are based on standard terms used in taxonomy and ecology. The suffix '-deme' does not imply that the plants in question form a population. The following is a list of the major categories of demes and their subtypes.

Denoting an association with a specific locality and/or habitat
Topodeme: occurring in a specified geographical area
Ecodeme: occurring in a specified habitat
Denoting phenotypic and/or genotypic difference
Phenodeme: differs from others phenotypically
Genodeme: differs from others genotypically
Plastodeme: differs phenotypically but not genotypically
Denoting reproductive behaviour
Gamodeme: individuals that interbreed naturally
Autodeme: composed of predominantly autogamous individuals
Endodeme: composed of predominantly closely interbreeding dioecious plants
Agamodeme: composed of predominantly apomictic plants
Denoting variational trends
Clinodeme: one of a series of demes, which collectively show a specified trend, or cline

Morphological similarities and differences that could be used to delineate species are difficult to describe objectively. Plant species may be genetically very similar yet reproductive barriers exist, preventing hybridisation. Indeed, some taxa such as *Musschia* (Campanulaceae) differ radically in morphology from their putative relatives, yet are closely related genetically. This is an example of morphology being out of phase with the plant's genome.

In practice, depending upon context, it is necessary to recognise several different kinds of species:
- successional species (palaeospecies),
- microspecies (agamospecies),
- biological species (genetical species),
- taxonomic species (morphological species; phenetic species),
- biosystematic species (ecospecies; coenospecies).

Hybridisation between morphologically highly distinct entities is particularly common in some groups such as the orchids. Allopolyploidy can lead to multiple origins of a new species (polytopic origin), while, among asexual populations, distinct phenotypes may persist indefinitely, for example *Limonium* (Plumbaginaceae). All these examples make it clear that the BSC is difficult to apply to plants. As with animals, distinctive allopatric populations, especially if they also have distinctive ecological requirements, are usually treated as species, but the evidence is generally inferential. Transplant experiments such as those done by Clausen, Keck and Hiesey cannot establish criteria for species boundaries although they can provide evidence for affinities.

For practical classification and identification purposes it is expedient to use the Taxonomic Species Concept provided one bears in mind that, like a phylogenetic tree, it only has an approximation to reality, though the taxonomic species corresponds precisely with the biological species. The development of molecular techniques has provided a measure of genetic distance between species, but at what percentage of difference in gene sequence or genetic markers should the boundary be placed? Anyway, if we were to recognise and name species on the basis of differences in DNA sequences, then the whole edifice of classification would collapse. Nevertheless, the recognition of differences in plant populations at the genetic level is profoundly important for conservation purposes, so the student of plant evolution has, simultaneously, to operate within several relatively independent frameworks.

In 1926 I called attention to another important similarity, which it seems to me, greatly strengthens the comparison between plant community and organism – the remarkable correspondence between the species of a plant community and the genes of an organism, both aggregates owing their 'phenotypic' expression to development in the presence of all the other members of the aggregate and within a certain range of environmental conditions.

A. G. Tansley, The use and abuse of vegetational concepts and terms, *Ecology*, **16**: 3 (1935), 284–307.

8.3.6 Plant ecology

Ecological studies were also being developed independently both at the level of individual species (autecology) and the community (synecology). The description of vegetation as communities of organisms was made by the American ecologist Clements (1874–1945), and the English ecologist Tansley (1871–1955), although the development of vegetation during this period was largely through the study of succession towards a climax vegetation.

Within an autecological perspective, and following on from the pioneering efforts of Gaston Bonnier and C. Schroeter (1926), Clements and the Danish botanist Turesson (1892–1970) developed the study of botanical genecology. Through their studies of the adaptation of plants to environments, they were able to demonstrate the distinction between the effects of inherited genotypic differences between individuals, from plastic differences moulded by the environment.

Population genetics rather ignored the relationship between the genotype and the environment, but the analysis of variance and the study of heritability of continuously varying traits, especially by plant and animal breeders, re-emphasised the contextual nature of the phenotype. Each phenotypic trait is determined by the genotype and the environment (including both the external and internal environments) acting together. The same trait could have high heritability, and be changed by selection, or it could have low heritability, and be highly resistant to change.

From the 1930s to the 1950s, the nature and extent of variation within and between plant populations was investigated by Clausen, Keck and Hiesey. Their approach was orthodox in that they believed that local populations are the units of evolutionary change but they saw that variation is contextual to a given environment and not limited to average differences among individuals. Despite the fact that they regarded macro-evolutionary change as being in accord with environmentally correlated gene expression, their views were, somehow, out of step with the prevailing 'modern synthesis'. Their findings on genera such as *Mimulus* (Scrophulariaceae) and *Viola* (Violaceae) showed that speciation in plants is a much more complex phenomenon than the Biological Species Concept (BSC) promoted by Mayr and Dobzhansky.

8.3.7 Voices of dissent

The 'modern synthesis' became the orthodoxy but some botanists maintained a different tradition. In the newly emerging Soviet Union, N. I. Vavilov made voluminous observation on variation in plants at different levels in the taxonomic hierarchy, but especially at and below the species level in grasses of economic importance. Vavilov was particularly struck by the parallel series of variations which occur in diverse lineages of plants, particularly those which have closer genealogical relationships. From these observations he was able to formulate his famous 'Law of Homologous Series in Variation', and to predict the presence or absence of particular traits in populations. Vavilov was laying the foundation for population phenetics, which, in a sense, was ahead of its time owing to the lack of genetic techniques. But he drew attention to the phenomenon of repetitious variation across diverse lineages that could not be explained adequately by natural selection acting on random mutations.

Vavilov, who was influenced by William Bateson, was well aware of the implications of Goethe's pioneering efforts, but he never got the opportunity to develop his ideas. Tragically, in 1940, he was incarcerated in Saratov prison for daring to criticise Lysenko, and he died there three years later. Since his official rehabilitation, the extent of Vavilov's immense contributions to botany can be fully appreciated. His book, *Centers of the Origin of Cultivated Plants* (1926) remains a classic for the study of crop plants, but he is also recognised as the foremost biogeographer of his time during which he participated in over one hundred expeditions to almost every corner of the globe.

> *. . . genera more or less nearly related to each other are characterized by similar series of variation with such regularity that, knowing a succession of varieties in one genus . . . one can forecast the existence of similar forms and even similar genotypical differences in other genera.*
>
> Vavilov, 1922

In the momentum of the new orthodoxy, research on reaction norms and phenotypic plasticity was largely ignored since it was viewed as of minor importance in evolution, and was thought to have little, if any, genetic basis. Simultaneously, the rational morphologists were severely criticised or, at worst, ignored. However, botanical morphologists such as Arber, Troll, Zimmermann and Willis, empirical saltationists such as Goldschmidt and Schmalhausen, and the palaeontologist, Schindewolf, didn't exactly disappear overnight, but continued to hover on the periphery of the new orthodoxy. Richard Goldschmidt felt that natural selection was only part of the evolutionary story and relevant mostly to micro-evolutionary events. The bulk of his research focused on the causal factors of macro-evolution, the origin of species and phenotypic novelty, and resulted in his book, *The Material Basis of Evolution* (1940), from which the unfortunate phrase 'hopeful monster' originated.

There is more to Goldschmidt's legacy than this hopeless caricature. His work, especially on plants, convincingly demonstrated that plants can survive and reproduce, even when their morphology is far from the norm. He saw that different environments will produce different phenotypes, and he coined the term phenocopy for those plants whose phenotypes resemble the effects produced by known mutations. One of the effects of mutation on plant development that Goldschmidt recognised was the phenomenon of *homeosis*, which we discussed in Chapter 2. Many of Goldschmidt's ideas have relevance today in modern studies of ontogenetic contingency and phenotypic plasticity. In a modified form (Neo-Goldschmidtian) his theories have resurfaced in the writings of van Steenis and, more recently, those of Bateman and Dimichelle.

Schmalhausen emphasised the importance of changes in the ontogeny of organisms for the evolution of form involving, initially, alterations in the norms of reactions by direct environmental influence, and later by the mediation of genes. Variations in critical environmental parameters may invoke morphological changes that may be interpretable as adaptive or not. For example, the leaf morphology of many aquatic plants such as *Ranunculus fluviatilis* (Ranunculaceae) is correlated with their ability for gaseous exchange in air or in water.

Schmalhausen also drew the important distinction between those plants which display phenotypic plasticity and those which do not (i.e. 'normal' phenotypes or wild-type reaction norms). He referred to the latter as displaying the effects of stabilising selection. Unfortunately, he also used the term in another sense for a two-stage process whereby a plant is able to utilise other parts of its range of reaction norms to accommodate or harmonise itself to changed conditions, and eventually becomes selected for a new stable state. Schmalhausen hypothesised that the mechanism responsible for this reshuffling of reaction norms is differential allelic sensitivity due simply to the biochemistry or physiology of the plant. The overall effect is eventually incorporated into the more complex genetic regulatory system, a process which is essentially the same as the 'canalisation' of Waddington. The difference between the two stages of the process is that the first is a 'reaction' to the environment whereas the second is 'anticipatory' of the environment

8.3.8 The natural philosophy of plant form

Outside of mainstream botany in the first half of the twentieth century there were several developments that had their origins from pre-Darwinian times, and from the period immediately following the publication of *On the Origin of Species*. A structuralist programme was largely missing from the modern synthesis, but several individuals may be singled out as having had a major influence on structural botany, even though they are not always given due recognition for their contributions.

Wilhelm Troll (1897–1978), was who was a student of K. von Goebel, was an idealist and rigid typologist, and could be said to have adhered most closely to the natural-theology of the previous century, even to a Platonic world-view. He rejected common descent as the basic explanation for systematic categories because his types, which represented the fundamental order of Nature, were invariant. This was probably the greatest weakness in his whole research programme. Although his approach was empirical, he only accepted evolution as occurring through major saltational changes in the types. For Troll, typology was the predominant and fundamental procedure of systematics, whereas phylogeny merely provided a means of tracing genealogical lineages.

Troll's view was based on two static type categories: organisation, which is a metaphysical idea, and form, which is the perceivable form or phenomenology of the organism. He was inspired by Goethe's belief in the 'unity behind diversity', but he lacked a dynamic perspective. His morphology also reflects Goethe's dualistic view of a universal reference system (archetype) and its many manifestations (Gestalten). His organisation type (or Bauplan) was equivalent to Goethe's archetype, but differed in being discrete from other organisation types. Intermediates were not recognised, each organisation type had to be one thing or another, and within each organisation type, a variety of forms or 'Gestalten' could be recognised that differed only in proportions. This was Troll's 'Principle of Variable Proportions'.

His concept of homology, which was the natural outcome of his purely structuralist approach, was based on relative position rather than identity through phylogenetic relationship. He used the term 'Gestalt' for the outer appearance of forms whether analogous or homologous, and he held the view that Gestalten could not be subdivided into component parts without loss of identity. They are therefore beyond analysis. Because of the diversity of flowers and inflorescences and their repeated convergences and parallelisms across unrelated groups, Troll believed that Gestalt is independent of the underlying archetype and functional constraints. From an evolutionary perspective it is easy to see the difficulty that Troll must have experienced in explaining botanical phenomena. He hypothesised an 'Urge to Form' in order to explain analogous similarities among diverse plant groups, which could be interpreted as vitalism. Despite the inevitable negative consequences of Troll's rigid typological system, there were a few hidden gems. By carefully documenting in diagrammatic form the characters of a great wealth of plants, especially their inflorescences, he introduced a reference system for all parts of the plant thus providing the first general view of plant diversity and a scientific procedure for abstracting general rules from individual plants.

Walter Zimmermann (1892–1980), whose main interests being plant phylogeny and evolution, rejected metaphysical influences, and regarded archetypes as genealogically related natural groups whose fluid nature was established through phylogenetic analyses. He believed in a strict distinction between subject and object and in this respect he could be said to be more in line with an empirical scientific approach rather than the idealistic views of Troll, or even the 'gentle empiricism' of Goethe. For Zimmermann, rational analysis was the preferred scientific procedure. He distinguished between 'Natural Laws', which are intellectual abstractions and the bases for hypotheses, and 'Natural Regularities', which are observable phenomena. In contrast to Goethe's Urpflanze, Zimmermann's hypothetical archetypes were potentially real plants of the past. Zimmermann had a major impact on the thoughts of Willi Hennig and the subsequent development of cladistics

The upsurge in phylogenetic emphasis after Darwin, in combination with a strong appreciation of biological form and evolution, prompted a diversity of evolutionary morphological studies. In the late nineteenth century zoologists such as Ernst Haeckel were promoting the famous 'Ontogeny Recapitulates Phylogeny Hypothesis', while, in botany an associate of Haeckel in Jena, Eduard Strasburger, was continuing the pioneering work of Hofmeister. In Britain, F. O. Bower also did much to advance knowledge of alternation of generations, particularly in ferns. In his *Origin of a Land Flora* (1908), Bower adopted an evolutionary-adaptive perspective on alternation.

Agnes Arber (1879–1960) was perhaps one of the greatest visionaries of the botanical world in the twentieth century. She became a renowned plant morphologist and anatomist, historian of botany, botanical bibliographer, philosopher of biology, the first woman botanist to be elected as a Fellow of the Royal Society of London, and the first woman to receive the Gold Medal of the Linnean Society of London. She made original contributions to botany that, like her two main contemporaries in Germany (Wilhelm Troll and Walter Zimmermann), have largely been bypassed by the rise of molecular systematics and developmental genetics. Her approach to plant morphology was developmental and dynamic, but it also allowed the nature of the investigation process to be revealed. She treated plant archetypes as abstract conclusions from empirical observations, but had a dynamic concept of the archetype rather than a static one, a natural outcome of her developmental approach. She was no narrow typologist, but possessed an unsurpassed awareness of the holistic nature of plant form. Her emphasis thus differed somewhat from the typological approach of Troll, or the phylogenetic approach of Zimmermann.

It is the business of morphology to connect into a coherent whole all that may be held to belong to the intrinsic nature of a living being.
Agnes Arber, 1950

Arber was greatly influenced by J. W. Goethe, and translated his major botanical work *Versuch die Metamorphose der Pflanzen zu erklären* in 1946. Following Goethe she employed a method whereby the whole is encouraged to 'speak' to us through its expression in the parts. Although she was limited by her circumstances, and availed herself of only the most basic anatomical equipment, her observational skills, and her ability to integrate her observations to a coherent whole, were without equal. Her talents were creative, the encompassing vision of the artist, and the grace and erudition of a great writer, in combination with the analytical eye of the scientist. Among her many publications, the following four books remain classics of botanical writing, namely: *Herbals, their Origin and Evolution* (1912); *Water Plants* (1920); *Monocotyledons* (1925); *The Gramineae* (1934). In her later years, Arber increasingly turned to philosophy and metaphysics, which resulted in the publication of *Goethe's Botany* (1946); *The Natural Philosophy of Plant Form* (1950); *The Mind and the Eye* (1954); and *The Manifold and the One* (1957). The *Natural Philosophy of Plant Form* is probably her most important contribution to theoretical plant morphology.

The leaf is a partial shoot . . . which has an inherent urge towards the development of whole shoot characters.
Agnes Arber, 1941

The nature of the leaf was a central aspect of Arber's thoughts, and she elaborated a partial-shoot theory, which may be viewed as an re-expression of the ideas of Dresser and C. de Candolle that, in themselves, were undoubtedly influenced by Goethe. This theory emphasises the parallels between leaves and shoots, and, therefore, their part in a developmental continuum, in contrast with the view that leaves are independent units distinct from the shoot. Arber was interested in iterative growth in plants, an aspect of plant development that later came to be of profound importance in developmental genetics. She wondered if the reiteration of leaf form in leaflets of a compound leaf could have relevance to the relationship between shoot and leaf. The homologous variation that she identified across structures of a single plant (i.e. partial-shoot theory) is logically very similar to the homologous series of variation across species and genera that were established by Vavilov.

8.3.9 The pursuit of objectivity

During the post-war years until about the 1960s the modern synthesis, or New Systematics as it became known, expanded and consolidated its domination of evolutionary studies. A systematic development of autecological studies matured to produce the study of infraspecific variation that became known as biosystematics. Population genetics as well as chromosome studies could be applied to systematic inquiry. The analysis of variation required a rigour not found in the more traditional evolutionary or intuitive systematics, and this led to the development of more precise biometric techniques. The first serious challenges to traditional systematic methods came in the 1960s from numerical taxonomy or phenetics as it became known. At first this technique was used most successfully in bacteriology, while in botany it was most successfully applied in the analysis of reticulate relationships below the species level. Numerical taxonomy made no claims to detect phylogeny but aimed to provide a more objective and rational basis for taxonomic inference. Much of the early hopes for a thoroughly objective method of systematic analysis were soon dashed when it was realised that different results could be obtained from different clustering methods. In addition, the choice of characters used was a highly subjective exercise. For a short period botanists debated whether intuitive methods might be preferable to computerised phenetics.

Apomorphy = a derived or specialised character state.
Synapomorphy = a shared apomorphy.
Plesiomorphy = an ancestral or primitive character state.
Symplesiomorphy = a shared plesiomorphy.

Shortly thereafter, in the mid 1960s, Willi Hennig's work on cladistic methodology (*Grundzüge einer Theorie der phylogenetischen Systematik*) became better known in the English-speaking world and, since then, cladistics has virtually been the method of choice for the majority of plant systematists, and latterly in conjunction with the use of molecular characters. Cladistics aims to find groups or clades on the basis of shared derived characters or synapomorphies, whereas ancestral characters or plesiomorphies were considered to introduce 'noise' into the analysis and thus create erroneous results. Usually more than one cladogram is obtained in cladistic analysis and the choice of the best one is decided on the basis of some parsimony criterion. Homology is determined a posteriori as those similarities that are shared through recent ancestry.

One of the difficulties in cladistics is the determination of the polarity of the characters, whether they are ancestral or derived. This often remains subjective, although outgroup comparison is often used to determine the status of the character. The algorithms used in cladistic analyses are designed to produce a cladogram with fully resolved branches and this most commonly takes the form of a hierarchy of dichotomies. This may be the greatest weakness of the method because Nature does not necessarily evolve in a dichotomous fashion. Furthermore, the assumption of cladistic analyses is that the basal, most ancestral stem of the clade represents the total expression of informative characters knowable for the ancestors of the group as a whole since plesiomorphic characters are discounted. Subsequently, new characters or synapomorphies are considered to arise *de novo*, or they are considered to have arisen more than once in different lineages within the group. It is difficult to reconcile such a narrow way of interpreting genotypic expression with what is actually observable in Nature, where the whole organism expresses many aspects of its phylogenetic legacy, not just a few synapomorphies. This has to be accounted for in a classification, which must, affer all, be workable for its practitioners.

Fickle Dame Nature has very different ends in view from that of creating neat hierarchies of species and genera, which naturalists can file away tidily in cabinets with the least possible trouble.
G. L. Stebbins, 1950

However, cladistic analyses can introduce an additional and welcome rigour into systematics but there are some levels in the taxonomic hierarchy where its use is inappropriate. For example at the infraspecific level variation often has a reticulate or tokogenetic pattern due to hybridisation and polyploidy. Older methods of multivariate statistics (numerical taxonomy or phenetics) such as factor analysis of character correlations and cluster anlaysis of overall similiarity measures are more effective in revealing the complex nexus of relationships that exist within and between closely related species.

In the purist Hennigian interpretation of cladistics, a clade is considered to be monophyletic if it includes both the ancestor and all the descendants of that ancestor. We object to this narrow interpretation of monophyly, for which there was a perfectly acceptable pre-Hennigian term, namely 'holophyly'. Similarly, the equating of the term 'natural' with 'phylogenetic' is unfortunate.

Although cladistic methodology has proved outstandingly successful, at least in attracting the young, a difficulty arises where cladistic phylogenies are translated directly into classifications. So, while Mayr advocated a cladistic method of analysis, he rejected the adoption of a cladistic method of classification, preferring an evolutionary approach that utilises the degree of divergence between phylogenetic branches, and takes into consideration the level of ecological adaptation (grade of organisation). Although there is no reason why the results of a cladistic analysis should, with due care, not be brought to bear on a classification, we must remember that at best, our estimate of phylogenetic relationships is inferential, based on available data.

8.3.10 The triumph of molecular systematics

The use of computer-based techniques to apply cladistic methodology is at the heart of molecular systematics. However, an argument could be made that it was not cladistic methodology but the new kind of data, in vast quantity, from molecular sequences that has revolutionised systematics and enabled a phylogeny of plants to be detected (see the APG arrangement described in Chapter 5).

Nowadays, molecular techniques dominate almost all systematic studies of plants. The unfortunate spinoff of this development is that it requires expensive equipment and can be carried out only by relatively few in institutions such as universities and the larger museums. This has made plant systematics an ever more esoteric discipline, the province of the expert, and remote from the amateur.

The use of cladistics has led to a proliferation of plant families, with the knock-on effect that Linnaean nomenclature is now deemed inadequate by the most enthusiastic cladists. In addition, molecular cladograms may not be the best source for inferring phylogenies since we now know that different genes are evolving at different rates, and different cladograms may result from their use in phylogeny reconstruction.

Figure 8.16. An example of the strange relationships discovered by molecular phylogenetics is that between (a) *Platanus*, a woody tree, and (b) *Nelumbo*, a kind of water-lily.

Placing facts above ideas, which is characteristic of extreme empiricism, has an injurious influence on the development of plant morphology. The facts . . . should be interpreted, systematized, and generalized.
A. Takhtajan, In *Evolutionary Trends in Flowering Plants*, 1991.

In botany the evolutionary or eclectic approach to classification that was employed by botanists such as Cronquist and Takhtajan, Thorne and Dahlgren has, in the past decade, been completely overtaken by molecular systematics leading to the publication of revised phylogenies for all plants (see Figure 8.16). Nevertheless we must not forget that the work of these earlier botanists provided the foundation for the current explosive development of molecular systematics. Working in an older tradition they brought a lifetime of experience to bear on their interpretations of plant evolution. Dahlgren's use of diagrams ('Dahlgrengrams') foreshadowed the modern approach of mapping characters or geographic distributions on to gene phylogenies. Of course, some aspects of the earlier systems were mistaken, for example the Dillenidae of Cronquist is now understood not to be a group, but their works contain much of lasting value. Perhaps the most extraordinary mistake was Hutchinson's recognition of two major lineages of flowering plants, one for woody plants and one for herbs, that seems incomprehensible now, but even so it prefigures the recognition of the Rosids (containing the majority of woody plants) and the Asterids (dominated by herbs and subshrubs) in the APG system, and Hutchinson's books dealing with the genera of flowering plants are still classics.

The kind of changes in classifications that cladistic methods have brought about are well illustrated by the treatment of a long recognised family, the Scrophulariaceae *sensu lato*. This is now proved to include members of at least four distinct lineages, and its genera have been distributed to four different families, separating showy garden flowers once grouped together, such as *Antirrhinum*, *Pentstemon*, *Linaria* and *Nemesia* from *Verbascum* and *Digitalis*, and from *Mimulus*. Even more surprisingly, the radically redefined families in which these genera now reside include other very distinct genera. What is left in the Scrophulariaceae (*sensu stricto*) are the genera *Verbascum* and *Digitalis*, but now also included are genera formerly placed in two other families such as *Selago* (formerly in Globulariaceae) and *Buddleja* (formerly in Loganiaceae or Buddlejaceae). The newly circumscribed Plantaginaceae has a huge morphological diversity that includes not only familiar garden plants, showy annuals, biennials and hardy perennials such as *Linaria*, *Antirrhinum* and *Nemesia*, with four or five stamens, but also shrubby genera such as *Veronica/Hebe* with two stamens. More remarkably it also includes *Plantago*, which is mainly wind-pollinated, with its scapes of tiny flowers. Astonishingly it also includes the aquatics with very reduced flowers, *Callitriche* (formerly in its own family Callitrichaceae) with its unisexual flowers lacking a perianth, and *Hippuris* (formerly in its own family the Hippuridaceae), lacking a corolla altogether, and having only a single stamen arising from the top of the ovary! A difficulty is that the newly circumscribed families include such divergent genera that they are very difficult to define morphologically. It seems that all that their members share is a genetic lineage. This is exactly the situation that was forecast by early critics of cladistics, and calls into question what these new classifications are for. This kind of treatment is likely to make a botanist trained in an older tradition where taxonomic groups are based on resemblance turn apoplectic.

8.4 The green future

In this chapter we have tried to trace the major developments in botany since its inauguration as a science in the Renaissance, and to highlight the events that contributed most to our understanding of the plant world. To some extent these developments reflect the rise of modernity, and botany, no less than any other science, has been influenced by the changing politics and social order in Europe during the past 500 years or so. Those aspects of botany that we have emphasised, such as philosophy, classification, morphology, ecology and genetics, have been most central to how we know plants. Of course, there are other important areas of botany that we have scarcely touched upon, for example, some of the most profound advances in botany in the twentieth century were made by plant physiologists and biochemists. Without doubt some of the greatest advances involved the elucidation of photosynthesis and photorespiration.

Botanical advances depended not only on new ways of looking, thinking, and experimenting, and new ways of collecting and analysing data, but also on the development of new technology such as computers, gas chromatography, mass spectrometry and the electron microscope. How we conceptualise plants has also been crucial in the development of botany as a science and, whether we like it or not, Descartes' metaphor of the machine has been central to all our investigations of plant-life until the present day. It still holds sway, despite the fact that few would interpret the metaphor literally. The world of Newtonian mechanics lent credence to Darwin's theory of natural selection, which is still the dominant causal explanation of evolution among present day biologists.

There is also no doubt that reductionism as an analytical technique has been singularly successful in advancing the scientific quest but one wonders how much this has been at the expense of a more synthetic, holistic biology. Now the advent of DNA sequencing has opened the 'blueprint' of plants to our gaze: first, the chloroplast genome of the liverwort *Marchantia polymorpha* was sequenced, followed by the complete genome of the cyanobacterium *Synechocystis*, and then the apogee of the reductionist approach came in 2001 with the sequencing of the whole genome of *Arabidopsis*. More than 25 000 protein-encoding genes were reported, leading one over-enthusiastic researcher to trumpet, rather naively, that all that was needed now was to find out what their products did! It is as if the gene sequences are a manual on how a plant works, and all that is necessary to understand it.

The Platonic world view, revived in various guises by Bacon and Descartes, allowed nature to be objectified, to be seen as 'the other', and therefore to be manipulated and exploited. The pervasive and deeply sublimated biblical account in *Genesis* of our exile from the Garden of Eden has reinforced the strange belief that we are somehow not part of the natural world. Since we do not belong in Nature, we treat it shabbily.

The investigative, reductionist techniques, the *modus operandi* of an anthropocentric western science, have had a sorry impact on plants and other forms of wildlife. With the rise of agriculture-based civilisation and then of industrialised capitalism, the story has been one of exploitation, subjugation and destruction of the natural world. Science, in the service of the prevailing interests of the day, has often been the vehicle for the domination and control of Nature rather than for an understanding of it. One wonders how different it might have been, if instead of being dominated by men, science in general, and botany in particular, had been led by women.

The life of plants is intimately enmeshed with climate change, and the ecological crisis that we are witnessing today, with increasing levels of greenhouse-gas emissions, global warming and rising sea levels is undoubtedly linked to the ever-escalating industrial activities of humans, in combination with continued destruction of the world's vegetation, the major sink for carbon dioxide. These activities are intimately bound up with a globalising culture that identifies personal happiness with material abundance and consumption. As human populations rise inexorably and the environment is challenged evermore by our pollutants, we will rely more and more on the ability of plants to respond and to buffer our planet from extreme changes. As botanists we are acutely aware of the accelerating decline of plant species worldwide and the prospect of major losses of biodiversity. This will have significant impact, not only on our utilitarian use of plants, but also on our aesthetic sensibilities, and the legacy we leave for future generations. In addition, it will have a cascading affect on other components of the planetary ecosystem, especially leading to catastrophic loss of animal species.

All these impacts ultimately derive from the application of science to wrong technology. Many developed nations acknowledge the threats and have taken steps to develop alternative and sustainable ways of energy generation, although, as we have seen with the proliferation of wind-farms, such steps are often controversial, profit-motivated and, ironically, may be an additional hazard to the environment. Countries such as the USA, which is one of the greatest producers of greenhouse gases, has given greater priority to national interests, and has refused to sign up to international agreements to reduce such emissions. Meanwhile, over the next decade or so we can expect the level of carbon emissions to increase dramatically as developing nations such as China and India rush to build economies that depend on material production and consumption.

The greatest danger that the Earth faces is the export of capitalist models of production and consumerisms, and the globalisation of markets. This can rapidly erode local economies and result in a dependence on export potential controlled by foreign interests and world prices. Under such conditions, pristine areas of the Earth and indigenous peoples will always be under threat from development. We can no longer go on exploiting Nature as if it were inexhaustible. All this is costing the Earth, but how can we break the impasse? Have we the right to preach to developing nations about limiting their desire to bring material prosperity to their citizens, when we in the west are still drowning in a glut of uncontrolled consumerism, spurred on by the philosophies of economic growth and progress?

The criticisms levelled at science are part of a general disaffection with modernity. Although a growing number of people now realise that science is not a panacea for harmonious relations between humans and the environment, there are still many who cling to modernity's vision of Utopia, and believe that science will solve all our problems. The paradigm of modernity is still very much with us. For example, in the universities, reductionist science continues apace, but is increasingly funded and subverted by commercial interests. Unfortunately, many researchers in plant biology do not see the plant in context, as a living component of the environment, and few have any interest in plants outside of the laboratory. One of the most worrying developments to spawn from this kind of research has been the proliferation of genetically modified (GM) crops, especially when one recalls the disastrous impact of the previous 'Green Revolution' on developing countries, and the can of worms revealed by Rachel Carson's *Silent Spring*. Today, the agrichemical industry is still thriving and is locked into extensive programmes involving GM crops that work in combination with herbicides and pesticides. The toolkit of plant researchers now includes gene cloning, whole genomic sequences, expressed sequence tags, RFLPs, PCR, knock-outs, reverse genetics and transgenic plants as well as the Internet and GenBank, enabling almost any type of molecular manipulation to be carried out.

The views of several prominent post-modern 'Continental' theorists such as Jaques Derrida, Michel Foucault, Gilles Deleuze, Félix Guattari, and others, have relentlessly criticised the presuppositions of modernity. Influenced by the anthropoligist Claude Lévi-Strauss, and philosophers such as Nietzsche and Heidegger, such criticisms stem from the rejection of a theory of historical progress, and the belief that developments in society are really the outcome of relatively independent cultural influences. But despite their legitimate concerns about modernity, their ideas of political pluralism have done little to lessen the assaults on Nature that are still occurring worldwide, and have contributed to a culture of nihilism. Ironically, they have frequently been accused of being conservative and unwittingly supporting modernity's status quo.

Judging by the rapid degradation of plant-life worldwide, we urgently need solutions that have universal applicability. Although, ultimately, we require global solutions to the problems of modernity, we have to implement short-term 'holding-actions' by way of the establishment of a greater number of protected reserves, and the implementation of more effective protection laws. The oft-quoted phrase 'think globally, act locally' has important implications for the transformation to a more ecoliterate society. The activist Wendell Berry is a strong advocate of local action and believes that, since we are not living harmoniously with Nature, we must look for new local economic and political solutions to what are essentially local problems. Because we acquiesce in an exploitative system, we all too easily allow ourselves to be robbed of what is fundamental to our greater well-being, that is a healthy natural environment. But there is also an imperative for immediate global action, for many of the planetary problems are international in their making, for example deforestation, pollution of the oceans and over-fishing, carbon emissions, mining and oil-extraction.

Over the past few decades various global solutions have been proposed, mostly by groups such as Friends of the Earth, Greenpeace, and by radical ecologists who want more revolutionary action to halt the transformation of the Earth into consumer products. Such changes are to be distinguished from the 'mere tinkering' with ecological problems, as advocated by the reform environmentalists who generally favour restraint and wise use of natural resources without really altering the paradigm of progress through science and technology. For example, the influential and respected biologist, E. O. Wilson, in his book *Consilience*, has proposed a challenge to the prevailing world view by attempting a synthesis of all the ways of knowing, by the proposition of a grand conception encompassing the sciences, the arts, ethics and religion. Unfortunately, this synthesis is nothing less than a scientific credo. Wilson makes it plain that he would have his grand synthesis operating within a framework of existing scientific and industrial technology.

Although there are many shades of radical ecology, there is little doubt that they all call for a fundamental overhaul of our relationship with Nature, at both the cultural and personal level. In particular, Deep Ecology, or transpersonal ecology, is particularly interesting in that it calls for a radical shift from *anthropocentrism* to a more balanced ecological perspective of the world, within the framework of a wide range of philosophical viewpoints. Of course, such a paradigm shift to a socially harmonious and ecological era that should create the social and perceptual context for a new politics is a long term project. It is generally acknowledged that, only when sufficient numbers of individuals develop a wider identification with all life, and the multiple interconnectedness of the planetary ecosystem, will large-scale changes begin to take place. There is a growing number who allow their sense of 'self' to include others, as well as animals, plants and ecosystems. Plants do not just interact with the environment, they are so intimately connected with it that they cannot be separated from it. Surely it is time to learn from them and make a break from the antagonism and estrangement between humans and

Nature that exists today. The wider identification that may result from such a transpersonal shift in psyche should allow for more empathy with, and a more spontaneous care for, plants and animals, instead of treating them as commodities. It may also lead to a more reverential approach to Nature and a greater sense of ecoliteracy, which, fortunately, can still be found in small, self-sustaining communities worldwide.

However, there are also progressive and liberating aspects to modernity, and there is a danger that, in our angst over ecological catastrophes, we fail to appreciate the benefits that modernity has brought. A re-enchantment of the world is not inconsistent with aspects of contemporary science, and it would be unrealistic to turn the clock back. But the progress to a more ecoliterate society could involve a synthesis of both old and new values, and a realisation of what we have already lost. We need balance and the skill to make the right decisions, but these can only come about by a new sensitivity to our place in nature.

And what of botany? There is so much information coming at us that we are in severe danger of overload. Information has now become a worldwide commodity, but are we able to synthesise it into a meaningful programme? A four-year university course is hardly adequate to provide a proper basis for twentyfirst-century botany. General botany, including field and descriptive botany, is no longer taught, and the history of botany is completely ignored, while systematic and taxonomic research is now virtually the province of the molecular expert, leaving the gifted amateur with almost no influence whatsoever on how we should interpret the botanical world. Systematics is dominated today by molecular biology and cladistics. Ph.D. theses are produced, ostensibly about plant evolution, but often with scarcely any reference to the plant living in Nature, while botanical journals have become dense and boring, filled with papers on molecular research. Too much taxonomic research consists of a seemingly endless comparison of cladograms produced by different molecular and analytical methods; and the search for the phylogenetic utility of the classifications that are derived from the cladograms is inward looking, almost forgetting the plants themselves.

This rather depressing picture does not give much optimism for an ecologically enlightened future generation. Perhaps the idea of formal university training in botany is just a hangover from an outmoded modernity programme, and we need new ways of re-enchanting botany. One of the finest examples of an inspiring alternative botany can be found in the work of the French botanist, Francis Hallé, but this reflects a lifetime of experience that is rare in the university lecture theatre. Yet, it may be timely to re-examine the world of past botanists and possibly learn what they had to teach. One approach that may prove richly rewarding is the 'delicate empiricism' developed by Goethe and subsequently practised by Agnes Arber. The phenomenological approach to botany allows for the dynamic interplay between observer and the observed, a fertile ground for the

Deep Ecology – I could also call it 'Green' – the Green Movement is a movement where you not only do good for the planet for the sake of humans but also for the sake of the planet itself. That's to say that you start from the whole of the globe and talk about the ecosystems, trying to keep them healthy as a value in itself. That is to say, for their own sake, like you do things for your own children or for your own dog, not only thinking of the dog as an instrument for your pleasure. So, deep ecology starts from a philosophical or religious view that all living beings have value in themselves and therefore need protection against the destruction from billions of humans.
Arne Naess (Nancho Rep: W. David Kubiak, http://www.nancho.net/advisors/anaess.html)

The highest that man can attain in these matters is wonder; if the primary phenomenon causes this, let him be satisfied
Goethe

imagination, or 'reason in its most exalted mood' as Wordsworth described it. Again, this may be asking too much. Any new direction is more likely to spring from an increasingly ecoliterate public, and the gradual spread of an ecological postmodernism, as advocated by Charlene Spretnak in her visionary book *The Resurgence of the Real.*

> *. . . how to order the signs and the symbols so they will continue to form new patterns developing into new harmonic wholes so to keep life alive in complexity and complicity with all of being – there is only poetry.*
>
> Kenneth White (*Walking the Coast*)

There are encouraging signs that this is already happening, with a growing plurality of botanically related interests such as vegetarianism, aromatherapy, gardening, ethnobotany, botanical art, etc., and environmental spinoffs such as reforestation schemes (e.g. the Community Woodlands Project in central Scotland). Environmental topics form a major part of the courses offered at progressive institutions such as Schumacher College in Devon, England, and at Prescott College in Arizona, USA, while the popularity of experiential courses offered by community education, lifelong-learning and other such programmes reveals that there is a widespread desire to reconnect with Nature. The award of the Nobel Peace Prize in 2004 to Wangari Maathai, the Kenyan leader of the Green Belt Movement that has assisted women in planting more than 20 million trees on their farms and on school and church compounds, was hopefully a harbinger of a new appreciation of the place of humanity in Nature.

The potential rewards of a change of attitude and behaviour are huge. The evolution of all life on Earth has been a long slow dance, the Dance of Shiva; of creation and destruction, where signal, response and flourish, are incorporated into an elaborating process of metamorphosis, innovation and improvisation. At the heart of our estrangement from Nature is our disconnection from ourselves. We have forgotten how to dance. By developing a transpersonal psyche we can extend the boundaries of 'self' and cease to see the world in dualistic terms. By entering the dance, the duality between art and science, between science and religion, dissolves. By bringing a poetic dimension, the 'poetry of the cosmos', back into our lives we reconnect with ourselves.

Further reading for Chapter 8

Abram, D. *The Spell of the Sensuous* (New York: Vintage Books, 1996).

Arber, A. *The Natural Philosophy of Plant Form*, Facsimile edition (1970) (Darien, CN: Hafner Publishing Co., 1950).

Berry, W. *Life is a Miracle. An Essay Against Modern Superstition* (Washington, D.C.: Counterpoint, 2000).

Bortoft, H. *The Wholeness of Nature* (New York: Lindisfarne Books, 1996).

Brandon, R. N. *Concepts and Methods in Evolutionary Biology* (Cambridge: Cambridge University Press, 1996).

Capra, F. *The Web of Life. A New Scientific Understanding of Living Systems* (New York: Anchor Books/Doubleday, 1996).

Creath, R. and Maienschein, J. (editors) *Biology and Epistemology* (Cambridge: Cambridge University Press, 2000).

Depew, D. J. and Weber, B. H. *Darwinism Evolving* (Cambridge, MA: MIT Press, 1996).
Fox, W. *Toward a Transpersonal Ecology: Developing New Foundations for Environmentalism* (Boston: Shambhala, 1990).
Gould, S. J. *The Structure of Evolutionary Theory* (Cambridge, MA: The Belknap Press of Harvard University Press, 2002).
Hallé, F. *In Praise of Plants* (Portland: Timber Press, 2002).
Maturana, H. R. and Varela, F. J. *The Tree of Knowledge: The Biological Roots of Human Understanding*, revised edition (Boston: Shambhala, 1998).
Mayr, E. *The Growth of Biological Thought* (Cambridge, MA: The Belknap Press of Harvard University Press, 1982).
Toward a New Philosophy of Biology (Cambridge, MA: Harvard University Press, 1988).
Morton, A. G. *History of Botanical Science* (London: Academic Press, 1981).
Naess, A. *Ecology, Community, and Lifestyle* (New York: Cambridge University Press, 1989).
Rieppel, O. C. *Fundamentals of Comparative Biology* (Basel: Birkhäuser Verlag, 1988).
Seamon, D. and Zajonc, A. (editors) *Goethe's Way of Science: A Phenomenology of Nature* (New York: State University of New York Press, 1998).
Schlichting, C. D. and Pigliucci, M. *Phenotypic Evolution: A Reaction Norm Perspective* (Sunderland, MA: Sinauer Associates Inc., 1998).
Schmalhausen, I. I. *Factors of Evolution: The Theory of Stabilizing Selection* (Chicago: The University of Chicago Press, 1986).
Sober, E. *Philosophy of Biology* (Oxford: Oxford University Press, 1993).
Conceptual Issues in Evolutionary Biology (editor), second edition (Cambridge, MA: MIT Press, 1994).
From a Biological Viewpoint (Cambridge: Cambridge University Press, 1994).
Spretnak, C. *Resurgence of the Real. Body, Nature and Place in a Hypermodern World* (New York: Routledge, 1999).
Stevens, P. F. *The Development of Biological Systematics* (New York: Columbia University Press, 1994).
Wilson, E. O. *Consilience. The Unity of Knowledge* (New York: Alfred A. Knopf, 1998).

Index